Risk Assessment and Environmental Management

A Case Study in Sudbury, Ontario, Canada

We are proud to provide this complimentary copy of *"Risk Assessment and Environmental Management, A Case Study in Sudbury, Ontario, Canada"*

Glen Watson
Vale Ltd

Marc Butler
Xstrata Nickel

Risk Assessment and Environmental Management

A Case Study in Sudbury, Ontario, Canada

Christopher Wren, PhD
MIRARCO, Sudbury, Ontario, Canada

2012 • Leiden, The Netherlands

Maralte

Copyright © 2012 Maralte B.V. Leiden, The Netherlands

All rights reserved. No part of this publication may be reproduced, stored in a retrieval system, or transmitted, in any form or by any means, electronic, mechanical, photocopying, recording, or otherwise, without the prior written permission of the publisher.

Requests for permission should be directed to info@maralte.com, or you may send in your request online via www.maralte.com by selecting "Contact us".

Library of Congress Cataloging in Publication Data
A catalog record from the Library of Congress has been applied for.

British Library Cataloging in Publication Data
A catalog record from the British Library has been applied for.

First edition: 2012

Design and typography: 2 kilo design *(www.2kilo.nl)*

ISBN: 978-94-90970-01-7 *(softcover)*
      978-94-90970-04-8 *(electronic version)*
ISSN: 1879-8764 *(series)*

DOI: 10.5645/b.1

Abridged Contents

	Foreword	7
1.0	Introduction	9
2.0	Historical Smelter Emissions and Environmental Impacts	23
3.0	The 2001 Soil Survey and Selection of Chemicals of Concern	49
4.0	Distribution of Chemicals of Concern in the Study Area	71
5.0	Human Health Risk Assessment Problem Formulation	97
6.0	Sampling and Analyses to Fill Data Gaps for the Human Health Risk Assessment	117
7.0	Detailed Human Health Risk Assessment Methods	157
8.0	Human Health Risk Assessment Results	207
9.0	Ecological Risk Assessment Problem Formulation	255
10.0	ERA Objective #1: Are the Chemicals of Concern Preventing the Recovery of Terrestrial Plant Communities?	277
11.0	Ecological Risk Assessment: Evaluating Risks to Terrestrial Wildlife	343
12.0	Risk Management	381
13.0	Public Consultation and Risk Communication	403
14.0	A Comparison of Two Community Based Risk Assessments	421
	Index	449

Preface

This book is written for practitioners and students of risk assessment. The purpose is to share the information and experiences gained at Sudbury with anyone contemplating a large-scale risk assessment, and with students wishing to see theory applied to a case study. While this study was extensively reported on the project's website and in technical reports, we recognized that few people would access and navigate through all those pages. It was generally felt that condensing the work into one easily accessible Volume would be the best way to share the data and experiences with a wider audience.

This book provides chapters devoted to all components of the project including human health risk assessment, ecological risk assessment, risk management and risk communication. One of the strengths of the risk assessments was the tremendous amount of Sudbury specific data that were collected for the exposure assessment. Much of that information is summarized and provided in this text. We hope it will be of interest and possible use to other risk assessors who may not have the luxury of collecting extensive exposure pathway data. The final chapter provides a comparison of the Sudbury study with another community wide risk assessment at Port Colborne, Ontario. Using this comparison the authors were able to identify strengths and weaknesses of the two different approaches.

This is not a 'how to' book on risk assessment, nor is it a guidance document. It is simply a documentation of how we undertook a very large risk assessment in Canada. There was no regulatory guidance on how to conduct a study of this scale or magnitude, but we followed the basic risk assessment paradigm. I was encouraged by some of the authors to include a 'disclaimer' that we did the best we could with the information available at the time. Science continuously changes and new information comes to light. For example, the toxicity reference values (TRVs) we used may already be replaced by other values. We hope the reader can learn from our experience, and not judge us too harshly for things we could have done differently. It should also be emphasized that the studies and risk assessments were carried out by independent consultants, and the process or the results should not be taken to reflect policy, opinions or positions of either of the mining companies, regulatory agencies or City of Greater Sudbury that were involved in this work.

I was fortunate that the MIRARCO organization and the two mining companies (Vale, Xstrata Nickel) supported my time to condense and edit all the technical reports into this one book. Each chapter is followed by acknowledgements in partial recognition of all the people who made this work possible. However, particular recognition is due to a few individuals. Without the commitment, hard work and vision of Glen Watson (Vale) and Marc Butler (Xstrata Nickel) I believe the studies would have gone off track, or at least stalled, and the outcome could have been very different. The fact that we emerged as good friends after the studies were complete is more a testament to their resolve than my competence. It was a pleasure to work with Dr. Glenn Ferguson (Intrinsik Environmental Sciences Inc.) who provided excellent technical and project management leadership throughout the years. Lastly, I express my everlasting gratitude to my wife, Lisa Robin, for her unqualified support and friendship.

Christopher D. Wren, Ph.D. MIRARCO *(Mining Innovation Research and Rehabilitation Corp.)*, Laurentian University, Sudbury, Ontario, Canada.

Dedicated to Hilary, Matthew and Breelyn.
May you always continue to learn.

DOI: 10.5645/b.1.1

1.0 Introduction

Authors

Christopher Wren, PhD
MIRARCO
935 Ramsey Lake Road, Sudbury, ON Canada P3E 2C6
Email: cwren@mirarco.org

Glen Watson, M.Sc.
Vale
337 Power Street, Copper Cliff, ON Canada P0M 1N0
Email: Glen.Watson@vale.com

Marc Butler
Xstrata Nickel
Falconbridge, ON Canada P0M 1S0
Email: MButler@xstratanickel.ca

Table of Contents

1.1 Study Rationale 10
1.2 Background and Setting for the Study 11
1.3 Study Area 12
1.4 Study Organization 13
1.5 Risk Assessment Process 15
    1.5.1 Human Health Risk Assessment 16
    1.5.2 Risk Assessment Framework and General Approach 17
1.6 Acknowledgements 19
1.7 References 19
1.8 Appendix Chapter 1: Abbreviations 21

Tables
1.1 Chronology of Significant Events in the Sudbury Region (modified from Conroy and Kramer, 1995) 11

Figures
1.1 Location of the City of Greater Sudbury 13
1.2 Stakeholder Linkages within the Sudbury Soils Study 14
1.3 General Schedule for the Sudbury Soils Study 15
1.4 Factors Required for a Risk of Health Effects 16
1.5 Overview of the Risk Assessment Framework 17

1.1 Study Rationale

The Sudbury Basin is an area rich in mineral deposits, particularly in the nickel and copper ores that have drawn people to the region for the past 125 years. The industries that developed around these valuable metals have provided significant economic benefits to the city, and have made a substantial contribution to the Canadian mining sector which is recognized around the globe. The mining and smelting operations required to extract these minerals have also had significant environmental impacts. The phenomenon of 'acid rain' and lake acidification was first recognized around Sudbury in the 1960s (Gorham and Gordon 1960; Beamish and Harvey 1972; Martin 1986). Numerous studies have also documented elevated metal levels in soils and other parts of the Sudbury environment reaching out many kilometers beyond the city boundaries (Hutchinson and Whitby 1974; Freedman and Hutchinson 1980). However, in the last three decades, tremendous effort has been put into reducing smelter emissions and re-greening the Sudbury landscape, making Sudbury a model for ecological restoration (Gunn 1995).

Since 1971, the Ontario Ministry of the Environment (MOE), Vale (formerly Inco Limited) and Xstrata Nickel (formerly Falconbridge Limited) have conducted sampling programs to determine the concentrations of metals in soils and vegetation across the Sudbury region. Not surprisingly, those studies demonstrated that there are areas in and around Sudbury with elevated metal levels in the soil. These areas are generally close to the historic smelting sites of Copper Cliff, Coniston and Falconbridge, but elevated metal levels are found throughout the city. The smelter operation at Coniston was closed by Inco in 1971. Vale still operates the smelter at Copper Cliff, while the smelter in the town of Falconbridge is still operated by Xstrata Nickel. Although metals do occur naturally in all soils, the studies indicated that elevated metal concentrations in surface soil (the top 5 cm) were the result of atmospheric deposition from the local mining, smelting, and refining operations, including the original roast yards.

In 2001, the MOE released a report entitled *Metals in Soil and Vegetation in the Sudbury Area (Survey 2000 and Additional Historic Data)* (MOE 2001). That report reviewed and summarized the results of the previous 30 years of studies, comparing metal levels in local soils to the criteria listed in the MOE's *Guideline for Use at Contaminated Sites in Ontario* (MOE 1997). During this evaluation, the MOE identified that concentrations of nickel (Ni), cobalt (Co), copper (Cu) and arsenic (As)[1] in Sudbury soils exceeded the MOE criteria. In addition, the MOE review identified significant gaps in the existing soil quality data in terms of spatial coverage (geographic area) and changing sampling and analytical methods over the 30 year period, making direct comparison of much of the data impossible. Therefore, the 2001 MOE report made two primary recommendations:

- that a more detailed soil chemistry study be undertaken to fill data gaps; and
- that a human health risk assessment (HHRA) and ecological risk assessment (ERA) be undertaken.

Both Vale and Xstrata Nickel voluntarily accepted these recommendations, and in 2001, the Sudbury Soils Study was initiated. The study was overseen by a collaborative Technical Committee (TC), comprised of Vale, Xstrata Nickel, the MOE, the Sudbury & District Health Unit (SDHU), the City of Greater Sudbury, and the First Nations and Inuit Health Branch of Health Canada. The study organization and development are described in more detail later in this chapter in section 1.4. It must be noted that the MOE soil criteria published in the *Guideline for Use at Contaminated Sites in Ontario* (MOE 1997) are not 'action levels'. Therefore, exceeding the generic criteria does not automatically imply that clean-up is required (MOE 2001). However, soil contaminant concentrations in excess of the criteria listed in the guideline (MOE 1997) do trigger the need for further investigation and/or a risk assessment.

Although the impetus for the study was soil quality, all potential exposure pathways were considered for both the HHRA and ERA. To make the risk assessment as directly relevant to Sudbury as possible, a considerable amount of local or site-specific exposure data were gathered. This included data on the concentrations of metals in air, drinking water, food, indoor dust, and various other environmental media. These data gathering studies and results are described in Chapters 6 and 9. The purpose of gathering site-specific exposure data is to reduce the level of uncertainty in the predictions of risk. It can never be stated unequivocally that a particular exposure to a chemical poses absolutely no risk. Therefore, it becomes a matter of deciding what is the level of 'acceptable' risk. The degree of acceptable uncertainty is then closely associated with the level of acceptable risk. This becomes a complex discussion and is dependent upon many factors, including the particular receptor(s) under consideration. Of paramount importance are the consequences of an incorrect prediction of acceptable risk.

---

[1] Note: In chemical terms, As is considered a metalloid and not a metal. However, for ease of terminology, the elements considered in this study are referred to as metals.

Another point to make in this introduction is to differentiate between Risk Assessment and Risk Management; the two are often mistakenly interchanged. However, the difference is considerable and important. Risk Assessment (either HHRA or ERA) is a process for collecting, organizing, and analyzing information to estimate the likelihood of undesired effects on human (HHRA) or non-human (ERA) receptors (after Suter et al. 2000). If potential risk is identified or predicted, risk management initiatives can be undertaken to reduce the predicted risk. Risk management decisions will be made based on a wide range of considerations and input from the stakeholders involved. Public regulatory agencies have legislated mandates for the protection of human or environmental health, while proponents or potentially responsible parties have legal, ethical and economic considerations. The public at large has a wide range of societal values that are also imposed on risk management decisions.

1.2 Background and Setting for the Study

Conducting the HHRA/ERA represented a significant and logical step in the history of the City of Greater Sudbury. There have been tremendous social, technical and economic benefits from the mining and smelting activities at Sudbury, not just for the City, but for Canada as a whole. However, these benefits also came with considerable cost to the environment. Environmental degradation around Sudbury was apparent in the early 1900s, and by the early 1940s, government representatives were already meeting with local industry to discuss the smelter emissions and damage to the forests and local agriculture. Stringent regulations on emission levels began to be enforced in the early 1970s. By the late 1980s, significant ecological recovery was under way in the area due to reduced emissions from the smelters and restoration efforts by many groups and stakeholders. A detailed overview of the 're-greening' of Sudbury's landscape was provided in a book in the mid-1990s (Gunn 1995). That monograph and related activities began to change the global perception of Sudbury from an industry town plagued with extensive environmental degradation to a community committed to environmental protection. At the same time, active development of the natural mineral resources of the area continued. A number of key events and dates related to the smelters and environmental programs are summarized in Table 1.1.

The Sudbury landscape of today is the result of many factors and activities interacting over a period of nearly a century: sulfur dioxide fumigations, metal deposition, intense logging, wild fires, water and wind erosion, and enhanced frost action (Winterhalder 1984). Until relatively recently, it was assumed that vegetation damage in the Sudbury area was the direct result of sulfur dioxide emissions. However, during the late 1960s, there was increasing interest and examination of the effects of soil acidity (due to acid rain resulting from sulfur dioxide emissions) and elevated metal concentrations.

Table 1.1 **Chronology of Significant Events in the Sudbury Region**
(modified from Conroy and Kramer, 1995)

Events	Year
Formation of Sudbury Basin	~2 billion years ago
Ojibway, Huron and Ottawa First Nations settle in the area	~10,000 years ago
First European explorer (Champlain)	1750
First mapping of mineralization in Sudbury area	1856
Discovery of mineralization in Sudbury during construction of transcontinental railway	1883
First purchase of mining lands by Murray	1884
First use of corrosion-resistant nickel steel	1885
First smelting of Sudbury ores by roast heap	1888
Parent company of Inco Limited established	1902
First geological map of Sudbury Basin	1905
Ontario Royal Commission of Nickel	1915

Events	Year
Damages by Sulphur Fumes Arbitration Act proclaimed	1921
Founding of International Nickel of Canada Ltd (Inco)	1928
Formation of Falconbridge Nickel Mines Ltd	1928
First ambient SO_2 monitoring network established	mid-1940s
Formation of the Sulphur Dioxide Committee	1944
First acid plant established to produce elemental sulfur from SO_2	1958
Ontario issues control orders to reduce emissions from area smelters	1969
First provincial control order on Falconbridge	1969
Sudbury Environmental Enhancement Programme (SEEP) Committee initiated	1969
Tree planting and 're-greening' activities began	1969
First provincial control order on Inco to reduce emissions	1970
First published concerns of long-range atmospheric damage from Sudbury	1970
Completion of Inco's 'Superstack' and major emission controls	1972
Vegetation Enhancement Technical Advisory Committee (VETAC) initiated	1973
MOE – Sudbury Environmental Study	1973–80
First International Conference on Acid Rain at Ohio State University	1976
Regional Land Reclamation initiated	1978
Acidic Precipitation Conference, Muskoka, Ontario	1985
Ontario government initiates Countdown Acid Rain Program	1985
First published evidence of reversal of acidification, Sudbury lakes	1986
U.N. Local Government Honours Award to Sudbury	1992
Production of Canadian (CCME) Approaches to Ecological Risk Assessment at Contaminated Sites	1993
Development of MOE Generic Soil Clean-up Criteria (Guideline for Use at Contaminated Sites – Revised 1997)	1993
Legislated reductions in sulfur dioxide emissions achieved	1994
Publication of book Restoration and Recovery of an Industrialized Region (J. Gunn, ed.)	1995
The Sudbury Soils Study	2001–09
Inco is purchased by Vale and becomes Vale Inco	2006
Falconbridge is purchased by Xstrata and Sudbury operation is Xstrata Nickel	2007
Design of the Sudbury Biodiversity Action Plan	2009

The ecological effects of smelter emissions in Sudbury have been widely studied and reported. Now, with this study, the human health implications of metals in soil and the environment are also addressed. This is the first comprehensive study to examine the potential human and ecological risks associated with metals in air, soil, water, vegetation, and other environmental matrices in the Sudbury area.

1.3 Study Area

The City of Greater Sudbury is located in central northern Ontario (Figure 1.1). Based on the 2001 Census, the total population of the new City of Greater Sudbury was 155,219. Sudbury is considered the largest municipality in Ontario, with its total area encompassing 3,627 km² including 330 lakes, many wetlands and different types of natural habitat and vegetation communities (City of Sudbury 2005). The study area also encompasses two First Nations communities, Whitefish Lake First Nation in the southern portion of the study area, and Wanapitei First Nation, situated on the shore of Wanapitei Lake. The initial study area for the Sudbury Soils Study was defined as the area from which soil samples were collected during the 2001 Sudbury Regional Soils Project (see Chapter 3). The extensive 2001 soil survey was undertaken to address data gaps in the concentration of metals and metalloids for the Sudbury area that were identified by the MOE in their review of the historical soil quality data (MOE 2001).

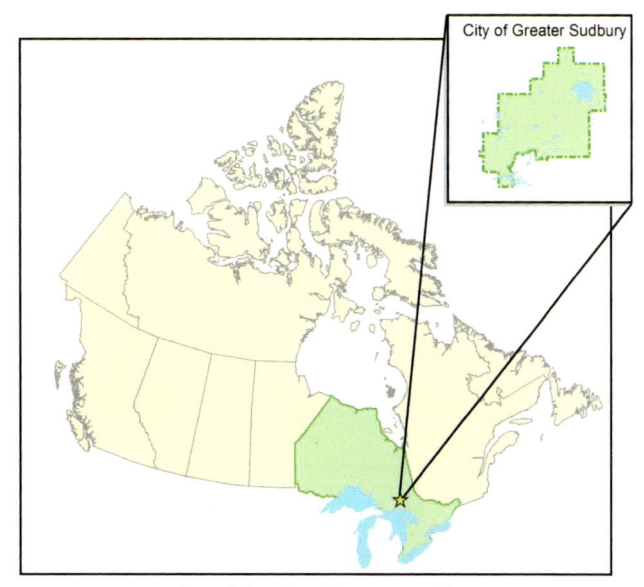

Figure 1.1   **Location of the City of Greater Sudbury**

1.4 Study Organization

A Technical Committee (TC) was formed in 2001 to oversee the Sudbury Soils Study and to provide technical guidance to the studies. The TC was comprised of the following partners:

- Vale
- Xstrata Nickel
- Ontario Ministry of the Environment
- Sudbury & District Health Unit
- The City of Greater Sudbury
- First Nations and Inuit Health Branch of Health Canada.

The study was funded by the two mining companies, Vale and Xstrata Nickel. A facilitator was retained to help organize and run the monthly TC meetings, with an administrative assistant to record developments and decisions at meetings. All decisions were made by consensus. A more complete description of the study organization is provided in Chapter 14. As the study progressed, additional committees, subcommittees and other stakeholders became involved. The relationship between these groups is conceptually illustrated in Figure 1.2. A Public Advisory Committee (PAC) was established early in the process to help coordinate contact with the public and to promote the flow of information between the TC and the public. A more complete description of the PAC and public consultation activities is provided in Chapter 13.

The overall vision of the Technical Committee for the HHRA and ERA was to develop:

> "A transparent process that provides a thorough scientifically sound assessment of the environmental and health risks to the Sudbury community, and effectively communicates the results so that future decisions are informed and valued."

The Sudbury Area Risk Assessment (SARA) Group was retained early in 2003 to undertake the HHRA and ERA. The SARA Group was an affiliation of several Ontario-based consulting firms specializing in different scientific disciplines required to carry out a study of this broad scope. These disciplines included toxicology, ecology, biology, GIS, statistics, communication specialists, epidemiology, project management, analytical chemistry, soil science, meteorology and air dispersion modeling to name a few.

The TC went to considerable effort to ensure the process was thorough and transparent. These efforts were driven, in part, by skepticism early in the process by members of the public that the companies funding the study would somehow be able to influence the results and outcome. Input from, and contact with, the community can occur at different levels, either through the PAC, directly to members of the TC, or to the SARA Group. An Independent Process Observer (IPO) was brought into the study to participate in all TC and PAC meetings and observe how decisions were achieved. The IPO selected for this role was Mr Franco Mariotti, a biologist by training, and a staff scientist at Science North in Sudbury. He was well known for promoting science education and advocating the virtues of Sudbury. As IPO, he had complete autonomy and published a regular IPO report on the study. The reports contained his observations and comments on the process of the study. His role was to help ensure that no one TC partner unduly influenced the process, and that community interests were being met.

Each of the stakeholders on the TC had its own technical specialists for reviewing reports and material provided by the SARA Group. Some members of the TC retained outside consultants to assist with their review of technical material. As an additional transparent check of the scientific quality of the data and reports, in 2004, the TC retained two additional experts, referred to as Scientific Advisors. The Scientific Advisors were Dr Stella Swanson for the ERA, and Dr Ronald Brecher for the HHRA (Figure 1.2). They were retained to provide advice and technical review of documents for the TC. They did not dialogue directly with the SARA Group and reported to the TC as a whole, not to any one member. In addition, the Scientific Advisors were available to the PAC for presentations or to answer questions.

Lastly, the draft reports generated by the SARA Group were subject to external scientific peer review by an Independent Expert Review Panel (IERP). Separate panels were convened to review the HHRA and ERA. In addition to reviewing technical material and submitting peer review comments, the two IERP panels met with the SARA Group to question and discuss the draft reports, methodology and preliminary results. The TC retained the services of TERA (Toxicology Excellence for Risk Assessment) Group from the United States to retain the individual panel members and to coordinate the flow of information. In this way, the IERP process was independent from both the TC and SARA Group.

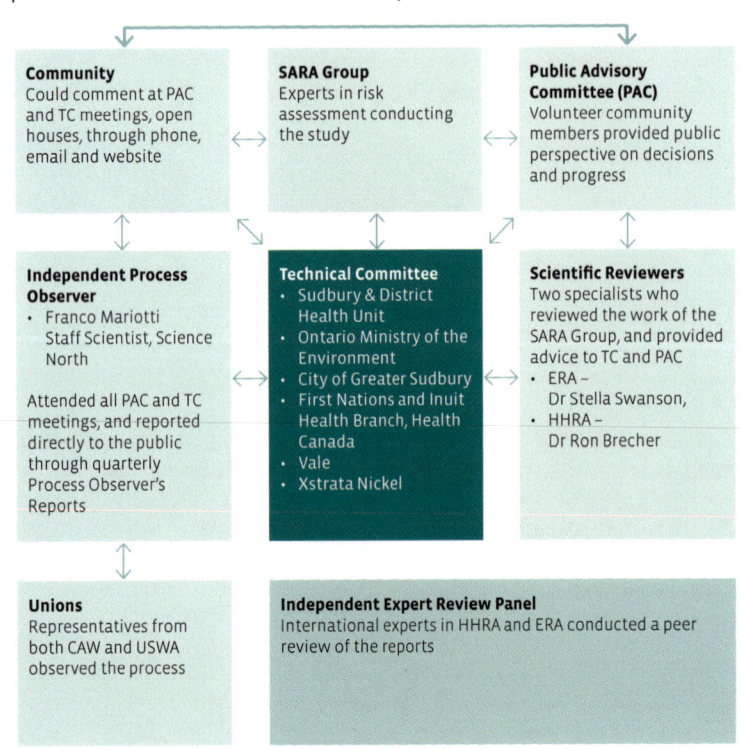

Figure 1.2   **Stakeholder Linkages within the Sudbury Soils Study**

The overall schedule for the studies is shown in Figure 1.3. The comprehensive soil sampling and analysis program was undertaken in the summer and fall of 2001. As mentioned previously, the data from that survey formed the basis for both the HHRA and ERA. The companies and the MOE worked through 2001 and 2002 to collect and chemically analyze the soil samples. The results were provided to the SARA Group to select the Chemicals of Concern (COC) for the risk assessments (see Chapter 3). The TC was also developed in 2001 and worked during this time to define the scope of the HHRA and ERA. A considerable number of investigations were conducted to collect Sudbury-specific exposure data for the HHRA (see Chapters 6 to 8) as well as field and laboratory studies to address specific objectives of the ERA (see Chapter 10). These were extensive and time consuming. Additional work on metal speciation and bioavailability in soil, indoor dust and air samples took place during 2005–2007 for the HHRA which added additional time to the project. It was decided early in the study to release the results of the HHRA to the public first as residents of Sudbury had expressed considerably more interest in human health over ecological health. Results of the HHRA were released to the public in May 2008, at three separate information sessions. Results of the ERA were released in March 2009, at one public information session.

2001

2001
- Extensive urban and regional soil sampling program
- Technical Committee formed

2002
- PAC formed, Scope of Study developed

2003
- SARA Group retained
- Data collection began for HHRA and ERA

2004-2005
- Sample collection and analysis continued
- Selection of Chemicals of Concern for HHRA and ERA

2006-2007
- HHRA and ERA Draft reports prepared and submitted to Technical Committee and to TERA for peer review
- Further sample analysis and data interpretation

2008
- Human Health Risk Assessment released to the public

2009
- Ecological Risk Assessment results released to the public
- Risk Management and Biodiversity Plan developed.
- Sudbury Soils Study concluded

2009

Figure 1.3   **General Schedule for the Sudbury Soils Study**

1.5 Risk Assessment Process

Several jurisdictions (e.g. Health Canada, MOE, United States Environmental Protection Agency [U.S. EPA], European Union) have published guidance documents outlining procedures for conducting a 'Site-Specific Risk Assessment', or SSRA. SSRA guidelines were developed to address concerns at relatively small, well-defined, properties where off-site contamination is not an issue. The broad geographic scope of the Sudbury Soils Study goes far beyond an SSRA. The term Community-Based Risk Assessment (CBRA) or Area Wide Risk Assessment was used in Ontario to studies of larger magnitude. To date, the Sudbury study remains one of the largest risk assessment ever completed in Canada. The ERA and HHRA for the Sudbury area were conducted in general accordance with the available Canadian, U.S. EPA, and MOE framework and regulatory guidance for SSRAs, with emphasis on those provided in the *Guideline for Use at Contaminated Sites in Ontario* (MOE 1997). The main advantage to conducting a CBRA is that it allows the estimation of potential health risks across a broad area, and can evaluate potential exposures from a variety of input sources (e.g. smelter emissions, locally grown foods, regional drinking water, etc.). However, this approach also requires considerably greater amounts of data specific to the study area. Another significant challenge is there is no specific regulatory guidance available governing area-wide risk assessment in Canada. It is also important to note that since the Sudbury study began, the provincial guidance documents on contaminated sites and risk assessment were revised in 2004 (MOE 2004; Ontario Regulation 153/04), and amended again in 2009 for implementation in 2011.

It is important to determine why a risk assessment is being conducted, or what 'triggered' the risk assessment. In the Sudbury case, the concentration of metal in soils exceeded the provincial government guidelines which prompted a 'regulatory' trigger. In these situations, a proponent can remediate to comply with the soil quality guidelines, or conduct a risk assessment. Both the HHRA and ERA followed similar general study approaches, although the details of each phase of the risk assessments were very different. Bother HRA and ERA evaluate the likelihood (or risk) of health effects following chemical exposures. They require consideration of the toxic properties of the chemicals, the presence of receptors, and the existence of exposure pathways to the receptors. When all three factors are present (i.e. chemicals, receptors and exposure pathways), there is a potential for adverse effects to occur if exposures to the chemicals are elevated above acceptable levels (see Figure 1.4). In HHRA, the primary receptors are people, while there can be a wide range of receptor types in ERA including plants, birds, mammals, fish, invertebrates, microbial communities, etc. Within human receptors, there are different age and sex categories, while ecological receptors are not generally distinguished on the basis of age or sex.

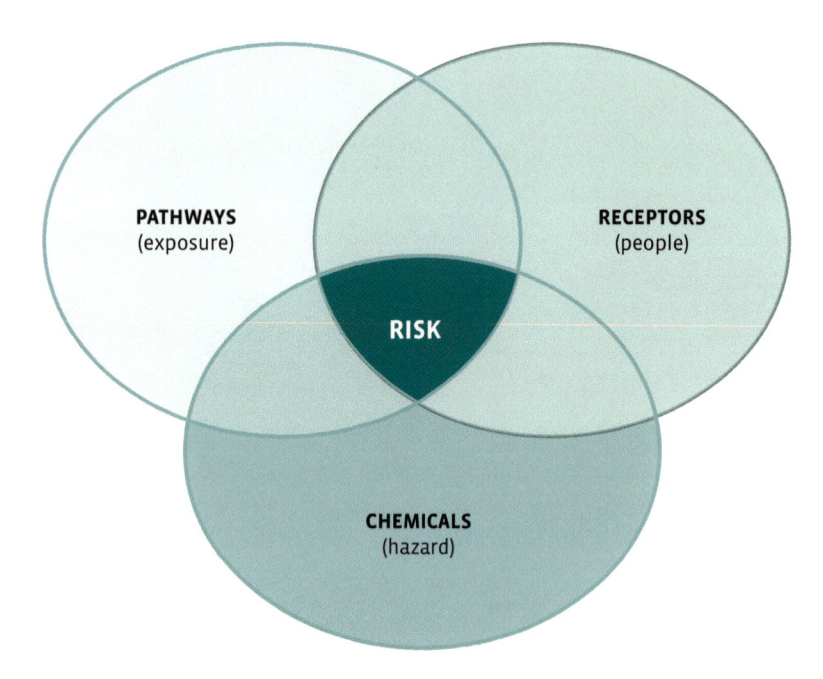

Figure 1.4   **Factors Required for a Risk of Health Effects**

1.5.1 Human Health Risk Assessment

An HHRA is a scientific study that evaluates the potential for the occurrence of adverse health effects from exposures of people (receptors) to chemicals of concern (COC) present in surrounding environmental media (e.g. air, soil, sediment, surface water, groundwater, food and biota, etc.), under existing or predicted exposure conditions. HHRA procedures are based on the fundamental dose-response principle of toxicology. The response of an individual to a chemical exposure increases in proportion to the chemical concentration in critical target tissues where adverse effects may occur. The concentrations of chemicals in the target tissues (the dose) are determined by the degree of exposure, which is proportional to the chemical concentrations in the environment where the receptor resides, works or visits.

The fundamental purpose of a HHRA is to estimate whether people working, living or visiting a given location are being exposed, or will be exposed to concentrations of chemicals that have the potential to result in adverse health effects. The assessment of potential occurrences of adverse health effects from chemical exposure is based on the dose-response concept that is fundamental to the responses of biological systems to chemicals, whether they are therapeutic drugs, naturally occurring substances, or man-made chemicals in the environment.

The Sudbury HHRA was conducted in accordance with the regulatory guidance provided by the MOE (*O. Reg. 153/04*; MOEE, 1997 and related documents) and Health Canada (1993, 2004), and is primarily based upon guidance developed as part of the U.S. EPA Superfund program (U.S. EPA 1989, 1992, 1997, 1999, 2001a,b, 2002, 2004). The current study is considered an area-wide risk assessment, as it evaluated a large geographical area, rather than an individual property (that would be considered a site-specific risk assessment or SSRA). It must also be recognized that a HHRA is not the same as a community health study. Community health studies may involve such tasks as questionnaires, interviews, medical records review and collection of biological tissue or fluid samples (e.g. blood, urine, hair) to measure human exposures directly. While these types of information can be valuable supplementary information for a human health risk assessment, they are not components of the established HHRA framework.

1.5.2 Risk Assessment Framework and General Approach

The current framework used to evaluate potential environmental and human health risks has evolved considerably from its roots in the 1983 regulatory document, *Risk Assessment in the Federal Government: Managing the Process*, released by the United States National Research Council (NRC 1983). Although there are slight variations between different jurisdictions with respect to how risk assessments are conducted, the key elements of the risk assessment framework are highly consistent across most agencies. The current HHRA and ERA follow the standard risk assessment framework (see Figure 1.5) that is composed of the following steps:

- Problem formulation
- Exposure assessment
- Hazard assessment
- Risk characterization.

Typically, where potential adverse impacts are predicted through risk characterization, an additional step providing risk management recommendations to address these concerns can be added, if necessary.

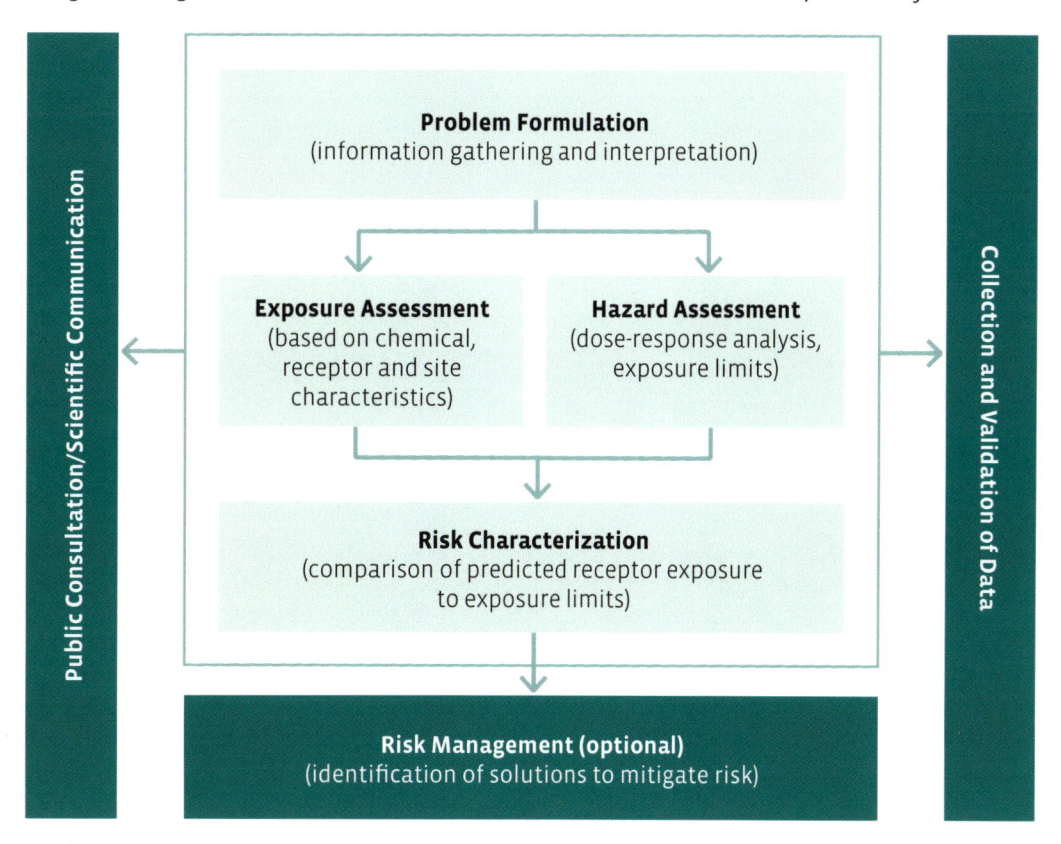

Figure 1.5 **Overview of the Risk Assessment Framework**

The risk assessment framework was applied to both the ERA and HHRA in three distinct phases:

- Phase 1 – Problem formulation
- Phase 2 – Sampling and analyses to fill identified data gaps
- Phase 3 – Detailed HHRA and ERA including exposure assessment, hazard assessment and risk characterization.

The phased approach allowed the risk assessment to proceed in a logical and sequential manner, and allowed for unresolved issues or major uncertainties to be addressed as they were identified. The phased approach was especially valuable for the HHRA as it permitted multiple iterations of the HHRA, such that various components of the HHRA were efficiently revisited and re-evaluated as new information became available. A brief introduction to and list of key activities within each phase is provided below.

Phase 1 – Problem Formulation

The Problem Formulation phase for the HHRA and ERA are described in Chapters 5 and 9, respectively. The key activities in Phase 1 included the following:

- completion of a thorough review of existing data for the Greater Sudbury Area (GSA)
- identification of the Communities of Interest (COI) for HHRA
- definition of the study area
- selection of Chemicals of Concern (COC)
- selection of Valued Ecosystem Components (VECs) for ERA
- development of the overall assessment methodology
- identification of the priority exposure pathways and development of conceptual site model
- identification and prioritization of data gaps and information requirements for Phases 2 and 3.

Once completed, the problem formulation became the blueprint for the detailed risk assessments that followed. Collection of samples and information to fill data gaps may expand, or analysis plans and approaches can change, which are additional reasons to leave these steps outside of the problem formulation stage. It was important to prepare and submit a distinct interim report on the results of the problem formulation for both the ERA and HHRA to the Technical Committee for approval.

Phase 2 – Sampling and Analyses to Fill Identified Data Gaps

The purpose of Phase 2 was to collect the data necessary to fill the data gaps identified in Phase 1 (if feasible), and to decrease the level of uncertainty in the risk assessment. Phase 2 included the development of sampling plans, sample collection and analyses, review of the new data, and incorporation of new information into the analysis or related databases.

Some guidance documents suggest that detailed analysis and sampling plans should be part of the Problem Formulation, however, for the Sudbury studies, the detailed analysis plans were incorporated into the next stages of the risk assessment methodology. Similarly, for this study, discussion of data quality objectives and areas of uncertainty were incorporated into the detailed risk assessment as it was difficult to imagine how uncertainty could be properly discussed until the data were thoroughly examined and analyzed.

Phase 2 for the Sudbury risk assessment included a wide range of activities to obtain site-specific information. For the HHRA, much of the site-specific data pertained to the exposure assessment such as collecting detailed information on the concentration of the COC in indoor dust, drinking water, ambient air, local foods and local food consumption patterns. Considerable effort was focused on speciation of Ni and As in soil, dust and air to enable species-specific exposure and toxicity issues to be addressed. In addition, the bioavailability/bioaccessibility of the COC were analysed in detail to predict risk based on the exposure scenarios more accurately. These activities are described in Chapters 6–8. The ERA Phase 2 activities included extensive measurement of ecological conditions in the surrounding Sudbury landscape as well as soil toxicity testing in the laboratory. These studies are described in Chapters 10 and 11.

Phase 3 – Detailed Risk Assessments

Once all identified data gaps had been addressed, the detailed risk assessments were conducted, employing both deterministic and probabilistic exposure analysis approaches to characterize the exposure, predicted risks, and overall uncertainty inherent within the assessment methodology and results. The following are some of the major items that were incorporated into the detailed Sudbury risk assessments:

- A quantitative exposure analysis that incorporated site-specific data
- A multi-pathway risk assessment model, based on published or peer reviewed literature. The model provided both deterministic and probabilistic results for all COC, receptors and scenarios evaluated
- A comprehensive toxicological profile for each COC
- Selection of toxicological criteria (i.e. toxicity reference values or exposure limits)
- The use of exposure and hazard data to characterize potential risks, under all plausible exposure scenarios, for all receptors and all COC
- A quantitative uncertainty and sensitivity analysis to identify those assumptions and exposure pathways that significantly contributed to the overall uncertainty inherent within the predicted risk estimates
- For the HHRA, the contribution to risk from each pathway and the cumulative risk from all pathways was evaluated
- For the HHRA, development of health-based, risk management objectives, if required
- For the ERA, identification of data and information that could be used in the ongoing and future re-greening and landscape restoration activities.

The detailed study methodology and results for the HHRA are presented in Chapters 5 to 8, and in Chapters 9 to 11 for the ERA.

1.6 Acknowledgements

The authors wish to acknowledge the many dedicated members of the Technical Committee that directed this study over an 8-year period. These included Dr Bruce Conard, Dr Mike Dutton, Dr Gord Hall, Denis Kemp, Brian Cameron, Dale Henry, Dr. Paul Welsh, Max Kaspers, Ido Vettoretti, Dr Penny Sutcliffe, Bruce Fortin, Bill Lautenbach, and Dr Stephen Monet. Recognition is given to Dick DeStefano for facilitating the TC meetings and Julie Sabourin for her administrative support. We especially recognize the patience and hard work of Ms Lindsay Boyle for keeping us all organized.

1.7 References

Beamish RJ, Harvey HH (1972) Acidification of the La Cloche Mountain lakes, Ontario, and resulting fish mortalities. J Fish Res Board Can 29:1131–1143

City of Sudbury (2005) Key Facts: City of Greater Sudbury. www.city.greatersudbury.on.ca/keyfacts/index.cfm?lang=en&app=keyfacts&label=Keyfacts_Accessed 17 April 2010

Conroy N, Kramer JR (1995) History of geology, mineral exploration and environmental damage. In: Gunn JM (ed) Restoration and recovery of an industrial region. Springer-Verlag, New York

Freedman B, Hutchinson TC (1980) Pollution inputs from the atmosphere and accumulation in soils and vegetation near a nickel-copper smelter at Sudbury, Ontario, Canada. Can J Bot 55(1):108–132

Gorham E, Gordon AG (1960) The influence of smelter fumes upon the chemical composition of lake waters near Sudbury, Ontario, and upon the surrounding vegetation. Can J Bot 38:477–487

Gunn JM (ed) (1995) Restoration and recovery of an industrial region. Springer-Verlag, New York, 358 pp

Health Canada (1993) Health risk determination. The challenge of health protection. ISBN 0-662-20842-0. Health Canada, Health Protection Branch, Ottawa, ON

Health Canada (2004) Contaminated sites program. Federal contaminated site risk assessment in Canada. Part I: Guidance on human health preliminary quantitative risk assessment (PQRA). Health Canada, Ottawa, ON.

Hutchinson TC, Whitby LM (1974) Heavy metal pollution in the Sudbury mining and smelting region of Canada. 1. Soil and vegetation contamination by nickel, copper and other metals. Environ Conserv 1:123–132

Martin HC (ed) (1986) Acidic precipitation. Proceedings of the International Symposium of Acidic Precipitation, Muskoka, Ontario, 15–20 September 1985. D. Reidel Publishing Co., Boston, MA

MOE (Ontario Ministry of the Environment) (1997) Guideline for use at contaminated sites in Ontario. Ontario Ministry of the Environment, Toronto

MOE (Ontario Ministry of the Environment) (2001) Metals in soil and vegetation in the Sudbury area (Survey 2000 and additional historic data). Ontario Ministry of the Environment, Toronto

MOE (Ontario Ministry of the Environment) (2004) Records of site condition. A guide on site assessment, the cleanup of brownfield sites, and the filing of records of site condition. Ontario Regulation 153/04 (O. Reg. 153/0), Soil, ground water and sediment standards for use under Part XV.1 of the Environmental Protection Act, 9 March 2004. Document 4728e.pdf. Ontario Ministry of the Environment, Toronto, ON

NRC (1983) Risk Assessment in the Federal Government: Managing the Process. U.S. National Research Council. National Academy Press, Washington, D.C.

Suter GW, Efroymson RA, Sample B, Jones D (2000) Ecological risk assessment at contaminated sites. Lewis Publishers, Boca Raton, FL

U.S. EPA (1989) Risk assessment guidance for Superfund. EPA/540/01. United States Environmental Protection Agency, Washington, DC

U.S. EPA (1992) Dermal exposure assessment: Principles and applications. Interim Report. Exposure Assessment Group Office of Health and Environmental Assessment, United States Environmental Protection Agency, Washington, DC

U.S. EPA (1997) Exposure factors handbook. Volumes I, II, and III. EPA/600/P-95/002Fa. Office of Research and Development, United States Environmental Protection Agency, Washington, DC

U.S. EPA (1999) Risk assessment guidance for Superfund: Volume 3. Part A, Process for conducting probabilistic risk assessment. Office of Solid Waste and Emergency Response, United States Environmental Protection Agency, Washington, DC

U.S. EPA (2001a) Risk assessment guidance for Superfund. Volume I: Human health evaluation manual. Part E, Supplemental guidance for dermal risk assessment. Interim Review Draft – for Public Comment. EPA/540/99/005. Office of Emergency and Remedial Response, United States Environmental Protection Agency, Washington, DC. www.epa.gov/superfund/programs/risk/ragse/index.htm. Accessed October 2003

U.S. EPA (2001b) Risk assessment guidance for Superfund. Volume 3: Part A – Process for conducting probabilistic risk assessment. EPA 540-R-02-002. Office of Emergency and Remedial Response, United States Environmental Protection Agency. www.epa.gov/superfund/RAGS3A/index.htm. Accessed April 20 2006

U.S. EPA (2002) Child-specific exposure factors handbook. EPA-600-P-00-002B. National Center for Environmental Assessment, Washington, DC

U.S. EPA (2004) Exposure scenarios. EPA/600/R-03/036. National Center for Environmental Assessment, Washington, DC

Winterhalder K (1984) Environmental degradation and rehabilitation in the Sudbury area. Laurentian Univ Rev 16(2):15–47

1.8 Appendix Chapter 1: Abbreviations

CAW, Canadian Auto Workers union
CBRA, Community Based Risk Assessment
CGS, City of Greater Sudbury
COC, Chemical of Concern
ERA, Ecological Risk Assessment
HHRA, Human Health Risk Assessment
IERP, Independent Expert Review Panel
MOE, Ministry of the Environment (Ontario)
PAC, Public Advisory Committee
SDHU, Sudbury & District Health Unit
SSRA, Site Specific Risk Assessment
TC, Technical Committee
TERA, Toxicology Excellence for Risk Assessment
USEPA, United States Environmental Protection Agency
USWA, United Steelworkers of America
VEC, Valued Ecosystem Component

2.0 Historical Smelter Emissions and Environmental Impacts

DOI: 10.5645/b.1.2

Authors

Glen Watson, M.Sc.
Vale
337 Power Street, Copper Cliff, ON Canada P0M 1N0
Email: Glen.Watson@vale.com

Monika Greenfield
Vale
337 Power Street, Copper Cliff, ON Canada P0M 1N0
Email: Monica.Greenfield@vale.com

Marc Butler
Xstrata Nickel
Falconbridge, ON Canada P0M 1S0
Email: MButler@xstratanickel.ca

Christopher Wren, PhD
MIRARCO
935 Ramsey Lake Road, Sudbury, ON Canada P3E 2C6
Email: cwren@mirarco.org

Table of Contents

2.1 Historical Smelter Emissions	25
2.1.1 The Beginning of Sudbury	25
2.1.2 Origin of the Mineral Deposits	25
2.2 History of the Sudbury Smelters	27
2.2.1 Early Smelting Processes	27
2.2.2 Summary of Major Technology and Emission Events in Sudbury	29
2.3 Evolution of the Mining Companies and Mineral Production	33
2.4 Smelter Emissions	35
2.4.1 Combined Emissions	35
2.4.2 Copper Cliff Smelter Emissions	36
2.4.3 Coniston Smelter Emissions	37
2.4.4 Xstrata Nickel Emissions at Falconbridge Smelter	38
2.4.5 Particulate Size	39
2.4.6 Wind Climate and Dispersion Patterns	40
2.5 Environmental Impacts Overview	41
2.5.1 The Original Forests and Early Logging	41
2.5.2 Roast Yards and Smelter Emissions	41

    2.5.3 Effects on Landscape: Barren and Semi-barrens 42
    2.5.4 Interacting Stressors 44
    2.5.5 Industrial Barren Landscapes 45
2.6 Acknowledgements 46
2.7 References 46
2.8 Appendix Chapter 2: Abbreviations 48

Tables

2.1 Major Events for Vale at Sudbury 30
2.2 Major Events for Xstrata Nickel Smelter at Falconbridge 32
2.3 Annual Production (Tonnes) from the Smelters in Sudbury 35
2.4 Metal and Sulfur Discharges (Tonnes) for 2001 for Vale and Xstrata Nickel 36
2.5 Total Annual Emissions from the Vale Smelter (Tonnes/Year) (Environment Canada 2010) 37
2.6 Annual Emissions (Tonnes Per Year) from Xstrata Falconbridge Smelter Primary Stack (Environment Canada 2010) 39
2.7 Historical Wind Characteristics 40

Figures

2.1 The Sudbury Basin – Satellite Image 26
2.2 A Roast Yard (Photograph Courtesy of Vale Archives) 28
2.3 O'Donnell Roast Yard, Circa 1915 (Photograph Courtesy of Vale Archives) 28
2.4 O'Donnell Roast Yard, Circa 1974 (Photograph Courtesy of Vale Archives) 28
2.5 SO_2 Generated from the Copper Cliff Smelter 36
2.6 Xstrata Nickel Smelter Annual SO_2 Emissions (from Xstrata Nickel 2011) 38
2.7 Resprouting birch trees in Untreated Barren Area (Photo Courtesy of the City of Greater Sudbury) 43
2.8 Map Showing Extent of Barren and Semi-Barren areas near the Sudbury Smelters 44
2.9 Factor Interactions Leading to the Formation of Barren Land (Redrawn from Winterhalder 1995, 1996) ..45

2.1 Historical Smelter Emissions

2.1.1 The Beginning of Sudbury

In 1824, the Hudson's Bay Company established a fur-trading post in the rugged land of pines, lakes, swamps and rock north of Georgian Bay, near what is now the City of Greater Sudbury. The region was sparsely populated by the Anishnabe people, who, with the opening of the post, came into contact with the French and English fur traders, missionaries, and government agents passing through what was near-trackless wilderness. Chicago's Great Fire of 1871 sparked the opening of the region for logging. Massive red and white pines were cut and floated down to Georgian Bay and Lake Huron, then rafted to sawmills in the northern United States for the rebuilding of the city (Wallace and Thomson 1993). The transcontinental railway development also put demands on the local forests because of the need for railway ties and trestles. Denudation of the forests resulted in the faster spread of man-made and natural fires. Lumbering remained dominant in the area until the 1920s (Winterhalder 1995).

While logging was a key industry over the next decades, it was the arrival of the railway surveyors that precipitated the birth of Sudbury. In 1883, the Canadian Pacific Railroad (CPR) established a camp at the junction between the line heading east from Sault Ste. Marie and the new line heading westward. The location of the camp, and consequently the city of Sudbury, was due to a surveying miscalculation. The surveyor, William Ramsey, accidentally plotted the rail line north of the lake that now bears his name rather than south. It is said that he got lost because his compass pointed to the ore body instead of north.

In February 1883, James Worthington, a senior CPR official, named Sudbury after his wife's birthplace in Suffolk, England. A cluster of shoddy, temporary buildings sprung up to house the 3350 men who worked for the CPR over the next two years (Wallace and Thomson 1993). When the last construction crew left in 1885, Sudbury could have melted back into the wilderness, as did most railway camps. But a CPR depot remained, and with it, the beginnings of the town.

William Ramsey's surveying error turned out to be a lucky stroke for Sudbury. In 1883, CPR work crews blasting a route through the rock discovered the mineral riches hidden beneath. Legend has it that CPR blacksmith Tom Flanagan noticed a rusty stain on rocks near what was to become Murray Mine, and Sudbury's future changed course forever. Flanagan's find was actually a re-discovery of the Sudbury basin copper and nickel deposits mentioned in an 1856 Geological Survey of Canada Report (Murray 1857). Even earlier missionary reports that Indians were mining copper north of Georgian Bay in the 1630s had also gone unnoticed (Wallace and Thomson 1993). Only in 1883, when the railway made the possibility of exporting minerals a reality, did the vast potential of the deposit demand the attention of the mining community.

The mineral deposits discovered at Sudbury are truly remarkable. Although Ni is one of the most common elements in the universe and is thought to make up a significant part of the earth's core, it is scarce in the crust in levels that are economic to extract. The Sudbury district contains one of the largest known nickel ore bodies on the planet. As described below, the extraction of Cu and Ni ores from the Sudbury basin started in the late 1880s. Past production and resources exceed 1548 million tonnes of ore at approximately 1.2% Ni, 1.1% Cu and 0.4 g/t Platinum Group Metals (PGM). It has been estimated that this has resulted in the production of 8.5 million tonnes of Ni and 8.4 million tonnes of Cu over the years. Total ore reserves are estimated at 1.6 billion tonnes (Patterson 2001).

2.1.2 Origin of the Mineral Deposits

Even after decades of research, the origins of the deposit are still under debate. The most popular theory among geologists is that of a meteorite impact, as presented in 1964 (Dietz 1964). The impact of a 10 kilometer-diameter meteorite is thought to have left a variety of shock phenomena, including shatter cones (half-cone-shaped fractures in the rock) and breccia (the welded broken rock characteristic of the area). After impact, large amounts of breccia fell into the crater. This theory is supported by recent research from the University of Toronto, which concluded that the Sudbury Igneous Complex is predominantly derived from a shock-melted lower crust rather than the average of the whole crust, as has been previously supposed. The researchers discovered a subtle but significant enrichment of iridium, an extremely rare metal found mainly in the Earth's mantle and in meteorites. Due to the low Mg and Ni content found in the samples they concluded that the iridium came from the meteorite itself rather than the Earth's mantle (Kelly 2004).

When Apollo 16 astronauts visited Sudbury in 1971, it was not to experience a barren 'moonscape' but because of the evidence of crater formation in the area. The astronauts practiced describing the rocks and geological features in preparation for reporting on the geology of the moon. Commander John Young of Apollo 16 described a lunar rock sample this way: 'It has a black fracture pattern running through the middle of it…It looks like a Sudbury breccia' (Stephenson 1979). While some meteorites do contain amounts of Fe and Ni, the quantity and sulfur content of the ore in the Sudbury area is too great to have come only from the meteorite. Proponents of the meteorite theory believe that the impact ruptured the earth's crust, triggering the flow of magma that formed the Sudbury Irruptive. The ore separated as hot sulfide droplets from the molten igneous rock erupted from the crater floor. Age dating of the igneous rock shows that it crystallized about 1.85 billion years ago (Gunn 1995). The elliptical shape of the basin is credited to the 30° angle at which the meteorite struck the earth.

Geologists skeptical of the meteorite theory point out that the shatter cones could also have been produced by volcanic activity. Possibly during the Keweenawan period up to 500 million years ago, a mass of molten matter deep within the earth was forced by subterranean pressure toward the surface. Checked by solid layers of sediment, this molten stream spread out between the base of the sediments and the rocks below causing considerable deformation in the sediments (LeBourdais 1953). This earlier theory can explain the presence of breccia. The Sudbury area contains the Murray fault system and the Onaping system, both of which present the possibility of volcanic activity. Science North in downtown Sudbury is built on and incorporates the Creighton fault in its architecture.

Sudbury's Ni ores are located in what is known as the Ni intrusive. Otherwise known as the Sudbury basin, the deposit is an oval, spoon-shaped, multi-layered body of igneous (volcanic) rock. At the surface, it measures 60 km long and 27 km wide. Figure 2.1 is a NASA satellite photograph using Landsat 7 color imagery taken in 2006. The elliptical impact zone is clearly visible with Lake Wanapitei as a large water-filled crater in the upper right. The floor of the basin is estimated to lie 10 to 15 km below the surface, and it contains deposits of Pb, Zn and Cu. The major Ni and Cu mineral deposits lie on the outer rim of the basin (Boldt 1967; Gunn 1995).

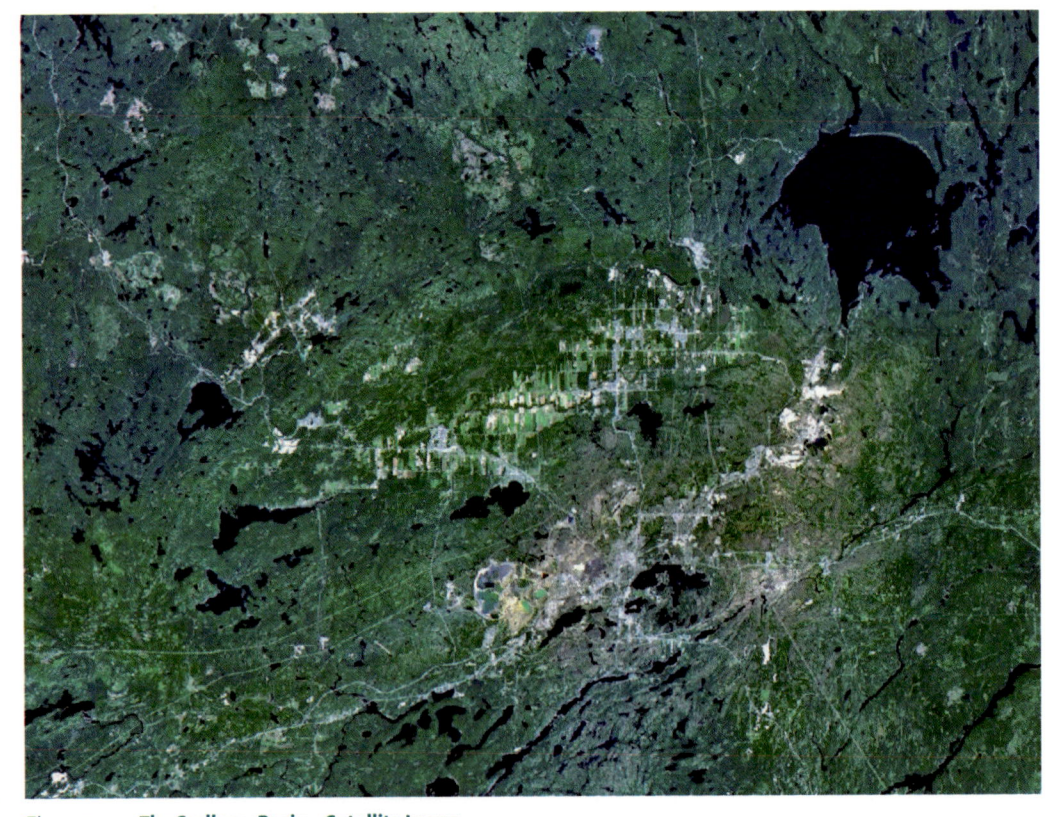

Figure 2.1   **The Sudbury Basin – Satellite Image**

2.2 History of the Sudbury Smelters

2.2.1 Early Smelting Processes

The primary goal of the smelting process has remained the same from the earliest attempts to the present day: to separate the minerals from the rock. In the case of the Sudbury ores, the biggest challenge lies in separating the pay metals (Cu, Ni) from the S, Fe and barren host rock to which they are melded. To reduce the amount of material being transported to the United States for refining, the need for effective on-site smelting arose. The infamous roast yards were the first stage in the smelting process. After a trial roast in December of 1886 proved its practicality, the first roast yard became operational in 1888. Between 1890 and 1930, approximately 28 million tonnes of ore were processed at the roast yards (Gunn 1995).

Production of feed for the Copper Cliff blast furnaces was carried out in open roast yards (Figure 2.2) until 1929. The ore was stockpiled, ignited using wood as fuel, and left to burn for extended periods of time. During roasting, about 50% to 70% of the S associated with the pyrrhotite was preferentially oxidized using air or oxygen and carried away in the exhaust gases as SO_2. The Fe was converted from a sulfide to an Fe oxide. Very little of the Ni and Cu sulfides were oxidized during roasting. The open burning precluded any form of air pollution control. The ore added to the roast yards consisted of various layers of coarse and fine ore piled to a height of about 4 m. The roasting reaction is exothermic; therefore, the wood was only necessary to raise the temperature to the ignition point, after which, the reaction was self-sustaining. There had to be enough porosity in the pile to allow sufficient air to pass through the mass, which enabled the sulfide in the ores to be oxidized to SO_2. Once the ore was raised to roasting temperatures by the wood fire, oxidation of the pyrrhotite generated sufficient heat energy to allow the roast to continue until the S in the ore had been reduced from approximately 25% to 7%. Approximately 6 months were required for a pile to be completely roasted.

Until the process was abandoned in 1929, 11 roast yards had operated, discharging clouds of acrid smoke across the land and nearby communities. Approximately 900,000 cords of wood were burned during this period, or about 27 football fields, all piled 100 feet deep (Gunn 1995). In 1916, the smaller yard in Copper Cliff was shut down due to numerous complaints and rebuilt as the largest of the roast yards, west of Creighton Mine, near the Vermillion River. The O'Donnell yard was 2286 m long by 52 m wide, and was highly mechanized for its time (Figure 2.3) while the smaller yards, including the one that is now Copper Cliff Park, were, for the most part, rehabilitated. The remains of the O'Donnell roast yard are still visible today (Figure 2.4). All of the former yards are located on property that is now or was formerly owned by Vale. Over the 40-year history of the roast yards, they released about 10 million tons of SO_2 at ground level, killing plants and acidifying soils (Gunn 1995). Open-bed roasting was a cheap but ultimately inefficient method, as it allowed some of the nickel and copper to be washed into the soil by rain. The process also required vast amounts of lumber, which led to a depletion of the supply by 1929.

The reaction took place under very low rates of oxidation; therefore, the localized air velocities would have been fairly low. This suggests that the distance that the non-volatile particulate emissions traveled from roast yards would likely have been limited. Ground-level concentrations of particulate matter in the air around the roast yards would have been substantial by today's standards. However, it is likely that the temperature in parts of the roast heap was above the vaporization temperature, allowing part of the semi-volatile metals including As (particularly oxidic compounds), Pb and Se to be released into the air.

Between 1890 and 1930, 28,040,068 tonnes of ore were processed, releasing an estimated 10 million tonnes of SO_2 (Laroche et al. 1979; Gunn 1995). The quantities of particulate emitted are unknown.

Risk Assessment and Environmental Management

Figure 2.2   **A Roast Yard (Photograph Courtesy of Vale Archives)**

Figure 2.3   **O'Donnell Roast Yard, Circa 1915 (Photograph Courtesy of Vale Archives)**

Figure 2.4   **O'Donnell Roast Yard, Circa 1974 (Photograph Courtesy of Vale Archives)**

The most significant damage caused by the roast yards was destruction of surrounding vegetation, and subsequent soil erosion. The O'Donnell roast yard area today has been naturally re-colonized by native plants, right up to the perimeter of the yard, because levels of metals in these areas were low. Only the roast yard itself remains toxic and barren because of the metals leached by rainwater into the soil. A more detailed review of the impacts of the roast yard's and early smelting activities is provided later in this chapter.

As the demand for Ni increased, lower grade ores (lower percentage of Ni and Cu) were mined. Higher amounts of metals could be produced from the same smelting process if the ore was milled into a powdered concentrate before being processed. To produce a concentrate, the ore was ground to a fine slurry in rod or ball mills. Magnetic separation and/or froth flotation was used to separate the tailings (waste containing little metal value) from the valuable minerals. The concentrate was fed to the smelters and the tailings were disposed of in piles or under water in ponds. Ores containing from 1.0% to 2.5% Ni could produce concentrates containing from 6% to 10% Ni. The Copper Cliff Mill produced a Ni concentrate containing 5% Ni and a Cu concentrate containing 30% Cu and 1% Ni. The Levack mill produced two concentrates; a Ni concentrate containing 9% Ni and 1% Cu and a Cu concentrate containing 30% Cu and 1% Ni (Boldt 1967). Sulfur dioxide and particulate matter (PM) were the principal air contaminants generated during roasting of concentrates.

2.2.2 Summary of Major Technology and Emission Events in Sudbury

The development and evolution of 3 smelters in relatively close proximity to each other in Sudbury can be confusing, and is perhaps of limited interest to an audience with a primary interest in risk assessment. However, the history of the smelters and the resultant atmospheric emissions are the basis for the Sudbury risk assessment and deserve to be documented as part of the story. To this end, some of the major events related to ore production, technology development and emission controls are summarized in Tables 2.1 and 2.2 for Vale and Xstrata Nickel, respectively. These tables provide the general reader with an appreciation of some of the many changes that have taken place during the past century and a quarter. The following section provides a brief description of the smelting processes at each of the three historical smelters.

2.2.2.1 Smelting at Copper Cliff

Smelting is a term used to describe a set of complex reactions that occur in a furnace in which sufficient heat is added to raise the temperature to melt the constituents. Flux is also added to mix with the impurities in the ores such as Mg, Al and silicates. The impurities form a slag that floats on the surface of the pool inside the furnace, and the metals partition into a matte. This matte is heavier and is tapped from the bottom layer of the furnace from a matte tap hole. The slag is skimmed off in a separate slag skim hole that is at a higher elevation. The Fe oxide produced during roasting partitions to the slag. The Sudbury ore is approximately 25% mixed Fe, Cu and Ni sulfide minerals (Bouillon 1995). The remainder is discarded as waste rock or tailings. The Cu and Ni minerals contain approximately equal amounts of metal and S. The Ni is distributed throughout the rock and is, therefore, more difficult to separate than the Cu. It is estimated that for every tonne of Ni produced, 8 tonnes of S will be processed (Bouillon 1995).

At Copper Cliff during the period from 1890 to 1929, smelting took place in blast furnaces. A new smelter close to the original site began operations in 1929 with reverberatory furnaces. According to Boldt (1967), the advantage of the reverberatory furnace was that the 'unit would smelt material of a fine particle size such as concentrates from flotation plants without blowing much of it from the furnace in exhaust gases'. By 1965, the Vale smelter at Copper Cliff consisted of 2 blast furnaces, 42 multi-hearth roasters (the roasting plant was divided into 7 batteries of 6 multiple hearth roasters, with each battery superimposed over a reverberatory furnace), 7 reverberatory furnaces, 2 oxygen flash smelting furnaces for Cu concentrates, and 24 converters, with 2 of the latter treating Cu matte. Electrostatic precipitators were used to control emissions from the roasters and converters.

In 1976, Vale replaced 8 of the multi-hearth roaster with a fluid bed roaster (FBR). A highly concentrated SO_2 stream was produced that could be tied into plants that converted the SO_2 to sulfuric acid or liquid SO_2. The installation of the particulate control devices on the roasting operation contributed to the reduction in particulate matter emissions after 1970, from 25,000 tonnes/year in 1940 to 15,000 tonnes/year in 1972. In 1994, the remaining multihearth roasters and reverberatory furnaces were shut down in favor of flash furnaces, which combine roasting and smelting in one unit and produce an off-gas containing a smaller volume and a higher percentage of SO_2 that has been economically processed to produce sulfuric acid. The flash furnaces also include highly effective controls for particulate and metal emissions. A summary of major events and technological changes for the Vale operations in Sudbury is provided in Table 2.1.

Table 2.1 **Major Events for Vale at Sudbury**

Year	Vale (including Copper Cliff, Mond Nickel – Coniston, & Mond Nickel – Victoria)
1886	The Canadian Copper Company (predecessor to Inco Ltd.) incorporated in Ohio, under the direction of Samuel Ritchie
1888	• Roast yard constructed at Copper Cliff. About 80 to 100 tonnes a day of calcine were treated. Eleven roast yards (which each ran between one and seventeen years) were operated in Sudbury between 1888 and 1929 • The Herreschoff No. 1 blast furnace was blown in at the East Smelter in Copper Cliff
1889	• Total production in 1889 was 8450 tonnes of matte containing 22.4% copper and 14.3% nickel • The second Copper Cliff blast furnace was blown in. 41,000 tonnes of ore were excavated, producing 8,450 tonnes of matte • Roast yards were relocated further away from populated areas
1890	Between 1890 and 1929, approximately 28,000,000 tonnes of ore were processed in Sudbury, resulting in roughly 11,000,000 tonnes of sulfur dioxide released to the atmosphere
1897	Copper Cliff ore begins to be pyrometallurgically processed into a matte
1899	Copper Cliff West smelter installed to replace the East smelter. Four blast furnaces were installed. 180 ft stack
1901	Mond Nickel starts nickel smelter in Victoria Mines, stack height 35 m (115 ft). Two blast furnaces and converters. Average capacity, 60,000 tonnes per year of ore
1902	Copper Cliff East smelter shut down
1903	Canadian Copper Co. and Orford Copper Co. unite to form International Nickel Co
1904	New Copper Cliff West smelter with 64 m (210 ft) stack
1905	Five blast furnaces at new west smelter (Copper Cliff) produced 2000 tonnes of nickel copper matte per day
1909	Mond Nickel Victoria plant capacity: 140,000 tonnes per year of ore
1911	Installation of reverbatory furnace at Copper Cliff
1913	• The second major smelter in Sudbury commenced operation when the Mond Nickel Company installed blast furnaces and converters in Coniston in 1913 • Victoria smelter shut down
1918	Coniston roast yards phased out
1919	Ore beginning to be pyrometallurgically processed into a matte at Coniston
1925	Inco tried to make sulfuric acid from converter gas in 1925, but the process was abandoned
1929	• Roasting of ores in open piles (roast yards) was stopped. All smelting carried out at smelters with stacks from 106 to 152 meters high • Inco unites with Mond Nickel • Reverberatory furnaces begin to replace blast furnaces at Copper Cliff
1950	• Emission controls consisted of settling chambers that would have removed about 50% of the particulate (mainly larger particle sizes) • In the period from 1900 to 1950, at all three smelters, more than 90% of the sulfur in the ore was emitted to the atmosphere
1952	New 454 tonnes per day oxygen flash furnace and 375 tonnes per day oxygen plant commissioned at Copper Cliff smelter. SO_2 produced was recovered as liquefied sulfur dioxide
1953	907 tonnes per day copper processed through the flash furnace, commenced construction of first unit of Iron Ore Recovery Plant (IORP)
1956	Started operation of first two units of IORP. Capacity: 907 tonnes per day of pyrrhotite producing 235,820 tonnes per year of iron ore pellets

Historical Smelter Emissions and Environmental Impacts

Year	Vale (including Copper Cliff, Mond Nickel – Coniston, & Mond Nickel – Victoria)
1958	• In 1958, the converter aisle at Copper Cliff consisted of 19 Pierce Smith converters each 10.6 m (35 ft) long by 4 m (13 ft) in diameter • Acid plant commissioned at IORP. Capacity: 363 tonnes per day H_2SO_4
1960	At Coniston, four blast furnaces were on site with only one or two running in the 1960s
1963	Major expansion of IORP to 2721 tonnes per day of pyrrhotite and 725,600 tonnes per year of iron ore pellets. Off-gases treated by 907 tonnes per day H_2SO_4 acid plant
1965	By 1965, the Inco Copper Cliff Smelter consisted of two blast furnaces, 42 multi-hearth roasters, 7 reverberatory furnaces, two oxygen flash smelting furnaces for copper concentrates, 24 converters. Electrostatic precipitators were used to control emissions from the roasters and converters
1967	Third acid plant commissioned at IORP, raising capacity to 2268 tonnes per day of H_2SO_4
1968	Expanded copper flash furnace
1969	Fluid bed roaster added to No. 1 reverberatory furnace with off-gases cleaned and SO_2 gases converted to sulfuric acid
1970	The installation of the particulate control devices on the roasting operation contributed to the sharp reduction in particulate matter emissions after 1970
1972	Coniston smelter shut down, concentrates diverted to the Copper Cliff smelter equipped with the 381-meter stack (this superstack had just recently replaced the three lower stacks)
1975	Development of nickel oxygen based flash furnace technology where SO_2 emissions were converted to sulfuric acid
1976	Inco replaced the multi-hearth roaster with a fluid bed roaster at No. 1 reverberatory furnace in 1976. A highly concentrated SO_2 stream was produced that could be tied into plants that converted the SO_2 to sulfuric acid or liquid SO_2
1977	Development of nickel oxygen based flash furnace technology where SO_2 emissions were converted to sulfuric acid
1980	• Additional precipitator capacity added to the Copper Cliff Nickel Refinery • A sharp transition took place during the 1980s and 1990s with the installation of acid plants at Copper Cliff to recover the SO_2, and less than 10% of the sulfur in the ore is now discharged to the atmosphere
1994	• Flash furnace installed at Copper Cliff Smelter • In 1994, the reverberatory furnaces were shut down in favor of flash smelting furnaces • Milling of ore was rationalized into one mill – Clarabelle, producing a bulk concentrate of Ni and Cu. This was accompanied by increased pyrrhotite rejection for additional containment of sulfur
1997	• CuS_2 was converted to Cu in a new converting process with N_2 bottom stirring. SO_2 was sent to the acid plant • Higher efficiency of Cu–Ni separation was achieved at the matte separation plant by using giant flotation cells and a flotation column
2001	• Cu anode production switched to the smelter from the Cu refinery. Chemical reduction of blister Cu was done using natural gas and steam blown into the melt with N_2 bottom stirring
2007	Fluid Bed Roaster Project, resulting in an approximate 30% reduction in SO_2 emissions and decreasing total emissions of Ni, Cu, As and Pb by approximately 80 tonnes/year
2010	Additional emission reductions proposed

2.2.2.2 Sintering at Coniston

Vale also operated a smelter in the town of Coniston. Here, sintering was used to roast concentrates over the period 1918 to 1972. The feed to a sinter plant consisted of concentrates that were blended with recycled sinter fines and other process materials. The bed was ignited on the surface with gas burners, and the flame front moved from the top to the bottom of the bed. A continuous series of individual pallets conveyed the bed over a series of wind boxes, through which air was drawn from the surface of the bed down through the pallets into the wind box below the pallets. The sintered product was discharged at the end of the strand as it passed over a breaker. Crushing, screening, and cooling produced a coarse feed to the blast furnace. Fine material was cooled and recycled back through the process.

Emissions from the sintering process arose from the handling of hot, dry, final sintered product; as well as the wind box gases that contained considerable particulate material (PM), metals, and SO_2. According to the US EPA AP-42 (US EPA 1995), the discharge of calcine from a roaster can generate 1.36 kg of particulate matter per tonne of concentrated ore processed, if not controlled. This 1.36 factor would be dependent upon the type of roaster but may have been applicable at Coniston for the discharge from roasters and sinter plants while being conveyed to the smelting furnaces. In 1972, the Coniston smelter was shut down and the concentrates were diverted to the Copper Cliff smelter, which by then was equipped with a 381-m stack. As a result, SO_2 and metal ground-level concentrations around Coniston were significantly reduced.

2.2.2.3 Smelting at Town of Falconbridge

Xstrata Nickel added a concentrator and a sinter plant in 1933 to treat lower grade ores. The sinter plant/blast furnace smelting process used at Falconbridge over the period from 1933 to 1978 was very similar to the operation at the Coniston Smelter. The sinter plant produced an agglomerate, or coarser material that had sufficient porosity to be processed in a blast furnace. In 1947, the sinter plant produced a waste gas volume of 170,000 cfm at 95°C containing 1% SO_2 (Jackson 1990). The original exhaust stack used for the blast furnace and converters at Falconbridge in 1930 was 175 feet high and 12 feet in diameter. In 1936, a 304-foot stack was added to handle the blast furnace and converter gases, and the sinter plant was tied into the 175-foot stack. The 304-foot (or 93 m) stack is still used today to disperse exhaust gases from the electric furnace, converter aisle, and acid plant tail gas. From 1965 to 1978, a 450-foot stack was used to better disperse the sinter plant emissions. The 450-foot stack still stands, but is not currently being used.

From 1932 to 1978, the smelter at Falconbridge used blast furnaces to smelt the sinter. A Cottrell electrostatic precipitator was installed in 1955 to clean the off-gases from the blast furnaces and converters. In 1978, the blast furnaces were shut down and switched to fluidized beds for roasting which fed into submerged arc electric furnaces for smelting. The No. 2 electric furnace still in operation at the Xstrata smelter currently produces an off-gas that contains about 1.0% SO_2, which is too low to be converted to sulfuric acid in an acid plant. In 1954, the grade of the concentrate produced by the Xstrata Nickel mill and fed to the smelter contained 4% Ni, and by 1978, the grade had been increased to 10% Ni; the mass of concentrate that had to be roasted, smelted, and converted to produce a tonne of matte at 50% Ni had been reduced by more than one-half over the period. Some of the technological changes related to production and emissions reduction are summarized in Table 2.2.

Table 2.2 **Major Events for Xstrata Nickel Smelter at Falconbridge**

Year	Xstrata Nickel
1928	Nickel mines incorporated to develop ore body outlined in Falconbridge township
1930	• The first matte from Falconbridge Smelter produced. Single blast furnace and two converters • 85% of sulfur in ore released to air. Dust collection by dust chamber
1933	• Added a concentrator and a sinter plant to treat lower grade ore • Concentrator built to produce fine concentrate • Two sinter strands added to make concentrate suitable for blast furnace feed
1936	• 3rd converter added, 304 ft high stack added for blast furnace and converter gases. Separate dust chamber for each stack • Production rate 327,783 tonnes of ore, 11,226,108 pounds of nickel
1937	3rd sinter strand added, 438,629 tonnes of ore treated
1942	Two additional sinter strands added
1943	Additional blast furnace added
1947	In 1947, the sinter plant produced a waste gas volume of 170,000 cfm at 95°C containing 1% SO_2

Historical Smelter Emissions and Environmental Impacts

Year	Xstrata Nickel
1953	3rd blast furnace, 4th converter added
1954	Sourcing of ore from west end; 3200 tonnes per day concentrate from east end and 1500 tonnes per day from west end. Pyrrhotite plant was commissioned to treat pyrrhotite from east end mill
1955	Four converters in use at the Falconbridge Smelter. An electrostatic precipitator was installed for cleaning gases from blast furnaces and converters but not the sinter plant
1958	New converter aisle added. 4th blast furnace added, No. 5 & 6 converter added, No. 4 converter shut down. No. 2 blast furnace shut down.
1963	No. 1, 2 & 3 converters shut down
1965	450 ft stack added for dispersion of sinter plant emissions. The previous 304 ft stack is still used today for the electric furnace, converter aisle and acid plant tail gas
1967	Strathcona mine and mill started production. 6000 tonnes per day capacity
1970	No. 1 blast furnace rebuilt and gases passed through cyclone and up 450 ft sinter stack
1971	6th sinter strand added
1972	• Pyrrhotite plant shut down • Construction starts on Smelter Environmental Improvement Project (SEIP)
1978	• Sinter plant and blast furnaces shut down. The remainder of the sinter plant replaced with fluid bed roaster and electric furnace. Fluid bed roaster off-gases to be tied into acid plant. • Blast furnaces shut down and switched to fluidized beds for roasting. All process gases now cleaned and the SO_2, metals and particulate emission reduction is significant
1985	No. 8 converter started up
1994	Furnace rebuild and production consolidation to a single unit
1998	Introduction of concentrate from the Raglan mine in northern Labrador
2004	• Furnace sidewall and roof rebuild • Introduction of concentrate from the Montcalm mine near Timmins, Ontario
2007	Commissioning of calciner
2008	Introduction of concentrate from Xstrata Nickel Australia

2.3 Evolution of the Mining Companies and Mineral Production

The first patent was issued in October 1884 to Thomas and William Murray, Harry Abbott, a CPR construction boss, and John Loughrin, a businessman and politician. The mining rights to the 310 acres that were to become the Murray mine sold for one dollar an acre (Wallace and Thomson 1993). By the turn of the century, independent prospectors had all but disappeared from the Sudbury landscape. The Ni and Cu proved to be very difficult to extract and refine, with very little high-grade ore accessible below the surface deposits. Tremendous amounts of capital became an essential ingredient in any bid to reap a profit from the rock. In his memoir, Aeneas McCharles spoke the hard truth:

> "The Sudbury district is not a poor man's camp. A few big companies are going to make all the money there. It takes large capital to work nickel mines, and if a prospector happens to find a good body of ore, the only thing he can do with it, is to try and sell it." (McCharles 1908)

In the 1880s and 1890s, the huge investment capital required to develop the Sudbury deposits was not available in Canada. The era of cheaper and more efficient transportation ushered in by the railway, coupled with technological developments in the refining process, attracted investment from beyond the nation's borders. The cycle of foreign ownership and foreign control over Sudbury's mining industry began before the first shaft was sunk. In 1885, Samuel Ritchie, an entrepreneur from Ohio, secured 97,000 acres of claims all over the district and formed a syndicate of Cleveland businessmen to begin mining development. On 25 May 1886, 25 or so workers began blasting at the Buttes (later Copper Cliff) mine and the mining of the Sudbury basin began in earnest (Stephenson 1979). The seeds of a mining dynasty were planted in January 1886 with the formation of the Canadian Copper Company of Cleveland, Ohio, with Ritchie as president. As the name suggests, Cu was still the main draw of the region, and the Ni content of the ore was seen almost as a nuisance.

Canadian Copper's local monopoly was broken just as the new century arrived: in 1899, prospector Rinaldo McConnell sold his stake to Ludwig Mond, a Swiss national living in England who had a different method of refining nickel ores. Mond Nickel began operations in 1900, when the European and American navies began to prepare for the brewing conflict. The fledgling company was not a direct threat to Canadian Copper, because it pursued mainly European markets. With time, it would become a key player in the Sudbury mining industry. The year 1901 saw the rise of monopolies in industries almost as new as the century. J. Pierpont Morgan unified various components of the U.S. steel industry, including coal, Fe mining, smelting, refining operations, wire, plate, and rail manufacture into the United States Steel Corporation. To keep control of the Ni supply from competitors, he formed the International Nickel Company (Inco) in New Jersey in 1902. For $10 million, Morgan and the financial interests behind U.S. Steel acquired Canadian Copper, Orford Nickel Corporation Ltd, the Socieété Miniere Caledonienne and a number of non-operating companies with mining rights in Sudbury and New Caledonia, the world's other main source of Ni at the time.

In 1918, International Nickel underwent a structural transformation to turn its Canadian subsidiary, Canadian Copper, into the International Nickel Company of Canada. In 1928, Inco merged with the rapidly growing Mond Corporation. Just prior to the merger, International Nickel became a 'Canadian' corporation by the exchange of shares between the former New Jersey parent and the Canadian subsidiary. Prior to this, Inco was an American company with a Canadian subsidiary – Canadian Copper. From this point on, Inco was a Canadian company with foreign subsidiaries.

Inco did not enjoy its monopoly position for long. A little over a week after the merger of Inco and Mond, a powerful new player entered the story: Falconbridge Nickel Mines Ltd. The development of the Falconbridge properties had an auspicious start. In his search for a Pb–Ni source for battery manufacture, Thomas A. Edison used a magnetic dip needle to discover the ore body that would become Falconbridge's first mine. However, in 1902 Edison abandoned the Falconbridge Township property after several failed attempts to sink a shaft in material resembling quicksand. Further drilling by the E.J. Longyear Company of Minneapolis in 1916–17 revealed extensive Ni-Cu deposits. The original Edison claim had long reverted to the Crown, so the Minneapolis interests staked the property. In 1928, the deposit came to the attention of Thayer Lindsley, a Harvard-educated roving geologist and the president of Ventures Limited, a company formed to support his mining development plans. Lindsley, who was to become known as an outstanding mine-maker, recognized the potential of the deposit, and arranged to purchase the Falconbridge properties in the name of a new company incorporated for the purpose, Falconbridge Nickel Mines Ltd. The selling price of $2.5 million was the highest ever paid for a mine in the Sudbury district. However, the purchase was made on more than a hunch; geological engineers estimated that 500,000–700,000 tons of ore existed above the 500 foot level (Stephenson 1979).

By 1930, the first shaft was down, a smelter ready to treat upward of 300 tons of ore a day was complete, and railway connections were in place. A new company town, complete with water supply, sewage disposal and a school sprang up in the shadow of the smelter. Simultaneously, the search for refining capabilities led to the purchase of the Kristiansand refinery in Norway. Kristiansand owned the European rights to the Hybinette refining process, which had become the industry standard. Falconbridge had learned from the failure of other fledgling mining companies that the only way to survive on Inco's turf was to have its own refining capabilities. In the years to come, the Falconbridge cargo of Ni–Cu matte en route to Norway, was said to be the most expensive on the ocean. Unlike the many companies that sank their fortunes into the Sudbury basin only to come away empty-handed, Falconbridge was to develop into a company more successful than even its visionary founder anticipated; by the 1970s, Falconbridge had become the third largest Ni producer in the 'free' world.

The economic face of mining in Sudbury continued to change even since the Sudbury Soils Study began. Worldwide demand for Ni and Cu, particularly in emerging markets such as India and China, drove the price of these two commodities to unprecedented levels in the period 2005–2009. This, combined with a growing trend toward globalization of major industries around the world, made both Inco and Falconbridge ripe for takeover by larger companies. The companies attempted a merger during 2005–2006, but this was delayed due to lengthy reviews by various anti-competition tribunals. Eventually, Falconbridge Ltd. was acquired by the publicly traded Xstrata Corporation based in Zug and London. The former Falconbridge operations became

Xstrata Nickel in October 2007. In the summer of 2006, the majority of shares of Inco Ltd. were acquired by the privately owned CVRD mining company of Brazil to form CVRD Inco. Late in 2007, the company name was formally changed to Vale Inco, and in 2010, the name was changed to simply Vale. After a century of development and change, the two iconic Canadian mining companies, Inco Limited and Falconbridge, no longer existed.

In the past, emissions to the atmosphere were closely related to mineral production and the amount of ore being processed. This changed dramatically in the 1960s when more stringent government regulations were put in place and technological advances were made by the industries that reduced emissions while production still increased. In 2008, production from the Xstrata smelter was about 65,000 tonnes of Ni and just under 18,000 tonnes of Cu (Table 2.3). For Vale, production in 2008 was roughly 85,000 tonnes of Ni and 115,000 tonnes of Cu.

Table 2.3   **Annual Production (Tonnes) from the Smelters in Sudbury**

	2004[1]	2005[1]	2006	2007	2008	2009
Xstrata Ni	52,595	63,093	61,066	67,576	64,906[2]	65,889[2]
Xstrata Cu	18,402	20,798	20,964	21,978	17,811[2]	18,560[2]
Vale Ni[4]	–	–	94,000	83,000	85,000	43,000[3]
Vale Cu[4]	–	–	109,000	113,000	115,000	42,000[3]

[1] Falconbridge (2005);
[2] Xstrata Nickel (2009);
[3] Production at Vale disrupted due to labor strike in 2009;
[4] Figures provided directly from Vale, November 2010

2.4 Smelter Emissions

2.4.1 Combined Emissions

The early methods of smelting included the use of roast beds, and later, roast yards. It is not possible to reliably know the actual amounts of particulate material (PM), metals and gases emitted during this period. However, estimates suggest that as much as 2.7×10^5 tonnes of SO_2 were emitted annually, together with many tonnes of heavy metal particulates (Holloway 1917), at the peak of the ore roast yard era between about 1895 and 1928. In 1928, the use of open roast beds was forbidden by an order from the Ontario Legislature. The open roast beds were supplemented with more efficient smelter facilities with smoke stacks. The 3 smelters in the Sudbury region were located at Copper Cliff, Coniston and Falconbridge. In the mid-1970s, Ni and Cu emissions from the three smelters were estimated at 1100 tonnes per year (Cox and Hutchinson 1981).

In 1976, the total Canadian SO_2 emissions were 6.0×10^6 tonnes with emissions from the Sudbury area representing about 25% of the national inventory (MOE 1976). Smelter emissions from the Sudbury area smelters have decreased significantly over the years. In 1995, the combined SO_2 emitted from the Vale Copper Cliff and the Xstrata Falconbridge smelters totaled 281,000 tonnes per year. Annual metal emissions, in 1995, were approximately 140 tonnes Cu, 10 tonnes Zn, 87 tonnes Ni, 52 tonnes Pb, 10 tonnes Cd, and 48 tonnes As (MOE 1976). The 2001 data for the Vale and Xstrata operations in Sudbury are presented in Table 2.4. The amount of trace elements released during smelter operations is a function several factors. These factors include the mineralogy of the ores being processed, the tonnage processed, the temperature of the smelting process, with the more volatile elements (e.g., As, Cd, Hg, Pb, Sb, Se, Tl, Zn) emitted at lower temperatures than the less volatile elements (e.g., Cu, Fe, Mn, Ni), and the efficiency of the emission control equipment at the facility (e.g., multi-cyclones, electrostatic precipitators or bag houses).

Table 2.4 **Metal and Sulfur Discharges (Tonnes) for 2001 for Vale and Xstrata Nickel**

Location	Cu	Zn	Ni	Pb	Cd	As	H$_2$SO$_4$	H$_2$S
Vale Central Mills	37.76	3.84	195.80	0.91	0.36	0.12	27.26	–
Copper Cliff Nickel Refinery	5.98	–	12.42	6.68	–	3.94	–	–
Copper Cliff Smelter Complex	109.64	18.96	64.19	63.40	4.84	52.91	1271.08	–
Vale Copper Refinery	43.42	1.20	0.50	2.92	0.05	2.54	0.99	–
Falconbridge Smelter	10.66	5.32	11.810	6.24	1.70	0.27	24.87	21.20

Data from Environment Canada (2002)

2.4.2 Copper Cliff Smelter Emissions

Past emissions of SO$_2$ and other substances from the Copper Cliff facility were substantial. Although SO$_2$ was not explicitly considered as a Chemical of Concern (COC) in the risk assessment, it was released in significant quantities from the smelters and was known to contribute to ecological damage. In addition, SO$_2$ releases were monitored quite accurately by the companies and government agencies relative to the release of metals from the stacks. Therefore, it is useful to examine the general trends in SO$_2$ releases which can be used as an indicator of expected trends in metal emissions. During the 1960s, the annual amount of SO$_2$ released to the atmosphere from the Copper Cliff facility exceeded 2×10^6 tonnes (Figure 2.5). Company records indicate that SO$_2$ emissions peaked at 5600 tonnes per day or about 2 million tonnes/year in the 1960s (Bouillon 1995) although other records suggest the emissions peaked in the 1970s. Sulfur dioxide is one of the most serious byproducts of smelting Ni ore, and consequently is one of Vale's biggest targets for emission reductions. Much of the company's SO$_2$ reduction efforts have been directed at sulfur removal before smelting. Since the Ontario government introduced the Countdown Acid Rain Program in 1986, the company had spent close to $1 billion to reduce SO$_2$ emissions at its Sudbury operations. By the year 2005, SO$_2$ emissions were less than 200,000 tonnes/year.

A new state-of-the-art $115 million facility, that began operating at the Vale Inco smelter late in 2006, has again reduced the amount of SO$_2$ that the company emits. In 2008, the Vale SO$_2$ release from the Copper Cliff facility was down to 148 kilotonnes (Environment Canada 2009). The facility uses unique fluid bed roaster (FBR) off-gas scrubbing technology to capture SO$_2$ from the smelter and convert it into sulfuric acid and liquid sulfur dioxide, which are both saleable products. In November 2010, Vale announced investment in the Atmospheric Emissions Reduction program. At an estimated cost of $1.2 to $2 billion dollars, the project will reduce SO$_2$ emissions by more than 80% from current levels.

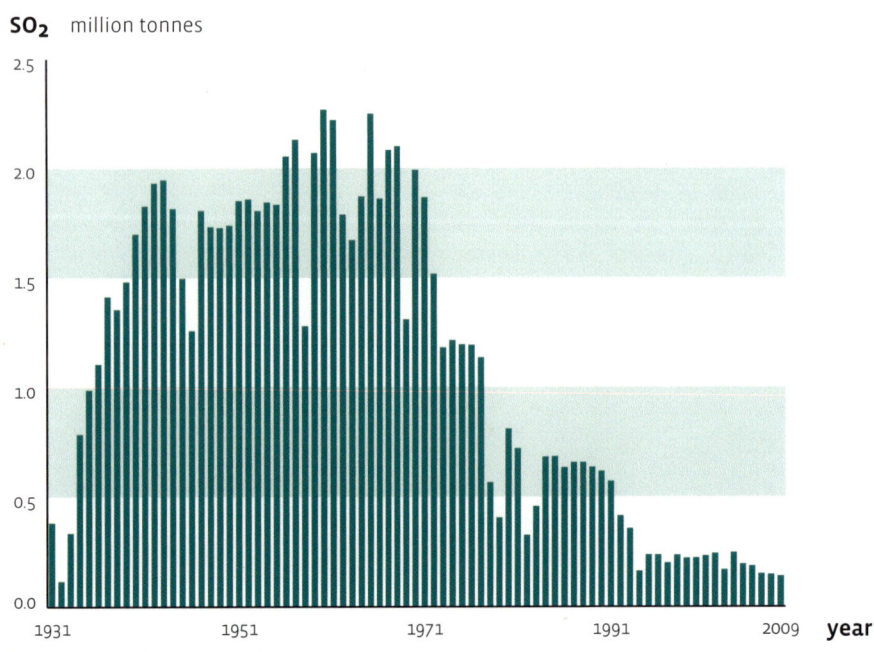

Figure 2.5 **SO$_2$ Generated from the Copper Cliff Smelter**

In addition to S-rich emissions, large quantities of metal-containing particulate materials are vented through the stacks. In 1976 and 1977, emissions from the Copper Cliff stack were estimated to be 1.0×10^4 tonnes, a reduction from the total Inco emissions of 3.4×10^4 tonnes in 1970. The particulate material (PM) emitted is primarily comprised of iron oxides, with significant amounts of Ni and Cu emissions. Total Canadian emissions for all particulates in 1972 were 2.12×10^6 tonnes with 1.42×10^6 tonnes originating from industrial processes (MOE 1976). The discharge of PM declined from a peak of 34,000 tonnes in 1970, to 1129 tonnes in 2008 (Environment Canada 2009).

Bouillon (2003) used emission factors to develop a history of PM emissions generated from smelting and related processes at the Copper Cliff smelter over the period from 1932 to 1996.. A sharp increase in emissions occurred in the 1930s, which was mainly related to an increase in production and the scale of the operations. The emissions peaked between about 1958-1967 at about 24,000 tonnes per year. Particulate emissions declined dramatically in the 1970s and 80s, with the advent of new technologies that included acid plants for recovery of sulfur emissions and electrostatic precipitators. In 2006, 2007 and 2008, particulate emissions from the Vale facility at Copper Cliff were 1259, 1284 and 1269 tonnes, respectively (Vale 2008).

The estimated emissions for select metals at the Copper Cliff facility are shown in Table 2.5 (data from Environment Canada 2010). Emission figures released by Vale are actually slightly higher for some metals. For example, Vale (2008) reported an atmospheric release of 38 tonnes of Pb in 2008, compared with 33 tonnes reported by Environment Canada, however, they are generally similar.

Table 2.5 **Total Annual Emissions from the Vale Smelter (Tonnes/Year) (Environment Canada 2010)**

Year	As	Cd	Co	Cu	Pb	Ni	TSP	Se	Zn	SO$_2$
2009	2.4	0.28	0.28	42	8.1	14	381	1.7	2.8	41,993
2008	10.7	1.1	1.1	92	33.1	37	1023	4.1	9.1	137,965
2007	10.9	1.8	0.65	96	22.8	31	1095	4.9	7.1	140,758
2006	24.7	2	0.8	87	22.5	41	2194	23	20	180,927
2005	34.8	1.9	1.1	101	19.5	50	2482	17	13	189,753
2004	62.6	4.3	1.9	118	42.5	121	3098	31	19	203,271
2003	47.5	5.7	1	70	32.8	51	1955	25	9.1	164,114
2002	58.2	5	1.8	223	77.3	86	2995	24	23	235,907
2001	53	4.8	1.8	110	63	64	NA	16	19	NA
2000	54	6.8	2.2	164	131	78	NA	NA	17	222,906
1999	64	4.3	NA	263	81	84	NA	NA	22	220,987
1998	50	7	NA	97	146	68	NA	NA	14	235,000
1997	53	NA	NA	107	51	144	NA	NA	10	200,003
1996	39	NA	2.5	205	52	122	NA	NA	7.3	236,041
1995	7.3	NA	5.6	107	68	418	NA	NA	16	236,033
1994	6.2	NA	4.4	87	65	333	NA	NA	16	NA

TSP, total suspended particulate matter.

2.4.3 Coniston Smelter Emissions

The Coniston Smelter was decommissioned in 1972 when the 'Super Stack' (381 m) was brought on line at the Vale Copper Cliff smelter. The Coniston smelter used essentially the same technology and gas cleaning technology as the Xstrata sinter plant/blast furnace did from 1933 to 1955. The sinter plant gases were cleaned in a settling chamber, until the strands were shut down in 1978. Vale reported the 1971 PM emission rate for Coniston at 2741 tonnes per year (Bouillon 2003). This was equivalent to 4.4 kg per tonne of concentrate treated. Historical Pb emissions for the Coniston smelter fluctuated between an estimated 10 and 30 tonnes per year (Bouillon 2003) from 1934 until the facility was closed in 1972.

2.4.4 Xstrata Nickel Emissions at Falconbridge Smelter

Sulfur dioxide emissions from the Xstrata smelter peaked in the late 1960s at about 370,000 tonnes per year (Figure 2.6). Government regulations were put into place in the early 1970s, and Xstrata was able to comply with these regulations, even as production increased. With improvements in technology, the SO_2 releases were down to approximately 28,000 tonnes per year in 2000, while the releases in 2008 and 2009 were reported to be 37,600 tonnes and 32,714 tonnes, respectively (Environment Canada 2009).

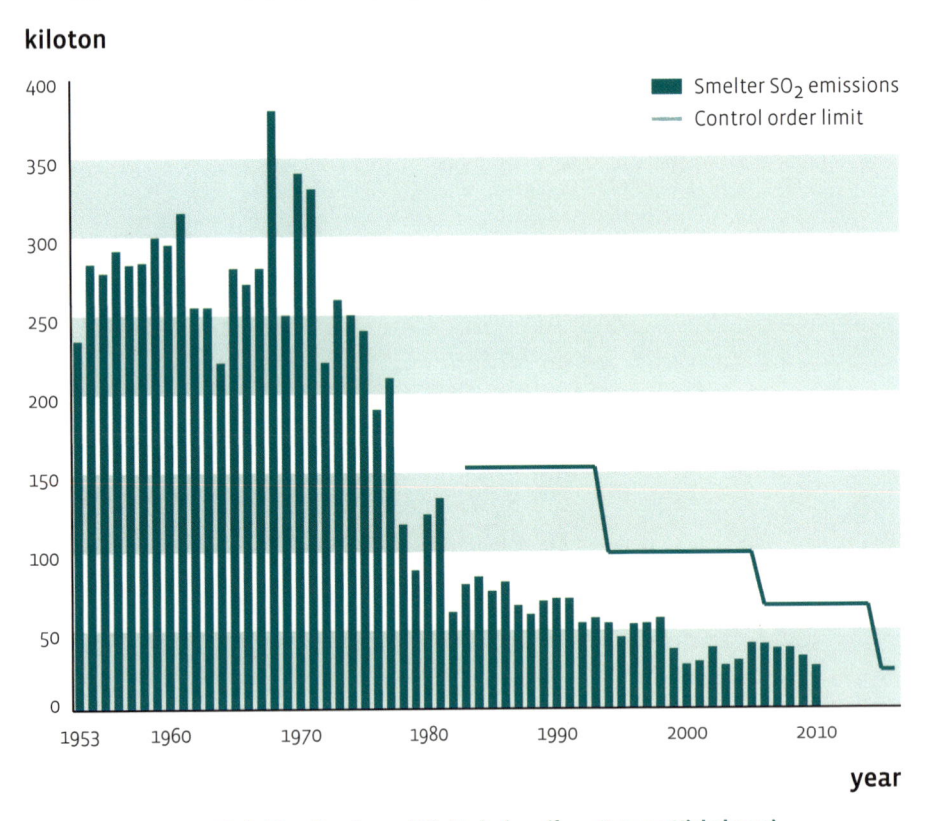

Figure 2.6 Xstrata Nickel Smelter Annual SO_2 Emissions (from Xstrata Nickel 2011)

The release of PM from the Xstrata smelter was estimated to be 12,820 tonnes in 1973 (MOE 1982) and this was significantly reduced with the closure of the sintering plant in 1978, which brought the release to less than 1000 tonnes by 1980. More recently, PM emissions in 2006 and 2007 were 266 and 363 tonnes, respectively (Environment Canada 2009). This represents about a 98% reduction in PM emissions since the 1970s.

The higher As content of ore processed at the Xstrata facility, and relatively high historical emission factors contributed to the prominence of As in the risk assessment in the community of Falconbridge (see also Chapter 8) due to elevated soil As levels. The emissions of the semi-volatile As exceeded the emission of Ni and Cu, which were at a much higher content in the materials being processed. In June 1977, a stack sampling study was carried out by the Ontario Ministry of the Environment at the sinter plant stack of the Xstrata smelter (MOE 1979). The report documents the emissions from the sampling of the 450-foot sinter plant stack as follows: Fe, 2.31 tonnes per day (693 tonnes per year); As, 0.44 tonnes per day (132 tonnes per year); Cu, 0.38 tonnes per day (114 tonnes per year); Ni, 0.27 tonnes per day (81 tonnes per year). Therefore, for the period before 1978, emissions of As from the Xstrata Nickel smelter sinter plant likely exceeded 100 tonnes per year. Arsenic emissions were estimated to be 11.7 tonnes in 1988, and now the annual release of As from the Xstrata facility is under 1 tonne per year. Emissions of select elements from 1994 to 2009 are presented in Table 2.6. Substances with low emission rates are not included in the Environment Canada (Table 2.6) reporting.

Table 2.6 **Annual Emissions (Tonnes Per Year) from Xstrata Falconbridge Smelter Primary Stack (Environment Canada 2010)**

Year	As	Cd	Co	Cu	Pb	Ni	TSP	Zn	SO$_2$
2009	0.83	1.1	1.1	7.2	5.1	13	983	4.9	32,714
2008	0.42	0.84	0.91	7.5	5.1	14	1631	4.4	38,064
2007	0.44	0.85	0.98	8.1	6	13	1792	4.8	37,750
2006	0.26	1.1	0.6	5.5	4.8	9.1	1507	3.9	40,445
2005	0.37	1.1	0.62	5.9	5.6	9.4	1630	na	40,839
2004	0.16	1.3	0.67	6.9	5.7	11	1047	4.1	30,212
2003	0.4	1.2	0.54	6.2	4.9	9.9	2029	4.4	27,133
2002	0.35	1.8	0.81	8.1	8.4	13	2463	5.3	38,300
2001	0.27	1.7	1.8	11	6.2	12	na	5.3	na
2000	0.13	na	0.82	6.1	3.5	14	na	4.3	na
1999	0.38	na	0.87	7.9	7.1	13	na	9.8	na
1998	0.14	na	1.7	10	9.3	17	na	8.3	na
1997	0.12	na	1.8	9.1	13	12	na	17	na
1996	na	na	2.2	9	8.1	14	na	10	na
1995	na	na	0.71	9.1	14	12	na	2	na
1994	0.052	1.3	0.51	4.4	7.7	11	na	1.1	na

TSP, total suspended particulate matter; na, not available.

A minor element balance completed for the Xstrata Nickel smelter in 2003 demonstrates that the collection efficiency in the Cottrell precipitator for non-volatile metals Ni, Cu and Co is about 99%, while the collection efficiency for semi-volatile metals such as Pb and Cd is about 95% (Hatch 2004).

2.4.5 Particulate Size

The size of the particles released from the smelter into the atmosphere is significant for a number of reasons. Larger particles will settle closer to their point of origin, while smaller particles will stay suspended in the atmosphere longer and travel greater distances. Different metals or metalloids also tend to associate with different particle sizes. And lastly, different particle sizes have different toxicological significance and were treated separately within the context of the human health risk assessment (HHRA) (see also Chapter 7.0). For the purpose of the risk assessment, PM was considered in three size fractions including: respirable particulate matter less than 2.5 μm in diameter (PM$_{2.5}$); respirable particulate matter less than 10 μm in diameter (PM$_{10}$); and total suspended particulate (TSP) matter less than 44 μm in diameter. The smaller the size fraction (PM$_{2.5}$), the further it is likely to travel into a person's respiratory system and make its way into the lung. Coarser particles (TSP) tend to be filtered out in the nasal cavity and not travel into the lungs.

Although there are limited data on the size of particles released, the particulate matter released by the Vale smelter is dominantly fine grained (80% by mass) with particles <3 μm, and particles from fugitive emissions (dust from roads, wind erosion of exposed surfaces, releases from material handling and on site storage) are relatively coarse grained at >2 μm. Metals such as Cu and Ni may be primarily associated with coarse particle sizes (>2.5 μm) with mass median diameters <9 μm (Chan and Lusis 1986). These authors further reported that Pb, Zn and As are most frequently associated with fine particles (<2.5 μm), typically with mass median diameters closer to 1 μm. Cumulative plots for coarse particle distributions of Cu indicate that 60–95% of Cu was associated with particles greater than 2.5 μm.

Metals, deposited from the atmosphere by wet and dry depositional processes, can accumulate in a variety of environmental media including soil, water and sediment. Concentrations of emitted metals typically decrease exponentially with distance from the source. For the purpose of monitoring or modeling atmospheric transport, PM is generally subdivided into a fine fraction (<2.5 μm) and a coarse fraction (>2.5 μm). Particulate matter may be primary or secondary. Primary particulate matter is emitted directly into the atmosphere, whereas secondary particulate matter is formed in the atmosphere through chemical and physical transformations. The principal gases involved in secondary particulate formation include SO_2, NO_x, volatile organic compounds and NH_3. Long-range transport of aerosols occurs mainly via particles with diameter less than a few micrometers. If an element is emitted in volatile form from a high temperature source such as a smelter, it will eventually condense on particles in the emission plume. Since smaller particles have a greater surface-to-mass ratio than larger particles, a preferential concentration of volatile chemical species occurs in the small particle fraction (Steines and Friedland 2006). Thus, elements that are entirely or partly emitted as volatile species will be available for long-range transport to a greater extent that those that are released in particulate form from the source.

Extremely fine particles (<0.1 μm) are formed mainly from the condensation of hot vapors during high temperature combustion processes and the nucleation of atmospheric species. These tiny nuclei-mode particles have a short atmospheric residence time as they nucleate and coagulate to yield larger particles. Particles 0.1–2.0 μm in diameter result from the coagulation of particles in the nuclei mode, and from the condensation of metal-rich vapors onto previously existing particles. Thus the growth of these particles may be continuous. This group of particles accounts for much of the particle mass in the atmosphere. Atmospheric removal processes are least efficient in this size range. These fine particles can remain in the atmosphere for days to weeks. Eventually, these particles are removed from the atmosphere by dry deposition and precipitation. Scavenging by precipitation accounts for 80–90% of the mass of particles in the accumulation mode (Wallace and Hobbs 1977).

2.4.6 Wind Climate and Dispersion Patterns

The dispersion of metals, gases and PM from the smelters is governed by the source characteristics of the material released as well as local meteorological conditions. For example, emissions from the Vale superstack may travel tens of kilometers before being deposited to the ground, whereas emissions from low-level sources including wind-blown dust from waste piles will more likely be deposited in the soils of the area immediately surrounding each facility.

In the Sudbury basin, the winds are most frequently from south-westerly and north/north-easterly directions. Long-term climate data for the Sudbury area were reviewed as part of the Sudbury risk assessment for the years 1971 to 2000 (SARA 2006). Winds from all directions are experienced over the course of time, but the winds are most frequently from the southwest and north (Table 2.7).

Table 2.7 **Historical Wind Characteristics**

	Jan	Feb	Mar	Apr	May	Jun	Jul	Aug	Sep	Oct	Nov	Dec	Year
Wind speed (km/h)	16.6	16.1	17.2	17.4	15.9	14.8	13.5	13.2	14.6	16.0	16.7	16.0	15.7
Predominant direction	SW	N	N	N	N	SW	SW	SW	S	S	SW	NW	SW

The wind and other meteorological factors have determined the fate of the emissions from the smelters and other operations since processing began at Vale and Xstrata Nickel, and thus it would be expected that the communities to the northeast and south of the smelting and other facilities would have experienced higher concentrations and subsequent fallout (deposition) of dust and COC than other areas in Sudbury. Local climate and wind patterns were studied as part of the risk assessment to help predict where historical emissions may have been most significant and also to help design an air monitoring program to measure metal levels in air as part of the HHRA exposure assessment (see Chapter 6.1). Another consideration in designing the air-monitoring program was locating monitoring stations to ensure adequate representation of populated areas in the City of Greater Sudbury expected to be influenced by current smelting operations. A preliminary dispersion modeling study was undertaken, based on current emissions information for the Copper Cliff and Falconbridge smelters for As, Co, Cu and Ni. These were the original COC when the study began.

2.5 Environmental Impacts Overview

The environmental effects of human activities, particularly emissions from smelters, in the Sudbury region have been measured and documented by numerous researchers and studies (Gorham and Gordon 1960; Dreisenger 1965; McGovern and Balsille 1972, 1975; Hutchinson and Whitby 1977; Rutherford and Bray 1979; Freedman and Hutchinson 1980a,b; Amiro and Courtin 1981; Cox and Hutchinson 1981; Pitblado and Amiro 1982; Taylor and Crowder 1983). As described earlier in this chapter, the atmospheric release of S gases and metals was significantly reduced during the mid-1970s and following years. These reductions led to rapid improvement in local environmental conditions which permitted restoration programs to be implemented and also led to the natural recovery of regional lake quality and some terrestrial vegetation communities. These recovery scenarios have been monitored and documented (McCall et al. 1995; Winterhalder 1995; Keller and Gunn 1995; Potvin and Negusanti 1995).

The cycle of landscape degradation and recovery has been the subject of comprehensive reviews (Courtin 1994; Winterhalder 1996, 2002) and it is not the intent of this chapter to describe in detail all the environmental studies that have taken place at Sudbury during the past several decades. Rather, this chapter provides an overview and basic understanding of the stressors that caused significant impacts to the surrounding vegetation, and a description of the current status of the landscape. This provides the platform for later chapters, in particular, the ecological risk assessment (ERA) discussed in Chapters 9–11. Although re-greening activities around Sudbury began 35 years ago and have been deemed an immense success, the reader should not be under the false impression that the ecosystem has recovered to anywhere near its full potential. Large areas remain devoid of soil and meaningful vegetation cover, while other areas that have recovered either naturally or through human intervention still do not represent a diverse, naturally sustaining vegetation community. To identify the factors that are limiting ecological recovery of the terrestrial vegetation was one of the objectives of the ERA and is described in Chapter 10. Chapter 12 presents a discussion of the landscape restoration programs implemented to reverse the environmental impacts that have occurred in the Sudbury basin.

2.5.1 The Original Forests and Early Logging

Sudbury lies within a vegetation zone known as the Great Lakes–St Lawrence Forest Region. The study area is in the Laurentian Upland section of this forest region. Prior to the influx of Europeans, the climax forest community in the northern section was characterized by extensive stands of red and white pine (*Pinus resinosa* and *P. strobus*) (Amiro and Courtin 1981). In the southern section, the pines were mixed with hardwoods such as sugar maple (*Acer saccharum*) and yellow birch (*Betula alleghaniensis*). White cedar (*Thuja occidentalis*) would have been present in the low lying wetlands and swamps. Other common tree species included black and white spruce (*Picea mariana* and *P. glauca*), balsam fir (*Abies balsamea*) and jack pine (*Pinus banksiana*).

Logging began about 1872, and lumber from the Sudbury area likely played a role in rebuilding Chicago after the fire of 1871. Larger pines were cut and floated down rivers into Lake Huron and across to saw mills in the northern United States. At first, the large red and white pine were harvested, then as these were depleted, other tree species were cut and removed. At one time, more than 11,000 men were employed in the logging and milling activities around Sudbury (Winterhalder 1995). The original forests were likely replaced by quickly growing successional species such as white birch (*Betula papyrifera*) and trembling aspen (*Populus tremuloides*).

The railway came through the area in 1883, which led to the discovery of mineral deposits as described earlier in this chapter. The advent of the railway led to more frequent forest fires, and it is also reported that early prospectors set fires to expose the underlying bedrock. Widespread impacts to the vegetation had begun which still persist 150 years later.

2.5.2 Roast Yards and Smelter Emissions

The basic goal of the smelting process is to separate the minerals from the rock. In the case of the Sudbury ores, the biggest challenge lay in separating the economic metals (Cu, Ni) from the S, Fe and barren host rock. To reduce the amount of material being transported to the United States for refining, the need for effective on-site smelting arose. The infamous roast yards were the first stage in the smelting process. After a trial roast in December 1886, the first roast yard became operational in 1888. Between 1890 and 1930, approximately 28 million tonnes of ore were processed at the roast yards (Gunn 1995). Eleven roast yards existed in the area between 1988 and 1929. Most trees in the surrounding area were felled for fuel for the roast yards, further contributing to the loss of regional forest cover. It was estimated that over 3.3×10^6 m^3 of wood was consumed in the 11 roast yards during their period of operation (Laroche et al. 1979).

The open roast beds emitted dense low-level fumes rich in SO$_2$ and particulate matter that would have killed nearby plants and vegetation, and acidified the soil. Although widespread vegetation damage is often attributed to the early roast yards, Winterhalder (2002) maintains that impacts from the roast yards were likely quite localized, and it was not until the smelters became operational that vegetation impacts became more widespread and permanent. The damage inflicted on gardens, crops and other vegetation was recognized by the turn of the 20th century but early public complaints appear to have been deflected in defence of the mining companies. For example, in 1915, the Ontario government passed legislation that all patents issued to settlers of land within a defined area include a clause exempting mining companies from liability due to smoke damage (Potvin and Negusanti 1995). The first piece of legislation related to impacts from emissions was the 'Damages by Sulfur Fumes Arbitration Act' which was proclaimed in 1921. The purpose of the Act was to facilitate the settlement of claims of damage to agricultural crops and vegetation. The effects of SO$_2$ to vegetation around Sudbury began to be recognized (Katz 1939) and a subcommittee was formed in 1944 by the Province of Ontario to investigate sulfur emissions and alleged forest damage in the Sudbury area. The subcommittee reported that 'severe burns' of tree foliage occurred as far as 35 km to the northeast, 20 km to the north, and 20 km to the south of the smelters (Murray and Haddow 1945).

The Province of Ontario responded to these growing environmental concerns by imposing control orders on the mining companies in 1969 (Xstrata Nickel, formerly Falconbridge) and 1970 (Vale, formerly Inco) to put annual limits on smelter SO$_2$ emissions. The purpose of the control programs was aimed at improving local air quality with little regard to long-range transport. The closure of the smelter at Coniston by Vale and construction of the 381 m 'super stack' by Vale at the Copper Cliff facility had the immediate result of reducing local SO$_2$ concentrations by 50% after 1972. Since then, SO$_2$ emissions have continued to decline (Figures 2.5 and 2.6) while government controls have become more stringent.

2.5.3 Effects on Landscape: Barren and Semi-barrens

As indicated above, the first concerns regarding emissions from the roasting yards and early smelters related to nearby agricultural crops. Following the early investigations in the 1940s, it appears there was a 10–15 year lag period with little documentation of further studies. Then, in the late 1950s, Linzon (1958) reported on the extreme sensitivity of white pine to SO$_2$ fumes. This species was absent within 19 km of the smelter and displayed decreased growth and increased mortality up to 40 km northeast of the smelters. Some tree species, such as red maple *(Acer rubrum)* and red oak *(Quercus rubra)* are more resistant, perhaps as a result of their deciduous nature, and survive to within 5 km of the smelters (Hutchinson and Whitby 1977).

In 1960, Gorham and Gordon (1960) reported severe impacts to vegetation within 3 km of the smelters, moderate damage up to about 7 km, and no obvious damage beyond 7–11 km from the smelters, although they noted white pine were absent up to 24 km from the smelters. Dreisinger and McGovern (1971) used mean atmospheric SO$_2$ concentrations to describe 'fumigation zones' around Sudbury. The first attempt to quantify the extent of smelter damage to the surrounding forests was undertaken by H. Struik, a forester with the Ontario Ministry of Natural Resources, using a series of aerial photographs from the 1940s to 1970. He mapped areas referred to as 'zones of site and vegetation stability' (Struik 1973), with vegetation being noticeably scarce or absent in concentric rings around each of the three smelters. Later, these areas were found to coincide with the vegetation community labelled as 'the barrens' by Amiro and Courtin (1981). Watson and Richardson (1972) recorded 102 km^2 around the Sudbury smelters as 'severely barren', and another 363 km^2 as having 'impoverished' vegetation. Hutchinson and Whitby (1977) reported that there was a 416 km^2 area around Sudbury that was devoid of vegetation and consisted of blackened rocks.

The barrens were characterized by bare and sparsely vegetated land, severely eroded and blackened hilltops and acidic (pH < 4.0) and metal contaminated soils (Freedman and Hutchinson 1980b; Amiro and Courtin 1981). The plant communities immediately outside the barren areas contained stunted vegetation, abundant open spaces and reduced species diversity. Amiro and Courtin (1981) identified 3 distinct types of vegetation communities in this next zone: maple transition community, birch transition community, and birch-maple community type. In structure, this last community was a mixed hardwood type with coniferous components that increase with distance from the pollution sources. It tends to include a variety of sites that do not fit into any other community type. These sites are located outside the birch and maple transition zones.

The larger area of impacted vegetation surrounding the barrens eventually became known as the 'semi-barrens'. It is unknown exactly when this category was developed but the term 'semi-barrens' was in use in the scientific literature by the mid-1990s (Gunn et al. 1995; McCall et al. 1995; Winterhalder 1995, 1996). In the 'semi-barren' area, conifers are generally absent and the forest consists of a near monoculture of stunted white birch *(Betula papyrifera)* (Amiro and Courtin 1981). Soil conditions are better in the semi-barren area (pH > 5.0), but many exposed hilltops are bare (Gorham and Gordon 1960) and many sensitive plant species such as epiphytic lichens (LeBlanc et al. 1972) are usually absent from this zone. Figure 2.7 shows exposed bedrock and re-sprouting white birch in a barren area that had not yet been treated or limed.

Figure 2.7   **Resprouting birch trees in Untreated Barren Area (Photo Courtesy of the City of Greater Sudbury)**

In the early 1970s, three significant events occurred that significantly reduced local SO_2 concentrations and metal deposition: smelter emissions were reduced following provincial control orders, the smelter at Coniston was closed in 1972, and the large 'super stack' at Copper Cliff went into operation in 1972. McCall et al. (1995) used a series of aerial photographs from 1970 to 1989 to examine changes in vegetation and recovery patterns in response to localized improved air quality. They reported very little natural recovery in the barrens area, although the total surface area of the barren zones shrank by 2554 ha during the 19 year period. But almost all changes (2113 ha) were attributed to the land reclamation program which began in 1978 and with the addition of lime, fertilizer, grass and legume seeds and planting of trees. Only the semi-barren area exhibited substantial natural recovery. The surface area of the semi-barren zone reportedly decreased by 22%, from 83,796 ha to 65,453 ha, as conifers moved into the periphery of the zone. The shrinkage in the semi-barren damage zone occurred mainly along the western and southern limits of the 1970 boundaries. Figure 2.8 provides a schematic map of the Sudbury area showing the spatial limits of the barrens and semi-barrens. Versions of this map have appeared in various publications including Winterhalder (1995), Gunn et al. (1995) and McCall et al. (1995). Although a significant reduction in the extent of the semi-barrens was reported by this analysis over 15 years ago, the approach used provides little information on the number and types of species present in the recovering areas. As a result of the detailed surveys conducted as part of the Sudbury ecological risk assessment, we now know that plant species diversity remains low in many of the reclaimed and natural recovery areas (see Chapter 10).

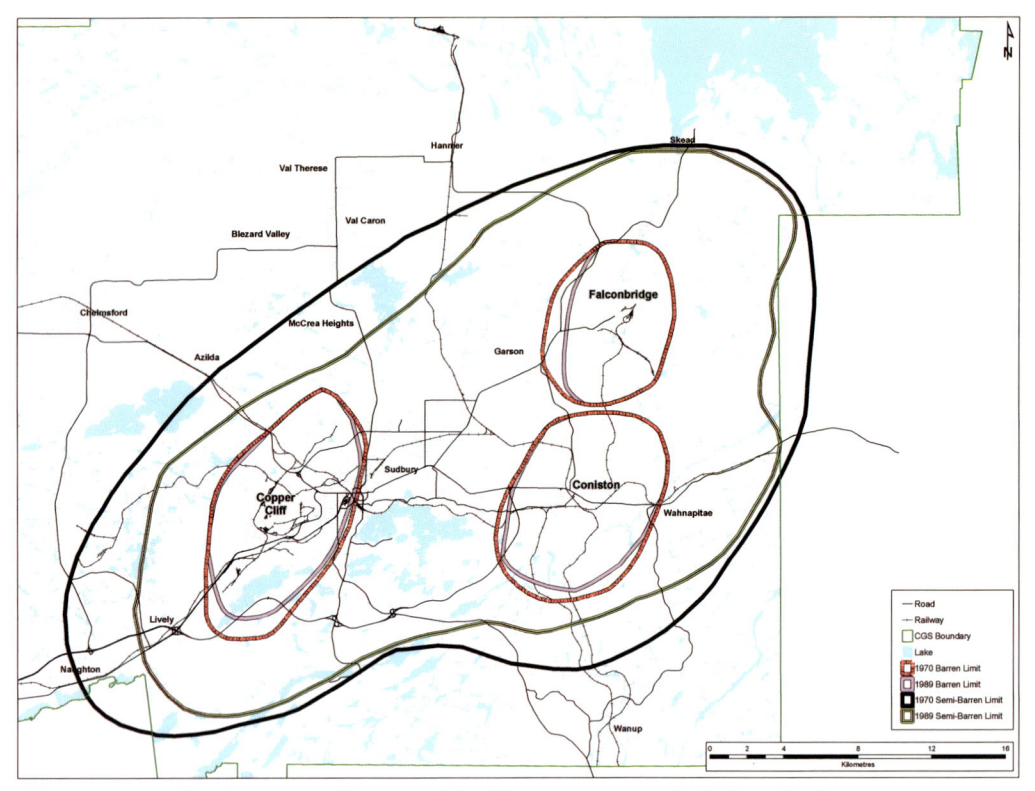

Figure 2.8   Map Showing Extent of Barren and Semi-Barren areas near the Sudbury Smelters

2.5.4 Interacting Stressors

For many years, it was thought that vegetation damage in the Sudbury area was the direct effect of SO_2 emissions from both the roast yards and smelter emissions up to the early 1970s. However, studies began to document elevated metal concentrations in soils surrounding the smelters and led researchers to suggest that metal accumulation was a 'complicating factor' with respect to vegetation impacts (Hutchinson and Whitby 1974). These authors also noted that increased acidity of the soil increased metal solubility and mobility resulting in phytotoxic conditions. Early soil toxicity testing demonstrated that the ambient soil conditions were directly toxic to plants (Whitby and Hutchinson 1974). The primary metals of interest from a toxicological perspective included Ni and Cu but later investigations noted the possible toxic effect of reduced soil pH and Al concentrations in the soil (Cox and Hutchinson 1981) as well as lower levels of Co (Winterhalder 1995). A professor at Laurentian University, Keith Winterhalder, seemed to be among the first to recognize that the Sudbury landscape was the result of many interacting factors including chemical, biological and physical stressors. Figure 2.9 represents a simple conceptual illustration developed by Professor Winterhalder in 1984 showing the relationship between many of these interacting factors that ultimately resulted in barren or semi-barren vegetation conditions.

The original logging activities for timber and fuel for the roast beds coupled with forest fires led to early clearing of the landscape. These caused increased erosion and soil loss from the hilltops and hillsides. During these processes, valuable organic material, leaf litter and soil nutrients were lost. The loss of vegetation and soil also reduced insulation to the earth's surface and resulted in severe microclimates. In these areas, frost action is enhanced leading to direct damage to vegetation, and the summer sun can cause scorching conditions on the exposed bedrock not amenable to plant growth. Later emissions of sulfur and metals from the roast beds and smelters and subsequent deposition in the surrounding landscape produced toxic conditions in the remaining soil. Decades of deposition of acidic material helped leach important nutrients and Ca from the soil, further exacerbating the toxic effect of metals. Thus, it has been a slow and complex process that resulted in the barren and impacted conditions covering many thousands of hectares of landscape. Understanding these factors and processes is necessary to help plan risk management strategies (see also Chapter 12). It is also important to recognize that these impacts took place over a period of more than a century, and that natural forest ecosystems evolved over many centuries. Thus, reversing these impacts will not be completed in a short timeframe of one or two generations.

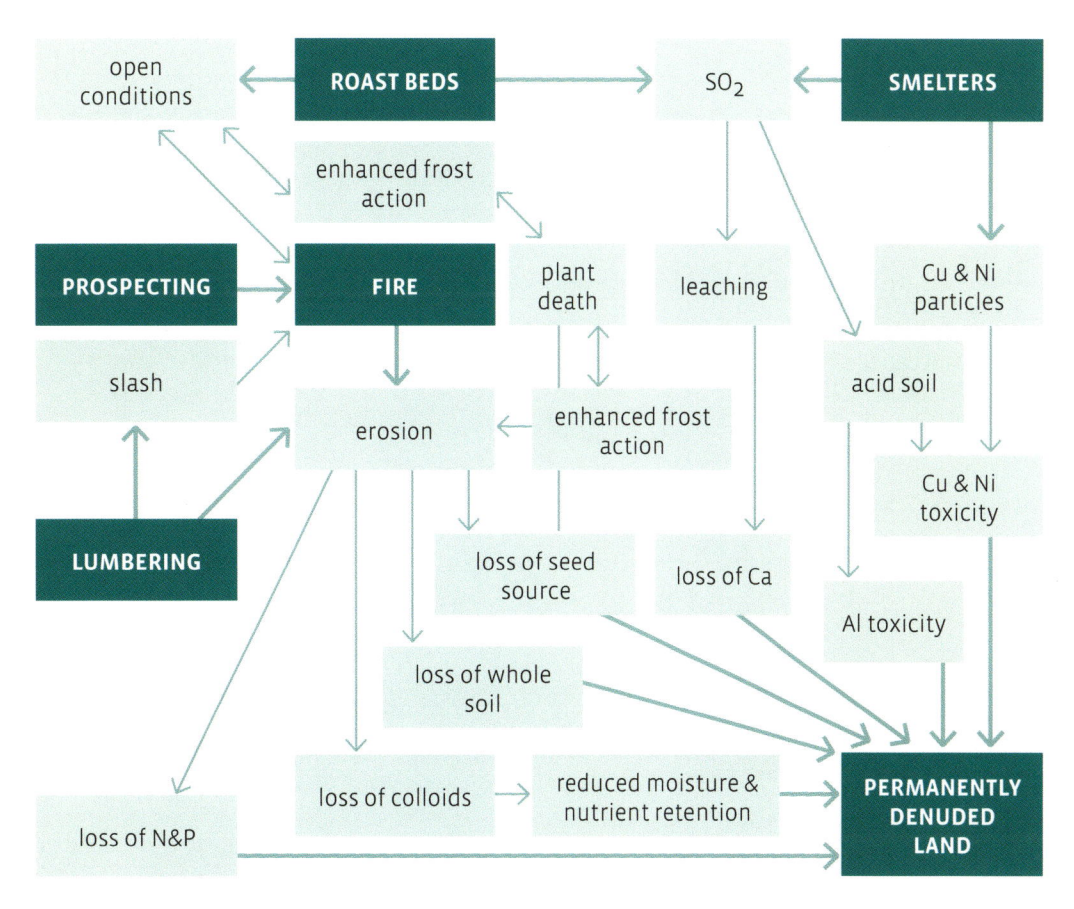

Figure 2.9 **Factor Interactions Leading to the Formation of Barren Land (Redrawn from Winterhalder 1995, 1996)**

2.5.5 Industrial Barren Landscapes

Sudbury is not alone in the world as an example of widespread vegetation impacts and destruction from industrial activities. Kozlov and Zvereva (2007) reviewed the available information on 36 industrial 'barrens' worldwide. Nearly all of them (33 of 36) were located adjacent to non-ferrous smelters and refineries, primarily those of Cu-, Ni-, Zn- and Pb-producing facilities. The authors considered industrial barrens as bleak open landscapes around point sources of industrial pollution due to deposition of airborne pollutants, with small patches of vegetation (cover usually 10% or less relative to control) surrounded by bare land. Yakovlev et al. (2008) described in some detail the environmental impacts from the Ni smelter at Norilsk, Russia, including an absence of trees within a 4 km radius of the smelter.

From their review of impacted landscapes, Kozlov and Svereva (2007) were able to identify several common factors and conditions among the sites. For example, the sites were located predominantly in mountainous or hilly landscapes, possibly because this is where the mineral ores occur, or because this type of topography is more susceptible to soil erosion. Even after emissions decline or stop, the soils remain contaminated and are subject to erosion processes. It was also noted that vegetation damage from smelters was usually preceded or accompanied by other human-induced disturbances, notably logging for mine timber and fuel. Vegetation destruction was often concluded by fires. These factors were all present at Sudbury.

In addition to fumigation, severely metal-contaminated soil was considered to be a significant factor contributing to barren landscapes. The absence of industrial barrens around power plants and Al smelters was given as possible evidence that the presence of metals strongly contributed to barren landscapes (Kozlov and Zvereva 2007). The metals also hamper natural recovery of vegetation although most industrial barrens display some signs of natural recovery. However, recovery may be very slow. For example, natural re-vegetation had not occurred in industrial barrens adjacent to a Zn smelter at Palmerton, PA, USA, for at least 50 years (Sopper 1989).

2.6 Acknowledgements

The authors would formally like to thank Dr Bruce Conard (formerly with Inco Ltd.) for his helpful review and knowledge of technology changes at the Vale smelter. We also thank Angie Robson (Vale) for providing updated information and Laura Taylor (AECOM) for providing a historical perspective of Sudbury in the original Sudbury Soils Study report.

2.7 References

Amiro BD, Courtin GM (1981) Patterns of vegetation in the vicinity of an industrially disturbed ecosystem, Sudbury, Ontario. Can J Bot 59(9):1623–1639

Boldt JR Jr (1967) The winning of nickel. D. Van Nostrand Company, Inc., Toronto

Bouillon D (1995) Developments in emission control technologies/strategies: A case study.
In: Gunn JM Restoration and recovery of an industrial region, pp. 275–286. Springer-Verlag, New York

Bouillon D (2003) Preliminary assessment of 100 years of Pb emissions from the Inco Sudbury operations. Internal Vale report

Chan WH, Lusis M (1986) Smelting operations and trace metals in air and precipitation in the Sudbury Basin. John Wiley & Sons, New York

Courtin GM (1994) The last 150 years; a history of environmental degradation in Sudbury.
Sci Total Environ 148:99–102

Cox RM, Hutchinson TC (1981) Environmental factors influencing the rate of spread of the grass Deschampsia ceaspitosa invading areas around the Sudbury nickel-copper smelters. Water Air Soil Pollut 16:83–106

Dietz R (1964) Sudbury structure: an astrobleme. J Geol 72:412–434

Dreisenger BR (1965) Sulphur dioxide levels and the effects of the gas on vegetation near Sudbury, Ontario. 58th Annual Meeting of Air Pollution Control Association, Toronto. Paper 65-121

Dreisenger BR, McGovern PC (1971) Sulphur dioxide levels and vegetation injury in the Sudbury area during the 1970 season. Ontario Dept. of Mines, Sudbury, 35 pp

Environment Canada (2002) National pollution release inventory. www.ec.gc.ca/inrp-npri

Environment Canada (2009) National pollution release inventory. www.ec.gc.ca/inrp-npri

Environment Canada (2010) National pollution release inventory. www.ec.gc.ca/inrp-npri

Falconbridge (2005) Falconbridge Annual Report 2005.
www.archive.xstrata.com/falconbridge/ourbusiness/nickel

Freedman B, Hutchinson TC (1980a) Pollutant inputs from the atmosphere and accumulation in soils and vegetation near a nickel-copper smelter at Sudbury, Ontario, Canada. Can J Bot 58:108–132

Freedman B, Hutchinson TC (1980b) Long term effects of smelter pollution at Sudbury, Ontario, on forest community composition. Can J Bot 58:2123–2140

Gorham E, Gordon AG (1960) Some effects of smelter pollution north-east of Falconbridge, Ontario. Can J Bot 38:307–312

Gunn JM (ed) (1995) Restoration and recovery of an industrial region. Springer-Verlag, New York

Gunn J, Keller W, Negusanti J, Potvin R, Beckett P, Winterhalder (1995) Ecosystem recovery after emission reductions: Sudbury, Canada. Water Air Soil Pollut 85:1783–1788

Hatch R (2004) 2003 Minor element balance for Falconbridge. Report prepared for Falconbridge Limited

Holloway ME (1917) Report of the Ontario Nickel Commission, with Appendix.
Legislative Assembly of Ontario, Toronto

Hutchinson TC, Whitby LM (1974) Heavy metal pollution in the Sudbury mining and smelting region of Canada, I. Soil and vegetation contamination by nickel, copper and other metals. Environ Conserv 1:123–132

Hutchinson TC, Whitby LM (1977) The effects of acid rainfall and heavy metal particulates on a boreal forest ecosystem near the Sudbury smelting region of Canada. Water Air Soil Pollut 7:421–438

Jackson JF (1990) Evolution of the Falconbridge smelter. Falconbridge Ltd, Sudbury Division

Katz M (1939) Effects of sulphur dioxide on vegetation. Bill 815. National Research Council of Canada, Ottawa, Canada

Keller W, Gunn JM (1995) Lake water quality improvements and recovering aquatic communities. In: Gunn JM (ed) Restoration and recovery of an industrial region, pp. 67–80. Springer-Verlag, New York

Kelly K (2004) Meteorite crash turned Earth inside out: New research paints a picture of what happened billions of years ago when a devastating meteorite crashed into the Earth. News @ the University of Toronto, 3 June 2004

Kozlov MV, Zvereva EL (2007) Industrial barrens: extreme habitats created by non-ferrous metallurgy. Rev Environ Sci Biotechnol 6:231–259

Laroche C, Sirois G, McIlveen WD (1979) Early roasting and smelting operations in the Sudbury region – an historical outline. Ontario Ministry of the Environment and Energy. Sudbury, Ontario. Cited in Winterhalder 1995

LeBlanc F, Rao DN, Comeau G (1972) The epiphytic vegetation of Populus balsamifera and its significance as an air pollution indicator in Sudbury, Ontario. Can J Bot 50:519–528

LeBourdais DM (1953) Sudbury basin: The story of nickel. The Ryerson Press, Toronto

Linzon SN (1958) Influence of smelter fumes on the growth of white pine in the Sudbury region. Department of Lands and Forest, Department of Mines, Ontario

McCall J, Gunn J, Stuik H (1995) Photo interpretive study of recovery of damaged lands near the metal smelters of Sudbury, Canada. Water Air Soil Pollut 85:847–852

McCharles A (1908) Bemocked of destiny: The actual struggles and experiences of a Canadian pioneer, and the recollections of a lifetime. W. Briggs, Toronto

McGovern PC, Balsille D (1972) Sulphur dioxide levels and environmental studies in the Sudbury area during 1971. Technical Report, Ontario Ministry of the Environment, Air Quality Branch, Sudbury, p. 33

McGovern PC, Balsille D (1975) Effects of sulphur dioxide and heavy metals on vegetation in the Sudbury area (1974). Ontario Ministry of the Environment, Northeast Region, Sudbury, p. 33

MOE (1976) Air pollution control directorate. Ontario Ministry of the Environment, Toronto

MOE (1979) Airborne investigations of the Inco and Falconbridge plumes 1976–1977. Ontario Ministry of the Environment Report No. ARB-TDA, pp. 60–79. Ontario Ministry of the Environment, Toronto

MOE (1982) Sudbury environmental study 1973–1980: Synopsis, p.119. Ontario Ministry of the Environment, Toronto

Murray A (1857) Report of Progress for the Year 1856; In Geological Survey of Canada, Reports of Progress 1853-1856, p. 145-188. Ottawa

Murray RH, Haddow WR (1945) First report of the subcommittee on the investigation of sulphur dioxide smoke conditions and alleged forest damage in the Sudbury region, February 1945. 67 p (cited in Winterhalder 2002)

Patterson JM (2001) Technical report on mineral properties in the Sudbury basin, Ontario. For Fort Knox Gold Resources Inc

Pitblado JR, Amiro BD (1982) Landsat mapping of the industrially disturbed vegetation communities of Sudbury, Canada. Can J Remote Sensing 8:17–28

Potvin RR, Negusanti JJ (1995) Declining industrial emissions, improving air quality, and reduced damage to vegetation. In: Gunn J (ed) Restoration and recovery of an industrial region, pp. 51–61. Springer-Verlag, New York

Rutherford GK, Bray CR (1979) Extent and distribution of soil heavy metal contamination near a nickel smelter at Coniston, Ontario. J Environ Qual 8:219–222

SARA (2006) Sudbury soils study, Volume I: Background and study organization. Report prepared by Sudbury Area Risk Assessment (SARA) Group for Soils Study Technical Committee. www.sudburysoilsstudy.com

Sopper WE (1989) Revegetation of a contaminated zinc smelter site. Landscape Urban Plan 17:241–250

Steinnes E, Friedland AJ (2006) Metal contamination of natural surface soils from long-range atmospheric transport: Existing and missing knowledge. Environ Rev 14:169–186

Stephenson R (1979) A guide to the golden age: Mining in Sudbury, 1886–1977. Laurentian University, Sudbury

Struik J (1973) Photo interpretive study to assess and evaluate the vegetational and physical state of the Sudbury area subject to industrial emissions. In: Sudbury Environmental Enhancement Program Summary Report, 1969–1973, pp. 6–11. Ontario Department of Lands and Forests, Sudbury, Ontario

Taylor JG, Crowder AA (1983) Accumulation of atmospherically deposited metals in wetland soils of Sudbury, Ontario. Water Air Soil Pollut 19:29–42

US EPA (1995) Compilation of air pollutant emission factors, Volume 1: Stationary point and area sources. AP 42, Fifth Edition. Office of Air Quality Planning and Standards, US EPA, Research Triangle Park, NC, USA

Vale (2008) Ontario operations sustainability report. Vale. 8 pp. www.nickel.vale.com/sustainability/reports/pdf/vale

Wallace JM, Hobbs PV (1977) Atmospheric science: An introductory survey, p. 467. Academic Press, Inc., New York

Wallace CM, Thomson A (eds) (1993) Sudbury: Rail town to regional capital. Dundurn Press Limited, Toronto

Watson WY, Richardson DH (1972) Appreciating the potential of a devastated land. Forest Chron 48:312–315

Whitby LM, Hutchinson TC (1974) Heavy metal pollution in the Sudbury mining and smelting region of Canada. II. Soil toxicity tests. Environ Conserv 1(3):191–200

Winterhalder K (1995) Early history of human activities in the Sudbury area and ecological damage to the landscape. In: Gunn JM (ed) Restoration and recovery of an industrial region, pp. 17–32. Springer-Verlag, New York

Winterhalder K (1996) Environmental degradation and rehabilitation of the landscape around Sudbury, a major mining and smelting area. Environ Rev 4:185–122

Winterhalder K (2002) The effects of the mining and smelting industry on Sudbury's landscape. In: Rousell DH, Jansons KJ (eds) The physical environment of the City of Greater Sudbury. Ont Geol Surv Spec Vol 6:145–174

Yakovlev AS, Pleekhanova IO, Kudryashov SV et al (2008) Assessment and regulation of the ecological state of soils in the impact zone of mining and metallurgical enterprises of Norilsk Nickel Company. Eurasian Soil Sci 41(6):648–659

Xstrata Nickel (2009) Xstrata Nickel Annual Report 2009. www.Xstrata.com/annualreport/2009/performance/nickel/operations

2.8 Appendix Chapter 2: Abbreviations

COC, Chemicals of Concern

CPR, Canadian Pacific Railroad

ERA, ecological risk assessment

FBR, Fluid Bed Roaster

HHRA, human health risk assessment

IORP, Iron Ore Recovery Plant

MOE, Ontario Ministry of the Environment

PGM, Platinum Group Metals

PM, particulate matter

SARA, Sudbury Area Risk Assessment Group

SEIP, Smelter Environmental Improvement Project

TSP, Total Suspended Particulate

DOI: 10.5645/b.1.3

3.0 The 2001 Soil Survey and Selection of Chemicals of Concern

Authors

Christopher Wren, PhD
MIRARCO, Laurentian University
935 Ramsey Lake Road, Sudbury, ON Canada P3E 2C6
Email: cwren@mirarco.org

Graeme A. Spiers, PhD
MIRARCO, Laurentian University
935 Ramsey Lake Road, Sudbury, ON Canada P3E 2C6
Email: gspiers@mirarco.org

Glenn Ferguson, Ph.D.
Intrinsik Environmental Sciences Inc.
6605 Hurontario Street, Suite 500, Mississauga, ON Canada L5T 0A3
Email: gferguson@intrinsikscience.com

Dave McLaughlin
Ontario Ministry of the Environment
125 Resources Road, Toronto, ON Canada M9P 3V6
Email: dave.l.mclaughlin@ontario.ca

Table of Contents

3.1 Background ... 51
3.2 Soil Sampling Study Design ... 52
   3.2.1 Regional Soil Survey ... 52
   3.2.2 Urban Soil Survey ... 53
   3.2.3 Falconbridge Soils Survey ... 53
3.3 Methodology ... 54
   3.3.1 Soil Sampling Protocol ... 54
   3.3.2 Sample Preparation and Analysis ... 54
3.4 Overview of Results ... 56
3.5 Selection of Chemicals of Concern (COC) ... 57
   3.5.1 Data Screening Process ... 57
   3.5.2 Selection Criterion #1: Parameter Must Be above MOE Guideline ... 59
   3.5.3 Selection Criterion #2: Parameter Must Be Present across the Study Area ... 62
   3.5.4 Selection Criterion #3: Parameter Must Show Origin from the Companies' Operations ... 64
3.6 Summary ... 69
3.7 Acknowledgements ... 69
3.8 References ... 69
3.9 Appendix Chapter 3: Abbreviations ... 70

Tables

- 3.1 Summary of Mean Method Detection Limits (mg/kg) in Soil ... 55
- 3.2 Summary of 2001 Soil Survey Results for 20 Inorganic Parameters ... 57
- 3.3 Initial Data Screening and Comparison with MOE Criteria ($n=8148$) ... 60
- 3.4 Summary of Secondary Data Screening and Evaluation ... 61
- 3.5 Correlation of Se to the Established COCs and Other Elements ... 63
- 3.6 Correlation between Pb, Cu, Ni and V in Soil (0–5 cm) ... 64
- 3.7 Mean Metal Concentration in the Regional Soil Survey, and Calculated Enrichment Factors for the Surface (0–5 cm) Layer (Adapted from CEM 2004) ... 65
- 3.8 Mean Metal Concentrations (mg/kg) in Urban-undisturbed Natural Soil Profiles ($n=14$ Samples per Depth) ... 66
- 3.9 Lead Levels (mg/kg) in Surface Soil (0–5 cm) Correlated to Distance from the Smelters ... 66

Figures

- 3.1 Study Area Boundaries and Soil Sampling Locations ... 52
- 3.2 Typical Soil Cores in Forested Area (Photo from CEM 2004) ... 54
- 3.3 Soil Data Screening Process for Selection of COC ... 59
- 3.4 Spatial Distribution of Ni in Samples from the Regional Soil Survey ... 62
- 3.5 Spatial Distribution of As in Samples from the Regional Soil Survey ... 63
- 3.6 Se Concentration vs. Distance within a 7 km Radius from the Copper Cliff Smelter ... 67
- 3.7 Pb Concentration vs. Distance within a 10 km Radius from the Copper Cliff Smelter ... 67
- 3.8 Backscatter Electron (BSE) Micrograph Showing Spherical Particle in Soil Associated with High Temperature Source (Photo from GeoLabs 2004) ... 68

3.1 Background

As described earlier in Chapter 1, in 2001, the Ontario Ministry of the Environment (MOE) released a report entitled *Metals in Soil and Vegetation in the Sudbury Area (Survey 2000 and Additional Historic Data)* (MOE 2001). That report reviewed and summarized the results of 30 years of environmental studies in the Sudbury area to examine historical trends, identify data gaps and to compare the metal levels in local soils to the MOE's *Guideline for Use at Contaminated Sites in Ontario* (MOE 1997). During the period 1971–1999, the MOE collected and analyzed soil samples from 92 locations. They also collected soils from 103 sites in 2000 for metal analysis. Other studies have also measured metal concentrations in Sudbury soils (Freedman and Hutchinson 1980; Dudka et al. 1995).

The detailed MOE review of existing soil quality information revealed that it was not possible to compare metal concentrations between many of the different studies due to differences in methodology including sampling and analytical protocols. In addition, spatial coverage was inconsistent and/or sample sizes were relatively small. Some studies measured select metals but not others. For example, Dudka et al. (1995) measured Cd, Co, Cu and Ni in samples of Sudbury soils, but not As, Pb or Se. Based on their 2000 study and analysis, the MOE recommended that a more detailed and broader soil study be undertaken to fill data gaps.

Through this historical review, the MOE also identified that the concentrations of As, Co, Cu and Ni in Sudbury soils exceeded the soil quality guidelines, and further investigation was needed. A comprehensive soil sampling and analysis program was undertaken in the summer and fall of 2001. The sampling program was divided between the participants as follows:

- Vale and Xstrata Nickel retained the services of Laurentian University's Centre for Environmental Monitoring (CEM) to collect soil samples in more remote and undisturbed areas to determine the spatial extent (geographic area) of the smelter 'footprint' and attempt to determine background concentrations of metals in the region (referred to as the 'regional' soils survey).
- The MOE collected soil samples from schools, daycares, parks and beaches across the Sudbury area, as well as from 439 residential properties (this is referred to as the 'urban' soils survey).
- Xstrata Nickel retained a private consulting firm (Golder Associates Ltd.) to collect soil samples on properties owned by the company within the Town of Falconbridge, as well as some surrounding municipal and crown lands (referred to as the Falconbridge soils survey).

During the sampling program, approximately 8400 soil samples were collected from 1190 locations throughout the study area. This makes it possibly the most comprehensive soil survey of its kind on record. Each sample was analyzed for 20 inorganic parameters as described below in Section 3.2. Soil samples were collected from different depths to provide a vertical profile of metal concentrations. In addition, numerous duplicate samples were collected for quality assurance and quality control purposes. Data from the three surveys (Golder Associates Ltd 2001; CEM 2004; MOE 2004a) were combined into a single database. The individual reports are available at www.sudburysoilsstudy.com as part of Volume I of the Sudbury Soils Study.

The CEM designed the regional soil sampling survey to collect data from rural and remote areas with undisturbed soils. The CEM study area was approximately 200 km × 200 km and encompassed the City of Greater Sudbury. The approximate boundaries of the sampling area are shown in Figure 3.1 along with sampling locations for the soil surveys. At the scale of Figure 3.1, all sample locations are not readily apparent.

This chapter provides a brief overview of the methods and results of the 2001 soil survey. The samples from these three sources were combined together into a database of soil metal concentrations in the Sudbury region which formed the basis of the risk assessment. These soils data were used for the selection of Chemicals of Concern (COC) which is described later in this chapter. The regional distribution of the COC in the study area and the distribution of the COC in the five Communities of Interest (COI) considered in the human health risk assessment are presented in Chapter 4.

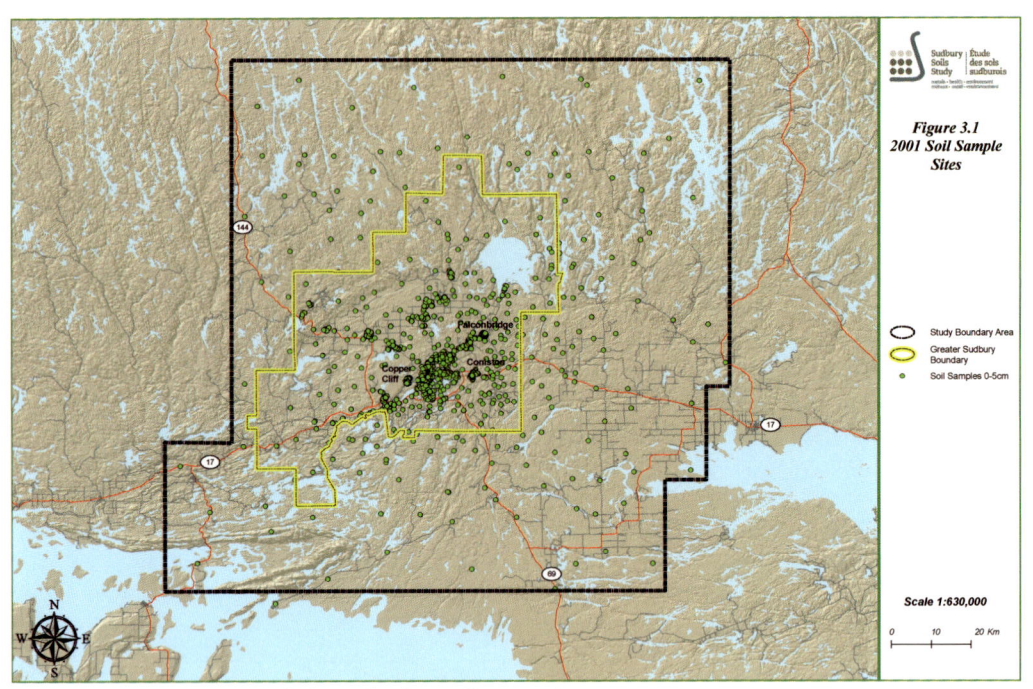

Figure 3.1   Study Area Boundaries and Soil Sampling Locations

3.2 Soil Sampling Study Design

3.2.1 Regional Soil Survey

The primary purpose of the regional soil sampling survey was to determine the spatial area of soil metal levels affected by the Sudbury smelters, and to establish regional soil chemical background levels. Soil data from the Sudbury region was deliberately avoided by the Ministry when it established background soil standards for Ontario because of the uncertainty of the spatial distribution of the impacts on soil chemistry from historical smelter emissions. The regional soil sampling survey conducted by the CEM was developed using a randomly stratified sampling plan, centered on the three historical smelters in Copper Cliff, Coniston and Falconbridge, with the center near the Copper Cliff smelter.

The final nested sampling grid covered an area approximately 200 km × 200 km in size (40,000 km²) which represents a study area almost the size of Switzerland (area 41,228 km²). An imaginary grid was overlain on the entire area. The cells of the stratified sampling plan ranged in size, with the smallest cells being in the zones of historically known smelter impact. The cells were 2, 4, 8 and 16 km square, respectively. The irregular 2 km grid cells extended approximately 8 km in each direction from the individual smelters. The 4 km grid cells extended 12 km out from the edge of the 2 km grid cells, with the 8 km grid cell area being squared off and designed to encompass the rest of the Greater City of Sudbury. The 16 × 16 km border grid was only one cell deep.

A series of sampling exclusion zones were developed to minimize any resultant data bias from contamination by recent erosion, sedimentation or flooding, as well as from non-smelting anthropogenic activities such as effects of road construction, road salt drift, and railway traffic fugitive dust. These exclusion zones included:

- industrial lands, such as tailings ponds, slag heaps, open pits
- wetlands
- lakes
- rivers and streams
- 200 m of the center line on primary roads
- 100 m of the center line on secondary roads
- 100 m of railway lines; and
- 100 m of major utility lines.

Soil samples were collected randomly from within each cell. Many of the soil sampling locations were remote and required helicopter access.

The regional survey collected soil cores following the MOE protocol described below (MOE 1993). In total, 386 sites were sampled as part of the regional soil survey. Core samples were sectioned to collect soils from 0 to 5 cm, 5 to 10 cm, and 10 to 20 cm depth. This provided a detailed examination of the vertical distribution of metals in the surface soils. The CEM survey made significant attempts to relate the geochemistry of surface soils to the bedrock mineralogy. A subset of samples was also taken at depth (i.e. over 80 cm deep) to determine the natural 'background' metal concentrations in Sudbury soils. This deep soil layer (parent material) was assumed to be unaffected by atmospheric deposition or other human sources.

3.2.2 Urban Soil Survey

For the urban soils survey, the MOE collected soil samples from four land uses: residential, schools, parks and agricultural. Three types of soil were sampled: soil, sand, and gravel. The division of these three soil types are as follows:

Soil

- Urban Soil (developed, grassed areas)
- Urban Garden Soil (residential vegetable gardens)
- Agricultural Soil (commercial market garden and berry farms)
- Undisturbed Natural Soil (undeveloped, naturally vegetated areas)

Sand

- Play Sand (material used around play structures, brought in for landscaping purposes)
- Beach Sand (from parks with beaches, tends to be naturally occurring)

Gravel

- Crushed Stone (used in the infields of baseball diamonds, tends to be brought in for landscaping purposes)
- Playground Gravel (used in many school playgrounds, tends to be brought in for landscaping purposes)

Sand and gravel were collected because, unlike grass-covered urban soil, these can come into direct contact with skin, increasing the risk of exposure. Soil, play sand, crushed stone and gravel samples were collected from each school and daycare within the City of Greater Sudbury. Soil and sand samples were also collected from the major parks and sports complexes within the City of Greater Sudbury.

The goal of the MOE program was to characterize the soil chemistry of the urban green spaces and residential areas of the City of Greater Sudbury. All schools, known commercial daycare centers and park areas were sampled. Residential property soil was sampled at about 10% of the houses in Copper Cliff, Coniston and Falconbridge, and a representative number of houses in the remainder of the City of Greater Sudbury. The breakdown of residential sampling locations in the risk assessment Communities of Interest was as follows:

- Falconbridge: 51
- Coniston: 75
- Copper Cliff: 74
- City of Greater Sudbury: 239

As part of the intensive urban soil survey, a total of 6734 soil samples were collected from 770 properties in the City of Greater Sudbury. This included 16 commercial agriculture properties, 146 schools, 169 parks and the 439 residential properties detailed above.

3.2.3 Falconbridge Soils Survey

Sample locations were limited to properties owned by Xstrata Nickel in the town of Falconbridge, as well as municipal and crown lands, to provide spatial coverage and representation of different terrain types including disturbed and natural (undisturbed) sites. A total of 33 sites were sampled, including three parks, 14 wooded areas, three residential yards, one school, two playgrounds, four grassy areas, three vacant lots, one gravel lot and two grass medians. Soil samples were collected, prepared and analyzed following the MOE protocol described below.

3.3 Methodology

3.3.1 Soil Sampling Protocol

Soil samples were collected according to the standard protocol in the government publication *'Field Investigation Manual, Part 1, General Methodology'* (MOE 1993). All sites for the regional survey were in stable landscape positions, with minimal evidence of erosion and with a full stable vegetation cover. Limed sites from the regional re-greening program were not selected as core sites for the sampling program. In the wooded areas, sampling was conducted within a 10 m quadrat, with the duff (leaf/grass litter) being lightly scraped away with a boot or hand. The UTM coordinates of the stake at the quadrat corner were taken with a GPS unit and recorded on standard site description forms.

All soil samples were collected with a hand-held soil corer (Figure 3.2) with 15 to 30 soil cores collected per site. Samples were collected along a grid, 'W' or 'X' pattern at each location or property, to ensure even coverage of the property. Each soil core was divided into three depth intervals (0 to 5 cm, 5 to 10 cm and 10 to 20 cm). The 0–5 cm samples from one site were mixed together to form a composite sample to represent each location. The same process was followed to create separate 5–10 cm and 10–20 cm composite samples. Duplicate composite soil samples were collected by performing the soil sampling procedure a second time.

The regional study also sampled parent material to aid in determining normal background levels of metals for the Sudbury Basin. A Dutch auger was used to remove the top 60 to 80 cm of soil, then soil parent material was collected using a bucket auger, gathering 25 to 30 cm depth of soil, and the soil sampling depth was recorded (e.g. from 85 to 112 cm).

Figure 3.2 **Typical Soil Cores in Forested Area (Photo from CEM 2004)**

3.3.2 Sample Preparation and Analysis

Samples were delivered to the laboratory where the sample bags were opened immediately, and the soil material was disaggregated to initiate air-drying to minimize chemical alteration due to anaerobic conditions within the sealed bags. The samples were then homogenized and further dried in weighed plastic containers to constant moisture, and weighed to enable bulk density calculation to quantify actual metal loadings to soils in subsequent data analysis.

The soil sample was then passed through a 2 mm mesh Fritsch Pulverizer, with the coarse material (>2 mm fraction) being weighed and stored for future analysis. The pulverized soil was split, using a stainless steel sample splitter, with a 200 g split being ground and sieved with 45 µm mesh and stored in labeled 250 ml plastic jars for shipment to the analytical facility.

3.3.2.1 Sample Analysis

The prepared soil samples were analyzed at the Environmental Analytical Services Division of SGS Lakefield Research Ltd., Lakefield, Ontario. The sample was mixed thoroughly to ensure sub-samples would be homogenous. Between 0.5 and 0.505 g of the sample was weighed into a Teflon sleeve and treated with 5 ml each of concentrated HNO_3 and HCl (*Aqua Regia*). The vessels were placed in a MARS 5 MAW2 Microwave Oven, put through a heat cycle and allowed to cool to less than 60°C. The contents were poured into 50 ml volumetric flasks and diluted to volume with deionized water. The solutions were analyzed by a combination of Inductively-Coupled Plasma-Optical Emission Spectrometry (ICP-OES), Inductively-Coupled Plasma-Mass Spectrometry (ICP-MS) and hydride generation-atomic emission spectrometry (HG-AES).

All soil samples were analyzed for the following elements:

aluminum (Al)	**antimony** (Sb)	**arsenic** (As)	**barium** (Ba)	**beryllium** (Be)
calcium (Ca)	**cadmium** (Cd)	**cobalt** (Co)	**copper** (Cu)	**chromium** (Cr)
iron (Fe)	**magnesium** (Mg)	**manganese** (Mn)	**molybdenum** (Mo)	**nickel** (Ni)
lead (Pb)	**selenium** (Se)	**strontium** (Sr)	**vanadium** (V)	**zinc** (Zn)

Table 3.1 presents the mean Method Detection Limits (MDL) for these elements that were achieved by SGS Lakefield Research Limited. The MDL is derived on an individual sample basis. It is calculated using the Instrument Detection Limit (IDL), an experimentally based value which is the minimum level at which the instrument can detect the element. The IDL is used to calculate the real MDL for soil, biota and tissue samples. Therefore, the MDLs are based on the actual mass of sample digested and the final volume. The calculation is as follows:

MDL = IDL × vol (digest)/mass(sample)

Every sample digested had different weights making the MDLs reported differ slightly between samples. Dilutions resulting from high concentrations of metals or matrix effects may produce MDLs that differ by one or two orders of magnitude.

Table 3.1 Summary of Mean Method Detection Limits (mg/kg) in Soil

Parameter	Mean Method Detection Limit (mg/kg)	Parameter	Mean Method Detection Limit (mg/kg)
Aluminum	2.5	Magnesium	1
Arsenic	5	Manganese	2
Barium	0.5	Molybdenum	1.5
Beryllium	0.5	Nickel	1
Calcium	10	Lead	1
Cadmium	0.8	Antimony	0.8
Cobalt	1	Selenium	1
Chromium	5	Strontium	10
Copper	1	Vanadium	2
Iron	5	Zinc	2.5

It is important to recognize that for the results described in the following sections, the MDLs for the quantification of As and Se in soil were 5 mg/kg and 1 mg/kg, respectively. For evaluation purposes, any value reported to be below the MDL was replaced with a value equal to half of its respective MDL.

The soil profile samples were analyzed by energy dispersive X-ray fluorescence spectrometry using the EMMA system in the CEM laboratories. Approximately 5 g of the 45 μm mesh sample was placed in a plastic tube sealed with a Mylar sheet and irradiated for 120 s. Data reduction was completed with propriety software, with data accuracy and precision being checked with selected NIST SRM materials.

Originally, one in 10 samples (10%) was analyzed for pH, electrical conductivity and total organic content (TOC). Subsequently, Laurentian University measured pH in all soils collected for the regional soil survey.

3.3.2.2 Quality Program

The soil samples were analyzed at SGS Lakefield Research Ltd., which is certified by the Standards Council of Canada (accredited ISO/IEC 17025 level) and the Canadian Association of Environmental Analytical Laboratories (CAEL). The calibration and testing activities at Lakefield followed the requirements of the ISO/IEC 9000 series standards. Quality control measures include duplicate samples, spiked blanks, spiked replicates, reagent/instrument blanks, preparation control samples, certified reference material analysis and instrument control samples. The Lakefield standard operating procedures required that, internally, at least 20% of samples analyzed were quality control samples.

In addition to the laboratory quality assurance/quality control (QA/QC) procedures, the Laurentian University CEM quality control program included submission of periodic splits of soil samples, and an analytical drift-monitoring sample collected from within the Sudbury region. This Internal Reference Material (IRM) was prepared from a 100 kg air-dried surface soil sample by sieving through the 2 mm mesh Fritsch Pulverizer, then through a 45 μm mesh sieve. The IRM sample was then tumbled for 24 h to ensure homogenization, and bottled in 250 ml plastic jars for storage and submission for analysis. Blind CEM IRM samples were submitted for every 30 samples. Reported laboratory values were considered accurate if their reported value was ±10% of the 'target' value of the blind reference material established by the analytical program.

Blind duplicates of the field samples were submitted every 20 samples. Reported laboratory values of the blind duplicate samples were considered precise and accurate if the values were within the 95% confidence intervals when plotted, with the slope of the regression line of the plotted data being between 0.95 and 1.05. The field duplicate accuracy and precision was evaluated in the same manner as the blind sample duplicates.

The overall QA/QC requirements of the Sudbury Soils Study meant that approximately 40% of all samples digested and analyzed were for data quality assessment and assurance, exclusive of the field duplicates which were required to allow for an assessment of landscape homogeneity. All IRM samples and duplicates were submitted in a randomized sequence relative to their duplicate, their geographic location, and their order of field collection. The laboratory analyses of duplicated splits of soil samples indicated very good reproducibility (precision). In general, there was also good agreement between the CEM laboratory analysis values and the reported values for the IRM. For the elemental analyses of the soil IRM, the majority of analyses fell within the ±10% criterion.

3.4 Overview of Results

Data from the three soil surveys conducted in 2001 were combined into a single integrated database in Microsoft Access format. The database was searchable using a range of variables and keyword locators for results presented by community, metal, land use, soil depth as well as other descriptors.

Summary statistics for the overall soil survey are provided in Table 3.2. It is important to note that Table 3.2 includes results from the three surveys including properties near the smelters as well as remote sample locations. The table incorporates soils sampled to a maximum depth of 20 cm. The regional soil survey conducted by Laurentian University collected over 250 samples at depth (>80 cm) considered to represent regional background or parent material metal concentrations. A discussion of the distribution of the elements in parent soil material is provided in Chapter 4.

Table 3.2 **Summary of 2001 Soil Survey Results for 20 Inorganic Parameters**

n=8148	Concentration in Soil (mg/kg)				
	Minimum	Median	Average	Maximum	95th Percentile
Al	2100	9800	10,400	39,000	18,000
As	2.5	6	16	620	61
Ba	9.8	47	56	720	120
Be	0.25	0.25	0.61	2	0.25
Ca	470	4000	5170	250,000	11,000
Cd	0.4	0.4	1	6.7	1.8
Co	1	9	14	190	42
Cr	9	31	34	1100	56
Cu	2.7	80	260	5600	1100
Fe	4400	15,000	16,300	110,000	26,000
Mg	350	2800	3070	26,000	5700
Mn	33	190	211	3300	360
Mo	0.75	0.75	1	21	1.8
Ni	7	95	264	3700	1100
Pb	1	16	35	790	130
Sb	0.4	0.4	0.48	8.1	1
Se	0.5	0.5	2	49	5
Sr	5	33	35	340	53
V	8	30	31	130	45
Zn	1.25	34	44	340	110

Sample size (n) is from combined soils database.
Soil depths combined: 0–5, 5–10, 10–20 cm. Note – some numbers adjusted to appropriate significant digits.

3.5 Selection of Chemicals of Concern (COC)

3.5.1 Data Screening Process

The focus of the risk assessment was particulate-associated metals and metalloids deposited in soils as a result of aerial deposition from smelter emissions. In addition, fugitive dust from tailings areas, waste rock piles and roasting beds may have contributed to localized metal levels in soils. Screening the soil chemistry results identifies those metals present in the study area that pose the greatest potential for exposure and risk to people and the terrestrial ecosystem.

It is common practice in risk assessment to focus detailed evaluation on those substances or chemicals that represent the greatest potential concern in the area under consideration. It is not practical or necessary to conduct a detailed risk assessment on every chemical present in a given study area. To identify chemicals that pose the greatest potential risk, data on the concentrations of chemicals in different environmental media are generally compared with regulatory guidelines or standards for the particular matrix. Chemicals, or metals in this study, with concentrations exceeding the regulatory guidelines are then typically carried forward for more detailed evaluation in a risk assessment. Chemicals with concentrations below the guidelines are generally 'screened out' and not considered further. In this study, metals data from the 2001 soil survey were compared with soil quality guidelines established by the Ontario Ministry of the Environment (MOE). Therefore, those chemicals present on site at concentrations in excess of the criteria are considered to be of most concern and were selected for further study in the form of a risk assessment.

Critical decisions concerning the primary COC selection criteria, list of analytes to be reviewed, sampling area (study area) and sampling methods were established by the study's Technical Committee based on a combination of regulatory, policy and scientific objectives. This is considered reasonable practice to help scope a study the magnitude of the Sudbury ecological risk assessment (ERA) and human health risk assessment (HHRA). As Suter (1993) points out, risk assessments performed for different assessment purposes will use different methods.

As discussed in Section 3.2 above, approximately 8400 soil samples from about 1190 sites were analyzed for 20 inorganic parameters. The results of the 2001 soils survey were merged into a comprehensive database and the soils data were screened against the following three criteria as determined by the Technical Committee.

1. Parameter must be above the Table A (potable groundwater) or Table B criteria (non-potable groundwater) published in MOE's *Guideline for Use at Contaminated Sites in Ontario* (the 'Guideline'; MOE 1997), depending on whether the specific area under study has surface or well water sources for potable water. In general, these criteria were developed "to protect against adverse effects to human health, ecological health and the natural environment" (MOE 1997); Note regarding soil quality criteria: When the Sudbury Soil Study was conducted, the data were interpreted using the regulatory soil quality criteria in effect at the time, which was the *Guideline for Use at Contaminated Sites in Ontario* (MOE 1997). On 9 March 2004 the Ontario government promulgated the Record of Site Condition Regulation, Ontario Regulation 153/04, *Soil, Ground Water and Sediment Standards for Use under Part XV.1 of the Environmental Protection Act* (MOE 2004b). Under O. Reg. 153/04 the former soil quality criteria of the 1997 *Guideline* became soil quality standards of the 2004 *Regulation*. Although the supporting Tables were re-structured, the numeric values of the standards remained the same as the original guidelines. For example, the background-based soil guidelines were found in Table F of the 1997 *Guideline* and in Table 1 in the 2004 *O. Reg. 153/04*. On 29 December 2009, the Ontario government further amended *O. Reg. 153/04*. Although most of the amendments related to streamlining risk assessments and obtaining a Record of Site Condition for remediated properties, many of the background and effects-based soil standards changed as well to reflect an update of the toxicological and geological literature for Ontario conditions. The amended regulation is *O. Reg. 511/09*. For this document, the authors have elected to use the 1997 *Guideline*, because that was the regulatory framework in place at the time the Sudbury Soil Study was conducted and the original data were evaluated.
2. Parameter must be present across the study area; and,
3. Parameter must have a scientific link to the companies' operations.

The overall selection process for COC for the HHRA and ERA is illustrated in Figure 3.3 and discussed in more detail below.

The use of the first criterion for COC screening requires a brief discussion of the MOE soil quality criteria and their purpose. Most of the MOE criteria were established to be protective of human and ecological health and the natural environment (MOE 1997). Plant toxicity values are typically lower than those reported for animals or the protection of human health. Therefore, many of the generic metals criteria in the MOE *Guideline for Contaminated Sites in Ontario* (MOE 1997) are based primarily on the effects of these metals on sensitive plant species, such as wheat or oats. Plant- or animal-based criteria have been developed for As, Ba, Cr, Co, Cu, Mo, Ni, Se, Sb, Zn and V. Soil criteria exist for different land uses (i.e. agricultural, residential/parkland, industrial/commercial), soil textures (i.e. fine, medium, coarse), and soil depths (i.e. surface, subsurface). Furthermore, the criteria are only applicable to surface soils with a pH range of 5.0 to 9.0 and to subsurface soils with a pH range of 5.0 to 11.0. This last fact is generally not an issue for soils in southern Ontario or agricultural areas but became a consideration in Sudbury where soils are naturally acidic and the soil pH is often less than 5.0.

The 2001 Soil Survey and Selection of Chemicals of Concern

2001 Soils Data

As, Al, Sb, Ba, Be, Cd, Ca, Cr, Co, Cu, Fe, Pb, Mg, Mn, Mo, Ni, Se, Sr, V, Zn.

↓

PRIMARY SCREENING CRITERIA

Parameter > Table A
Parameter across study area
Parameter linked to smelters

↓

Chemicals of Concern (COC)

As, Co, Cu, Ni, Pb, Se

↓

SECONDARY SCREENING CRITERIA

Soil pH < 5.0
Parameter > Table F
Parameter across study area
Parameter linked to smelters

↓

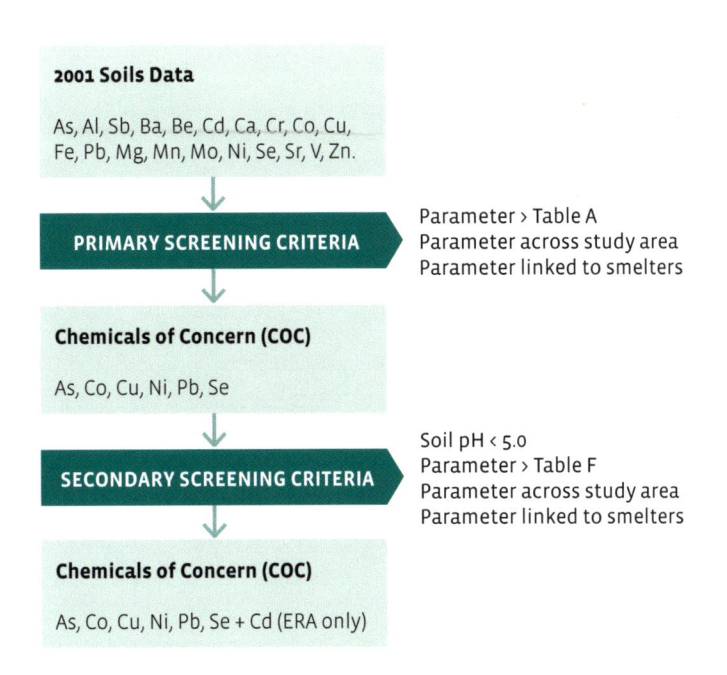

Chemicals of Concern (COC)

As, Co, Cu, Ni, Pb, Se + Cd (ERA only)

Figure 3.3 **Soil Data Screening Process for Selection of COC**

The MOE Table A and Table B values were developed to provide guidance for cleaning up contaminated sites and, at the time the Sudbury Soils Study was initiated, were not legislated regulations. Furthermore, the criteria are *not* action levels, where exceeding a particular criterion value would indicate immediate risk or that remediation or clean-up is required. The significance of the criteria to the Sudbury area was to provide triggers to identify the need for additional investigations (MOE 2001). In fact, metal concentrations higher than the generic criteria in soil samples collected up to the year 2000 were the impetus for the Sudbury Soils Study.

The values known as Table F or 'Ontario Soil Background Criteria' are the mean plus two standard deviations, which represents the 97.5th percentile of the concentrations of various chemical parameters detected in background soil in Ontario. These were derived from a province-wide soil sampling program to determine the range of background chemical concentrations in surface soil in Ontario resulting from either natural geological processes or human activity but remote from the direct influence of known sources of pollution.

3.5.2 Selection Criterion #1: Parameter Must Be above MOE Guideline

Results of metal analysis from approximately 8148 soil samples were compared with the provincial Table A guidelines (Table 3.3). The table indicates that the Table A guideline was exceeded for at least some soil samples for As, Be, Co, Cu, Ni, Pb and Se. In other words, these elements met Criterion #1 for selection of a COC for the risk assessment. In particular, a significant number of samples exceeded the guideline for As, Co, Cu and Ni.

The number of soil samples with Cr or Be levels exceeding the MOE criterion was considered exceptionally small (only 1 or 2) relative to the total number of samples. Therefore, Cr and Be were dropped from further consideration.

The number of samples with Pb and Se meeting or exceeding the Table A criterion was also relatively small (< 1.5% of total number of samples) but these were carried through for further evaluation.

Soil quality criteria for the protection of human and ecological health have not been established in Ontario, Canada, or the U.S. for several of the inorganic parameters analyzed. These include Al, Ca, Fe, Mg, Mn, and Sr (Table 3.3). These elements are common components of the earth's crust, with Al, Ca, Fe, and Mg comprising approximately 8%, 3%, 5%, and 2%, respectively (McQuarrie and Rock 1991). No further evaluation was considered necessary for these common elements.

Table 3.3 **Initial Data Screening and Comparison with MOE Criteria[1] (n=8148)[2]**

Metal	Concentration in Soil (mg/kg)			MOE Generic Criteria (mg/kg)		
	Minimum	Average	Maximum	Table A Criterion	No. of samples above Table A	No. of samples equal to or above Table A
Al	2100	10,434	39,000	NC	–	–
As	<5	16	620	20	1431	1493
Ba	9.8	56	720	750	0	0
Be	<0.5	0.61	2	1.2	2	2
Ca	470	5165	250,000	NC	–	–
Cd	<0.8	1	6.7	12	0	0
Co	1	14	190	40	460	483
Cr	9	34	1100	750	1	1
Cu	2.7	260	5600	150	2839	2948
Fe	4400	16,327	110,000	NC	–	–
Mg	350	3065	26,000	NC	–	–
Mn	33	211	3300	NC	–	–
Mo	<1.5	1	21	40	0	0
Ni	7	264	3700	150	3105	3187
Pb	1	35	790	200	108	129
Sb	<0.8	0.48	8.1	13	0	0
Se	<1.0	2	49	10	91	113
Sr	5	35	340	NC	–	–
V	8	31	130	200	0	0
Zn	1.25	44	340	600	0	0

[1] Table A criteria for coarse textured soil (potable groundwater condition); with pH 5.0–9.0 for surface soil and pH 5.0–11.0 for subsurface soil; and, intended for residential/parkland uses.
[2] Sample size is from combined soils database, with soil depths of 0–5, 5–10, and 10–20 cm.
NC = No Criterion.

3.5.2.1 Secondary Data Screening with Criterion #1

As mentioned previously, use of the Table A soil quality criteria in the *Guideline for Use at Contaminated Sites in Ontario* (MOE 1997) only applies to soils with a pH range of 5.0 to 9.0. For soils with a pH outside this range, Table F criteria, or background levels, should be used as alternate screening values (MOE 1997).

Generally, the pH of soil samples in Ontario is not a limitation of applying the MOE guidelines. However, soil pH in northern Ontario is known to be naturally low (pH < 5.0) with soil pH further reduced in the study area due to the smelter emissions and sulfur deposition.

During the original 2001 survey, soil pH was measured in only one in every 10 samples, or 10%. There were 229 soil samples from the urban soil survey of residential properties for which pH results existed. Of these, 224 (98%) had a pH >5.0. This higher pH range was attributed to homeowners amending residential soils with lime, organic matter and fertilizers and to builders landscaping with clay-textured soils to promote urban lawn growth. Therefore, it was valid to screen the urban soils data against the Table A criteria. Low soil pH relative to the MOE generic criteria was considered to be an issue primarily in samples from rural, unamended sites, and not urban residential properties.

Samples from the regional, or more remote, soils survey were all re-analyzed for pH which was not undertaken at the beginning of the study. Of 365 surface soil (0 to 5 cm) samples analyzed, soil pH was <5.0 in 347 (95%) of the samples. Therefore, the regional soils database was re-screened using Table F as a criterion, for those samples with soil pH <5.0. The results of this secondary screening exercise are summarized in Table 3.4.

Table 3.4 **Summary of Secondary Data Screening and Evaluation**

Parameter	Table A criterion (mg/kg)	Table F Criterion (mg/kg)	No. of samples with values > Table F	No. of samples with pH < 5 and values > Table F[1,2]	Max value (mg/kg)	95th Percentile (mg/kg)
Al	NC	NC	–	–	38,000	20,000
As	20	17	388	161	410	80
Ba	750	210	2	1	720	130
Be	1.2	1.2	0	0	1	0.25
Ca	NC	NC	–	–	24,000	9500
Cd	12	1	152	29	4.1	1.7
Co	40	21	238	58	150	45
Cr	750	71	42	7	1100	67
Cu	225	85	699	241	5000	1100
Fe	NC	NC	–	–	74,000	28,000
Mg	NC	NC	–	–	10,000	5400
Mn	NC	NC	–	–	3300	480
Mo	40	2.5	53	9	17	2.3
Ni	150	43	954	286	3649	1100
Pb[3]	200	120	57	3	790	120
Sb[3]	13	1	40	6	4.4	0.9
Se	10	1.9	468	183	27	6
Sr	NC	NC	–	–	110	52
V	200	91	0	0	74	50
Zn	600	160	10	0	310	97

[1] Samples with pH values = 1201 (values are from original and duplicate samples taken from 0–5, 5–10, and 10–20 cm soil depths). Samples with pH < 5 = 401.

[2] Only values that have an associated pH value.

[3] No value reported for some samples.

NC = No Criterion.

This secondary screening step identified five additional elements that exceeded the Table F values in soil samples with pH <5.0 (sample size exceeding Table F in parentheses): Ba (1), Cd (29), Cr (7), Mo (9) and Sb (6). Of these, four of the elements (Ba, Cr, Mo and Sb) exceeded Table F in a very small number of samples. In addition, the 95th percentile concentration of these elements was less than Table F. Therefore, these elements were excluded from further consideration as candidate COC.

Cadmium exceeded Table F in 29 soil samples with pH <5.0, and the 95th percentile concentration (1.7 mg/kg) was marginally greater than the Table F value (1.0 mg/kg). Therefore, Cd was retained as a COC for detailed assessment in the ERA, but not the HHRA. The reasons for this decision were twofold:

1. Low soil pH was associated almost exclusively with samples collected from rural and remote sites, therefore more applicable to ecological receptors than humans.
2. The maximum Cd concentration was less than half of the Table A criteria which are protective of human health (i.e. 12 mg/kg for residential/parkland, and 3 mg/kg for agricultural land, based on exposure to grazing animals).

In summary, as a result of the secondary data screening for COC selection, Cd was added to the list of candidate COCs. After evaluation of Criterion #1, the following elements were carried forth for further comparison with Criteria #2 and 3: As, Co, Cd (ERA only), Cu, Ni, Pb and Se.

3.5.3 Selection Criterion #2: Parameter Must Be Present across the Study Area

The distribution of samples with metal concentrations exceeding the Table A values was plotted on maps of the study area. A large number of samples contained elevated levels of As, Co, Cu or Ni and these were widely distributed throughout the study area, although samples with higher concentrations tended to be closer to the smelters (see also Chapter 4). For example, Figure 3.4 shows the distribution of soil samples with Ni concentrations greater than the MOE Table A value. Arsenic levels tended to be higher in samples closer to the Falconbridge smelter (Figure 3.5). This was not unexpected since the ore type processed at this location is known to contain elevated levels of As (Chapter 2). There was also a high statistical correlation of the concentration of these elements in the soil samples which would be expected if they were all emitted from the smelter.

Therefore, As, Co, Cu and Ni clearly met Criterion #2 in that they were present across the study area. The reason to include Criteria #2 and #3 as screening steps was to eliminate any possible chemical or metal that originated from another point source in the study area not associated with the mining and smelting operations. It was clear that As, Co, Cu and Ni were emitted in significant quantities from the smelters and contributed to elevated soil concentrations. There was still some uncertainty related to Pb and Se, therefore, more detailed examination of the spatial distribution of these elements was undertaken.

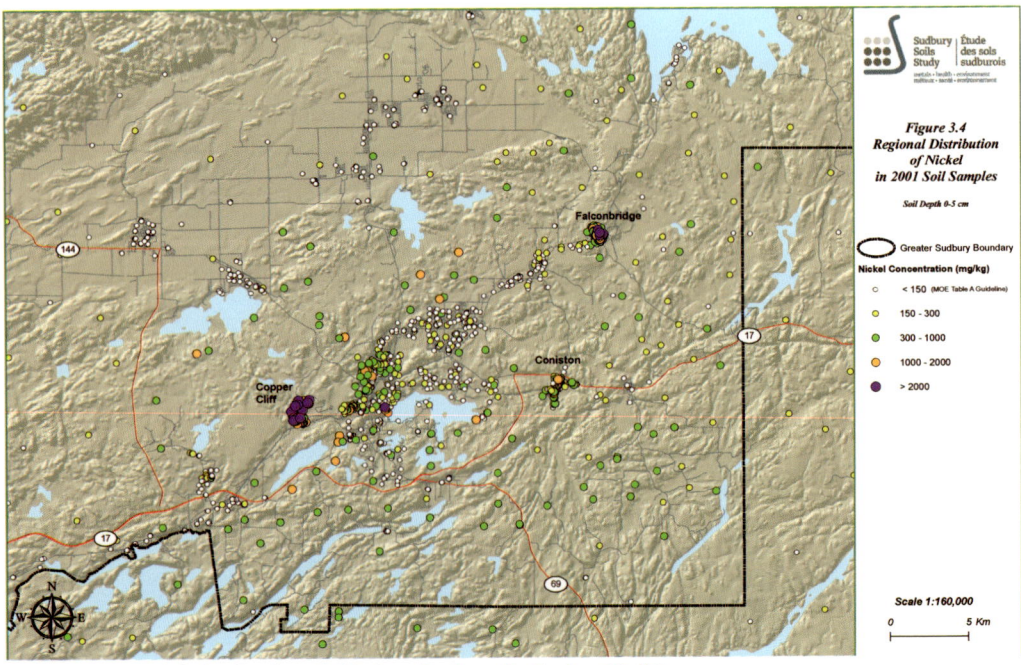

Figure 3.4 **Spatial Distribution of Ni in Samples from the Regional Soil Survey**

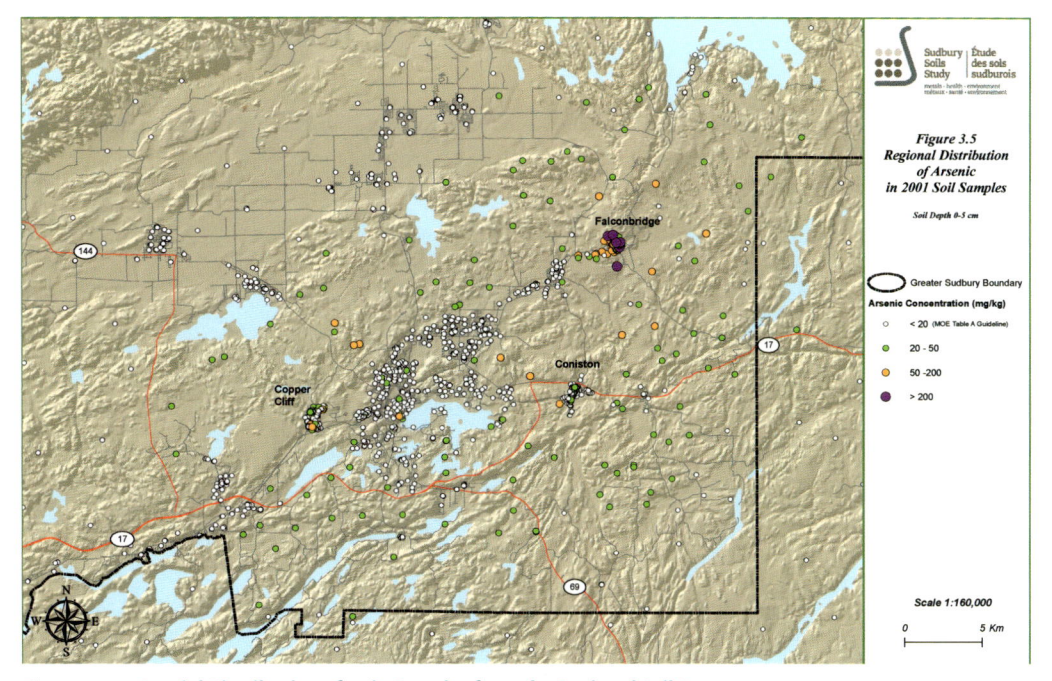

The 2001 Soil Survey and Selection of Chemicals of Concern

Figure 3.5   Spatial Distribution of As in Samples from the Regional Soil Survey

Selenium fulfilled Criterion #1 in 113 (1.4%) out of 8148 soil samples. Of these, 103 samples (91.1%) of the samples with Se in excess of or equal to the Table A guidelines were located in Copper Cliff adjacent to the Vale smelter. In addition, the majority of the samples with Se levels above or equal to the MOE Table A occurred in surface soils (0–5 cm, 101 out of 113 samples), while 10 samples occurred at 5–10 cm, and two samples at 10–20 cm. The elevated Se levels were, therefore, confined primarily to the uppermost soil layer (0–5 cm) near the smelter suggesting atmospheric deposition was the likely source.

An additional comparison between Se in surface soil and other COCs associated with smelter operations was performed. The premise was that if Se follows a distribution pattern in soil similar to the established COCs already associated with smelter operation, then the results would suggest that the source of Se was also due to smelter operations.

Spearman's correlation coefficients were calculated to examine the correlation between the level of Se and concentration of other established COCs associated with the smelter (Table 3.5) as well as parameters not associated with the smelters (i.e. Be, V and Al). Metal concentrations used in the analysis were from the 0-5 cm soil layer.

Table 3.5   **Correlation of Se to the Established COCs and Other Elements**

	Se	As	Co	Cu	Ni	Pb	Be[2]	V[2]	Al[2]
Correlation Coefficient[1] (r)	1.000	0.76	0.76	0.82	0.81	0.76	−0.072	0.16	0.19

[1] $\alpha=0.01$; one-tailed test; $N=1536$.
[2] Not a COC.

There was a significant positive correlation between the concentration of Se and the other COCs (r values >0.7, P<0.01) in soils. In contrast, there was a negligible correlation between the concentration of Se and other elements (Be, V, Al) not associated with the smelters as a source. These results indicate there was no consistent association between the levels of Se and metals not associated with smelter operations.

The results of the analysis show that Se fulfills Criterion #2; it existed within the study area although elevated levels were predominantly in the Copper Cliff area.

Lead in soil exceeded the Table A value in 129 samples. These were all in samples located adjacent to the smelters (see below), although over half of the elevated values occurred near the Vale smelter at Copper Cliff. An additional comparison between Pb and other COCs associated with smelter operations was performed based on the premise that if Pb follows a distribution pattern similar to the established COCs associated with smelter operations, then the results would suggest that the source of Pb was also due to smelter operations.

Spearman's coefficient was used to correlate the levels of Pb and other established COCs associated with the smelter, namely Cu and Ni. The same parameters used in the first correlation also apply here (i.e. $n=1536$, $\alpha=0.01$, and one-tailed test). The results of the analysis are summarized in Table 3.6. For comparison purposes, the correlation between soil Pb and V concentrations was examined, as V is an element not related to the smelting process.

Table 3.6 **Correlation between Pb, Cu, Ni and V in Soil (0–5 cm)**

Smelter source	Correlation coefficient test values (r)[1]		
	Cu	Ni	V[2]
Copper Cliff ($n=1536$)	0.923	0.929	0.179
Coniston ($n=370$)	0.847	0.853	−0.183
Falconbridge ($n=367$)	0.930	0.887	0.301

[1] Test type = one-tailed, $\alpha=0.01$.
[2] Not a COC.

There was a significant, positive correlation between levels of Pb and the two other COCs ($r>0.8$). The correlation between Pb and V was almost not significant ($r=0.179$, $P<0.01$). These results indicate no consistent association between Pb levels and elements not associated with smelter operations.

3.5.4 Selection Criterion #3: Parameter Must Show Origin from the Companies' Operations

To further determine whether soil metal levels were related to atmospheric smelter emissions, three additional lines of evidence were examined within this criterion:

1. Vertical distribution of metals in soil, with metal enrichment in the surface layers suggesting atmospheric deposition as the source.
2. Concentration of metals relative to distance from the smelter, with soil concentrations decreasing with increasing distance from the smelters indicating the smelters as the source.
3. A comparison of the mineralogy of Pb in soil and air filter samples from the smelter stacks.

These three lines of evidence are discussed further in the following sections.

3.5.4.1 Metal Enrichment in Regional Soil Surface Layers

The mean concentrations from all sites of *Aqua Regia* extractable metal(loids) for all soil layers sampled during the regional survey program are summarized in Table 3.7. The data show the enrichment of Sb, As, Co, Cu, Pb, Mo, Ni, Se and Zn in the surface layers (0–5 cm) which may be indicative of particulate fall-out from regional smelting operations. The surface organic soil layers (LFH horizons) effectively act as a filter retaining the fall-out and preventing translocation to underlying mineral horizons, an observation also described for the Kola region of Russia (Nikonov et al. 1999; Koptsik et al. 2003). The designation LFH refers to the fresh plant detritus (L = litter) on the soil surface, the partially decomposed organic layer (F = fermentation) and the well-decomposed organic layer (H = humus). The retention of the anthropogenic metals in this filter zone is further suggestive of a relatively low solubility and bioavailability of metals in the high-temperature particulates.

To evaluate if the metal(loid) content of the 0–5 cm layer was derived from natural or anthropogenic sources, an approximate enrichment factor (EF) was calculated for the above elements using the following equation: $EF = ((Metal)/(Al)_{LFH})/((Metal)_{pm}/(Al)_{pm})$. Enrichment factors ranging between 0.5 and 2 can be considered in the range of natural variability, whereas ratios greater than 2 indicate enrichment from anthropogenic inputs (Shotyk et al. 2000; Hernandez et al. 2003). The enrichment factors calculated in this study are normalized to Al as a reference element because Al is relatively immobile in the soil, and there is minimal indication of Al additions to the soils from the industrial sector.

The calculated EF values do suggest a strong anthropogenic influence in the concentrations of the metal(loid)s Sb, As, Cu, Pb, Mo, and Se in the 0–5 cm layers of regional soils. Cadmium, although at concentrations below detection limits for many of the soil parent material samples, may also be enriched in the surface (0–5 cm) layers of the regional soils. The lack of enrichment for elements such as Cr, Co and Zn in the surface layers (0–5 cm) suggests that there has been minimal anthropogenic output of these elements from the industrial activity within the region. This enrichment of specific elements in the 0–5 cm depth layers suggests that the LFH horizons which dominate the 0–5 cm layer act as a filter preventing the translocation of the aerosol particles to the deeper soil horizons.

Table 3.7 **Mean Metal Concentration in the Regional Soil Survey, and Calculated Enrichment Factors for the Surface (0–5 cm) Layer (Adapted from CEM 2004)**

Depth	Al (%)	Fe (%)	Mg (%)	As (mg/kg)	Cd (mg/kg)	Cr (mg/kg)	Co (mg/kg)	Cu (mg/kg)	Pb (mg/kg)	Mn (mg/kg)	Ni (mg/kg)	Se (mg/kg)	V (mg/kg)	Zn (mg/kg)
0–5 cm	1.00	1.59	0.18	14.81	0.36	38.7	12.5	261.4	50.0	306	263	2.2	31.1	39.1
5–10 cm	1.34	1.68	0.23	9.72	0.01	38.9	7.2	101.2	15.0	232	81.5	0.6	37.8	31.6
10–20 cm	1.72	1.92	0.32	3.80	0.00	44.0	7.3	49.7	8.9	215	50.6	0.2	40.3	33.7
Parent material	1.78	2.28	0.62	1.11	NA	56.4	8.9	26.4	5.9	293	36.1	0.1	45.3	29.7
0.5/Parent	0.56	0.70	0.29	13.34		0.69	1.39	9.92	8.46	1.04	7.28	37.09	0.69	1.31
EF (C-0.5)	0.6	0.7	0.3	13.3		0.7	1.4	9.9	8.5	1.0	7.3	37.1	0.7	1.3

Enrichment factor (EF) = $((M)/(Al)_{LFH})/((M)_{pm}/(Al)_{pm})$.

The lack of a significant EF for Co was somewhat surprising, as many hundreds of soil samples exceeded the MOE soil guideline for Co, and it was clearly considered as a COC for the risk assessment.

The concentrations of As, Cu and Pb were also higher in the 5–10 cm layer than in the deeper soils. Copper is known to be chelated in soil by the potentially mobile soil humic acids which would help explain some translocation from the 5–10 cm layer to the next layer in these undisturbed forested soils. The possibility of vertical translocation of specific elements further supports the conjecture of Nriagu et al. (1998) that release of anthropogenic metals from regional soils may affect dissolved metal levels in regional lakes for perhaps hundreds of years. The evidence for translocation also points to the potential slow solubility and potential bioavailability of the same anthropogenic metals currently stored as particulates in the surface layers of regional soils.

Fourteen soil core samples were also collected as part of the urban-residential soil sampling program in 2001 to examine evidence of atmospheric deposition of metals from the smelters (MOE 2004a). Undisturbed areas within the city were chosen because development and landscaping in urban areas have altered most soils, both physically and chemically, through the processes of adding, grading, removing, mixing and/or other activities that may have occurred repeatedly over time. Undisturbed soils provide a better picture of atmospheric deposition.

The results are presented in Table 3.8. The data show that the highest concentrations of each of the candidate COC occur within the surface (0–5 cm) layer, and generally decrease with depth. This information indicates that atmospheric deposition is the likely prominent source of metals to soils in the study area.

Table 3.8 **Mean Metal Concentrations (mg/kg) in Urban-undisturbed Natural Soil Profiles (n=14 Samples per Depth)**

Soil depth (cm)	As	Co	Cu	Ni	Pb	Se
0–5	29	37	660	983	63	1.9
5–10	8.9	13	168	191	18	1.0
10–20	3.5	10	52	69	8	0.7

The next line of evidence involved examining the concentration of metals with distance from the smelter. This was undertaken for all elements but is illustrated here using only Se and Pb.

A two-variable regression and correlation analysis was performed to evaluate the possible relationship between concentration of selenium and distance from the smelter. The correlation coefficient for selenium versus distance from the smelter was $r=-0.601$ ($P<0.01$) which implies a significant, negative correlation between Se and distance from the smelter. In other words, Se levels in the surface soils decrease with increasing distance away from the smelter. This relationship is illustrated in Figure 3.6, which shows that most of the samples containing Se exceeding the MOE Table A value occurred at a distance of less than 5000 m from the smelter. Beyond 5000 m, most samples contained Se levels below the Table A value.

The results of these analysis including vertical soil profiles, spatial distribution, correlation analysis and distance from the smelter all indicate that Se is related to the smelting operations and meets Criterion #3 to be considered a COC.

The correlation of Pb concentration in surface soils versus distance from the smelters was also examined. The results of the analysis are summarized in Table 3.9.

Table 3.9 **Lead Levels (mg/kg) in Surface Soil (0–5 cm) Correlated to Distance from the Smelters**

Smelter source	Correlation coefficient (r)[1]	P-value
Copper Cliff (n=1536)	−0.475	0.000
Coniston (n=370)	−0.159	0.000
Falconbridge (n=367)	−0.292	0.000

[1] Test type=one-tailed, $\alpha=0.01$.

The correlation coefficient for Pb versus distance from the Copper Cliff smelter was $r=-0.475$ with $P<0.001$ which denotes a significant negative correlation between Pb and distance from the smelter. The correlation is also significant, although weaker, for the other two smelters. The relationship between Pb in surface soils and distance from the Copper Cliff facility is illustrated in Figure 3.7. This graph indicates that all of the samples with Pb above the MOE Table A value were less than 2000 m from the Copper Cliff smelter.

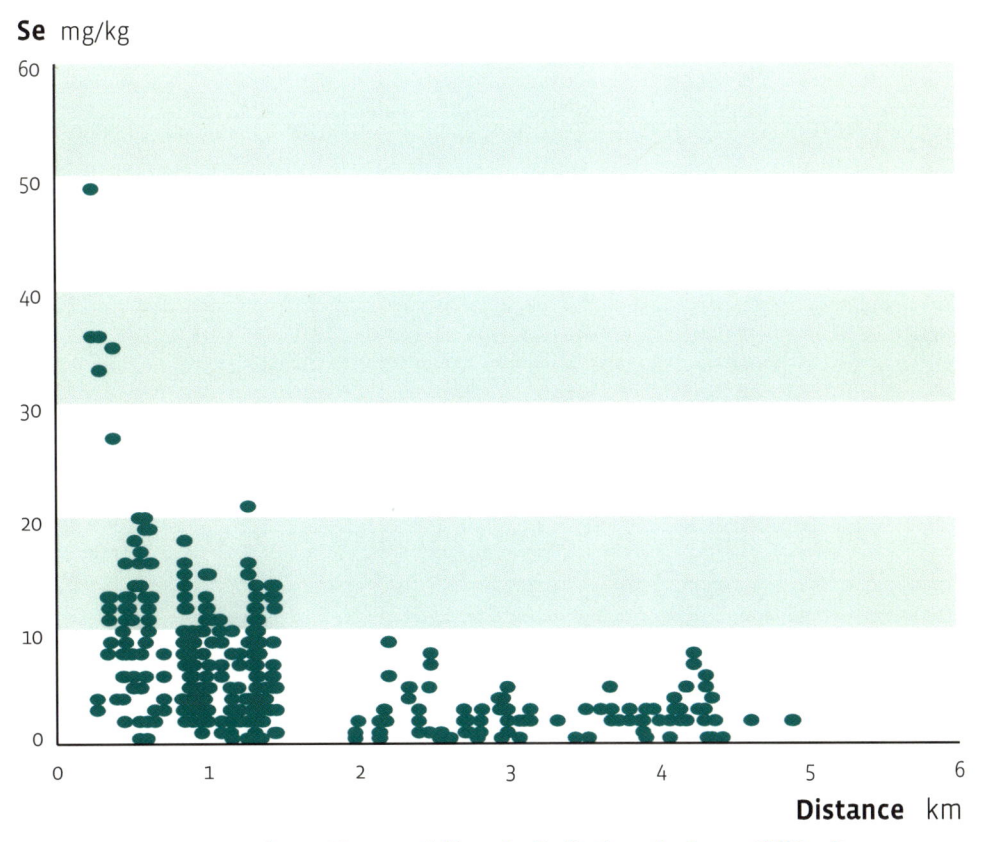

The 2001 Soil Survey and Selection of Chemicals of Concern

Figure 3.6 Se Concentration vs. Distance within a 7 km Radius from the Copper Cliff Smelter

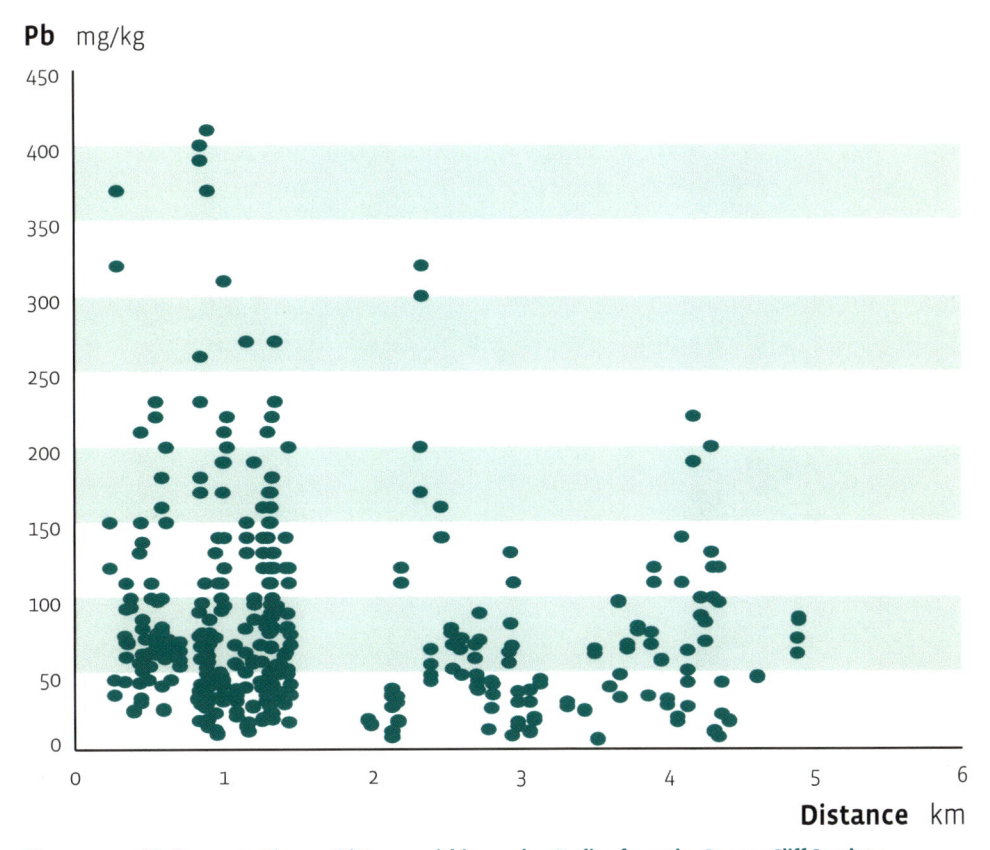

Figure 3.7 Pb Concentration vs. Distance within a 10 km Radius from the Copper Cliff Smelter

Lead Mineralogy

The third line of evidence within this criterion was a comparison of the mineralogy of Pb in soil and air filter samples from the smelter stacks. The objectives of this evaluation were to (a) characterize the mineralogy of Pb in Sudbury soil samples and (b) compare those results with particles on air filter samples collected from inside the Vale and Xstrata smelter stacks. This was considered an important step since there can be other sources of Pb in older residential communities unrelated to smelter emissions, particularly the historic use of Pb-based paint and leaded gasoline.

Characterization was completed on 22 soil samples collected from several areas in the Sudbury region, but especially from properties near the three smelter sources: Copper Cliff, Coniston and Falconbridge. The concentration of Pb in these samples ranged from 66 to 640 mg/kg. In addition, two filter stack samples were obtained from the Vale smelter (located in the town of Copper Cliff) and one filter stack sample from the Xstrata Nickel smelter (town of Falconbridge) for detailed examination.

Epoxy mounts of the filter material were prepared without any separation. Energy dispersive (ED) analyses were completed using a scanning electron microscope (SEM). Wavelength dispersive (WD) analyses were completed using an electron microprobe (EMP). All electron micrographs and element X-ray maps were collected using the SEM. ED analyses were performed using an accelerating voltage of 20 kV, beam current of 2 nA, count time of 100 s at a working distance of 38 mm. Pure metal standards and natural oxides were used as standards, using in-house characterization values of the oxide standards for quality control. Backscatter electron (BSE) micrographs were collected using the same analytical conditions.

Soil samples from Copper Cliff (near the Vale smelter) contained spherical and sub-spherical lead-bearing morphologies. Silica and iron-rich 'stack spheres' were common in all soil samples (Figure 3.8), indicating deposition of particles from high-temperature emissions sources. Lead-bearing spheres were not observed in soils from locations other than Copper Cliff.

Lead occurred in the soil samples in both oxide and metallic form. No other elements were detected in the metallic phase of lead. Of the lead-bearing particles analyzed, 56% had detectable levels of Ni and Cu. There was a positive correlation between Pb and Cu, and between Pb and Ni.

Particles from the stack filter samples were generally less than 10 μm in diameter, which is the minimum particle size for ED analysis. Particles from both Vale and Xstrata stack filters contained almost perfectly spherical 'stack spheres' which are formed as a result of high temperatures encountered during the smelting process. However, the chemistry of the particles was different between samples from Vale and Xstrata.

Lead phases were common in the filter samples from the Vale stack. The Pb phases occurred both within the spheres and along the sphere rim (Figure 3.8). No Pb phases were detected in the filter sample from the Xstrata stack. Sulfur, Fe, Cu and Ni were the primary elements detected in particles from the Xstrata filter samples.

The findings of the SEM work indicate that Pb in the Sudbury soils is present in both oxide and metallic form. The presence of 'stack spheres' in all of the soil samples is consistent with deposition from a high-temperature source. Air filters from the Vale (Copper Cliff) smelter included spherical particles that contained significant levels of Pb, in conjunction with Ni and Cu. The one stack filter sample from Xstrata Nickel (community of Falconbridge) did not contain detectable concentrations of Pb.

These filter observations are consistent with the spatial distribution of Pb in the Sudbury area, where elevated soil Pb levels occurred primarily near the Copper Cliff smelter.

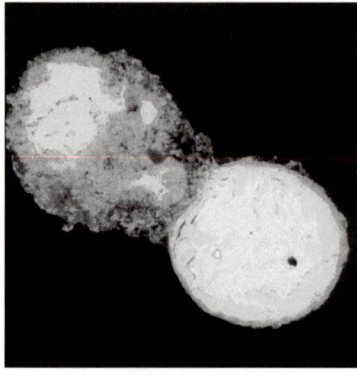

Figure 3.8 **Backscatter Electron (BSE) Micrograph Showing Spherical Particle in Soil Associated with High Temperature Source (Photo from GeoLabs 2004)**

3.6 Summary

Data from approximately 8148 soil samples were compared with three criteria for the selection of COC to be used for the detailed risk assessments. The concentrations of As, Co, Cd, Cu, Ni, Pb and Se exceeded the Ontario soil quality guidelines in many samples. The concentrations of these elements were spatially related to the smelter sources, and concentrations were elevated in the upper soil layers, indicating atmospheric deposition as the likely source. Detailed statistical and mineralogical examination of the data and samples provided further evidence that these elements were related to smelter emissions. The screening process identified these elements as COC for further evaluation in the risk assessments.

3.7 Acknowledgements

Many people contributed to the soil collection and initial report preparation. Field crews and laboratory assistants without whom this report could not have been started include: Caroline Hawson, Miriam Kaliomaki, Alan Lock, Duncan Quick, Chris Peloso, Francois Prevost, Ryan Post, Jacqueline Richard, Chantal Rosen, Paula Takats, Dana Willson. Special thanks go to Francois Prevost for data display and statistical analysis; Caroline Hawson for the literature review and quality control.

Day Aviation provided excellent field support. Acknowledgement is also given to personnel of the Ontario Ministry of the Environment, especially Brain McMahon and Rusty Moody. Personnel from Golders Associates, namely Sam Gauvreau and Natalie Boudreau, provided support and discussions during the sample design and sample collection phases. Dr David Pearson, Laurentian University, helped guide the field program and provided thoughtful insight in reviewing the report.

The characterization of the mineralogy of Pb in soil samples from the Sudbury area was conducted by Ms Linda Semenyna, Geosciences Laboratories, Ministry of Northern Development and Mines (MNDM), Sudbury.

We would also like to express our gratitude to Steven Gautreau for excellent graphics, mapping and additional statistical analysis in the final reports and in this book.

3.8 References

CEM (2004) Metal levels in the soils of the Sudbury smelter footprint. Report prepared by the Centre for Environmental Monitoring (CEM). Laurentian University, Sudbury

Dudka S, Ponce-Hernandez R, Hutchinson TC (1995) Current level of total element concentrations in surface layer of Sudbury soils. Sci Total Environ 162:161–171

GeoLabs (2004) Mineralogy report: Characterization of lead particles in Sudbury soils and comparison of particles with suspected source material. Report to C. Wren and Associates for the Sudbury Soils Study. Geosciences Laboratories. Ministry of Northern Mines and Development, Sudbury

Golder Associates Ltd. (2001) Town of Falconbridge soil sampling program, comprehensive Falconbridge survey. Report prepared by Golder Associates Ltd, Sudbury

Freedman B, Hutchinson TC (1980) Pollutant inputs from the atmosphere and accumulation in soils and vegetation near a nickel–copper smelter at Sudbury, Ontario, Canada. Can J Bot 58:108–132

Hernandez L, Probst A, Probst JL, Ulrich E (2003) Heavy metal distribution in some French forest soils: evidence for atmospheric contamination. Sci Total Environ 312:195–219

Koptsik S, Koptsik G, Livantisova S, Eruslankina L, Zhmelkova T, Vologdina Zh (2003) Heavy metals in soils near the nickel smelter: chemistry, spatial variation, and impacts on plant diversity. J Environ Monit 5:441–450

McQuarry D, Rock A (1991) Chapter 2 – Atoms and molecules. In: General chemistry, 3rd edn. WH Freeman and Company, New York

MOE (1993) Field investigation manual, Part 1, General methodology.
Ontario Ministry of the Environment, Toronto

MOE (1997) Guidelines for use at contaminated sites in Ontario. Ontario Ministry of the Environment, Toronto

MOE (2001) Metals in soil and vegetation in the Sudbury area (Survey 2000 and additional historic data). Ontario Ministry of the Environment, Toronto

MOE (2004a) City of Greater Sudbury 2001 urban soil survey. Report No. SDB-008-3511-2003. Ontario Ministry of the Environment, Toronto

MOE (2004b) Ontario Ministry of the Environment Record of Site Condition Regulation, Ontario Regulation 153/04 (O. Reg. 153/0) Soil, ground water and sediment standards for use under Part XV.1 of the Environmental Protection Act, 9 March 2004

Nikonov VV, Lukina NV, Frontas'eva MV (1999) Trace elements in Al-Fe-Humus podzolic soils subjected to aerial pollution from the apatite-nepheline production industry. Eurasian Soil Sci 32:1331–1339

Nriagu J, Wong HKT, Lawson G, Daniel P (1998) Saturation of ecosystems with toxic metals in Sudbury basin, Ontario, Canada. Sci Total Environ 223:99–117

Shotyk W, Blaser P, Grunig A, Cheburkin A (2000) A new approach for quantifying cumulative anthropogenic, atmospheric lead deposition using peat cores from bogs: Pb in eight Swiss peat bog profiles. Sci Total Environ 249:281–295

Suter GW (1993) Ecological risk assessment. Lewis Publishers, Boca Raton, Florida, 550 pp

3.9 Appendix Chapter 3: Abbreviations

BSE, backscattered electron

CAEAL, Canadian Association of Environmental Analytical Laboratories

CEM, Centre for Environmental Monitoring (Laurentian University)

COC, chemicals of concern

COI, communities of interest

ED, energy dispersive

EF, enrichment factor

EMP, electron microprobe

ERA, ecological risk assessment

HG-AES, hydride generation atomic emission spectrometry

HHRA, human health risk assessment

ICP-MS, inductively-coupled plasma-mass spectrometry

ICP-OES, inductively-coupled plasma-optical emission spectrometry

IDL, instrument detection limit

IRM, internal reference material

LFH Horizons, refers to surface organic soil layers ranging in thickness from 2 to 15 centimetres. The designation LFH refers to the fresh plant detritus (L = litter) on the soil surface, the partially decomposed organic layer (F = fermentation) and the well-decomposed organic layer (H = humus).

MDL, method detection limits

MOE, Ontario Ministry of the Environment

n, sample size

NIST SRM, standard reference materials

QA/QC, quality assurance/quality control

SEM, scanning electron microscope

TOC, total organic content

WD, wavelength dispersive

DOI: 10.5645/b.1.4

4.0 Distribution of Chemicals of Concern in the Study Area

Authors

Graeme A. Spiers, PhD
MIRARCO, Laurentian University
935 Ramsey Lake Road, Sudbury, ON Canada P3E 2C6
Email: gspiers@mirarco.org

Christopher D. Wren, PhD
MIRARCO, Laurentian University
935 Ramsey Lake Road, Sudbury, ON Canada P3E 2C6
Email: cwren@mirarco.org

Dave McLaughlin
Ontario Ministry of the Environment
125 Resources Road, Toronto, ON Canada M9P 3V6
Email: dave.l.mclaughlin@ontario.ca

Table of Contents

4.1 Sudbury Geology and Soils 73
    4.1.1 Geology 73
    4.1.2 Physiography 73
    4.1.3 Soils 73
4.2 Regional Distribution of Metals in Sudbury Soils 74
    4.2.1 Metal Distribution in Parent Materials 74
    4.2.2 Regional Distribution of COC in Surface Soils 77
    4.2.3 Zonation of Metal Enrichment in the Sudbury Smelter Footprint 83
4.3 Distribution of COC within the Communities of Interest (COI) 85
    4.3.1 Copper Cliff 85
    4.3.2 Coniston 87
    4.3.3 Falconbridge 88
    4.3.4 Hanmer 89
    4.3.5 Sudbury Center 91
    4.3.6 Spatial Distribution of COC within a Community 92
4.4 References 94
4.5 Appendix Chapter 4: Abbreviations 95

Tables

4.1 Summary of the Concentration of Major and Trace Elements in Parent Soil Materials ($n=254$) 74
4.2 Summary of COC Concentrations (mg/kg) in Regional Surface Soils ($n=385$) 77
4.3 Summary of Soil Metal Levels (mg/kg) Reported in Other Mining-related Risk Assessments 78

4.4 Mean Concentration of Metal(loid)s (mg/kg) in the 0 to 5 cm Layer of Soils within Concentric Zones around the Sudbury Smelter Region. ...84
4.5 Summary of Metal Concentrations (mg/kg) in Copper Cliff ...86
4.6 Summary of Metal Concentrations (mg/kg) in Coniston ...88
4.7 Summary of Metal Concentrations in Falconbridge (mg/kg) ...89
4.8 Summary of Metal Concentrations in Hanmer (mg/kg) ...91
4.9 Summary of Metal Concentrations (mg/kg) in Sudbury Center ...92

Figures
4.1 Regional Distribution of Arsenic (mg/kg) ...78
4.2 Regional Distribution of Cobalt (mg/kg) ...79
4.3 Regional Distribution of Copper (mg/kg) ...80
4.4 Regional Distribution of Nickel (mg/kg) ...81
4.5 Regional Distribution of Lead (mg/kg) ...82
4.6 Regional Distribution of Selenium (mg/kg) ...83
4.7 Graphs Illustrating the Concentrations of Individual Metal(loid)s with Distance (km) from the Smelter Zone Centroid (from CEM 2004) ...84
4.8 Distribution of Ni within Town of Copper Cliff ...93
4.9 Distribution of Pb within Town of Copper Cliff ...93

4.1 Sudbury Geology and Soils

4.1.1 Geology

Metals constitute a natural component of soils, with concentrations dependent on soil mineral composition and geochemical history. Localized, strongly elevated metal concentrations have been created in surface soils by atmospheric deposition in the neighborhood of metal extraction facilities. Soils, more than any other sampling medium, reflect the total historical metal accumulation from the point source, but modified to varying degrees by soil-forming processes and erosion. The metals originating from anthropogenic sources in soil may behave differently from the naturally present metals depending on mineral phase.

The Sudbury Geological Structure is an elliptical unit produced by a meteorite colliding with the southern part of the Superior Province (see Figure 2.1). The melted rocks form the world's largest impact-related melt sheet. The melt sheet of the Sudbury Igneous Complex is composed, from base to top, of norite, quartz gabbro and granophyre. As host to the ore deposits, its mineralogy and geochemistry are important in discriminating between anthropogenic contamination and influences of bedrock in the geochemistry of soils. The Sudbury Igneous Complex has been divided into the Main Mass (norite, quartz gabbro, and granophyre) and Sublayer (Contact Sublayer and Offset Sublayer) (Dressler et al. 1991). Mineralization of importance occurs in the Sublayer and the Offset dikes.

4.1.2 Physiography

The Sudbury area includes parts of the Abitibi Uplands, Penokean Hills and Cobalt Plain of the James Physiographic Region and a small part of the Laurentian Highlands within the Laurentian Physiographic Region (Barnett and Bajc 2002). The Abitibi Uplands north of Sudbury are underlain by Archean crystalline rocks and comprise broad, rolling, bedrock-controlled surfaces rising gently toward the Abitibi Uplands southern boundary to a maximum elevation of 450 m. Locally, relief may be as much as 90 m along deeply incised canyons (Dredge and Cowan 1989). The Laurentian Highlands is an old erosion surface consisting of low, rounded knobs and ridges; locally, relief may be as much as 30 to 50 m. Elevation of this area is up to 300 m (Barnett and Bajc 2002).

The Sudbury Basin consists of an oval central low, the 'Valley', rimmed by a zone of high relief ridges to the north, east and south. The rocks of the Sudbury Igneous Complex and the Onaping Formation form these ridges. The Valley is a plain exhibiting low relief interrupted by bedrock ridges caused by the broad folding of the Chelmsford Formation. The plain slopes to the south-west with a drop in elevation of about 40 m over 39 km (Barnett and Bajc 2002). Local relief in the Valley is about 15 m and in places as much as 30 m. Some bedrock ridges reach 320 m above sea level.

4.1.3 Soils

The characteristics of soil are greatly influenced by the nature of the parent material, together with weathering and erosion processes. The soil mineralogical and chemical composition in the Sudbury area will, therefore, reflect the bedrock geology of the region, the up-ice geology, the organic input from the flora and fauna of the region, and exogenous materials such as particulate matter from both long- and short-range transport processes. Sudbury area soils belong to five orders of the Canadian Soil Classification System (Agriculture Canada Expert Committee on Soil Survey 1987): Luvisolic, Gleysolic, Podzolic, Brunisolic, and Organic (Gillespie et al. 1983).

The focus of this study was on well-to-imperfectly drained soils developed on the regional glaciogenic sediments. The well-to-imperfectly drained undisturbed soils of the Sudbury region are characterized by having organic (LFH) horizons ranging in thickness from 2 to 15 cm. The designation LFH refers to the fresh plant detritus (L=litter) on the soil surface, the partially decomposed organic layer (F=fermentation) and the well-decomposed organic layer (H=humus). These poorly studied LFH horizons, initially composed almost entirely of organic matter, are crucial sinks for the aerosolic particles (Spiers et al. 2002) from both local and long-range sources, acting both as filters to prevent particle translocation to lower horizons, and as exchange surfaces to absorb dissolved metals in precipitation and water through flow. Colloidal soil organic matter, for example, strongly adsorbs Cu, Zn, Fe and other transition metal ions, by acting as a chelating agent (Bohn et al. 2001).

In a study in the Falconbridge area, Golder Associates Ltd. (2001) reported total C content ranges from a low of 0.16% to a peak of 10.1% in the 0–5 cm layer of the sampled soils. The study documented carbonate content ranging from below detection limit to a high of 0.89% in the same 0–5 cm layer, with the higher levels being in areas which were either landscaped or limed. Gundermann and Hutchinson (1995) reported that the organic C content of soils in the 0–5 cm layer in the Coniston smelter area decreased over the period between 1972 and 1992, probably because of soil erosion. The latter study also reported a concomitant decrease in water extractable metal content from 74, 33 and 52 mg/kg Ni, Cu, and Al to 2, 2, and 3 mg/kg, Ni, Cu and Al, respectively, suggesting that the decrease is strongly linked to the erosion of surface organic matter. Hazlett et al. (1983) reported organic C content of soils around the Coniston smelter in the range 0.1% to 19.4%, with the high values being for the LFH horizons and the lowest values for C horizons. A study of soil samples collected for the 0–20 cm layer from throughout the Sudbury region some 12 years ago, documented organic carbon content ranging from about 0.5% to 2.1% (Adamo et al. 2002).

4.2 Regional Distribution of Metals in Sudbury Soils

4.2.1 Metal Distribution in Parent Materials

The sites selected for the regional study were sampled to a depth of greater than 80 cm, wherever possible, using bucket augurs to obtain soil samples assumed to be unaffected by recent industrial activities. These samples, referred to as parent material samples, were obtained from over 70% (*n*=254) of the sites visited during the sampling program. The analytical data obtained from these samples represent the first known attempt to establish the pre-industrial levels of metals in regional soils, with the data providing an excellent indication of regional background levels of *Aqua Regia* extractable metal(loid)s.

The concentrations of soil parameters measured in the parent material are presented in Table 4.1. Also presented in Table 4.1 for comparison purposes are mean values for soils of the Canadian Shield from an earlier study (McKeague et al. 1979) and Ontario Ministry of the Environment (MOE) Table F values, which are considered to represent Ontario background soil concentrations.

The soils in the Sudbury region are formed on primarily coarse textured tills and glacio-fluvial materials which are mineralogically dominated by quartz and feldspars, with minor amounts of heavy and clay minerals. As the heavy and clay mineral fraction are the sources for the metals of interest to the current studies, it is not surprising to observe that both the mean concentration and 95th percentile of most elements measured in the parent material are less than the generic Ontario background level (Table F values). The Table F criteria in the MOE *Guideline* reflect the mean plus two standard deviations of the entire provincial background soil data base, which is approximately the 97.5th percentile. Therefore, if the Sudbury Regional soil survey data are normally distributed, the mean concentrations should be less than the MOE generic background levels. In fact, the only two elements of the Sudbury regional sampling that have 95th percentile values greater than Table F are Cr and Ni, perhaps reflecting some incorporation of local metal-rich bedrock in the glacial detritus of the soil parent materials.

Table 4.1 **Summary of the Concentration of Major and Trace Elements in Parent Soil Materials (*n*=254)**

Parameter	Arithmetic mean	Range	95th UCL	Standard error	Standard deviation	Shield soils[1]	MOE Table F[2]
Al (%)	1.78	0.21–9.1	1.9	0.067	1.07	6.7	NG
Ca (%)	0.78	0.1–5.8	0.8	0.067	1.11	1.8	NG
Fe (%)	2.27	0.21–7.8	2.4	0.068	1.09	2.5	NG
Mg (%)	0.67	0.04–3.8	0.7	0.034	0.54	0.53	NG
As (mg/kg)	1.11	<dl–98	4.1	n/a	6.5	n/a	17
Ba (mg/kg)	98.4	13–390	120	n/a	80.5	n/a	NG
Be (mg/kg)	0.15	<dl–1.1	0.4	n/a	0.2	n/a	NG
Cd (mg/kg)	<dl	<dl	<dl	n/a	n/a	n/a	1

Parameter	Arithmetic mean	Range	95th UCL	Standard error	Standard deviation	Shield soils[1]	MOE Table F[2]
Cr (mg/kg)	56.4	12–130	59.1	1.63	26.04	19	71
Co (mg/kg)	8.9	2–38	9.5	0.29	4.64	19	21
Cu (mg/kg)	26.4	<dl–270	28.8	1.73	27.65	12	85
Mn (mg/kg)	293	23–1800	342	11.35	180.8	417	NG
Mo (mg/kg)	0.11	<dl–3.1	0.8	0.03	0.46	n/a	2.5
Ni (mg/kg)	36.1	8.5–163	41.9	1.32	21.0	12	43
Pb (mg/kg)	5.9	1–47	7	0.24	3.85	20	120
Se (mg/kg)	0.06	<dl–2	0.06	0.02	0.31	0.18	1.9
Sr (mg/kg)	43.8	11–80	45.7	1.08	17.27	n/a	NG
V (mg/kg)	43.3	6.7–220	47.4	1.29	20.61	n/a	91
Zn (mg/kg)	29.7	5.4–160	31.9	1.16	18.5	57	160
pH (n=35)	5.6	3.7–7.8	NC	0.15	0.9	n/a	n/a

Based on values for 254 samples; pH reported for subset of 35 samples (CEM 2004).
UCL = upper confidence level.
NG = no guideline available.
NC = not calculated.
n/a = not available.
<dl = below detection limit.
[1] Values for soils of the Canadian Shield (McKeague et al. 1979).
[2] Table F considered Ontario background concentrations (MOE 1997).

Although the bedrock in the Sudbury basin is known to be locally highly mineralized, this is not reflected in higher background soil concentrations relative to the Ontario generic criteria, possibly because of dilution with upstream rock materials due to glaciations. Furthermore, the base metal-rich mineral phases hosted in the sulfide-rich units of the regional bedrocks are relatively soft, and may thus have been transferred and dissolved from the surficial materials as a result of glacial activity and weathering. In fact, the true natural 'background' surface soil metal concentrations in the mineralized areas of the Sudbury basin are similar to those documented in other regions of the Canadian Shield. The results of the parent material analysis and comparison with Table F indicate that background metal concentrations in the Sudbury area are not higher than levels considered as background for other parts of Ontario. This is an interesting conclusion because the MOE specifically avoided the Sudbury basin area when collecting the samples that were used to calculate the 1997 Table F background-based soil guidelines because of the suspicion that the surficial soils may be mineralized above normal background levels.

In the following discussion, the values obtained for Sudbury parent soils are compared with data for parent materials from a variety of sources, including 'background' Ontario soil concentrations.

4.2.1.1 Arsenic

The maximum value for As in the parent samples was 98 mg/kg, with minimum values below detection limits. The arithmetic mean concentration of As in the parent materials of the Sudbury region was 1.1 mg/kg. Of the 254 parent material samples analyzed, only 24 had an As concentration above detection limits by the methodology used in this study. The MOE Table F background concentration limit for As (17 mg/kg) for all non-agricultural use for surface materials was exceeded in six samples, centered on the Copper Cliff smelter. Since the Table F guideline is essentially the 98th percentile, if the sample population is normally distributed, about two out of every 100 samples may have a concentration greater than Table F. Therefore, for a population size of 254 samples, one would expect about five samples to exceed the normal background range, which is close to what was observed. Although comparative data for the levels of As for a similar study in Shield soil materials is not available, mean levels for uncontaminated soils of Southern Ontario are documented at 5.2 mg/kg, while that for U.S. soils is 7.4 mg/kg, with a range from 1 to 97 mg/kg (Shacklette et al. 1971; Gough et al. 1988). For comparison, Henderson et al. (2002) estimated the background level of As for soils of the Rouyn-Noranda area approximately 250 km north-east of Sudbury at 6 mg/kg. In a study of background concentrations for soils formed in tills underlain by Precambrian bedrock formations in the Flin Flon region of Manitoba, McMartin et al. (1999) documented As mean concentration as 20 mg/kg.

4.2.1.2 Cadmium

Cadmium was below the method detection limit (MDL) for the 254 samples of regional soil parent materials. Henderson et al. (2002) estimated the background level of Cd for soils of the Rouyn-Noranda area at 1 mg/kg, a level in contrast to a measured mean background concentration of 0.3 mg/kg for soils formed in tills in the Flin Flon region of Manitoba (McMartin et al. 1999).

4.2.1.3 Cobalt

With a range in concentration from 2 to 38 mg/kg, the arithmetic mean Co level (8.9 mg/kg) in soil parent materials of the Sudbury region was lower than those described by McKeague et al. (1979) for Shield soils. However, the mean value is similar to that defined for U.S. soils (10 mg/kg) by Shacklette et al. (1971) and Gough et al. (1988). Several sample sites in the Sudbury area had concentrations above the MOE Table F background Co guideline of 21 mg/kg for all non-agricultural uses for surface materials.

4.2.1.4 Copper

The overall mean level of Cu (26.4 mg/kg) was similar to that defined for U.S. soils (25 mg/kg) by Shacklette et al. (1971) and Gough et al. (1988), but double that documented by McKeague et al. (1979) for samples of Shield soils. One site in the Kelley Lake delta, with a Cu level of 270 mg/kg, skewed the data for the regional distribution of Cu. This site is probably enriched with Cu as a result of erosion from the mining and smelter operations further up the Junction Creek watershed. This is also the only site in the region with soil Cu levels above the MOE Table F guideline (85 mg/kg) for surface materials. Interestingly, the data obtained in the current Sudbury study were similar to the estimation of background concentration of Cu at 30 mg/kg for soil of the Rouyn-Noranda area approximately 250 km north-east of Sudbury (Henderson et al. 2002). The measured mean background concentration of Cu in soils formed in tills in the Flin Flon region of Manitoba was 122 mg/kg (McMartin et al. 1999).

4.2.1.5 Lead

The mean regional value for Pb (5.9 mg/kg) in Sudbury area soil parent materials was much below that documented by Dudas and Pawluk (1980) for Prairie soils, by McKeague et al. (1979) for soils of the Shield region and by Shacklette et al. (1971) and Gough et al. (1988) for soil forming materials of the conterminous U.S. (20 mg/kg). The Pb levels for all parent material sample sites in this Sudbury area study had concentrations below the MOE Table F background lead concentration limit of 120 mg/kg for all non-agricultural uses for surface materials. The Table F soil Pb guideline is considerably higher than most of the parent material soil Pb data because the MOE guideline is calculated from samples of surface soil. Although known industrial point sources of contaminants were obviously avoided during the background sampling, the ubiquitous use of leaded gasoline for seven decades in the 1900s resulted in the widespread distribution of Pb in the Ontario airshed, and subsequently deposition of Pb to surface soil. Therefore, the background soil Pb guideline for Ontario reflects a significant societal anthropogenic signal. Furthermore, the data obtained in the current Sudbury study are much lower than the estimate of background concentration of Pb at 80 mg/kg for soil of the Rouyn-Noranda area approximately 250 km north-east of Sudbury (Henderson et al. 2002). The measured mean background concentration of Pb in soils in the Flin Flon region of Manitoba was 8 mg/kg (McMartin et al. 1999).

4.2.1.6 Nickel

The arithmetic Ni level (36 mg/kg) for the Sudbury region was higher than those levels documented in the review by McKeague et al. (1979) for agricultural soils of the Shield (12 mg/kg) and in conterminous U.S. soils (20 mg/kg) (Shacklette et al. 1971; Gough et al. 1988). The range in parent material Ni concentrations was from 9 to 163 mg/kg, with the higher concentrations being in the soil parent materials near the Kelly Lake delta that is the receiver of the sediment load from the major mineral extraction operations of the region. The high values were, however, below the MOE Table F guideline of 43 mg/kg for surface materials where soil pH is 5.0 to 11.0. The measured mean background concentration of Ni in soils formed in tills in the Flin Flon region of Manitoba was 67 mg/kg (McMartin et al. 1999), approximately twice as high as in the soils of the Sudbury region. The data obtained in the current Sudbury study were much lower than the estimation of background concentration of Ni at 1 mg/kg for the humus layer of soils of the Rouyn-Noranda area approximately 15 km north-east of Sudbury (Henderson et al. 2002).

4.2.1.7 Selenium

The arithmetic Se level (0.06 mg/kg) for the Sudbury region was lower than documented in the review by McKeague et al. (1979) for agricultural soils of the Shield (0.18 mg/kg). The reports of Shacklette et al. (1971) and Gough et al. (1988) for soils of the U.S. did not document levels for Se. The values reported in this study were considerably below the MOE Table F guideline of 1.9 mg/kg for surface materials where soil pH is 5.0 to 11.0.

4.2.2 Regional Distribution of COC in Surface Soils

A primary objective of the regional soil survey was to measure the spatial distribution (geographic area) of metals in surface (0 to 5 cm) soils to determine the potential 'footprint' of particulate airborne emissions from the Sudbury smelters. The analytical methods and detection limits for the 2001 Soils Study are described in Chapter 3. A summary of metal concentrations in regional surface soils (0–5 cm layer) is provided in Table 4.2 with further discussion below. Data from other studies and other smelter locations are provided for comparison with the Sudbury results. Note the following discussion is for regional samples only; a discussion of metal levels found in urban residential soils, which tend to include samples closer to the smelters, is provided in Section 4.3 below.

Table 4.2 **Summary of COC Concentrations (mg/kg) in Regional Surface Soils (n=385)**

Parameter	Minimum	Maximum	Mean	Standard deviation
Arsenic	ND	305	14.8	21.3
Cadmium	ND	3.2	0.36	0.54
Cobalt	2.0	78.5	12.5	10.3
Copper	6.1	3850	261	314
Nickel	14.0	2900	263	296
Lead	3.5	194	50.0	25.0
Selenium	ND	17.0	2.2	2.1

ND = below analytical detection limit.

A series of maps is provided that illustrate the concentration of each of the COC (As, Co, Cu, Ni, Pb, Se) in surface soils to a distance of approximately 20 km from the smelter centers. In general, elevated concentrations of metals were centered on the three historic smelting centers of Coniston, Copper Cliff and Falconbridge. This is clearly illustrated in Figures 4.1 to 4.6. Although the actual soil sampling extended over a vast area approximately 200 km × 200 km, the major influence of the smelters is much more localized. The actual extent of effects from smelter emissions on regional background concentrations is thought to extend out about 120 km. This is discussed in more detail in Section 4.2.3 below.

4.2.2.1 Arsenic

The maximum value for As in the 0–5 cm samples was 305 mg/kg, with minimum values below detection limits. The regional arithmetic mean value was 14.8 mg/kg, which compares with an upper crustal average of 4.8 mg/kg (Rudnick and Gao 2003). The Table F Ontario background concentration is 17 mg/kg and the MOE Table A effects-based generic guideline is 20 mg/kg. The Table A limit was exceeded in 113 samples. The elevated As concentrations are centered on the Coniston, Vale and Xstrata smelters (Figure 4.1).

It is apparent that large portions of the Sudbury area between the smelters of Copper Cliff and Falconbridge contain soils with As concentrations <20 mg/kg. This may be attributed to residents importing, vertically mixing or amending soil in their yards. This pattern also appears for the other COC. Soils with elevated As values are localized within a few kilometers of the smelters. Soils with As levels more than 200 mg/kg (10× the Table A value) only occur in the town of Falconbridge.

During surveys from 1976 to 1997 in the Sudbury area, the MOE recorded a maximum soil As level of 510 mg/kg in 1976 in a sample approximately 2 km north of the Xstrata smelter (MOE 2001). Table 4.3 contains soil metal data for some other mining related sites in Canada and Australia. The maximum As level in soil surrounding the smelter at Trail, British Columbia was reported to be 130 mg/kg (Intrinsik 2007), while the maximum soil As level at the Deloro site in Ontario was 605 mg/kg (Cantox 1999). The MOE (1991) detected soil As levels ranging up to 234 mg/kg in residential areas of Port Hope and up to 1045 mg/kg in Wawa (MOE 1999). Residential communities in both of these Ontario towns were impacted by historic smelting emissions.

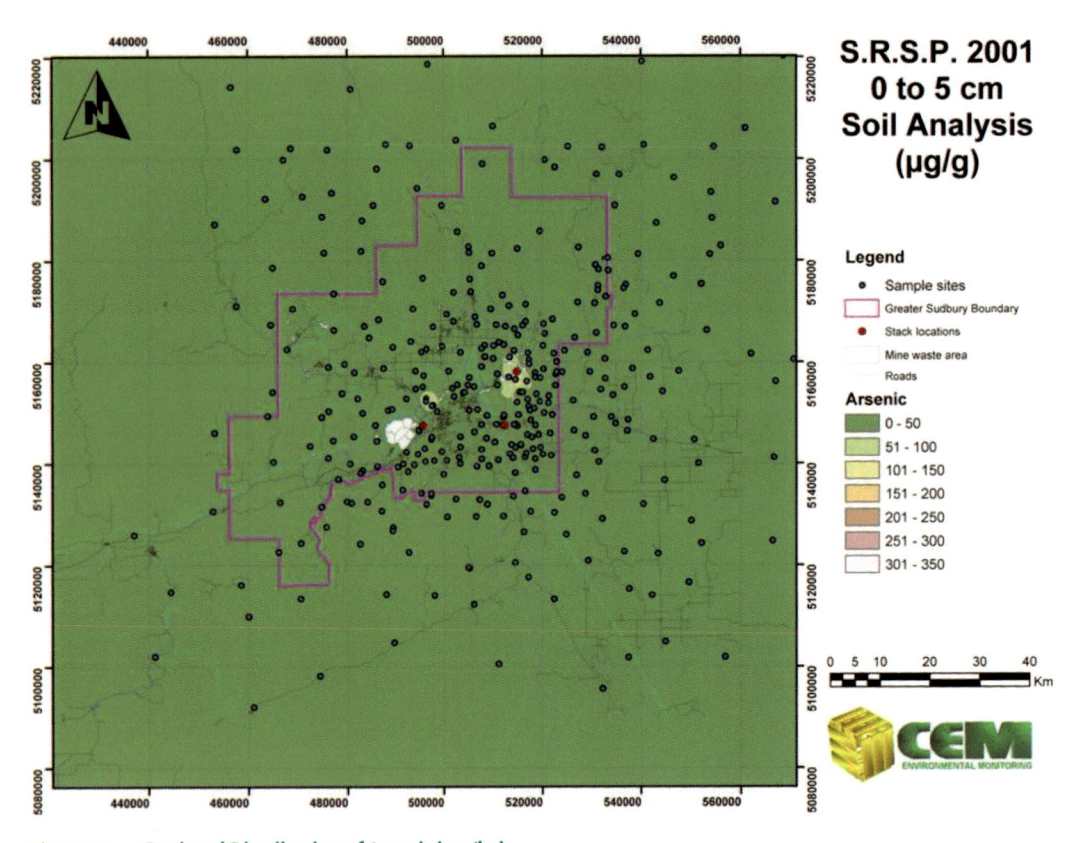

Figure 4.1   **Regional Distribution of Arsenic (mg/kg)**

Table 4.3   **Summary of Soil Metal Levels (mg/kg) Reported in Other Mining-related Risk Assessments**

Location (Reference)	Statistic	As	Co	Cd	Cu	Pb	Ni	Se
Port Colborne, Ontario (MOE 2002)	Range n=>100	ND–350	5–262	ND–5.1	4–2720	6–1800	19–17,000	0.1–19.4
	Mean	16	51	1.2	240	217	2508	2.4
Trail, British Columbia (Intrinsik 2007)	Range n=364	1.2–130		0.1–27.6	2–326	2–3330		
	Mean	17.7		3.2	23	207		
Deloro, Ontario (Cantox 1999)	Range n=147	2.4–605	5.1–340		6–115	3.5–655	8.8–195	
	Mean	111	57		29	122	44	
Mt Isa, Australia (Taylor et al. 2010)	Range n[1]=60			0.7–12	31–12,100	8–5700		
	Mean			1.8	342	105		

n = sample size; ND = Not detected.
[1] For fine texture soils.

4.2.2.2 Cobalt

Elevated concentrations of Co in surface soils are not widely distributed throughout the study area (Figure 4.2). Soils with Co concentrations above Table A for residential land use and coarse soils are primarily confined to the immediate vicinity of the three smelting centers of Copper Cliff, Coniston and Falconbridge. Several sample sites in the Sudbury area have concentrations above the Ontario soil quality guideline of 40.0 m/kg.

With a range in concentration from 2 to 78 mg/kg, the arithmetic mean Co levels (12.5 mg/kg) in soil surface layers of the Sudbury region are slightly lower than those described by McKeague et al. (1979) for the surface mineral horizon of Shield soils. The mean Co level in this study was very similar to that (11 mg/kg) reported by Dudka et al. (1995) in a survey of 73 locations around Sudbury (maximum 113 mg/kg). The mean value was also similar to that defined for U.S. soils (10 mg/kg) by Shacklette et al. (1971) and Gough et al. (1988). Yakovlev et al. (2008) reported Co levels ranging from 384 to 1229 mg/kg in soil samples collected near the Norilsk Nickel Company smelter in Russia, which are substantially higher than those levels found in Sudbury.

The MOE collected and analyzed soil samples at 92 locations around Sudbury on a regular basis from 1971 to 1999. During those years, the maximum soil Co level recorded was 788 mg/kg in a sample collected in 1971 approximately 5 km west of the Copper Cliff smelter (MOE 2001). The maximum soil Co level detected in 1999 was much lower at 57 mg/kg (MOE 2001).

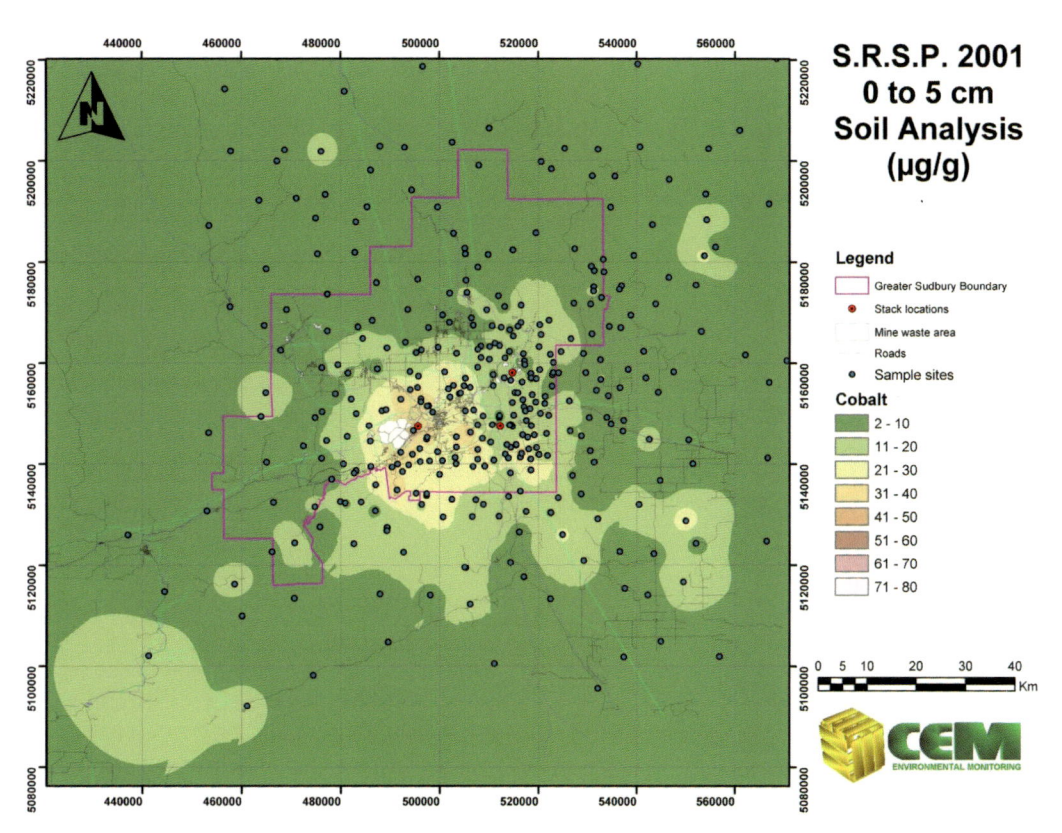

Figure 4.2   **Regional Distribution of Cobalt (mg/kg)**

4.2.2.3 Copper

The overall mean level of Cu (261.4 mg/kg) for the 0 to 5 cm layer of Sudbury regional surface soils was substantially greater than those defined for U.S. soils (25 mg/kg) by Shacklette et al. (1971) and Gough et al. (1988), and for surface layers of Canadian Shield soils (11 mg/kg) as documented by McKeague et al. (1979). With a range from 6 to 3850 mg/kg, the distribution of Cu in surface soils, as illustrated in Figure 4.3, shows a classic wind driven ellipsoidal pattern centered on the regional smelter complex. The soil concentrations of Cu are elevated above Table A for residential land use and coarse soils over a relatively wide area (Figure 4.3). Similar to As, many locations within the City Centre had Cu levels below Table A. Copper concentrations over 2000 mg/kg were confined to the immediate vicinity of the Copper Cliff and Falconbridge smelters.

During their regular surveys (1971 to 1999) around Sudbury, the maximum soil Cu level reported was 2800 mg/kg in a sample less than 1 km west of the Copper Cliff smelter (MOE 2001). Many soil samples contained Cu in excess of 1000 mg/kg. A subsequent survey in 2000 generally detected lower soil Cu levels with maximum values less than 1000 mg/kg (MOE 2001). Another survey of Sudbury soils (n = 73) reported a mean Cu concentration of 116 mg/kg, with a maximum value of 1891 mg/kg (Dudka et al. 1995).

Taylor et al. (2010) reported that soil Cu levels ranged from 31 to 12,100 mg/kg in soils surrounding the Xstrata Nickel smelter at Mt Isa City, Australia (Table 4.3), while Yakovlev et al. (2008) reported Cu levels up to 15,000 mg/kg in soil around the Norlisk Nickel smelter in Russia. These are considerably higher than maximum soil Cu levels found near the Trail smelter (326 mg/kg) or at the Deloro site in Ontario (115 mg/kg) shown in Table 4.3. The measured mean humus concentration of Cu in soils formed in tills within 5 km of the Flin Flon region of Manitoba was 1970 mg/kg (McMartin et al. 1999).

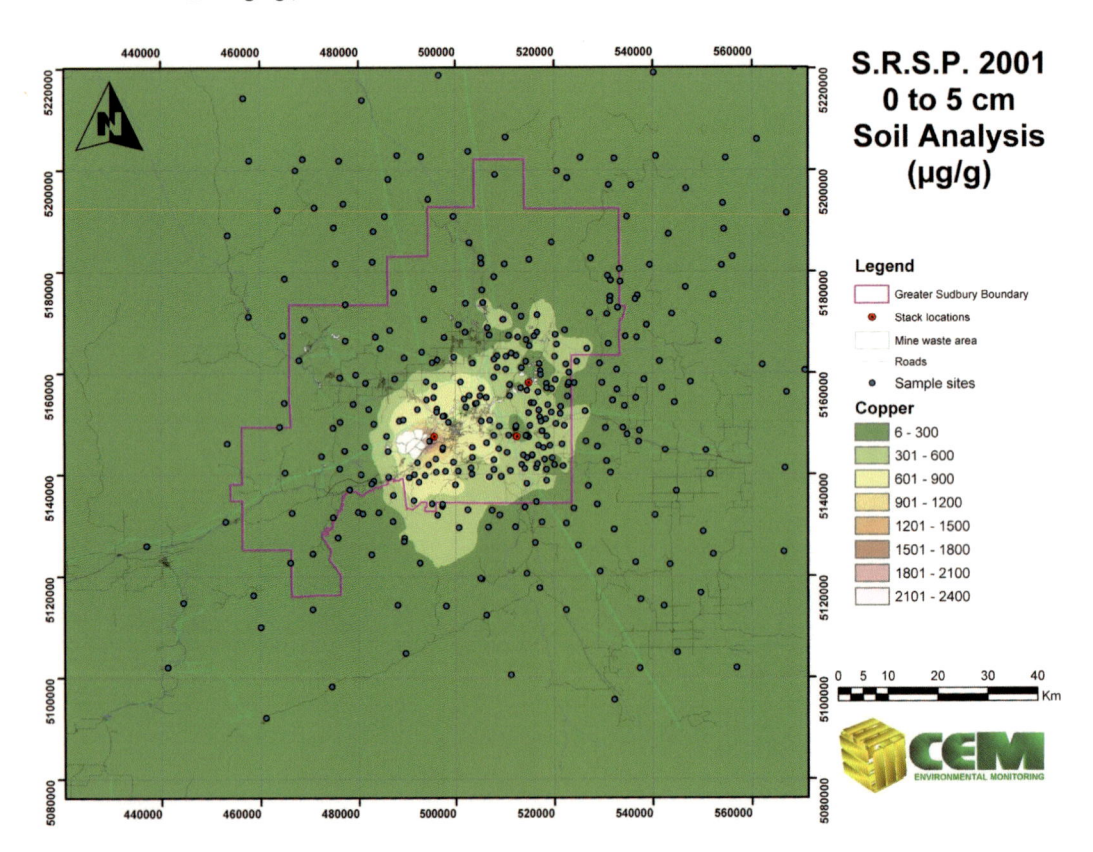

Figure 4.3   **Regional Distribution of Copper (mg/kg)**

4.2.2.4 Nickel

The distribution of Ni levels in surface soils very closely resembles the pattern of Cu (Figure 4.4). Many locations within the City Centre have Ni levels below the provincial soil quality guideline. This can likely be attributed to importing, amending or vertically mixing soils in residential or park properties.

The arithmetic mean Ni level (263 mg/kg) for the Sudbury region was much higher than those levels documented in the review by McKeague et al. (1979) for agricultural soils of the Shield (12 mg/kg) and in the USGS documented levels for the conterminous U.S. soils (20 mg/kg) (Shacklette et al. 1971; Gough et al. 1988). The range of the 0 to 5 cm soil layer Ni concentrations was from 14 to 2900 mg/kg, with the higher concentrations found near the smelter operations (Figure 4.4). The Ni concentrations for the 0 to 5 cm layers in the core of the study area were considerably above the MOE Table A (150 mg/kg) guidelines for surface materials where soil pH is 5.0 to 11.0. The maximum Ni levels measured in this survey were higher than those recorded by the MOE during their regular soil surveys of Sudbury. The previous maximum Ni value was 2300 mg/kg reported in a 1997 sample less than 1 km east of the Copper Cliff smelter (MOE 2001). The maximum value reported from the 1999 and 2000 surveys was 1000 mg/kg (MOE 2001). Another survey of Sudbury soils ($n=73$) reported a mean Ni concentration of 105 mg/kg, with a maximum value of 2149 mg/kg (Dudka et al. 1995).

For comparison, the measured mean background concentration of Ni in the humus layers of soils formed in tills in the Flin Flon region of Manitoba was 7 mg/kg (McMartin et al. 1999). In Rouyn, on the other hand, Henderson et al. (2002) reported levels of Ni in the humus layers of 20.5 mg/kg, with a range of 7 to 82 mg/kg, levels much lower than in the soils of the Sudbury region. Yakovlev et al. (2008) reported Ni levels up to 2915 mg/kg in soils surrounding the Norilsk smelter in Russia, while Ni levels in residential soils in Port Colborne, Ontario, ranged from 19 to 17,000 mg/kg, with a mean Ni concentration of 2508 mg/kg (MOE 2002). In undisturbed soil of deciduous woodlots between 1 and 2 km east (downwind) of the Port Colborne Ni refinery, the Ni levels ranged up to 33,000 mg/kg (Jacques Whitford Limited 2004).

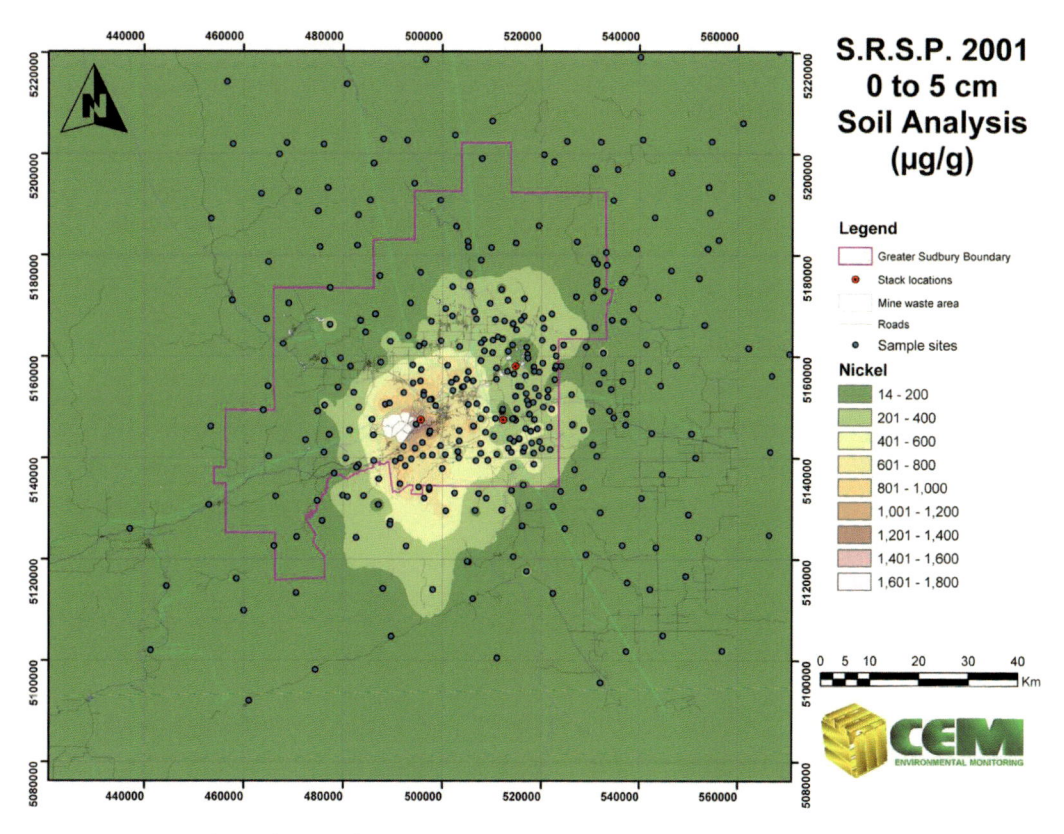

Figure 4.4   **Regional Distribution of Nickel (mg/kg)**

4.2.2.5 Lead

The distribution of Pb in surface soils was somewhat more patchy than the other elements (Figure 4.5). Soils containing Pb above the Ontario soil quality guideline (200 mg/kg) were primarily confined to the three smelting centers, with the largest density of these samples in proximity to the Copper Cliff smelter. Samples containing 100 to 200 mg/kg were primarily situated between the smelting centers, with samples containing 50 to 100 mg/kg being more widespread.

With a range in concentrations from 3.5 to 194 mg/kg, the mean regional value for Pb (50 mg/kg) in Sudbury area soil (0 to 5 cm) was higher than that documented by Dudas and Pawluk (1980) for Prairie soils, by McKeague et al. (1979) for soils of the Shield region (20 mg/kg) and by Shacklette et al. (1971) and Gough et al. (1988) for conterminous U.S. soils (20 mg/kg). Lead levels in soils surrounding the smelter at Trail, British Columbia, ranged from 6 to 3330 mg/kg (Intrinsik 2007).

In previous surveys of Sudbury, the MOE reported a maximum soil Pb level of 1000 mg/kg near the Copper Cliff smelter in 1986. A subsequent survey in 1999 did not reveal any soil samples exceeding background (120 mg/kg, MOE 1997), while the maximum soil Pb level in 2000 was only 160 mg/kg, with the Table A effects-based guideline of 200 mg/kg (MOE 1997) being exceeded at only two sites (MOE 2001).

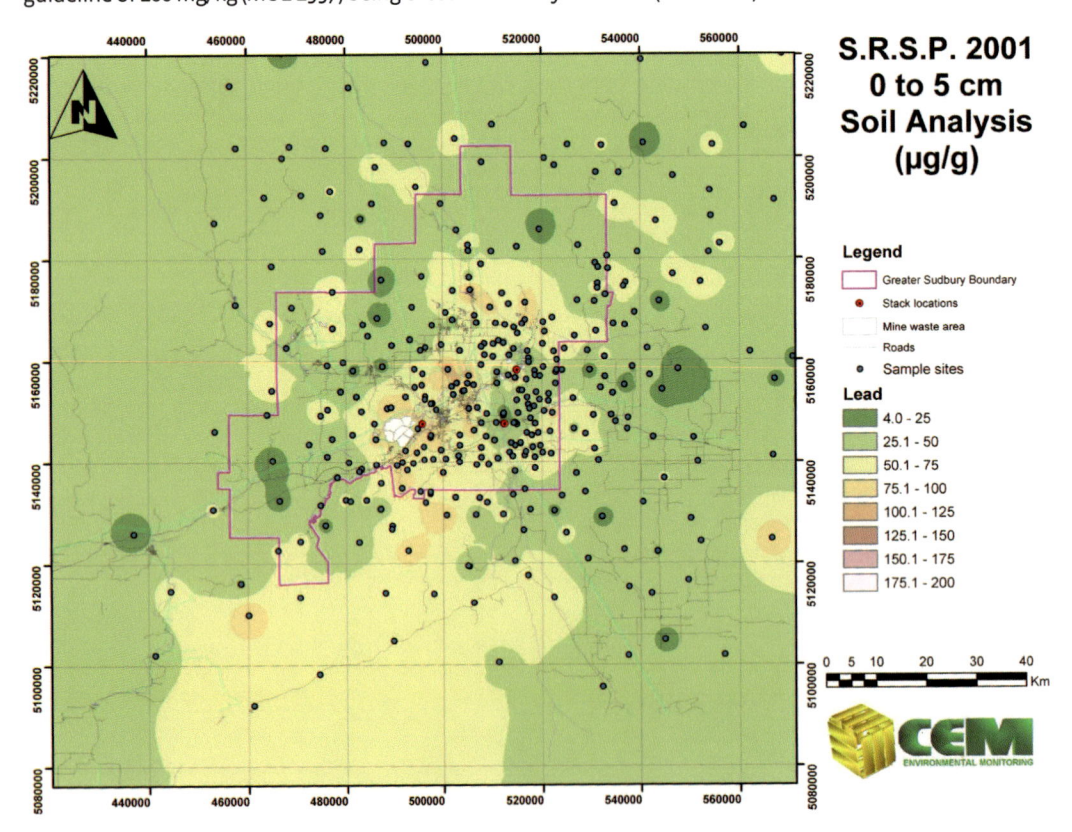

Figure 4.5   **Regional Distribution of Lead (mg/kg)**

4.2.2.6 Selenium

The soil concentrations of Se were generally low throughout the study area (Figure 4.6). The primary exception was soil samples in the immediate vicinity of Copper Cliff, which were elevated above the provincial Table A effects-based soil quality guideline (10 mg/kg). The distributional map for Se (Figure 4.6) also illustrates the classic ellipsoidal nature characteristic of aerosol deposition of a point source origin, with a locus near the Copper Cliff smelting operations. The highest concentration zone was immediately to the north-east of the Copper Cliff operations. During their 1971–1999 regular surveys, the MOE reported that Se concentrations rarely exceeded the Table A guideline, and were only elevated near the Copper Cliff smelter, with two soil samples containing Se levels of 11 and 33 mg/kg MOE (2001). During the survey in 2000 (n = 103 sites), the MOE reported that Se only rarely exceeded background (1.9 kg/kg) values and the maximum measured was just 3.7 mg/kg. However, it appears that Se was not measured in many of the samples from that year for some reason.

The arithmetic mean selenium level (2.2 mg/kg) for the 0 to 5 cm layer of undisturbed soils the Sudbury region was higher than documented in the review by McKeague et al. (1979) for agricultural soils of the Shield (0.18 mg/kg). The Sudbury Se levels were similar to the mean Se concentration (2.4 mg/kg) in soil reported in Port Colborne, Ontario (MOE 2002). Few of the other studies reported soil Se concentrations.

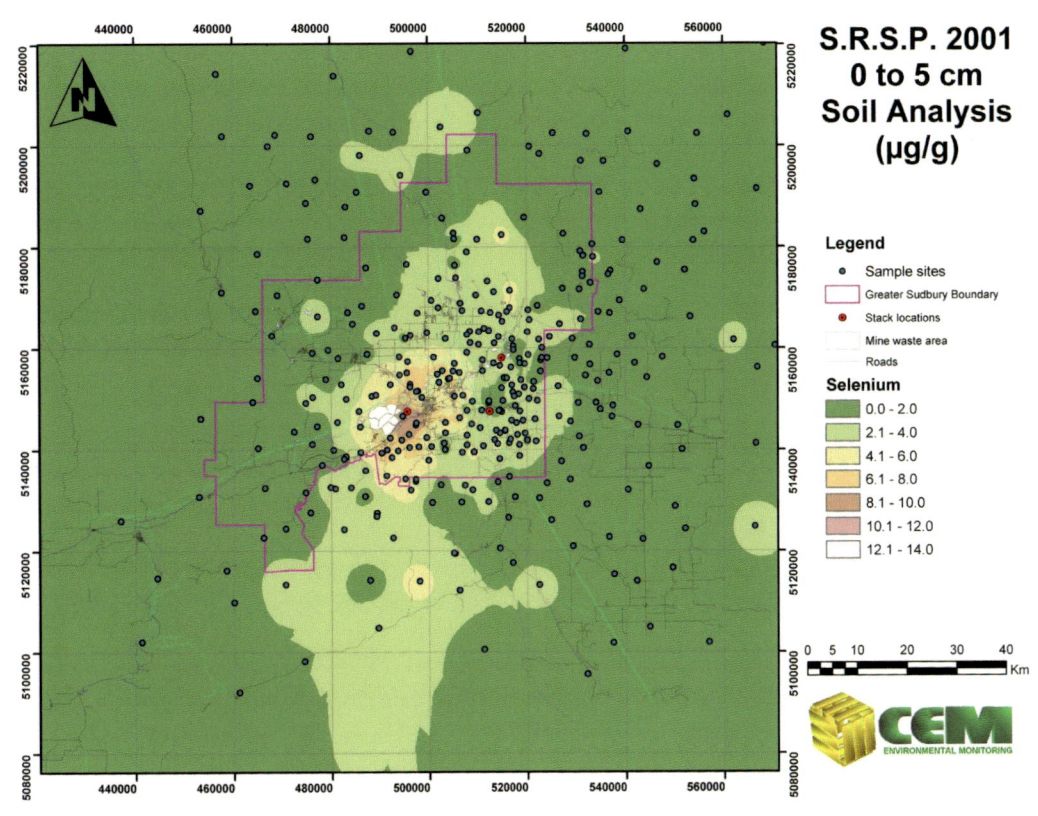

Figure 4.6   **Regional Distribution of Selenium (mg/kg)**

4.2.3 Zonation of Metal Enrichment in the Sudbury Smelter Footprint

The regional geochemical maps (Figures 4.1 to 4.6) indicate that the loading of the aerosol particular fallout from the regional smelters follows an ellipsoid, with a dominant southwest–northeast axis. The graphs in Figure 4.7 illustrate the concentrations of the individual anthropogenic metal(loid)s along a gradient from the center of the smelter zone toward regional background approximately 120 km from the heart of Sudbury. This estimate of the regional impact of smelter emissions agrees well with those estimated using metal accumulation on lichen thalli as deposition indices (Tomassini and Neiboer 1976). This estimate also compares with a calculated distance to regional background in humus of between 50 and 110 km (Zoltai 1988; McMartin et al. 1999; Henderson et al. 2002) for the smelter in the Flin Flon area of Manitoba, and a zone of 40 to 50 km for the Horne Smelter in the Rouyn-Noranda area of Quebec (Henderson et al. 2002). Goodarzi et al. (2001) documented enrichment of a series of six elements (As, Cd, Cu, Hg, Pb and Zn) in soils of the Trail, British Columbia area to a distance of 26 km from the smelter. Using associated moss bag studies, the authors concluded that the enrichment of Hg and As in the regional soils was not attributed to smelter activity. These authors emphasized the need for high quality atmospheric deposition data to supplement, and help to explain, data obtained from regional soil survey data.

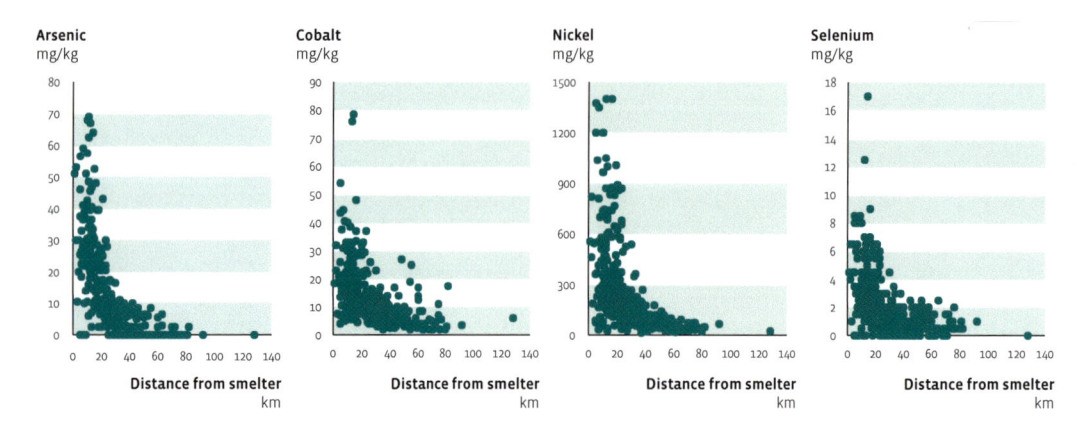

Figure 4.7 Graphs Illustrating the Concentrations of Individual Metal(loid)s with Distance (km) from the Smelter Zone Centroid (from CEM 2004)

The data in Table 4.4 summarize the mean concentrations of 20 elements in the 0 to 5 cm layer of the soils of this study in a series of circular zones around the center of smelter activity in the Sudbury region. The circular zonation does not exactly mimic the ellipsoidal zone suggested by wind rose and extrapolated map concentration data (Figures 4.1 to 4.6) but does provide an indication of the decrease in anthropogenic metal concentration in the surface soils with distance, to a final distance of 100 km at the borders of the current study zone.

The non-smelter emitted elements (Al, Ca, Mg, Mn, Ba, Be, Cr, Sr, V and Zn) tend to generally exhibit a similar concentration in the surface layer throughout the zones of the region, with no obvious enrichment pattern in the 0 to 5 cm layer. Some of these elements (Ca, Mn and Sr) actually are depleted in the surface layer nearer to the center of the smelter zone. This depletion perhaps reflects the effects of earlier higher levels of soil acidification from the high sulfur dioxide washout to regional soils before the implementation of the modern control systems. These control systems have resulted in the sulfur dioxide from the smelting process being converted into sulfuric acid, a valuable by-product of the mineral extraction process.

The metal(loid)s which are influenced by anthropogenic or smelter processes, on the other hand, show distinct concentration drops in the 0 to 5 cm layer with distance from the smelter zone, a characteristic response to point source emissions documented in most studies (e.g. Zoltai 1988; McMartin et al. 1999; Nikonov et al. 1999; Goodarzi et al. 2001; Henderson et al. 2002; Koptsik et al. 2003). The last zone (60 to 100 km) may exaggerate the drop in concentration with distance because the data are predominantly from the outer portion of the survey region with the 16 km cells at the extreme corners of the square.

Table 4.4 Mean Concentration of Metal(loid)s (mg/kg) in the 0 to 5 cm Layer of Soils within Concentric Zones around the Sudbury Smelter Region.

	Parent Material Influence Metals							
	Aluminum	Calcium	Magnesium	Manganese	Chromium	Strontium	Vanadium	Zinc
0–5 km	8564	2036	1510	220	37	23	26	35
5–15 km	9980	2649	1918	265	38	28	31	39
15–30 km	10,054	3286	1853	308	40	35	31	40
30–60 km	9730	3154	1773	344	38	35	32	39
60–100 km	8720	3025	1556	343	36	33	29	41

Distribution of Chemicals of Concern in the Study Area

Anthropogenic Influence Metal(loid)s								
	Arsenic	Cadmium	Cobalt	Copper	Iron	Lead	Nickel	Selenium
0–5 km	30	0.52	23	545	19,145	62	582	4.4
5–15 km	30	0.41	19	511	19,352	58	450	3.3
15–30 km	14	0.52	14	283	16,205	53	307	2.3
30–60 km	3.1	0.19	7.2	82	13,789	43	104	1.1
60–100 km	0.57	0.17	4.6	33	11,620	40	47	0.7
Parent material	1.11	NA	8.9	26	22,800	5.9	36	0.06

4.3 Distribution of COC within the Communities of Interest (COI)

The following sections summarize soil metal concentrations in the five COI (Copper Cliff, Coniston, Falconbridge, Hanmer, and Sudbury Centre) that were selected for detailed examination in the Human Health Risk Assessment (HHRA) (see Chapter 5). It was necessary to develop summary statistics for each of the COI to calculate community-specific exposure values in the HHRA. This is described in more detail in Chapter 7.

4.3.1 Copper Cliff

A review of the metal concentrations for the six COC in Copper Cliff is presented in the following text. In this community, a total of 315 soil samples were analyzed from the 0 to 5 cm soil depth, with 290 soil samples analyzed from both the 5 to 10 cm and 10 to 20 cm depths. A summary of the COC concentrations in the soils of this community is provided in Table 4.5.

4.3.1.1 Arsenic

The concentration of As reported at all three soil depths ranged from 2.5 to 101 mg/kg; note the minimum value represents one-half of the MDL for As. At the 0 to 5 cm soil depth, the mean As concentration was 17.6 mg/kg. The maximum As concentration for this depth was 72 mg/kg, and the 95th percentile concentration was 44.3 mg/kg. The deepest soil sampled (10 to 20 cm) exhibited a maximum As concentration of 99 mg/kg, with a mean concentration at this depth of 23.6 mg/kg. In a detailed HHRA in a community near a smelter operated by Vale in Port Colborne, Ontario, the maximum soil As in residential soil was reported to be 350 mg/kg (MOE 2002), while soil As levels in Deloro, Ontario, ranged from 2.4 to 605 mg/kg (Cantox 1999) and in Wawa, soil As levels ranged up to 1045 mg/kg (MOE 1999).

4.3.1.2 Cobalt

The mean Co concentration in the uppermost soil layer was 32.4 mg/kg with a range from 6 to 110 mg/kg. The 5 to 10 cm soil depth exhibited a Co range of 3 to 70 mg/kg. The mean and 95th percentile concentrations at this depth were 23.3 mg/kg and 47.1 mg/kg, respectively. The 10 to 20 cm soil depth showed Co concentrations varying from 5 to 46 mg/kg, with a mean concentration of 19.9 mg/kg. Cobalt levels in Sudbury soils are slightly lower than those reported for other Ontario locations at Port Colborne (mean 51 mg/kg) and Deloro (mean 57 mg/kg).

4.3.1.3 Copper

The Cu concentrations for the 0 to 5 cm soil depth ranged from 26 to 5600 mg/kg. The reported mean and 95th percentile concentrations were 1367 mg/kg and 3430 mg/kg, respectively. Soil from the 10 to 20 cm depth showed a range of Cu concentrations from 25 to 2000 mg/kg. The mean Cu concentration at this depth was 590.0 mg/kg, whereas the 95th percentile concentration was 1300 mg/kg. Copper concentrations in Sudbury soils tended to be higher than those reported for other Ontario locations (Port Colborne, Deloro, see Table 4.3) and in soils surrounding the smelter at Trail, British Columbia (soil Cu range 2 to 326 mg/kg). However, higher soil Cu levels were reported surrounding the Norilsk Nickel smelter in Russia (range 3300 to 15,000 mg/kg; Yakovlev et al. 2008) and the Xstrata smelter at Mt. Isa in Australia (range 31 to 12,100 mg/kg; Taylor et al. 2010).

4.3.1.4 Nickel

The range of Ni concentrations in the uppermost soil layer (0 to 5 cm) was 24 to 3700 mg/kg, with an average concentration of 978 mg/kg. In comparison, the 95th percentile concentration for this depth was 2500 mg/kg. The mean Ni concentration in the 5 to 10 cm soil depth was 720 mg/kg. The Ni concentration range at this depth was 40 to 3100 mg/kg, with a 95th percentile concentration of 1755 mg/kg. Soil Ni levels tended to be lower in the deeper soils but were still elevated relative to background concentrations. The 10 to 20 cm soil depth demonstrated Ni concentration values from 27 to 1900 mg/kg. The mean concentration in this lowermost depth was 582 mg/kg, with a 95th percentile concentration of 1200 mg/kg.

In contrast, soil Ni concentrations in residential soils from Port Colborne, Ontario, ranged from 19 to 17,000 mg/kg (MOE 2002). The distribution of soil Ni concentrations in this study was quite skewed as the mean soil Ni level was 2508 mg/kg, much lower than the maximum value but still higher compared to Sudbury soils. Yakovlev et al. (2008) reported Ni levels ranged from 870 to 2915 mg/kg in soils collected around the Norilsk smelter in Russia.

4.3.1.5 Lead

Lead concentrations in the upper soil layer ranged from 3 to 410 mg/kg. The mean Pb concentration in the upper 5 cm of soil was 83.0 mg/kg and the 95th percentile concentration was 220 mg/kg. The Pb concentrations at the 5 to 10 cm depth ranged from 7 to 330 mg/kg, with an average concentration of 71.9 mg/kg. The reported 95th percentile concentration was 190 mg/kg at this soil depth. The deepest soil sampled (10–20 cm) demonstrated an average Pb value of 85.9 mg/kg. The maximum soil Pb concentration was actually found at this depth (610 mg/kg) which might be attributed to soil mixing, or possibly a source of Pb other than the smelters.

Much higher soil Pb concentrations have been reported in other locations (Table 4.3) compared to the Sudbury values. For example, maximum soil Pb concentrations reported in Port Colborne (1800 mg/kg), Trail, B.C. (3330 mg/kg) and Mt. Isa (5700 mg/kg) are all above anything found in the Sudbury area. In a risk-based characterization of soils from an urban setting in the UK, Hooker and Nathanail (2006) reported that Pb concentrations ranged from 27 to 2853 mg/kg ($n=454$). Soil Pb levels in older residential communities in Toronto range up to 3150 mg/kg (MOE 2002).

4.3.1.6 Selenium

The minimum Se concentration for all three soil sample depths was 0.5 mg/kg; note that this value represents one-half of the MDL for Se. The mean Se value for the 0 to 5 cm soil range was 7.5 mg/kg, with a 95th percentile concentration of 19 mg/kg. The maximum Se concentration at this depth was 49 mg/kg. The mean concentration in the 5 to 10 cm soil depth was 3.9 mg/kg, with a 95th percentile concentration of 8 mg/kg. The maximum Se concentration at this depth was 14 mg/kg. The mean Se concentration in the 10 to 20 cm soil depth was 2.8 mg/kg, with a maximum concentration of 11 mg/kg. The 95th percentile concentration reported for this depth was 6 mg/kg.

Table 4.5 **Summary of Metal Concentrations (mg/kg) in Copper Cliff**

Depth (cm)	Statistic	Arsenic	Cobalt	Copper	Nickel	Lead	Selenium
0–5	Min	2.5	6	26	24	3	0.5
$n=315$	Max	72	110	5600	3700	410	49
	Mean	17.6	32.4	1367	978	83.0	7.5
	95th percentile	44.3	79.3	3430	2500	220	19
5–10	Min	2.5	3	26	40	7	0.5
$n=290$	Max	101	70	2800	3100	330	14
	Mean	22.2	23.3	779	720	71.9	3.9
	95th percentile	53.6	47.1	1655	1755	190	8
10–20	Min	2.5	5	25	27	5	0.5
$n=290$	Max	99	46	2000	1900	610	11
	Mean	23.6	19.9	590	582	85.9	2.8
	95th percentile	55.6	36.55	1300	1200	240	6

4.3.2 Coniston

A review of the metal concentrations for the six COC in Coniston is presented in the following text. In this COI, a total of 324 soil samples were analyzed from the 0 to 5 cm soil depth, with 304 and 288 soil samples analyzed from the 5 to 10 cm and 10 to 20 cm depths, respectively. A summary of the COC concentrations in the soils of this community is provided in Table 4.6.

4.3.2.1 Arsenic

The minimum As concentration reported at all three soil sampling depths was 2.5 mg/kg (one-half of the MDL for As). The mean As level in the topmost soil layer was 10.3 mg/kg, with a maximum As concentration of 66 mg/kg and a 95th percentile concentration of 33 mg/kg. The maximum As concentration in the 5 to 10 cm soil range was 53 mg/kg. The 95th percentile concentration was 28.9 mg/kg, whereas the mean concentration was 10.6 mg/kg at this depth. The 10 to 20 cm soil layer demonstrated a mean As concentration of 9.9 mg/kg (range 2.5 to 55 mg/kg) with a 95th percentile concentration of 23.7 mg/kg.

4.3.2.2 Cobalt

Cobalt concentrations ranged from 3 to 74 mg/kg in the 0 to 5 cm soil layer, with a mean concentration of 16.0 mg/kg. The mean Co value in the 5 to 10 cm soil layer was 13.0 mg/kg (range 3 to 46 mg/kg). At the 10 to 20 cm depth, the minimum and maximum Co concentrations were 4 and 43 mg/kg, respectively. The mean concentration in this layer was 11.9 mg/kg, with the 95th percentile concentration reported to be 22 mg/kg.

4.3.2.3 Copper

Concentrations of Cu in the topmost soil layer (0 to 5 cm) ranged from 8.3 to 1200 mg/kg, with a mean value of 236.2 mg/kg. The 95th percentile concentration in this soil layer was 782.5 mg/kg. The mean Cu concentration measured at the 5 to 10 cm soil depth was 213.6 mg/kg, with the concentration range encompassing 8.2 to 920 mg/kg. The Cu concentration range at the 10 to 20 cm depth was 17 to 1100 mg/kg. The mean concentration in this layer was reported to be 210.5 mg/kg, with a 95th percentile concentration of 506.5 mg/kg.

4.3.2.4 Nickel

The mean Ni content in the 0 to 5 cm soil layer was 320.7 mg/kg, with a 95th percentile concentration of 1100 mg/kg. The Ni concentration range in the topmost soil layer was 16 to 1900 mg/kg. The 5 to 10 cm soil depth demonstrated a mean Ni concentration of 282.4 mg/kg, with a Ni concentration range of 14 to 1200 mg/kg reported. The 10 to 20 cm soil range exhibited Ni concentrations from 22 to 1400 mg/kg with a mean concentration at this soil level of 262.4 mg/kg.

4.3.2.5 Lead

The minimum Pb concentration in the uppermost (0 to 5 cm) and middle (5 to 10 cm) soil layers was reported at 2 mg/kg. In the upper 5 cm of soil, the maximum Pb concentration was 400 mg/kg, and had a reported mean of 47.9 mg/kg. The 5 to 10 cm soil depth displayed a mean Pb content of 40.2 mg/kg. The maximum Pb level at this depth was 270 mg/kg, with a 95th percentile concentration of 130 mg/kg. The mean Pb concentration in the 10 to 20 cm soil depth was 43.0 mg/kg. Lead values at this soil level spanned a range of 3 to 280 mg/kg, with a 95th percentile concentration of 146.5 mg/kg.

4.3.2.6 Selenium

The minimum and 95th percentile Se concentrations at all three soil depths were equivalent at 0.5 mg/kg and 3 mg/kg, respectively. The minimum concentration represents one-half of the MDL for Se. The mean Se content of the 0 to 5 cm soil depth was 1.1 mg/kg, with a maximum Se concentration of 5 mg/kg. The maximum Se level in the middle soil depth (5 to 10 cm) was 4 mg/kg, and the mean concentration at this level was reported to be 1.0 mg/kg. The deepest soil layer (10 to 20 cm) exhibited a maximum Se level of 9 mg/kg, with a mean Se concentration equal to that of the middle soil layer at 1.0 mg/kg.

Table 4.6   **Summary of Metal Concentrations (mg/kg) in Coniston**

Depth (cm)	Statistic	Arsenic	Cobalt	Copper	Nickel	Lead	Selenium
0–5	Min	2.5	3	8.3	16	2	0.5
n=324	Max	66	74	1200	1900	400	5
	Mean	10.3	16.0	236.2	320.7	47.9	1.1
	95th percentile	33	44.85	782.5	1100	140	3
5–10	Min	2.5	3	8.2	14	2	0.5
n=304	Max	53	46	920	1200	270	4
	Mean	10.6	13.0	213.6	282.4	40.2	1.0
	95th percentile	28.85	30.85	637	895.5	130	3
10–20	Min	2.5	4	17	22	3	0.5
n=288	Max	55	43	1100	1400	280	9
	Mean	9.9	11.9	210.5	262.4	43.0	1.0
	95th percentile	23.65	22	506.5	653	146.5	3

4.3.3 Falconbridge

A review of the metal concentrations for the six COC in Falconbridge is presented in the following text. In this COI, a total of 311 soil samples were analyzed from the 0 to 5 cm soil depth, with the exception of Pb for which 310 samples were analyzed. In the 5 to 10 cm and 10 to 20 cm depths, 286 and 282 samples were analyzed, respectively (Table 4.7).

4.3.3.1 Arsenic

The minimum As level across all three soil depths was 2.5 mg/kg, which represents one-half of the MDL for As. The maximum As concentration in the 0 to 5 cm soil range was 400 mg/kg, with a mean concentration of 65.6 mg/kg. The mean As concentration in the middle (5 to 10 cm) soil layer was 95.0 mg/kg, and the 95th percentile concentration was 307.5 mg/kg. The maximum As value at this soil depth was 570 mg/kg. The 10 to 20 cm soil depth demonstrated a mean As concentration of 76.1 mg/kg, with a maximum As concentration of 620 mg/kg. This was the highest As value recorded in the study.

4.3.3.2 Cobalt

The range of Co concentrations at the shallowest soil depth (0 to 5 cm) was 4 to 190 mg/kg, with an average Co level of 45.0 mg/kg. The mean Co concentration in the 5 to 10 cm depth was 36.3 mg/kg, with a range of 1.9 to 150 mg/kg. The 10 to 20 cm soil layer exhibited a mean Co concentration of 21.6 mg/kg, and a 95th percentile of 49 mg/kg. The Co concentration range at the lowest soil depth was 2.2 to 110 mg/kg.

4.3.3.3 Copper

The maximum Cu concentration was 3000 mg/kg at both the 0 to 5 cm and 5 to 10 cm soil depths; these depths also had identical 95th percentile concentrations of 1800 mg/kg. The minimum Cu concentration in the topmost soil level (0 to 5 cm) was 10 mg/kg, with the mean concentration at this level reported to be 706.9 mg/kg. The mean concentration in the 5 to 10 cm depth was 639.4 mg/kg, with a range of 9.5 to 3000 mg/kg. The mean soil Cu level in the 10 to 20 cm soil depth was 340.9 mg/kg, and the maximum Cu level at this soil depth was 2000 mg/kg.

4.3.3.4 Nickel

The minimum Ni concentrations in the shallowest and deepest soil depths were equivalent at 17 mg/kg; the maximum soil concentrations at both of these depths were 3700 and 2500 mg/kg, respectively. The mean soil Ni concentration in the 0 to 5 cm depth was 751.0 mg/kg. The 5 to 10 cm soil depth exhibited a concentration range of 22 to 3100 mg/kg. The average concentration at this level was 700.0 mg/kg. The deepest soil layer (10 to 20 cm) was reported to have a mean concentration of 401.7 mg/kg.

4.3.3.5 Lead

The mean Pb concentration in the 0 to 5 cm soil layer was 72.1 mg/kg, with a range of 2 to 370 mg/kg. The 95th percentile concentration in this uppermost soil layer was 190 mg/kg. The Pb range in the 5 to 10 cm layer of soil was 3.6 to 340 mg/kg. At this depth, the mean and 95th percentile Pb concentrations were 67.2 and 197.5 mg/kg, respectively. The 10 to 20 cm soil layer demonstrated a mean Pb concentration of 44.3 mg/kg with a range from 3.1 to 790 mg/kg.

4.3.3.6 Selenium

The minimum Se concentration reported for all three soil layers was 0.5 mg/kg which represents one-half of the MDL for Se. The average Se level in the upper soil layer (0 to 5 cm) was 2.5 mg/kg, with a maximum of 12 mg/kg. The maximum Se concentration exhibited by soil in the 5 to 10 cm depth was 11 mg/kg, while the mean and 95th percentile concentrations were 2.6 and 6.75 mg/kg, respectively. The average Se level in the 10 to 20 cm soil depth was 1.9 mg/kg.

Table 4.7 **Summary of Metal Concentrations in Falconbridge (mg/kg)**

Depth (cm) and number of samples (n)	Statistic	Arsenic	Cobalt	Copper	Nickel	Lead	Selenium
0–5	Min	2.5	4	10	17	2	0.5
n=311	Max	400	190	3000	3700	370	12
	Mean	65.6	45.0	707	751	72.1	2.5
	95th percentile	190	110	1800	1950	190	6
5–10	Min	2.5	1.9	9.5	22	3.6	0.5
n=286	Max	570	150	3000	3100	340	11
	Mean	95.0	36.3	639	700	67.2	2.6
	95th percentile	308	96	1800	1975	198	6.7
10–20	Min	2.5	2.2	9.5	17	3.1	0.5
n=282	Max	620	110	2000	2500	790	8
	Mean	76.1	21.6	341	402	44.3	1.9
	95th percentile	230	49	900	1095	140	5

4.3.4 Hanmer

A review of the metal concentrations for the six COC in Hanmer is presented in the following text. Hanmer was chosen to represent a reference community, as it has similar demographics and is geologically similar to the smelter communities but the residential soils do not appear to be influenced by smelter emissions. In this COI, a total of 85 soil samples were analyzed from the 0 to 5 cm soil depth, while 30 and 28 samples were analyzed from the 5 to 10 cm and 10 to 20 cm depths, respectively. A summary of the COC concentrations in the soils of this community is provided in Table 4.8.

4.3.4.1 Arsenic

The minimum As level reported at all three soil depths was 2.5 mg/kg, representing one-half of the MDL. The maximum As value was 25 mg/kg at the 0 to 5 cm soil depth and 8 mg/kg at both of the deeper soil depths. The mean As concentrations in the three soil depths were: 4.2, 3.2, and 2.8 mg/kg, decreasing correspondingly with soil depth. Similarly, the 95th percentile values decreased with soil depth and were reported to be concentrations of 15, 7, and 4 mg/kg. The mean soil As concentrations from all three sample depths were substantially below the MOE background guideline of 17 mg/kg, so the influence of historic smelter emissions on As soil quality in Hanmer has been negligible.

4.3.4.2 Cobalt

The Co concentration range in the 0 to 5 cm soil layer ranged from 4 to 33 mg/kg. The mean Co value at this depth was 7.6 mg/kg, while the 95th percentile concentration was 17.8 mg/kg. The Co range in the middle soil layer (5 to 10 cm) was 2 to 6 mg/kg, with a mean value of 4.1 mg/kg. The 10 to 20 cm soil depth demonstrated a mean Co concentration of 4.3 mg/kg, with a range of only 3 to 5 mg/kg. The mean soil Co concentrations from all three sample depths were substantially below the MOE background guideline of 21 mg/kg, so the influence of historic smelter emissions on Co soil quality in Hanmer has been negligible.

4.3.4.3 Copper

The mean Cu concentration in the uppermost 5 cm of soil was 54.0 mg/kg with a range of 13 to 330 mg/kg. The mean concentration in the 5 to 10 cm soil layer was 25.0 mg/kg with a Cu range of 11 to 54 mg/kg. The 10 to 20 cm soil depth range showed Cu concentrations varying from 3.8 to 34 mg/kg, with a mean concentration of 17.4 mg/kg. The mean soil Cu concentrations from all three sample depths were less than the MOE background guideline of 85 mg/kg. However, some samples of surface soil did exceed normal background levels and there was a distinct vertical concentration gradient with Cu levels decreasing rapidly with depth. Therefore, the Cu soil quality in Hanmer has been influenced by historic smelter emissions, but the impact has been relatively insignificant.

4.3.4.4 Nickel

The Ni concentration in the uppermost soil segment (0 to 5 cm) varied from 16 to 297 mg/kg, and had a reported mean of 55.8 mg/kg. The 95th percentile concentration exhibited at this soil level was 217 mg/kg. The 5 to 10 cm soil depth displayed a Ni concentration range from 17 to 50 mg/kg with a mean Ni level of 34.4 mg/kg. The average Ni concentration in the 10 to 20 cm soil depth was 29.4 mg/kg with a range from 14 to 56 mg/kg. Some soil samples from all three depths exceeded the MOE background standard of 43 mg/kg and the mean Ni concentration in surface soil exceeded the guideline. When considered in combination with a decreasing vertical concentration gradient, it is apparent that the soil Ni quality in Hanmer has been noticeably influenced by historic smelter emissions.

4.3.4.5 Lead

In the top 5 cm of soil, a mean Pb level of 14.6 mg/kg was reported, with a 95th percentile concentration of 54.4 mg/kg. The Pb range in this upper soil layer was 2 to 79 mg/kg. The 5 to 10 cm soil layer, with a total Pb range of 5 to 44 mg/kg, exhibited a mean of 10.5 mg/kg. The 95th percentile value for this depth was 29.6 mg/kg. The deepest soil (10 to 20 cm) was observed to have a Pb range of 4 to 19 mg/kg, with an average Pb level of 6.9 mg/kg. The mean soil Pb concentrations from all three sample depths were substantially below the MOE background guideline of 120 mg/kg and well within levels normally found in urban communities in Ontario. Also, because of the historic use of leaded gasoline, it is common to find higher Pb levels in surface soil, and so a vertical concentration gradient is not uncommon in urban areas in the province. Therefore, it is difficult to conclude if historic smelter emissions have had any measurable impact on soil Pb quality in Hanmer.

4.3.4.6 Selenium

All three soil levels demonstrated minimum Se values of 0.5 mg/kg, which represents one-half of the MDL for Se. The 5 to 10 cm and 10 to 20 cm soil depths also exhibited a value of 0.5 mg/kg for mean, maximum, and 95th percentile values. The top soil layer was reported to have a maximum Se concentration of 3 mg/kg, with average and 95th percentile concentrations of 0.7 and 2 mg/kg, respectively. The mean soil Se concentration from all three sample depths was substantially below the MOE background guideline of 1.9 mg/kg and the vertical concentration gradient was weak and inconsistent. Therefore, it is difficult to conclude if historic smelter emissions have had any measurable impact on Se soil quality in Hanmer.

Table 4.8   **Summary of Metal Concentrations in Hanmer (mg/kg)**

Depth (cm)	Statistic	Arsenic	Cobalt	Copper	Nickel	Lead	Selenium
0–5	Min	2.5	4	13	16	2	0.5
n=85	Max	25	33	330	297	79	3
	Mean	4.2	7.6	54.0	55.8	14.6	0.7
	95th percentile	15	17.8	198	217	54.4	2
5–10	Min	2.5	2	11	17	5	0.5
n=30	Max	8	6	54	50	44	0.5
	Mean	3.2	4.1	25.0	34.4	10.5	0.5
	95th percentile	7	5	36.1	44.6	29.6	0.5
10–20	Min	2.5	3	3.8	14	4	0.5
n=28	Max	8	5	34	56	19	0.5
	Mean	2.8	4.3	17.4	29.4	6.9	0.5
	95th percentile	4.1	5	30	40.9	10.6	0.5

4.3.5 Sudbury Center

A review of the metal concentrations for the six COC in the Sudbury Center is presented in the following text. In this COI, a total of 1129 soil samples were analyzed from the 0 to 5 cm soil depth, while 643 and 607 samples were analyzed from the 5 to 10 cm and 10 to 20 cm depths, respectively. The COC concentrations in the soils for this community are summarized in Table 4.9.

4.3.5.1 Arsenic

Minimum As concentrations in the three soil layers were 2.5 mg/kg which represents half of the MDL for As. The mean As concentration in the upper 5 cm of soil sampled was 6.0 mg/kg, with a 95th percentile concentration of 18 mg/kg. The maximum As concentration in this upper soil level was 65 mg/kg. In the middle (5 to 10 cm) soil range, the maximum detected As level was 39 mg/kg. The mean and 95th percentile concentrations at this depth were 6.5 and 17 mg/kg, respectively. The deepest soil layer (10 to 20 cm) contained an average of 5.8 mg/kg As, with a maximum and 95th percentile concentration of 67 and 14.7 mg/kg, respectively.

4.3.5.2 Cobalt

The minimum Co content of all three soil depths was equivalent at 3 mg/kg. The upper soil layer (0 to 5 cm) had a reported mean Co value of 11.2 mg/kg and a maximum concentration of 100 mg/kg. A mean concentration of 9.0 mg/kg Co was reported in the 5 to 10 cm soil depth, with a maximum value of 36 mg/kg. The 95th percentile concentration in this soil layer was 17 mg/kg. The 10 to 20 cm soil range contained a maximum Co level of 28 mg/kg, with a mean concentration of 8.5 mg/kg.

4.3.5.3 Copper

The 0 to 5 cm soil depth demonstrated a Cu range of 6.2 to 1800 mg/kg. The average and 95th percentile Cu concentrations were 155 and 590 mg/kg, respectively. The middle soil layer (5 to 10 cm) contained a mean Cu value of 122 mg/kg, with a range from 12 to 1100 mg/kg. The deepest soil layer (10 to 20 cm) had a reported Cu mean concentration of 94.4 mg/kg, while its 95th percentile concentration was 270 mg/kg. Concentrations of Cu in the 10 to 20 cm layer ranged from 11 to 530 mg/kg.

4.3.5.4 Nickel

Nickel concentrations in the topmost soil layer (0 to 5 cm) varied from 11 to 3284 mg/kg, with a mean value of 172 mg/kg. The 95th percentile concentration at this depth was 596 mg/kg. The mean Ni concentration in the 5 to 10 cm layer was 140 mg/kg with a range from 18 to 970 mg/kg. The deepest soil layer sampled (10 to 20 cm) contained a mean Ni level of 115 mg/kg, with a range of 15 to 820 mg/kg. The 95th percentile concentration for this soil depth was 300 mg/kg.

4.3.5.5 Lead

The reported mean Pb level in the 0 to 5 cm layer was 26.4 mg/kg, with a 95th percentile value of 101 mg/kg and a range from 1 to 320 mg/kg. The minimum Pb value reported at both of the lower soil depths (5 to 10 cm and 10 to 20 cm) was 2 mg/kg. The mean Pb level in the middle (5 to 10 cm) soil range was 24.8 mg/kg and was 21.9 mg/kg in the 10 to 20 cm depth layer. The maximum Pb concentration reported at the lowest soil depth (10 to 20 cm) was 470 mg/kg.

4.3.5.6 Selenium

Minimum Se levels reported at all three soil depths were 0.5 mg/kg which represents half of the MDL for Se. Maximum, mean, and 95th percentile Se concentrations in the 0 to 5 cm soil range were 13, 1.1, and 3.6 mg/kg, respectively. The mean Se concentration observed in the middle soil layer (5 to 10 cm) was 0.9 m/kg. The maximum and 95th percentile Se concentrations at this depth were 5 and 2 mg/kg, respectively. Mean and 95th percentile values in the 10 to 20 cm soil layer were 0.8 and 2 mg/kg, respectively, with Se concentrations peaking at 4 mg/kg.

Table 4.9 Summary of Metal Concentrations (mg/kg) in Sudbury Center

Depth (cm)	Statistic	Arsenic	Cobalt	Copper	Nickel	Lead	Selenium
0–5	Min	2.5	3	6.2	11	1	0.5
n=1129	Max	65	100	1800	3284	320	13
	Mean	6.0	11.2	155	172	26	1.1
	95th percentile	18	28	590	596	101	3.6
5–10	Min	2.5	3	12	18	2	0.5
n=643	Max	39	36	1100	970	310	5
	Mean	6.5	9.0	122	140	25	0.9
	95th percentile	17	17	379	429	84	2
10–20	Min	2.5	3	11	15	2	0.5
n=607	Max	67	28	530	820	470	4
	Mean	5.8	8.5	94.4	115	21.9	0.8
	95th percentile	14.7	15	270	300	76	2

4.3.6 Spatial Distribution of COC within a Community

The distribution of each COC was mapped within each of the COI. Figures 4.8 and 4.9 illustrate the spatial distribution of Ni and Pb, respectively, as examples within the community of Copper Cliff. Each soil sample location is color coded to indicate a different concentration range. Visual examination of the COC distribution at this scale revealed no obvious trends along streets, or in relation to the smelter source. Several geostatistical approaches were also used to examine the spatial distribution of each metal within each of the COI to determine whether there were any apparent trends or patterns. The lack of a pattern at this localized scale was verified by the geostatistical analysis. This indicates that the distribution of metals within a community was essentially random. This was not unexpected since topsoil for residential properties has been extensively moved about and imported from different locations for landscaping. Therefore, metal levels at a particular property cannot be used to predict with confidence metal concentrations in soil on neighboring properties. This determination was important for the HHRA where it was necessary to select soil exposure point concentrations for the human exposure model (see Chapter 7). The lack of a predictable pattern also has implications if further soil sampling is required for risk management.

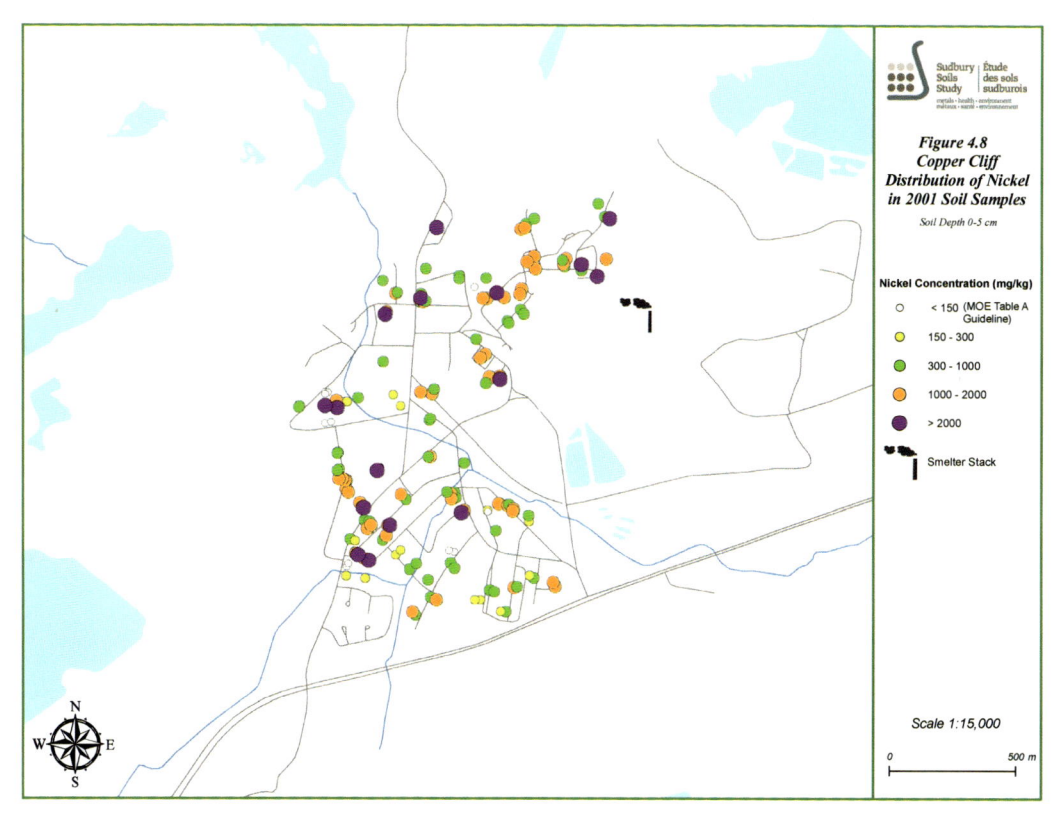

Distribution of Chemicals of Concern in the Study Area

Figure 4.8   **Distribution of Ni within Town of Copper Cliff**

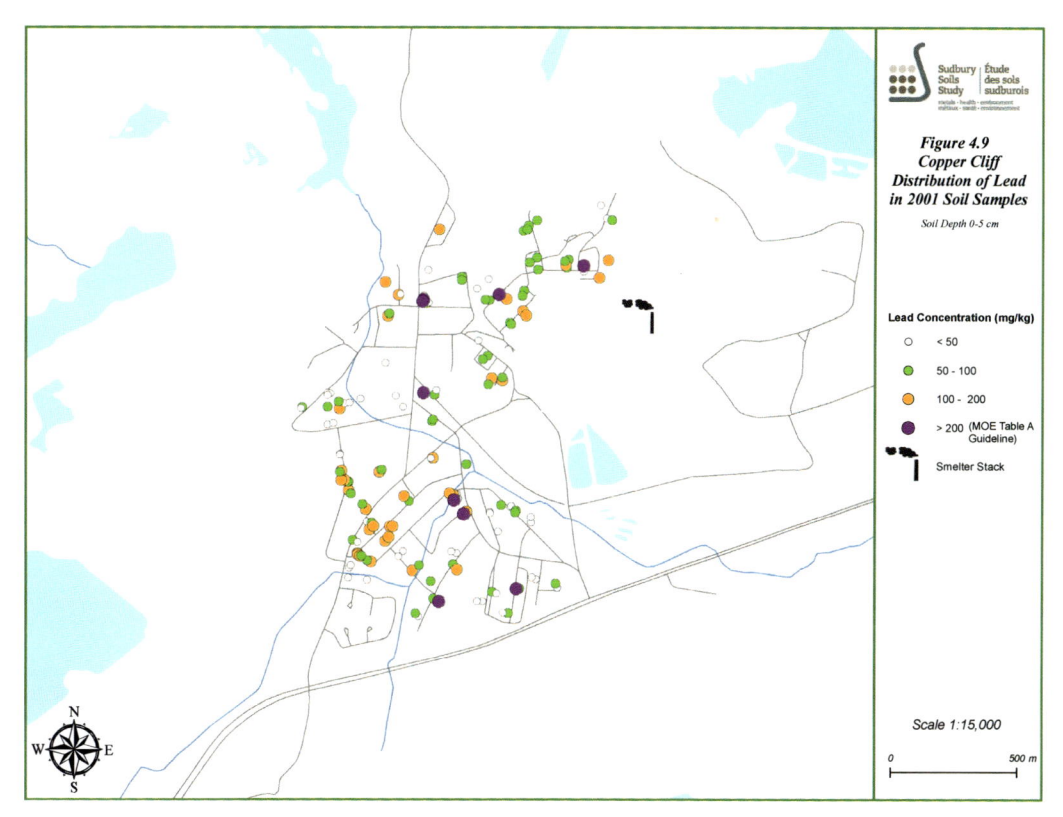

Figure 4.9   **Distribution of Pb within Town of Copper Cliff**

4.4 References

Adamo P, Dudka S, Wilson MJ, McHardy WJ (2002) Distribution of trace elements in soils from the Sudbury Smelting area (Ontario, Canada). Water Air Soil Pollut 137:95–116

Agriculture Canada Expert Committee on Soi1 Survey (1987) The Canadian system of soil classification, 2nd edn. Agriculture Canada Publication 1646. Agriculture Canada, Ottawa, Ontario, 164 pp

Barnett PJ, Bajc AF (2002) Quaternary geology. In: Physical environment of the City of Greater Sudbury. Special Volume 6, pp. 57–86. Ontario Geological Survey, Ontario

Bohn HL, McNeal BL, O'Connor GA (2001) Soil chemistry. John Wiley & Sons, Inc., New York, 307 pp

CANTOX (1999) Deloro Village exposure assessment and health risk characterization for arsenic and other metals. Final Report to Ontario Ministry of the Environment. Cantox Environmental, Toronto

CEM (2004) Metal levels in the soils of the Sudbury Smelter footprint. Report prepared by Centre for Environmental Monitoring (CEM), Laurentian University, Sudbury

Dredge LA, Cowan WR (1989) Quaternary geology of the southwestern Canadian Shield. Volume v.K–1 (no. 1), pp. 214–249. Geological Society of America (Geological Survey of Canada), Ottawa

Dressler BO, Gupta VK, Muir TL (1991) The Sudbury structure. In: The geology of Ontario. Special Volume 4, Part 1, pp. 593–625. Ontario Geological Survey, Ontario

Dudas MT, Pawluk S (1980) Natural abundance and mineralogical partitioning of trace elements in selected Alberta soils. Can J Soil Sci 62:763–771

Dudka S, Ponce-Hernandez R, Hutchinson TC (1995) Current level of total element concentrations in surface layer of Sudbury soils. Sci Total Environ 162:161–171

Gillespie JE, Acton CJ, Hoffman DW (1983) Soils of Sudbury area. Soil Survey Report. Ontario Institute of Pedology, Ontario

Goodarzi F, Sanei H, Duncan WF (2001) Monitoring the distribution and deposition of trace elements associated with the zinc-lead smelter in the Trail area, British Columbia, Canada. J Environ Monit 3:515–525

Golder Associates Ltd. (2001) Town of Falconbridge soil sampling program comprehensive Falconbridge Survey. Unpublished report prepared for Falconbridge Ltd. Golder Associates Ltd, 14 pp

Gough LP, Severson RC, Shacklette HJ (1988) Element concentrations is soils and other surficial materials of Alaska. Professional Paper 1458. United States Geological Survey

Gundermann DG, Hutchinson TC (1995) Changes in soil chemistry 20 years after the closure of a nickel-copper smelter near Sudbury, Ontario, Canada, pp. 559–562. Elsevier, Amsterdam, New York

Hazlett PW, Rutherford GK, van Loon GW (1983) Metal contaminants in surface soils and vegetation as a result of nickel/copper smelting at Coniston, Ontario, Canada. Reclam Reveg Res 2:123–127

Henderson PJ, Knight RD, McMartin I (2002) Geochemistry of soils within a 100 km radius of the Horne Cu smelter, Rouyn-Noranda, Québec. Geological Survey of Canada Open File Report 4169. Geological Survey of Canada

Hooker PJ, Nathanail CP (2006) Risk-based characterization of lead in urban soils. Chem Geol 226:340–351

Intrinsik (2007) Ecological risk assessment for Teck Cominco Operations at Trail, British Columbia. Final report. Terrestrial Risk Modeling Level of Refinement #3. Intrinsik Environmental Sciences, Calgary, Alberta, 143 pp

Jacques Whitford Limited (2004) Community based risk assessment, Port Colborne, Ontario. Ecological Risk Assessment – Natural Environment. Jacques Whitford Limited

Koptsik S, Koptsik G, Livantisova S, Eruslankina L, Zhmelkova T, Vologdina Zh (2003) Heavy metals in soils near the nickel smelter: chemistry, spatial variation, and impacts on plant diversity. J Environ Monit 5:441–450

McKeague JA, Desjardins JG, Wolynetz MS (1979) Minor elements in Canadian soils. Agriculture Canada, Research Branch, 75 pp

McMartin I, Henderson PJ, Nielsen E (1999) Impact of a base metal smelter on the geochemistry of soils of the Flin Flon region, Manitoba and Saskatchewan. Can J Earth Sci 36:141–160

MOE (1991) Assessment of human health risk of reported soil levels of metals and radionuclides in Port Hope. Hazardous Contaminants Branch, November 1991, ISBN 0-7729-9065-4, Ontario Ministry of the Environment, Toronto

MOE (1997) Guidelines for use at contaminated sites in Ontario. Ontario Ministry of the Environment, Toronto

MOE (1999) Studies of the terrestrial environment in the Wawa Area, 1998–1999. Phytotoxicology and Soil Standards Section, Standards Development Branch. Report No. SDB-047-3511-1999. Ontario Ministry of the Environment, Toronto

MOE (2001) Metal levels in soil and vegetation in the Sudbury area (Survey 2000 and additional historic data). Report SDB-045-3511-2001. Ontario Ministry of the Environment, Toronto

MOE (2002) Soil investigation and human health risk assessment for the Rodney Street community, Port Colborne. www.ene.gov.ca/envision/techdocs/4255e. Ontario Ministry of the Environment, Toronto

Nikonov VV, Lukina NV, Frontas'eva MV (1999) Trace elements in Al-Fe-Humus podzolic soils subjected to aerial pollution from the apatite-nepheline production industry. Eurasian Soil Sci 32(12):1331–1339

Rudnick RL, Gao S (2003) Composition of the continental crust. In: Rudnick RL (ed) The Crust, Vol. 3 Treatise on Geochemistry (eds. Holland HD and Turekian KK). Elsevier-Pergamon, Oxford, pp. 1–64

Shacklette HJ, Hamilton JC, Boerngen JG, Bowles JGM (1971) Elemental composition of surficial materials in the conterminous United States. Professional Paper 574-D. United States Geological Survey

Spiers GA, Pawluk S, Dudas MJ (1984) Authigenic mineral formation by solodization. Can J Soil Sci 64:515–532

Spiers, GA, Pearson DAB, Prevost F (2002) Distribution of Anthropogenic Metals in Soils of the Sudbury Smelter Footprint: Presented at the 5th International Nickel Conference, Murmansk, Russia

Taylor MP, Mackay AK, Hudson-Edwards KA, Holz E (2010) Soil Cd, Cu, Pb and Zn contaminants around Mount Isa city, Queensland, Australia, Potential sources and risks to human health. Appl Geochem 25:841–855

Yakovlev AS, Plekhanova IO, Kudryashov SV, Aimaletdinov RA (2008) Assessment and regulation of the ecological state of soils in the impact zone of mining and metallurgical enterprises of Norilsk Nickel Company. Environ Soil Sci 41(6):648–659

4.5 Appendix Chapter 4: Abbreviations

COC, chemicals of concern

COI, community(ies) of interest

HHRA, human health risk assessment

LFH horizons, refers to surface organic soil layers ranging in thickness from 2 to 15 centimeters. The designation LFH refers to the fresh plant detritus (L = litter) on the soil surface, the partially decomposed organic layer (F = fermentation) and the well-decomposed organic layer (H = humus)

MDL, method detection limit

MOE, Ontario Ministry of the Environment

n, sample size

ND, not detected; below analytical detection limit

S.R.S.P., Sudbury regional soils project

UCL, upper confidence level

DOI: 10.5645/b.1.5

5.0 Human Health Risk Assessment Problem Formulation

Authors

Glenn Ferguson, PhD
Intrinsik Environmental Sciences
6605 Hurontario Street, Suite 500, Mississauga, ON Canada L5T 0A3
Email: Gferguson@intrinsikscience.com

Elliot Sigal
Intrinsik Environmental Sciences
6605 Hurontario Street, Suite 500, Mississauga, ON Canada L5T 0A3
Email: Esigal@intrinsikscience.com

Table of Contents

5.1 Introduction	99
5.2 Site Characterization	100
5.3 Identification of COC	100
5.4 Receptor Identification and Characterization	101
5.4.1 Communities of Interest (COI) for the Sudbury Soils Study	101
5.4.2 Human Receptors within the COI	102
5.4.3 Selection of Dietary Consumption Rates	105
5.4.4 Local First Nation Wild Game and Fish Consumption Rates	105
5.5 Identification of Exposure Pathways	107
5.5.1 Pathway Analysis	107
5.5.2 Resuspended Dust	110
5.5.3 Mother's Milk	111
5.6 Exposure Scenarios	111
5.7 HHRA Conceptual Model	112
5.8 Screening Level Risk Assessment (SLRA)	113
5.9 Identification of Data Gaps	113
5.10 References	114
5.11 Appendix Chapter 5: Abbreviations	115

Tables

5.1 Receptor Characteristics – Preschool Child (7 months to 4 years)	104
5.2 Local Game and Fish Consumption Rates (times consumed per year)	106
5.3 Comparison of Reported Game and Fish Consumption Rates (g/day)	107
5.4 Exposure Pathways Evaluated/Screened	108
5.5 Exposure Pathways Excluded from Assessment	109
5.6 Comparison of Estimated Indoor Dust and Outdoor Air Concentrations	110

Figures
5.1 Location of the Communities of Interest (COI) for the Sudbury HHRA 102
5.2 Conceptual Model for the Sudbury HHRA 113

5.1 Introduction

The fundamental purpose of a human health risk assessment (HHRA) is to estimate whether people working, living or visiting a given location are being exposed, or will be exposed to concentrations of chemicals that have the potential to result in adverse health effects. The assessment of potential occurrences of adverse health effects from chemical exposure is based on the dose–response concept that is fundamental to the responses of biological systems to chemicals, whether they are therapeutic drugs, naturally occurring substances, or man-made chemicals in the environment. Thus, an HHRA evaluates the likelihood (or risk) of health effects following chemical exposures. It requires consideration of the toxic properties of the chemicals, the presence of receptors, and the existence of exposure pathways to the receptors. When all three factors are present (i.e. chemicals, receptors and exposure pathways), there is a potential for adverse health effects to occur if exposures to the chemicals are elevated above acceptable levels (See Figure 1.1 in Chapter 1).

The Sudbury HHRA was conducted in accordance with the regulatory guidance provided by the Ontario Ministry of the Environment (MOE 2004; MOEE 1997) and Health Canada (1993, 2004), and is primarily based upon guidance developed as part of the U.S. EPA Superfund program (U.S. EPA 1989, 1998, 2002, 2004a,b). The current study was considered an area-wide risk assessment, as it evaluated a large geographical area, rather than an individual property (that would be considered a site-specific risk assessment or SSRA). The main advantage to this approach is that it allowed the estimation of potential health risks across a broad area, and evaluated potential exposures from a variety of input sources (i.e. smelter emissions, locally grown foods, regional drinking water). However, this approach also required considerably greater amounts of data specific to the study area.

It must also be recognized that a human health risk assessment is not the same as a community health study. Community health studies may involve such tasks as questionnaires, interviews, medical records review and collection of biological tissue or fluid samples (i.e. blood, urine, hair) to measure human exposures directly. While these types of information can be valuable supplementary information for a human health risk assessment, they are not routine components of the established HHRA framework. The HHRA framework was applied in three distinct phases:

- **Phase 1** – Problem formulation and screening level risk assessment
- **Phase 2** – Sampling and analyses to fill identified data gap, and
- **Phase 3** – Detailed human health risk assessment (HHRA).

The phased approach allowed the HHRA to proceed in a logical and sequential manner, and allowed for unresolved issues or major uncertainties to be addressed as they were identified. A phased approach also allowed for multiple iterations of the HHRA, such that various components of the HHRA were efficiently revisited and re-evaluated as new information became available. This chapter describes the Problem Formulation step, while Phase 2 is discussed in Chapter 6, and Phase 3 is discussed in Chapters 7 and 8.

The first step in the HHRA was an information gathering and interpretation stage that planned and focused the approach of the study to critical areas of concern for the area being evaluated. The problem formulation defined the nature and scope of the work to be conducted, permitted practical boundaries to be placed on the overall scope of work and ensured that the assessment was directed at the key areas and issues of concern. This step was critical to the success of the risk assessment. Careful planning during the problem formulation step reduced the need for significant modifications once the risk assessment began. The data gathered and evaluated in this step provided information on the physical layout and characteristics of the study area, possible exposure pathways, potential human receptors, chemicals of concern (COC) and any other specific areas or issues of concern to be addressed. The key tasks evaluated within the problem formulation step included the following:

Site characterization – delineation of study area, and review of available site data to identify factors affecting the availability of contaminants to potential receptors, such as location and medium of contamination.

Identification of chemicals of concern (COC) – identification of the primary COC based on site environmental monitoring data.

Receptor characterization – identification of 'receptors of concern', including those with the greatest probability of exposure to chemicals from the site and those that have the greatest sensitivity to these chemicals.

Identification of exposure pathways – consideration of various factors that influence the means by which receptors come into contact with COC in environmental media, including: chemical-specific parameters, such as speciation, characteristics of the site, and physiology and behavior patterns of the receptors.

The outcome of these tasks formed the exposure scenarios which were the basis of the approach taken in the risk assessment and defined the scope of the HHRA. Stakeholder consultation was a critical component of the problem formulation step. During the problem formulation phase, there was considerable dialog with members of the Sudbury community, local stakeholders and the TC (Technical Committee) overseeing the study to help focus and define the scope of the risk assessment. Input from the public was sought during a series of workshops, open houses and public lectures. The extent of the public participation and stakeholder consultation is described in Chapters 13 and 14.

5.2 Site Characterization

Site characterization typically includes the following activities:

- establishment of spatial and temporal boundaries
- site visit(s) and reconnaissance
- review of past site reports/investigations
- interviews with persons knowledgeable about the site
- description of the physical characteristics of the site (i.e. geology, hydrogeology, and general topography), and
- consideration of historical and potential future land uses.

Delineation of Study Area

The temporal and spatial boundaries of the study area were determined by several factors. The temporal boundary of the study area was defined as the current conditions in the Greater Sudbury Area (GSA). The spatial boundary of the study area included the politically defined borders of the City of Greater Sudbury and the surrounding area from which soil samples were collected for metal analyses in 2001. Remote and undisturbed areas surrounding the City of Greater Sudbury were included in the study area to determine the geographical area impacted by the smelters and to provide background metal concentrations. Refer to Chapter 3 for a detailed discussion of the soil sampling program undertaken in 2001.

There are benefits to evaluating human health risks over a large defined area versus considering numerous small parcels of land within the GSA (i.e. several site-specific risk assessments or SSRAs). The SSRA approach was designed to evaluate risks in an individual property setting, and would be impractical given the size of the Sudbury study area and scope of the current assessment. Unlike the SSRA approach, an area-wide risk assessment allowed for:

- consideration of relevant geographic areas for key receptors of concern
- assessment of possible health risks to a broader population base
- development of risk management options that do not stop at a specific property's borders, and
- potential development of a single risk management plan, if required, rather than the development of numerous separate plans.

The area-wide approach does require the collection of a significant amount of data, from a variety of exposure media, across a larger geographical area. This results in significantly greater time and cost for completion of a study with this scope. The area-wide approach to risk assessment was used at other mining and smelter-impacted communities in Canada, such as Trail, B.C., and Port Colborne and Deloro, Ontario (Hilts et al. 2001; JWEL 2004; CEI 1999).

5.3 Identification of COC

It is common practice in HHRA to limit the number of chemicals evaluated to those that, due to their environmental concentrations, distribution, or chemical and toxicological properties, have the greatest potential to contribute to health risks to individuals residing in the study area. However, it is important to note that the identification of a substance as a COC does not automatically lead to the conclusion that the substance is, in fact, a contributor to health risk. Rather, the appropriate conclusion is that those substances identified as COC should be the subject of further evaluation. This is done because it is impractical in terms of time and cost to conduct a risk assessment for every chemical that has been found to occur in a particular area. In addition, the concentrations of many chemicals associated with a particular site may be similar to chemical concentrations found naturally in the area, rather than as a result of current or former activities on a site. The following three criteria were established by the TC as requirements for a particular chemical to be considered a COC in soil:

- chemical concentration in soil must be above the MOE Table A soil guideline for residential/park land use (MOEE, 1997)
- chemical must be present at elevated levels in soils across the study area, and
- chemical must be scientifically demonstrated to originate from the local mining/smelting operations.

Candidate COC were selected from a list of 20 inorganic parameters, which were measured in soil samples collected in the 2001 sampling program. As a result of statistical screening of the 2001 soil database, As, Co, Cu, Pb, Ni and Se were identified as COC for further assessment in the HHRA (see also Chapter 3). Metal concentrations in additional media (air, water, food, dust) subsequently collected to fill identified data gaps (see Chapter 6) were also compared with regulatory and study-specific screening criteria (i.e. vegetable garden survey screening criteria). However, no additional COC were included beyond those already established using soil screening criteria. Therefore, the final COC carried forward for detailed assessment in the HHRA were: As, Cu, Co, Pb, Ni and Se.

5.4 Receptor Identification and Characterization

5.4.1 Communities of Interest (COI) for the Sudbury Soils Study

As part of the analyses in Phase 1 of the study, the TC clearly identified Copper Cliff, Coniston and Falconbridge as the three primary COI based on the location of smelters in these towns. The definition of a COI was closely related to the objectives of the HHRA. Two of the key goals of the Sudbury HHRA were to:

- assess whether human health in the study area was currently, or in the future will be, adversely affected by elevated metal levels in the soil, and other local environmental media, impacted by historic and/or present day particulate smelter emissions; and
- establish soil intervention levels (site-specific remediation goals or clean-up criteria) and/or other risk management options protective of human health, now and into the future, if required.

The definition of COI was related to the second objective in that the intervention levels and risk management activities must be applied on an area-wide basis. However, in cases where there are distinct differences in a particular community, a unique intervention level may be established or other risk management activities implemented for that community. Differences that could influence an intervention level or implementation of a particular risk management activity for a specific community can include, but are not limited to, the following: unique secondary media concentrations (those not directly related to soil concentration, i.e. air, water, fish, food); unique exposure pathways; unique receptor behaviors (increased sensitivity, increased exposure), and, unique media characteristics (pH, soil organic carbon content, etc.).

The most likely of these factors to influence the ultimate soil intervention levels or other risk management activities was secondary media concentrations (i.e. vegetable gardens, potable water). For example, since different communities in the GSA have different drinking water sources (surface and ground), exposures related to the drinking water pathway differed between many of the communities. Additionally, air monitors were installed in a number of the communities within the GSA allowing community-specific air concentrations to be measured. Typically, as secondary media concentrations vary, the allocation of exposure available for the soil intervention level will vary, potentially resulting in community-specific intervention levels.

In addition to the three primary COI (Copper Cliff, Coniston, and Falconbridge), Hanmer and Sudbury Centre were selected based on the criteria described previously and the unique characteristics of each of the two communities (Figure 5.1). Sudbury Centre represented the greatest population of residents in the GSA. This COI was clearly defined by municipal borders and was located in the core of the city in the middle of current and historical smelter activities. Hanmer was also clearly defined geographically, and although part of the GSA, it is not in close proximity to the smelter operations. Hanmer has similar physiography and population demographics as the other COI but metal concentrations in soil were not notably affected by smelter emissions (Chapter 4). Therefore, Hanmer was used to represent a reference for comparison with the other COI.

Based upon the evaluation of available data, the following communities were selected as COI for ongoing assessment in the HHRA: Copper Cliff, Coniston, Falconbridge, Sudbury Centre and Hanmer.

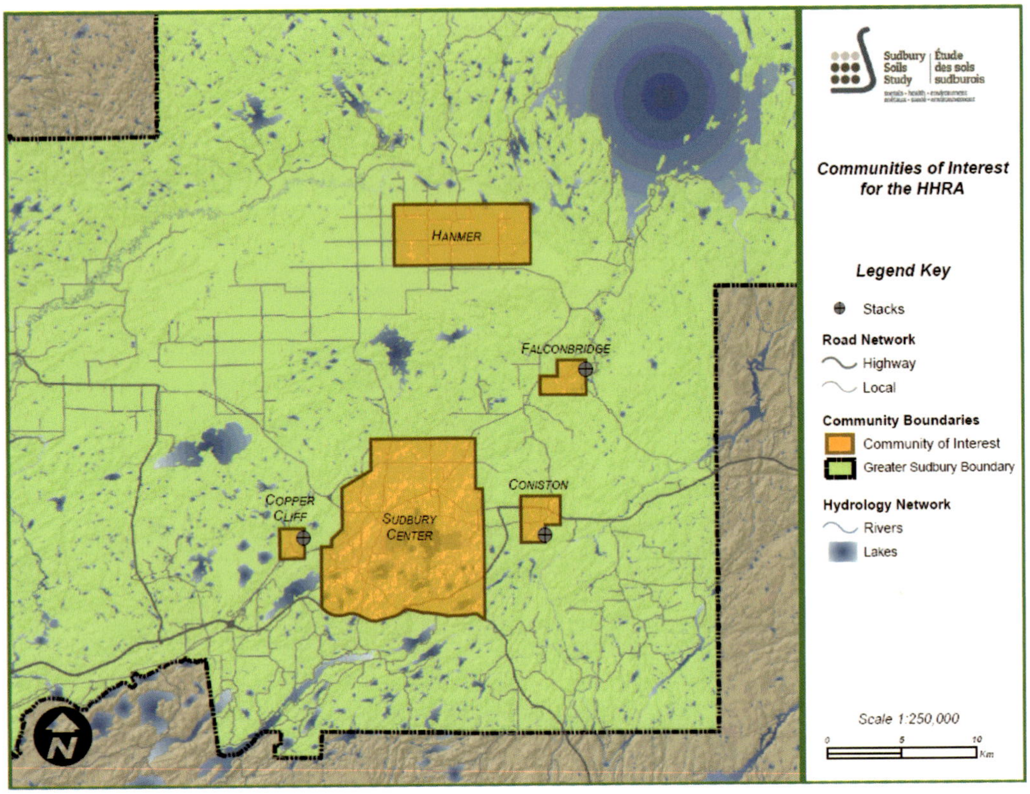

Figure 5.1 Location of the Communities of Interest (COI) for the Sudbury HHRA

Two First Nation communities also exist within the general study area, Whitefish Lake and Wanepetei, but being more distant from the smelters, metal concentrations in soils were low in these communities. However, a number of First Nations members live within the City of Greater Sudbury while also participating in traditional activities of fishing and hunting which might increase their exposure to the COC. Potential exposures of First Nation community members were addressed by evaluating risks related to the unique activities and lifestyle of First Nation peoples living within the established COI, rather than assessing either Whitefish or Wanapitei communities as a separate COI. Therefore, exposures experienced by First Nation members living in one of the established COI, but partaking in traditional First Nation activities, would be expected to be greater than those experienced by individuals living in either of the two established First Nation communities. A background community/typical Ontario resident (TOR) was also evaluated for comparative purposes, based on available data from various regulatory and published scientific sources.

5.4.2 Human Receptors within the COI

A human receptor was considered as any person who resided, visited, or worked in the area being investigated and was, or potentially was, exposed to the COC. It is important to note, however, that occupational exposure to the COC and associated risk was not evaluated in this HHRA. General physical and behavioral characteristics specific to the receptor type (body weight, breathing rate, amount of food consumed, etc.) were used to estimate the amount of chemical exposure received by each receptor. Due to differences in physiological characteristics and activity patterns between children and adults and between males and females, the exposures received by a female child, a male child, a female adult or a male adult are different. Consequently, the potential risks estimated for the same COC may differ depending on the receptor chosen for evaluation.

Human receptors are typically selected such that the most sensitive and/or most exposed individuals are represented. Consideration was given to such characteristics as body weight, breathing rate, dietary habits and daily activity pattern (time spent at a work place, and time spent at home, either indoors or outdoors). Chemical sensitivity, as a function of either physiological maturity or personal afflictions, which could compromise an individual's ability to cope effectively with otherwise harmless levels of exposure, was also considered. For example, because small children are in a state of rapid growth and still immature in terms of development, they are often more sensitive than adults to certain chemicals in their environment. For the current comprehensive risk assessment, male and female receptors in each of the following five life stages were evaluated to predict risks associated with exposure to COC:

- infant (0 to <6 months)
- preschool child (7 months to 4 years)
- child (5 years to 11 years)
- adolescent (12 to 19 years), and
- adult (20 years and over).

The toddler was considered to be the life stage most exposed to chemical in soil due to their habit of playing and crawling on the ground or floors, and hand to mouth activities. Since females have a longer expected life expectancy than males, the female toddler was selected as the most sensitive human receptor for this risk assessment. Chemicals considered to be carcinogenic were evaluated over a lifetime of exposure as the development of cancer is a long-term process that may take many years to manifest. For this reason, a special type of receptor called a 'lifetime' or 'composite' receptor was selected for evaluation of potential carcinogenic risks. This receptor is a 'composite' of all relevant life stages for which exposures were evaluated. Health risks associated with exposure to carcinogenic compounds were expressed as an estimate of excess or incremental lifetime cancer risk (ILCR) resulting from exposures to a particular source. Thus, risks associated with carcinogenic compounds were predicted using the average daily dose over a human receptor's entire life span.

To evaluate potential exposure to COC, it was necessary to characterize the physiological and behavioral characteristics of each receptor group evaluated. Several published sources were considered in the selection of these parameters. The Compendium of Canadian Human Exposure Factors for Risk Assessment (Richardson 1997), Health Canada (2005 pers. commun.), and the U.S. EPA's Exposure Factors Handbooks (U.S. EPA 1997a,b) were used as the primary sources of the receptor parameter data for the HHRA. These sources were used in other HHRAs that have been critically reviewed and accepted by regulatory agencies across Canada and the United States. The Compendium of Canadian Human Exposure Factors for Risk Assessment relies on data from published and reliable Canadian sources, such as Health Canada, Statistics Canada, and the Canadian Fitness and Lifestyles Research Institute. Where insufficient data were available in these sources to characterize relevant activity patterns and/or behavioral/physiological characteristics of a certain receptor group appropriately, other appropriate sources such as the U.S. EPA Exposure Factors Handbooks (U.S. EPA 1997a,b) and Burmaster (1998) were used to supplement the receptor parameter dataset.

For the current assessment, the lengths of the 'summer' and 'winter' exposure periods in Sudbury were conservatively assumed to be 8 and 4 months, respectively, for all receptors. The difference between the two periods pertains to the degree to which an individual may be exposed to soils. During the 'winter' period, the ground is either frozen or covered by snow, minimizing the degree to which one can come in direct contact with impacted soils. A greater amount of clothing is also worn in the winter period due to the temperature, further limiting exposures to soil. While it is likely that the winter period may be considered longer than 4 months in Sudbury, it was conservative to err on the side of a longer 'summer' period (which includes spring and fall), due to the greater potential for direct exposure to impacted soils.

Detailed physiological and behavioral parameters for male and female receptors at each life stage (infant, preschool child, child, adolescent, adult) were used in the HHRA. Since the preschool child was selected as the most sensitive receptor for the Sudbury HHRA, only detailed characteristics for this receptor are presented in Table 5.1. Full receptor characteristics for all life stages were provided in the original report (www.sudburysoilsstudy.com//volumeII/chapter2). Receptor-specific data presented in Table 5.1 provides both the central tendency estimate (CTE) and reasonably maximally exposed (RME) exposure scenarios, as well as the underlying mean and standard deviation of the overall receptor characterization dataset. The normal probability distribution functions (PDFs) represent uncertainty around the arithmetic mean and all other PDFs represent variability of the sample population. In Table 5.1, the upper 95% confidence limit (95 UCL) on the arithmetic mean was used to characterize chronic food intake rates. Default data used by the U.S EPA's Integrated Exposure Uptake Biokinetics (IEUBK) model (U.S. EPA 1994a) was employed to develop outdoor soil and indoor dust ingestion rates. The IEUBK model uses a default outdoor/indoor 45:55 split which applies 55% of the total soil and dust ingestion rate to indoor dust with the remaining 45% being applied to soil. Equivalent average body weights (arithmetic mean values reported by Richardson (1997) were used for both central tendency estimate (CTE) and reasonably maximally exposed (RME) exposure scenarios.

Table 5.1 Receptor Characteristics – Preschool Child (7 months to 4 years)

Receptor parameter	Female[a] Mean	SD	CTE[c]	RME[d]	PDF[b]	Male[a] Mean	SD	CTE[c]	RME[d]	PDF[b]	Reference
Body weight (kg)	16.4	4.5	16.4	16.4	L	16.5	4.6	16.5	16.5	L	Richardson 1997
Amount of air inhaled (m³/day)	8.8	2.4	8.5	11.9	L	9.7	2.7	9.4	13.3	L	Richardson 1997
Amount of soil ingested (g/day)	0.036	na	0.036	0.036	na	0.036	na	0.036	0.036	na	Health Canada 2004
Amount of dust ingested (g/day)	0.044	na	0.044	0.044	na	0.044	na	0.044	0.044	na	Health Canada 2004
Total skin surface area (m²)	na	na	0.69	0.69	na	na	na	0.69	0.69	na	Burmaster 1998
Amount of drinking water ingested (L/day)	0.6	0.4	0.5	1.09	L	0.6	0.4	0.5	1.09	L	Richardson 1997
Amount of milk and dairy consumed (g/kg per day)	44.5	38.8	28.7	46.7	N	45.1	30.0	38.1	47.2	N	Health Canada 2005 pers. commun.
Amount of meat and eggs consumed (g/kg per day)	6.2	5.2	5.7	6.5	N	6.2	5.8	5.1	6.6	N	Health Canada 2005 pers. commun.
Amount of fish and shellfish consumed (g/kg per day)	3.0	2.6	2.5	3.8	N	4.4	4.4	2.6	5.5	N	Health Canada 2005 pers. commun.
Amount of root vegetables consumed (g/kg per day)	7.4	5.3	7.1	9.5	N	7.9	6.9	5.90	8.5	N	Health Canada 2005 pers. commun.
Amount of other vegetables consumed (g/kg per day)	4.7	2.9	5.3	6.3	N	4.8	5.2	3.1	6.3	N	Health Canada 2005 pers. commun.
Amount of fruits and juices consumed (g/kg per day)	17.8	13.6	14.5	20.8	N	16.9	12.8	14.2	17.9	N	Health Canada 2005 pers. commun.
Amount of cereal and grains consumed (g/kg per day)	11.7	8.7	9.6	13.5	N	12.2	10.6	9.1	13.4	N	Health Canada 2005 pers. commun.
Amount of sugar and sweets consumed (g/kg per day)	4.0	1.8	4.6	6.7	N	3.7	4.9	1.9	4.4	N	Health Canada 2005 pers. commun.
Amount of fats and oils consumed (g/kg per day)	1.8	1.1	2.1	2.4	N	0.87	0.79	0.7	1.2	N	Health Canada 2005 pers. commun.
Amount of nuts and seeds consumed (g/kg per day)	1.0	0.7	0.9	1.4	N	0.9	0.79	0.7	1.2	N	Health Canada 2005 pers. commun.
Exposure frequency – summer (days/year)	243	na	229	243	na	243	na	229	243	na	Assumed
Exposure frequency – winter (days/year)	122	na	122	122	na	122	na	122	122	na	Assumed
Time spent outdoors (min/day)	91	83	67.2	182.2	L	91	83	67.2	182.2	L	Richardson 1997

na: Not applicable.
– Not provided.
[a] Whole body surface area was calculated using body weight from Richardson (1997) and model from Burmaster (1998).
[b] N- Normal Probability Distribution Function (PDF), L- Lognormal PDF.
[c] With the exception of body weight, all parameters representing the central tendency estimate (CTE) were characterized using 50th percentile values to represent the central tendency.
[d] With the exception of body weight and food intake rates, all parameters representing the reasonably maximally exposed (RME) individual were characterized using upper percentile (i.e. 90 to 95th percentile) values.

5.4.3 Selection of Dietary Consumption Rates

The dietary consumption rates used to evaluate both Sudbury residents and the typical Ontario resident (TOR) were based upon anonymized data collected in the 1970 to 1972 Nutrition Canada Survey on food use and biometrics which were published as summaries in the Richardson (1997) document. The raw data from the 1970 to 1972 Nutrition Canada Survey, which has previously undergone QA/QC by both Statistics Canada and Health Canada, was provided by Health Canada (2005 pers. commun.) and was used in the current HHRA. This data set is somewhat dated. Health Canada is currently completing a more current consumption survey; however, the new consumption data were not available while this assessment was being completed (Health Canada 2006 pers. commun.). The 1970 to 1972 Nutrition Canada Survey represents information for Canadian populations (unlike similar surveys conducted in the United States) and was recommended by recent Health Canada guidance for risk assessments undertaken in Canada (Health Canada 2004). However, to ensure that the older data did not grossly underestimate potential consumption rates, the Nutrition Canada survey raw data (Health Canada 2005) were compared to the United States Department of Agriculture 1994 to 1996 Continuing Survey of Food Intakes by Individuals and the 1994 to 1996 Diet and Health Knowledge Survey three-year nationwide food consumption survey data set (USDA 1997, 1998).

During a number of workshops at the beginning of the study, the risk assessment team was frequently told by members of the public that Sudbury residents consumed significant quantities of local foods including home grown vegetables, fish, wild game and blueberries. However, no quantitative information was available that could be used in the exposure assessment. To address this data gap, a food consumption survey was conducted with the resulting data considered as part of the exposure assessment. The key research questions addressed by the survey were:

- What types of local foods do residents consume?
- What approximate quantities of local foods do residents consume?
- What are the sources of local food consumed by residents?

The survey was designed to collect information on consumption patterns from population sub-groups predicted to have higher levels of local food consumption (gardeners, hunters, First Nation residents), and to obtain broad information from the general public. More details on the methodology are provided in Chapter 6; however, the results are presented here as they were used as part of the receptor characterization within the Problem Formulation stage.

5.4.4 Local First Nation Wild Game and Fish Consumption Rates

Local First Nation people living in one of the five COI participate in a variety of hunting and fishing activities, and therefore, may consume significant amounts of local wild game and fish. Such local foods may contain different levels of the COC than store-bought foods. First Nation members are often considered a high exposure population subgroup due to potentially higher chemical exposures via game and fish consumption arising from subsistence hunting and/or fishing (Richardson 1997). As the existing reserve areas were outside the primary area of influence of smelter emissions, only those First Nation individuals living off-reserve and in one of the identified COI were evaluated for potential health risks. Therefore, while it was assumed that these individuals may hunt and/or fish to a greater degree than the general population, they were not assumed to be subsistence hunters and/or fishers for the current assessment.

Data regarding the local consumption patterns of the First Nation population subgroup were collected as part of this survey. U.S. EPA (1997a,b) reviewed several studies which indicate that fish consumption among the Native American general population was similar (maximum 50% to 100% higher) to that of the general non-Native population. However, they note that the fish consumption rate for subsistence Native populations was much higher than that of either the general Native or non-Native populations. The Sudbury food consumption survey reported the number of times per year that respondents from the general population of the GSA, hunters and anglers living in the GSA, and residents of the Whitefish Lake First Nation reserve consumed local game and fish species. The total game and total lake fish consumptions of the Whitefish First Nation respondents were higher than those of the general population, but lower than those of the hunters and anglers (Table 5.2).

Table 5.2 **Local Game and Fish Consumption Rates (times consumed per year)**

Category	General population (n=1,226) CTE[a]	General population (n=1,226) RME[b]	Whitefish First Nation (n=218 households) CTE	Whitefish First Nation (n=218 households) RME	Hunters and anglers (n=70) CTE	Hunters and anglers (n=70) RME
Moose	4.5	6.7	18.5	41.5	24.6	48.6
Deer	4.3	8.5	16.5	39.5	10.8	20.8
Grouse	2.8	5.73	4.1	8.1	12.5	27.5
Total game	**11.6**	**20.9**	**39.1**	**89.1**	**47.9**	**96.9**
Walleye	5.4	12.1	14.2	36.2	18.2	42.2
Trout	4.7	10.7	–	–	–	–
Pike	5.9	21.3	10.2	20.2	17.6	46.6
Perch	4.0	10.5	–	–	21.5	48.5
Bass	–	–	9.2	19.2	–	–
Total fish	**20**	**54.6**	**33.6**	**75.6**	**57.3**	**137.3**

– While respondents indicated some consumption of these fish, statistics representing serving sizes could not be calculated from the available data.
[a] CTE: Central Tendency Estimate, mean of self-reported values for eaters only, both sexes combined, and all age groups.
[b] RME: Reasonable Maximum Estimate, mean plus standard deviation of reported values.

The survey results in Table 5.2 indicate that the local game and fish consumption patterns of the First Nation population subgroup may place them at increased risk relative to the general population, but not relative to the hunters and anglers subgroup. This evidence indicates that risk management measures protective of the hunters and anglers subgroup would also be protective of the First Nation population subgroup. However, several limitations with the survey data should be noted:

- Local food consumption rates were self-reported and not confirmed by any other survey method.
- After non-consumers of local fish and game were excluded, the number of respondents was relatively small.
- Data from both sexes and the different age groups were combined.
- Local game and fish consumption rates for residents of the Whitefish Lake First Nation reserve (which is not a COI) could only provide reasonable estimates of the same parameters for First Nation individuals living in one of the COI.

Given the above limitations of the survey data, there was a possibility that the dietary habits of First Nation individuals living in the COI could differ from those reported for the Whitefish First Nation in the Local Food Consumption Survey. For that reason, further information with regard to the local game and fish consumption habits of First Nation populations was considered in a weight-of-evidence approach. Wild game and fish consumption data for Native Canadian individuals were compiled in the Compendium of Canadian Human Exposure Factors for Risk Assessment (Richardson 1997). Comparison of the Richardson data with those obtained from the Sudbury survey showed that the consumption rates reported by Richardson (1997) were much higher than any of the rates obtained in the Sudbury survey (Table 5.3).

Table 5.3 **Comparison of Reported Game and Fish Consumption Rates (g/day)**

Consumption category	Sudbury general population[a] CTE	Sudbury general population[a] RME	Whitefish First Nation[a] CTE	Whitefish First Nation[a] RME	Sudbury hunters and anglers[a] CTE	Sudbury hunters and anglers[a] RME	Richardson[b] CTE	Richardson[b] RME	U.S. EPA[c] CTE	U.S. EPA[c] RME
Total game	7	13	24	55	30	60	270	570	63	125
Total fish	12	34	21	47	36	85	220	400	70	170

[a] Survey data were converted from units of times consumed per year to g/day by assuming a serving size of 227 g and dividing by 365 days in a year.

[b] Richardson (1997) data were collected in 1971 and 1972 from Native Canadians (both Inuit and Amerindian).

[c] U.S. EPA (1997a) estimated intakes for U.S. Native American subsistence fishing populations; and U.S. EPA (1997b) intake values for home-produced meat among eaters only of the general population of the Northeast.

It should be noted that the Richardson (1997) study was based on data collected in the early 1970s. In addition, wild game and fish consumption of Native Canadians may vary across geographic regions. The data reported by Richardson (1997) indicate that the local game and fish consumption rates of some Native Canadian populations may exceed those of the Sudbury General Population, and the hunters and anglers. However, it is unlikely that the Richardson data apply to First Nation individuals living in the Sudbury area because of the age of the data, the inclusion of Inuit respondents (the Inuit are high game and fish consumers), and the likelihood that a significant proportion of respondents were subsistence hunters and/or fishers. Other studies reporting fish consumption by anglers and First Nation groups around the Great Lakes were also reviewed (Kearney and Cole 2003; Gibson 1990). Volume II (Food Ingestion Factors) of the U.S. EPA's Exposure Factors Handbook (U.S. EPA 1997a) was another source of information on food consumption. The U.S. EPA (1997a) provides recommended fish intake values for Native American populations relying on subsistence fishing which were much higher than any of the intake values derived from the Sudbury Local Food Consumption Survey.

For the purpose of the Sudbury HHRA, the First Nation and hunters and anglers population subgroups were considered to differ from the general population only in their local food intake rates. The majority of the evidence suggested that the game and fish intake rates of First Nation individuals who were not subsistence hunters and/or fishers did not exceed those of hunters and anglers. Based on the weight of evidence, it was concluded that any risk management measures that are protective of the hunters and anglers population subgroup would also be protective of the First Nation subgroup.

5.5 Identification of Exposure Pathways

5.5.1 Pathway Analysis

People can come into contact with chemicals in their environment in a variety of ways, depending on their daily activities and use of local resources. The means by which a person comes into contact with a chemical in an environmental medium is referred to as an *exposure pathway*. The means by which a chemical enters the body from the environmental medium is referred to as an *exposure route*. There are three major exposure routes through which chemicals can enter the body: inhalation, ingestion and dermal absorption (i.e. uptake through the skin). For each of these major exposure routes, there are a number of potential sources of chemical exposure or exposure pathways:

- Inhalation of air, vapors, and dust through the lungs.
- Ingestion of soil, dust, drinking water, garden produce, food and accidental ingestion of water and sediments.
- Dermal absorption from soil, dust and water contact with skin.

Exposure pathways may require direct contact between receptors and media of concern (i.e. incidental ingestion of soil), or may rely on indirect pathways which require movement of the chemical from one environmental medium to another (i.e. the uptake and/or transfer of a chemical from soil into home garden vegetables which are then ingested by an individual). The potential for adverse health effects is dependent, in part, upon the existence of an exposure pathway. In other words, if there is no possible exposure to a chemical, regardless of its concentration or toxic potency, then there is no potential for the development of adverse health effects from that chemical. However, it is important not to overlook relevant and/or potentially significant exposure pathways within the HHRA process.

As a result of chemical deposition, human exposure may also occur through indirect pathways. Individuals may come in contact with these chemicals in secondary media as the chemicals move through the environment after release. Indirect exposure may include inhalation and ingestion of chemicals present in dusts generated from soil, soil ingestion itself (particularly for children), and ingestion of garden produce grown in the study area. Given the abundance of wildlife and fish habitat in the study area, and that dietary preferences of many of the Sudbury residents (and First Nation communities, in particular) included wild fish and game, the risk assessment also considered potential exposures to the COC arising from the consumption of wild plants, fish and game animals obtained from within the study area.

Sudbury residents also utilized surface and groundwater resources within the study area that may be impacted by the COC. However, some of these potential pathways contribute minimally to the overall exposure to Sudbury residents (dermal exposure to water while showering or swimming) when compared to other direct exposure pathways (consumption of potable water). Table 5.4 lists all of the theoretically possible pathways of exposure considered in this risk assessment. Only those pathways considered to be of greatest significance were evaluated in the quantitative exposure assessment. The relative contributions of the various pathways were not expected to be equal. Several of the pathways made only a very small contribution to the total estimated daily intake while others were more important. The benefits gained by inclusion of minor pathways into the quantitative phase of the assessment were very limited.

Table 5.4 **Exposure Pathways Evaluated/Screened**

Potential exposure pathways	Exposure route		
	Inhalation	Ingestion	Dermal
Air	■ indoor air ■ outdoor air		☐
Soil (outdoors)	☐ resuspended soil	■	■
Dust (indoors)	☐ living spaces ☐ attic dust ☐ resuspended dust	■ living spaces ☐ attic dust	■ living spaces ☐ attic dust
Potable water		■	☐ showering/bathing
Surface water		☐	☐ swimming
Food (market basket)		■	
Food (local)		■ fish ■ game ■ vegetables ☐ livestock ■ berries ■ fruit ☐ mushrooms	
Food (home grown)		■ vegetables ■ fruits	
Mother's milk		☐	
Infant formula		■	
Transplacental transfer		☐	

Note: ☐ indicates that the pathway was excluded (see Table 5.5 for detailed explanation).
■ indicates that the pathway was quantitatively evaluated in detailed HHRA.

Further discussion of those pathways considered less significant and, therefore, not carried forward into the detailed HHRA, is provided in Table 5.5.

Table 5.5 **Exposure Pathways Excluded from Assessment**

Exposure route	Exposure pathway	Basis for exclusion
Inhalation	Attic dust	Intermittent and short-term concern only
Inhalation	Indoor dust	Particulates entering buildings are likely to settle out and not be inhaled. For the purpose of the risk assessment, it was assumed that airborne particulates, as measured by the air monitoring study, will be inhaled throughout the day, both indoors and outdoors (U.S. EPA, 1998)
Inhalation	Outdoor dust	Assumed to be captured *via* the air monitors that collected data for TSP, PM_{10} and $PM_{2.5}$.[a] For the purpose of the risk assessment, it was assumed that airborne particulates, as measured by the air monitoring study, will be inhaled throughout the day, both indoors and outdoors
Inhalation	Resuspended dust	U.S. EPA recommends that inhalation of resuspended dust be evaluated only if site-specific exposure setting characteristics indicate that this is potentially a significant pathway (U.S. EPA, 1998)
Ingestion	Livestock	Not considered due to the limited grazing lands within the areas of Sudbury directly impacted by the facilities. Results of livestock sampling in the GSA indicated that local livestock COC levels were similar to those found elsewhere in Ontario
Ingestion	Food (local mushrooms)	Local mushrooms were considered as part of the overall above ground vegetable food group (not individually). It should be noted that the Nutrition Canada food consumption survey indicated that 0.3% of the Canadian population typically consume raw mushrooms
Ingestion	Surface water (swimming)	Ingestion of surface water was not considered to be a significant exposure pathway because the estimated ingestion rate while swimming is 50 mL/h (U.S. EPA 1989), which amounts to only 150 mL/day if a child swims 3 h/day. This will result in an approximately 5% annual increase in water-borne exposures if a child swims 3 h/day for 4 months of the year in a local lake
Ingestion	Sediment	Ingestion of sediment while wading or playing on the beach was evaluated quantitatively as part of the screening process. Given the low potential risk and the intermittent nature of these exposure, this pathway was not considered significant and not evaluated further
Ingestion	Attic dust	Intermittent and short-term concern only
Ingestion	Mother's milk	No published methodology is available for consideration of mother's milk exposures to inorganic compounds. Maternal transfer exposures are considered by the IEUBK model for Pb but no such model was available for the other COC
Dermal	Air	Not a significant exposure route for inorganic compounds, including COC
Dermal	Potable water (showering/bathing)	Not a significant exposure route for the COC (U.S. EPA 2004a)
Dermal	Surface water (swimming)	Not a significant exposure route for the COC (U.S. EPA 2004a). Assuming a child swims 3 h/day for 4 months of the year in a local lake, swimming exposure is less than 10% of potable water exposure. Considered insignificant by U.S. EPA
Dermal	Sediment	Given the low potential risk and the intermittent nature of these exposure, this pathway was not considered significant and not evaluated further
Transplacental transfer		Maternal exposure can be important for establishing the body burden of certain metals before birth, as a result of transplacental transfer. Transplacental exposures cannot be directly estimated from environmental measurements, and such assessments require the use of pharmacokinetic models. A few models of transplacental transfer of Pb in humans have been developed; models for other metals were not available for use in risk assessment (U.S. EPA 2004b). The IEUBK lead model was used as part of this assessment to ensure that transplacental transfer of lead was considered in the HHRA

[a] $PM_{2.5}$, respirable particulate matter less than 2.5 μm in diameter;
PM_{10}, respirable particulate matter less than 10 μm in diameter;
TSP, total suspended particles.

5.5.2 Resuspended Dust

As noted above, U.S. EPA recommends that inhalation of resuspended dust be evaluated only if site-specific exposure setting characteristics indicate that this is potentially a significant pathway (U.S. EPA 1998). Therefore, the potential impacts of resuspended dust were evaluated to determine whether it was an important exposure pathway for the current assessment. The assumption that 100% of indoor dust (PM_{10}) originated from settled dust was compared with the assumption that indoor airborne dust is equivalent to outdoor air/dust. To conduct this comparison, it was assumed that indoor airborne dust (PM_{10}) levels were 29.8 µg/m³ (the typical indoor suspended dust level reported in Pellizzari et al. 1999) and that, conservatively, 100% of this dust originated from settled dust. Settled dust concentrations were estimated based on the regression relationship between outdoor soil and indoor dust derived as part of the dust study (refer to Chapter 6). Table 5.6 provides a comparison between the estimated indoor dust concentrations (resuspended) and the measured outdoor air (dust) concentrations used in the risk assessment (the 95% upper confidence limit on the mean (95% UCLM)). The use of the measured outdoor air concentration to represent indoor dust concentrations is conservative compared to predicted indoor dust levels based on resuspension of settled dust (assuming concentrations in indoor dust are equivalent to those present in outdoor air would lead to higher COC concentrations than those predicted based upon resuspended dust). As such, the current assessment conservatively assumed that indoor air concentrations of the assessed COC were equivalent to those measured outdoors, and resuspended dust was not assessed separately.

Table 5.6 **Comparison of Estimated Indoor Dust and Outdoor Air Concentrations**

COC	COI	95% UCLM [soil] (µg/g)	Settled dust concentration (µg/g)	Indoor dust (PM_{10}) level (2.98×10^{-5} g/m³)	Indoor airborne dust concentration (µg/m³)	Outdoor air concentration (µg/m³)
As	Coniston	12.1	16.8	2.98×10^{-5}	5.0×10^{-4}	2.4×10^{-3}
Co	Coniston	18.4	42.6		1.3×10^{-3}	9.0×10^{-4}
Cu	Coniston	315.5	619.2		1.8×10^{-2}	1.6×10^{-2}
Ni	Coniston	432.8	668.7		2.0×10^{-2}	1.2×10^{-2}
Pb	Coniston	52.0	127.4		3.8×10^{-3}	8.0×10^{-3}
Se	Coniston	1.3	2.7		8.0×10^{-5}	3.4×10^{-3}
As	Copper Cliff	19.0	18.5	2.98×10^{-5}	5.5×10^{-4}	5.0×10^{-3}
Co	Copper Cliff	33.4	59.7		1.8×10^{-3}	2.5×10^{-3}
Cu	Copper Cliff	1370.0	842.8		2.5×10^{-2}	8.1×10^{-2}
Ni	Copper Cliff	976.1	896.1		2.7×10^{-2}	6.0×10^{-2}
Pb	Copper Cliff	97.9	150.2		4.5×10^{-3}	2.2×10^{-2}
Se	Copper Cliff	7.5	15.4		4.6×10^{-4}	5.5×10^{-3}
As	Falconbridge	78.6	25.3	2.98×10^{-5}	7.5×10^{-4}	2.4×10^{-3}
Co	Falconbridge	56.5	80.6		2.4×10^{-3}	2.5×10^{-3}
Cu	Falconbridge	1005.5	789.8		2.4×10^{-2}	2.6×10^{-2}
Ni	Falconbridge	1071.5	926.7		2.8×10^{-2}	2.8×10^{-2}
Pb	Falconbridge	82.3	143.6		4.3×10^{-3}	1.5×10^{-2}
Se	Falconbridge	3.1	6.3		1.9×10^{-4}	3.4×10^{-3}

COC	COI	95% UCLM [soil] (µg/g)	Settled dust concentration (µg/g)	Indoor dust (PM$_{10}$) level (2.98×10^{-5} g/m³)	Indoor airborne dust concentration (µg/m³)	Outdoor air concentration (µg/m³)
As	Sudbury Centre	7.2	14.9	2.98×10^{-5}	4.4×10^{-4}	6.1×10^{-3}
Co	Sudbury Centre	11.3	32.1		9.6×10^{-4}	9.7×10^{-3}
Cu	Sudbury Centre	204.0	565.0		1.7×10^{-2}	1.7×10^{-1}
Ni	Sudbury Centre	210.1	515.5		1.5×10^{-2}	9.5×10^{-2}
Pb	Sudbury Centre	35.9	115.7		3.4×10^{-3}	2.5×10^{-2}
Se	Sudbury Centre	1.3	2.7		7.9×10^{-5}	9.2×10^{-3}
As	Hanmer	4.3	13.3	2.98×10^{-5}	4.0×10^{-4}	5.6×10^{-3}
Co	Hanmer	6.5	23.6		7.0×10^{-4}	7.0×10^{-4}
Cu	Hanmer	67.0	447.2		1.3×10^{-2}	9.9×10^{-2}
Ni	Hanmer	67.9	343.3		1.0×10^{-2}	1.2×10^{-2}
Pb	Hanmer	19.2	98.4		2.9×10^{-3}	9.8×10^{-3}
Se	Hanmer	0.7	1.4		4.1×10^{-5}	4.0×10^{-3}

95% UCLM = 95% upper confidence limit on the mean.

5.5.3 Mother's Milk

A review of the literature indicated no published methodology for consideration of mother's milk exposures to inorganic compounds. In the specific case of exposure to Pb, maternal transfer exposures are considered by the IEUBK model (U.S. EPA 1994a,b) but no such model was available for the other COC. As an alternative to mother's milk, infants may consume formula. A simple exposure comparison indicates that formula exposures to As, Co, Cu and Ni are similar to or higher than estimated mother's milk exposures. Given that formula exposures are likely similar to or higher than mother's milk exposures, and that mother's milk exposures cannot be reliably estimated, consumption of mother's milk was not considered further in the current assessment. Consumption of infant formula was considered as part of the overall market basket estimated daily intake (EDI) route-of-exposure.

5.6 Exposure Scenarios

Exposure scenarios describe the situations and conditions in which receptors may be exposed to COC in environmental media. In developing an exposure scenario, a variety of factors were considered, including human access to specific areas or environmental media, physical activities/behavioral patterns, time spent in contact with exposure media, other potential sources of exposure to COC, lifestyle factors (i.e. hunting, fishing and other uses of natural resources) and the existence of sensitive sub-populations or sensitive locations within the community (i.e. children at schools, playgrounds; the elderly in nursing homes). Evaluation of different exposure scenarios can also be performed in an iterative manner to assess the efficacy of various remedial options during the risk management phase of a project. In general, while all receptors may potentially be subject to the same or a similar set of exposure pathways and environmental concentrations, the magnitude of exposure experienced by an individual via those pathways is, to some extent, dependent on the behavioral and physical characteristics of that individual.

For the Sudbury HHRA, a number of exposure scenarios were developed based on the applicability of particular activities and behavior patterns to certain groups or subpopulations. Consideration was given to provide a sufficient level of protection for GSA residents, as well as a sufficient level of sensitivity to enable, if necessary, the potential comparison of exposure estimates to biological monitoring data (such as urinary As levels). The exposure scenarios were deliberately selected to be conservative in nature (reasonable worst case) to ensure that potential exposures to chemicals and the resultant risks were neither overlooked nor underestimated.

5.6.3.1 Selected Exposure Scenarios

- Typical Ontario resident (background): In this scenario, it was assumed that exposure to COC occurred at typical Ontario background (or ambient) levels. Background exposures from the general food basket survey were incorporated into all final exposure estimates for all receptors in this scenario.

- Typical GSA resident: This scenario considered individuals currently living in the GSA. This included both receptors that would likely be away from home for regular scheduled periods due to either work or school, and those receptors who do not leave the home area on a regularly scheduled basis (stay-at-home receptors/primary care-givers). It was assumed that 'stay-at-home' receptors (i.e. infants, preschool children and adults that care for them, retired persons, etc.) would occasionally leave the home area for such reasons as shopping, visiting, vacations, etc. The amount of time in different areas of the GSA as well as the time spent indoors and outdoors during summer/winter months was evaluated as a function of receptor age class. A complete sensitivity analysis of exposure pathways was conducted to identify the most critical pathways with respect to significance of contribution and associated uncertainty.

The following additional scenarios were designed to account for certain subgroups within the overall Sudbury population that were of potential concern due to specific behavioral or activity patterns:

- First Nation residents: Local First Nation people living in the study area participate in a variety of hunting and trapping activities and, therefore, can consume significant amounts of local wild game and fish. To account for these specific behaviors, wild game and fish consumption patterns for the typical Sudbury resident were adjusted to address those potentially higher dietary exposures.

- Recreational hunters/anglers: Local residents who participate in seasonal hunting and fishing activities can also consume significant amounts of local wild game and fish. As with First Nation residents, the wild game and fish consumption patterns for the typical Sudbury resident were adjusted to address these potentially higher dietary exposures.

One exposure scenario that was not evaluated is that of local residents who were additionally exposed to the COC at their workplace (in the mines or at the smelting/processing facilities). Assessment of occupational exposures was outside the scope of the HHRA. Occupational exposure studies were addressed by the Vale and Xstrata Nickel medical departments and the various Joint Occupational Health and Safety Committees as well as the Ontario Ministry of Labour.

5.7 HHRA Conceptual Model

A primary tool in the Problem Formulation step was to develop a conceptual model that illustrated all potential human receptors and the related exposure pathways. Typically, each relevant pathway was identified, including those that may be eliminated through screening or risk management measures. Therefore, the conceptual model should be able to distinguish between routes of exposure that would potentially exist without risk management measures and routes that are expected to exist under conditions which include risk management measures. Figure 5.2 provides a conceptual model for the Sudbury HHRA. For the exposure assessment scenarios, risk management measures were not considered, and all potential pathways of exposure were considered available to each of the receptors (Sudbury residents).

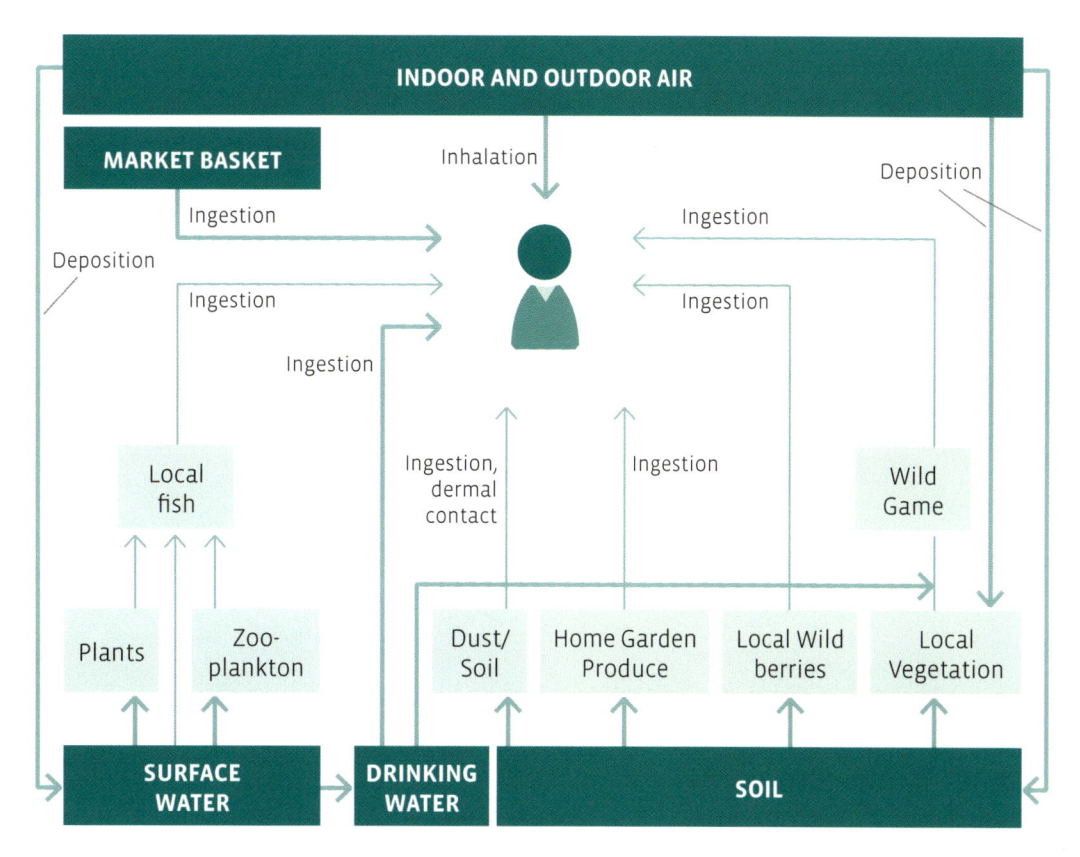

Figure 5.2   Conceptual Model for the Sudbury HHRA

5.8 Screening Level Risk Assessment (SLRA)

Typically, a SLRA is a cursory, deterministic, quantitative HHRA, employing a multitude of conservative and non-site-specific assumptions. The results of the SLRA are generally considered a significant overestimation of actual exposures and related health risks. As a result, those communities, chemicals, receptors and pathways that are not identified by the SLRA as being of concern to human health can confidently be removed from further detailed assessment.

For the Sudbury risk assessment, much of the work to screen and select COC was completed in Phase 1 of the study, and was outlined in the original study terms of reference upon which the current risk assessment was built. Furthermore, it was always the expectation of the TC, and members of the Sudbury community at large, that a full detailed HHRA would be conducted for each of the COC. Therefore, a formal quantitative SLRA was not conducted by the study team. Rather, components of an SLRA were evaluated qualitatively to help guide the detailed risk assessment.

5.9 Identification of Data Gaps

An important part of the Problem Formulation step involved the identification of key data gaps which required further sampling and/or analysis before conducting the detailed HHRA. These data gaps represented areas of uncertainty in the data or assumptions used within the HHRA. The key areas of uncertainty identified during the Problem Formulation were:

- speciation of Ni and As in soil, dust and air to enable metal species-specific exposure and toxicity issues to be addressed
- the bioavailability of COC from key environmental media (dust, soil)
- the concentration of COC in outdoor air

- food consumption patterns within each COI, particularly the unique First Nation communities
- concentrations of COC in local fish and livestock
- concentrations of COC in local wildlife
- concentrations of COC in private potable water sources
- concentrations of COC in indoor dust, and
- concentrations of COC in garden produce.

These and other issues were addressed in Phase 2 of the HHRA and are discussed further in Chapter 6.

5.10 References

Burmaster DE (1998) Lognormal distributions of skin area as a function of body weight. Risk Anal 18(1):27–32

CEI (1999) Deloro village exposure assessment and health risk characterization for arsenic and other metals. Final report. December 1999. www.ene.gov.on.ca/envision/sudbury/cantox_deloro/index.htm. Accessed 20 December 2006

Gibson RS (1990) Principles of nutritional assessment. Oxford University Press, New York.

Health Canada (1993) Health risk determination. The challenge of health protection. ISBN 0-662-20842-0. Health Canada, Ottawa, Ontario

Health Canada (2004) Federal contaminated site risk assessment in Canada. Part I: Guidance on human health preliminary quantitative risk assessment (PQRA). Environmental Health Assessment Services. ISBN: 0-662-38244-7. Health Canada, Ottawa, Ontario

Health Canada (2005) Personal communication with Mark Richardson, Senior Health Risk Assessment Specialist, Health Canada. 17 May 2005

Health Canada (2006) Personal communication with Maya Villeneuve P. Dt, Head, Nutrition Survey Section, Health Canada. 11 January 2006

Hilts SR, White ER, Yates CL (2001) Identification, evaluation and selection of remedial options. Trail lead program. January 2001. www.thec.ca/docs/TF%20Final%20Report.pdf. Accessed 15 November 2005

JWEL (2004) Human health risk assessment. Port Colborne community based risk assessment, Draft report. Prepared for: Inco Limited. Jacques Whitford Environmental Limited, St John's, NL.

Kearney JP, Cole DC (2003) Great Lakes and inland sport fish consumption by licensed anglers in two Ontario communities. J Great Lakes Res 29(3):460–478

MOE (2004) Ontario Ministry of the Environment Record of Site Condition Regulation, Ontario Regulation 153/04 (O. Reg. 153/0) Soil, Ground Water and Sediment Standards for Use under Part XV.1 of the Environmental Protection Act, 9 March 2004. Ontario Ministry of the Environment, Toronto, Ontario

MOEE (1997) Guidelines for use at contaminated sites in Ontario. February 1997. Ontario Ministry of Environment and Energy, Toronto, Ontario

Pellizzari ED, Clayton CA, Rodes CE, Mason RE, Piper LL, Fort B, Pfeifer G, Lynam D (1999) Particulate matter and manganese exposures in Toronto, Canada. Atmos Environ 33:721–734

Richardson GM (1997) Compendium of Canadian human exposure factors for risk assessment. O'Connor Associates, Ottawa, Ontario

USDA (1997) Data tables: Results from USDA's 1994–96 continuing survey of food intakes by individuals and 1994–96 diet and health knowledge survey. U.S. Department of Agriculture, Agricultural Research Service. Online: ARS Food Surveys Research Group Table Set 10. www.ars.usda.gov/Services/docs.htm?docid=7716#individualsurveyyears. Accessed 13 February 2006

USDA (1998) Data tables: Food and nutrient intakes by region, 1994–96. U.S. Department of Agriculture, Agricultural Research Service. Online: ARS Food Surveys Research Group Table Set 10. www.ars.usda.gov/Services/docs.htm?docid=7716#individualsurveyyears. Accessed 13 February 2006

U.S. EPA (1989) Risk assessment guidance for superfund, Volume 1, Human health evaluation manual (Part A). Office of Emergency and Remedial Response. EPA/540/1-89/002. U.S. EPA, Washington, DC

U.S. EPA (1994a) Technical support document: Parameters and equations used in the integrated exposure uptake biokinetic (IEUBK) model for lead in children. The Technical Review Workgroup for Lead. EPA/540/R-94/040. U.S. EPA, Washington, DC

U.S. EPA (1994b) Guidance manual for the integrated exposure uptake biokinetic model for lead in children. EPA/540/R-93/081. U.S. EPA, Washington, DC

U.S. EPA (1997a) Exposure factors handbook, Volume II – Food ingestion factors. Update to the original handbook released in May 1989. EPA/600/P-95/002Fa. U.S. EPA, Washington, DC

U.S. EPA (1997b) Exposure factors handbook. Volume I – General factors. Office of Research and Development. EPA/600/P-95/002Fa. U.S. EPA, Washington, DC

U.S. EPA (1998) Human health risk assessment protocol for hazardous waste combustion facilities. Volume One. Multimedia Planning and Permitting Division, Office of Solid Waste, Center for Combustion Science and Engineering. Region 6. EPA/530-D-98-001A. U.S. EPA, Washington, DC

U.S. EPA (2002) Child-specific exposure factors handbook. EPA/600-P-00-002B. U.S. EPA, Washington, DC

U.S. EPA (2004a) Risk assessment guidance for superfund, Volume I: Human health evaluation manual (Part E, Supplemental guidance for dermal risk assessment) final. Exhibit B-4. Office of Superfund Remediation and Technology Innovation. U.S. EPA, Washington, DC

U.S. EPA (2004b) Framework for inorganic metals risk assessment. Risk Assessment Forum. U.S. EPA, Washington, DC

5.11 Appendix Chapter 5: Abbreviations

COC, chemical of concern

COI, community of interest

CTE, central tendency estimate

EDI, estimated daily intake

GSA, Greater Sudbury Area

HHRA, human health risk assessment

IEUBK, integrated exposure uptake biokinetic (model)

ILCR, incremental lifetime cancer risk

MOE, Ministry of the Environment (Ontario)

PDF, probability distribution function

PM, particulate matter

QA/QC, quality assurance/quality control

RME, reasonable maximum estimate

SD, standard deviation

SLRA, screening level risk assessment

SSRA, site specific risk assessment

TC, technical committee

TOR, typical Ontario resident

TSP, total suspended particulates

USDA, United States Department of Agriculture

USEPA, United States Environmental Protection Agency

95% UCLM, 95% upper confidence limit on the mean

6.0 Sampling and Analyses to Fill Data Gaps for the Human Health Risk Assessment

DOI: 10.5645/b.1.6

Authors

Glenn Ferguson, PhD
Intrinsik Environmental Sciences
6605 Hurontario Street, Suite 500, Mississauga, ON Canada L5T 0A3
Email: Gferguson@intrinsikscience.com

Elliot Sigal
Intrinsik Environmental Sciences
6605 Hurontario Street, Suite 500, Mississauga, ON Canada L5T 0A3
Email: Esigal@intrinsikscience.com

Chris Bacigalupo
Intrinsik Environmental Sciences
6605 Hurontario Street, Suite 500, Mississauga, ON Canada L5T 0A3
Email: Cbacigalupo@instrinsikscience.com

Christopher Wren, PhD
MIRARCO
Laurentian University
935 Ramsey Lake Road, Sudbury, ON Canada P3E 2C6
Email: cwren@mirarco.org

Table of Contents

6.1 Introduction	119
6.2 Air Monitoring Program	119
6.2.1 Overview and Approach	119
6.2.2 Air Monitoring Study Results	121
6.2.3 Conclusions of the Air Study	122
6.3 Sudbury Locally-Grown Food Consumption Survey	123
6.3.1 Purpose and Approach	123
6.3.2 Survey Respondent Profiles	124
6.3.3 Survey Results	125
6.4 COC in Private Drinking Water Supplies	125
6.5 Evaluation of COC in Indoor Dust	126
6.6 Bioavailability/Bioaccessibility	130
6.6.1 Overview of Bioaccessibility	130
6.6.2 Approach for Bioaccessibility Testing in this Study	131
6.6.3 Bioaccessibility Results	132
6.6.4 Recommended Relative Absorption Factors	133

6.7 Speciation of the COC in Environmental Media ... 134
   6.7.1 Introduction ... 134
   6.7.2 Sample Selection ... 135
   6.7.3 Sequential Leach Analysis ... 135
   6.7.4 Mineralogical Procedure ... 138
   6.7.5 Discussion and Conclusions ... 141
6.8 Levels of COC in Locally Grown Vegetables ... 142
6.9 Levels of COC in Local Fish and Livestock ... 147
   6.9.1 Local Fish Survey ... 147
   6.9.2 Livestock ... 147
6.10 Falconbridge Urinary Arsenic Study ... 148
   6.10.1 Background to the Study ... 148
   6.10.2 Arsenic Metabolism ... 149
   6.10.3 Communities ... 149
   6.10.4 Study Approach ... 149
   6.10.5 Participation Rate ... 150
   6.10.6 Study Results ... 150
6.11 Acknowledgements ... 151
6.12 References ... 151
6.13 Appendix Chapter 6: Abbreviations ... 154

Tables

6.1 Site Locations and Parameters for Air Quality Monitoring Network ... 120
6.2 Summary of COC Concentrations (µg/m^3) in Air PM$_{10}$ samples ... 122
6.3 Private Drinking Water Survey Concentrations (µg/L) ... 126
6.4 Summary of Mean Indoor Dust Concentrations by Community of Interest ... 128
6.5 Summary of Mean Outdoor Yard Soil Concentrations by Community of Interest ... 128
6.6 Concentrations of Metals in Indoor Dust from Elementary Schools ... 128
6.7 Summary Statistics of Soil/Dust Concentration Values ... 129
6.8 Summary of Best Fit Linear Regression Equations for Each COC ... 129
6.9 Summary of Bioaccessibility Results (%) ... 132
6.10 Summary of Recommended Relative Absorption Factors (RAF) ... 133
6.11 Tessier Leach Fractions and Methodology (from SGS 2005) ... 136
6.12 Results of Sequential Leach Analyses in Soils (data from SGS 2005) ... 136
6.13 Results of SEM Analyses of Ni Species at Sudbury Centre West Station Air Filters (SGS 2006) ... 140
6.14 Summary of Proposed Ni Species Fingerprints ... 142
6.15 Range of COC Concentrations (µg/g dry wt.) in Garden Soils (0–15 cm layer) ... 143
6.16 Metal Concentrations (µg/g, wet wt.) in Above-Ground Commercial Vegetables ($n=18$) ... 143
6.17 COC Concentrations in Fruits and Wild Blueberries (µg/g wet wt.) ... 144
6.18 COC Concentrations (µg/g wet wt.) in Above-Ground Vegetables by Community ... 145
6.19 Summary of COC Concentrations (µg/g wet wt.) in Below-Ground Vegetables by Community ... 146
6.20 COC Concentrations in Fish Tissues (µg/g wet wt.) ... 147
6.21 Concentrations of COC in Tissue Samples (µg/g wet wt.) from Beef Cattle ... 148
6.22 Summary of Participants in the Urinary As Exposure Study ... 150
6.23 Summary of Inorganic and Total Urinary As Concentrations (µg/L) ... 151

Figures

6.1 Air Monitoring Sites for the Sudbury Soils Study ... 121

6.1 Introduction

The purpose of Phase 2 of the HHRA was to collect the necessary data to fill the information gaps identified during the Problem Formulation phase (see Chapter 5) to decrease the level of uncertainty in the risk assessment. Phase 2 included the development of sampling plans, sample collection and analyses, review of the new sampling data and incorporation of new information into the spatial analysis or related databases. Based on the review of available data and analysis of data gaps, the following issues were addressed in Phase 2:

- a comprehensive air monitoring program to measure the concentrations of Chemicals of Concern (COC) in outdoor air
- detailed dietary (i.e. food consumption survey, including local fish and wild game) and behavioral data (i.e. participation in gardening, hunting, and fishing activities)
- concentrations of COC in private potable water sources
- concentrations of COC in indoor dust
- the bioavailability of COC in soil and dust media (investigation of bioavailability/bioaccessibility in soil and dust for each COC
- speciation of COC in soil and air samples to enable species-specific exposure and toxicity issues to be addressed
- concentrations of COC in private residential garden produce
- concentrations of COC in local fish and livestock, and
- exposure and uptake of As from soil to residents in Falconbridge.

An overview of the methodology and results for each of these studies is provided in the following sections.

6.2 Air Monitoring Program

6.2.1 Overview and Approach

Given the historic and ongoing atmospheric emissions from the two active smelting facilities in the study area, it was very important to have an accurate measure of airborne concentrations of the COC to which Sudbury residents were exposed. Inhalation and breathing represent an important exposure pathway of COC to the human receptor. Some historic and ongoing ambient air monitoring data were available within the area but the monitoring stations maintained by the two mining companies at the start of the risk assessment were limited to small geographical areas (i.e. Copper Cliff, Falconbridge, and Sudbury Centre) and did not provide adequate estimation of ambient conditions in all of the communities of interest (COI). In addition, the routine monitoring programs did not include analysis of all of the COC included in this risk assessment. This was considered a significant data gap for the HHRA. Therefore, an extensive air monitoring program was established to collect samples of the air that would represent the inhalation pathway for Sudbury residents.

The air monitoring network followed the Canadian National Air Pollution Surveillance Program (NAPS) 6-day schedule and was conducted between October 2003 and September 2004, inclusive. Three size fractions of particulate matter (PM) were collected on quartz fiber filters using high volume (approximately 1630 m^3 of air per day – termed 'hi-vol') and low volume (approximately 24 m^3 of air per day – termed 'lo-vol') samplers. The three size fractions of PM sampled were:

- respirable particulate matter less than 2.5 μm in diameter (PM$_{2.5}$)
- respirable particulate matter less than 10 μm in diameter (PM$_{10}$), and
- total suspended particulate matter less than 44 μm in diameter (TSP).

These size fractions were relevant to the HHRA because they represent particulate matter that could be retained in the nose (TSP), upper lung (PM$_{10}$) and lower lung (PM$_{2.5}$). The particulate size considered to be of the most toxicological significance in the HHRA was PM$_{10}$ (i.e. this fraction was used as the primary inhalation component for the assessment modeling), while concentrations detected within the PM$_{2.5}$ size fraction also provide useful qualitative information for the assessment. To establish representative ratios between the different size fractions, samples of all three size fractions were collected as part of this study. The TSP, PM$_{10}$ and PM$_{2.5}$ ratios were determined in part to help differentiate the sources of the samples collected (i.e. smelting versus non-smelting operations in the Greater Sudbury Area, such as blown dust from tailings piles). It was not considered necessary to install three monitors (i.e. one for each size fraction) at every sampling location. Rather, a plan was adopted to apply the ratios from the sites with three monitors to sites where only PM_{10} was measured.

Ten monitoring sites were chosen for the air quality monitoring survey (Table 6.1 and Figure 6.1). These included two existing sites operated by the Ontario Ministry of the Environment (MOE) (Copper Cliff and Falconbridge), seven new sites within the Greater Sudbury area and one background site. It took the study team several months to design a program that would provide adequate monitoring data to be used in the HHRA exposure assessment. Preliminary dispersion modeling was undertaken to identify locations of exposure to stack emissions in the different communities. Power, security, access, and unobstructed air flow to the site were some of the additional conditions considered when choosing site locations.

Table 6.1 **Site Locations and Parameters for Air Quality Monitoring Network**

Site location	Parameters measured at each site
Copper Cliff (pumphouse on Nickel Street)	TSP, PM_{10} and $PM_{2.5}$
Falconbridge (Edison Building)	TSP, PM_{10} and $PM_{2.5}$
Sudbury Centre West (Travers Street, Catholic School Board yard)	TSP, PM_{10} and $PM_{2.5}$
Garson (Public Works Building yard)	TSP, PM_{10} and $PM_{2.5}$
Walden (Jesse Hamilton School, adjacent to SO_2 monitor)	PM_{10}
Coniston (on hill adjacent to communication tower)	PM_{10}
Hanmer (pumphouse on Notre Dame Road)	PM_{10}
Sudbury Centre South (Algonquin Public School)	PM_{10}
Skead (Bowland Bay Road, adjacent to SO_2 monitor)	PM_{10}
Windy Lake Provincial Park, Onaping (near works yard)	TSP, PM_{10} and $PM_{2.5}$

At the laboratory, each air filter was cut into strips for analysis. Extra filter strips were cut from the Copper Cliff and Falconbridge samples and were distributed to the MOE and Xstrata Nickel laboratories for independent analysis. This procedure served as one quality assurance measure (i.e. the comparison of results obtained from different laboratories for the same filter). Differences in analytical methodologies must be taken into consideration, as well as the possibility of uneven distribution of particulate matter over the surface of the filters caused by additional handling. All sampling units were subject to a full calibration once every 3 months, or more often if equipment was replaced or other issues arose. The MOE performed an audit on all of the units at the onset of the study, before any samples were collected. A second audit was performed in March (after 6 months of operation) and the last audit was performed at the conclusion of the study. All units were given a pass designation by the MOE at each auditing session.

Laboratory analysis of the samples collected on the quartz filters included total particulate matter for the TSP, PM_{10} and $PM_{2.5}$ samples and a multi-metal scan. Some of the samples collected were also submitted for metal speciation for COC as discussed below in Section 6.7. The requirement for speciation was based upon the relative differences in respiratory toxicology of the various chemical species (i.e. soluble versus insoluble forms). One blank filter was also submitted for analysis with every 10 sample filters, as an additional quality assurance measure.

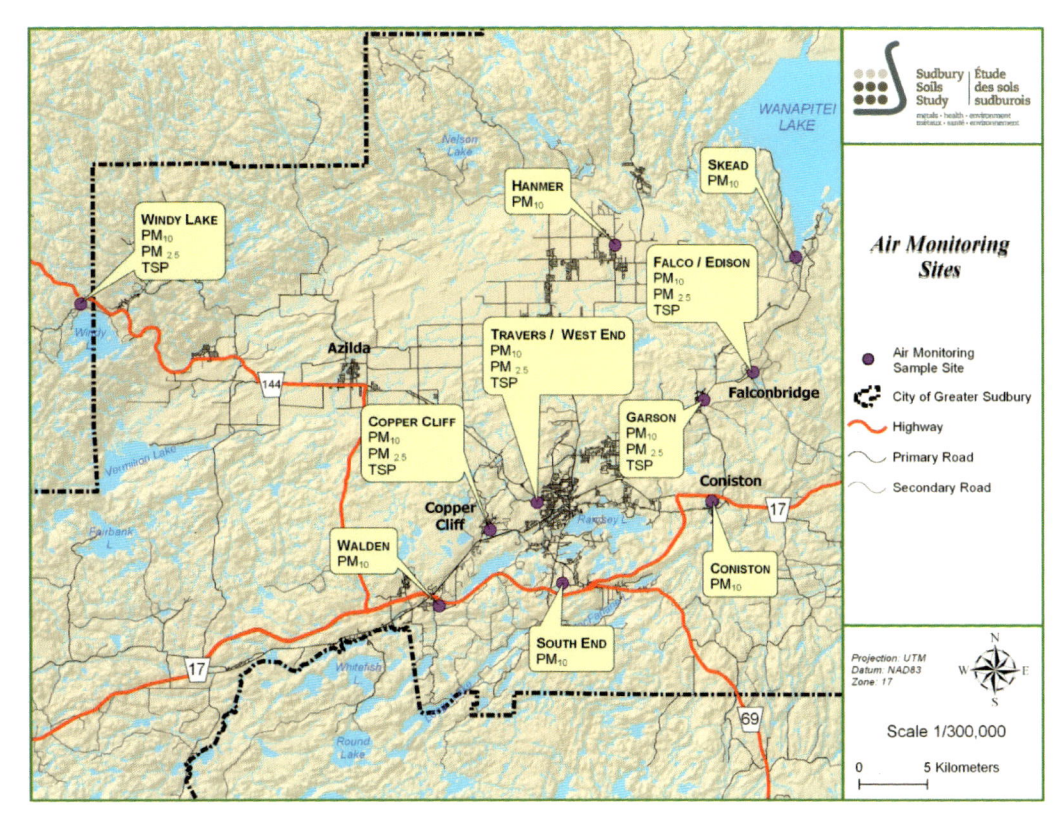

Figure 6.1   Air Monitoring Sites for the Sudbury Soils Study

6.2.2 Air Monitoring Study Results

Arsenic Concentrations

In general, As concentrations were at or below the detection limits, even at the 75th percentile level across all size fractions and across all of the monitoring sites (Table 6.2). The exception was the Sudbury Centre West station, where there was a significant number of values ranging from 0.002 to 0.008 µg/m³ (PM$_{10}$ fraction). This was well above the detection limit but still at least two orders of magnitude lower than the provincial Ambient Air Quality Criterion (AAQC) level of 0.3 µg/m³. The median value was at the detection limit, indicating that there were 'non-detectable' levels measured during the majority of the sampling days at all stations. The highest 95th percentile level was measured at the Sudbury Centre West site, which was three times greater than the concentration at any of the other monitoring stations.

Cobalt Concentrations

The measured Co concentrations at the 10 sites were very consistent throughout the duration of the survey. The highest concentrations were measured at the Sudbury Centre West station, followed by significantly lower, but still measurable, levels at Copper Cliff and Falconbridge. The majority of Co concentrations were several orders of magnitude less than the 24-h AAQC of 0.1 µg/m³ for Co. The one exception was the absolute maximum concentration of 0.06 µg/m³ (PM$_{10}$ fraction) measured at the Sudbury Centre West station, which approached but remained below the AAQC. Since Co was not widely sampled at monitoring stations around Ontario, no comparison values were available.

Copper Concentrations

Fairly low (compared to AAQC), but detectable levels of Cu were measured at all of the 10 monitoring stations. The highest concentrations were at the Sudbury Centre West station, followed by Copper Cliff and Hanmer. The other sites reported concentration distributions with 95th percentiles all less than 0.05 µg/m³. The highest 24-h value for the year was 1.05 µg/m³ (PM$_{10}$ fraction) at the Sudbury Centre West station, which was still well below the AAQC of 50 µg/m³ for Cu. As a point of comparison, the arithmetic mean of Cu concentrations (PM$_{10}$) at monitoring stations throughout Ontario between 1998 and 2002 was 0.02 µg/m³ (MOE 1998, 1999, 2000, 2001, 2002).

Lead Concentrations

Results of the monitoring program indicate that, except for occasional excursions, all stations recorded consistently low levels of Pb (compared to the AAQC). Occasional higher values were observed in Copper Cliff, and the Sudbury Centre West station. This would indicate that Copper Cliff experienced some relatively elevated values of Pb, but not as often as the moderately high levels experienced at the Sudbury Centre West station. The 25th and 75th percentiles and median values were all below 0.025 µg/m³ (PM$_{10}$ fraction). The highest single concentration of 0.13 µg/m³ was measured at the Sudbury Centre West station. As a point of comparison, the arithmetic mean of Pb concentrations (PM$_{10}$) at monitoring stations throughout Ontario between 1998 and 2002 was 0.01 µg/m³ (MOE 1998, 1999, 2000, 2001, 2002). All Sudbury stations in this study recorded concentrations well below the provincial AAQC level of 2.0 µg/m³.

Nickel Concentrations

Relatively low concentrations of Ni were recorded at most stations (compared to the AAQC) with the exception of the Sudbury Centre West station, which reported considerably higher distributions than the other stations, and Copper Cliff, which was moderately higher than the other locations (Table 6.2). The maximum single Ni concentration measured during the year was 0.87 µg/m³ (PM$_{10}$ fraction) at the Sudbury Centre West station. As a point of comparison, the arithmetic mean of Ni concentrations (PM$_{10}$) at monitoring stations throughout Ontario between 1998 and 2002 was 0.007 µg/m³ (MOE 1998, 1999, 2000, 2001, 2002). All levels were below the 24-h AAQC of 2.0 µg/m³.

Selenium Concentrations

The concentration of Se was essentially non-detectable, and was measured only very occasionally at levels above the detection limit. The Sudbury Centre West station recorded the highest single concentration during the year (0.08 µg/m³ on the PM$_{10}$ fraction) whereas the other sites did not record much higher than the detection limit for most of the time. Since Se was not widely sampled at monitoring stations around Ontario, no broad comparison data were available. All values were well below the 24-h AAQC level of 10.0 µg/m³.

6.2.3 Conclusions of the Air Study

Relatively low concentrations of all metals/metalloids and particulate matter were measured in all of the 1220 hi-volume air samples collected from October 2003 to September 2004, inclusive. The provincial air quality limits were exceeded 14 times (five times for the AAQC, three times for the Interim AAQC and six times for the Canada-Wide Standard), with some of these exceedances attributed to natural sources. Data from the survey were incorporated into the HHRA, representing the ambient air concentrations inhaled by residents of the various communities. Data from both Sudbury Centre West and Sudbury Centre South monitoring stations were used to represent exposure to residents of Sudbury Centre as a whole. Use of the Sudbury Centre West station to represent typical airborne concentrations for the entire Sudbury Centre COI would be a very conservative approach, given the proximity of this site to both the Copper Cliff smelter and the nearby slag piles and would likely overestimate actual exposure to the broader area. Table 6.2 provides a summary of the PM$_{10}$ air concentrations used for the human exposure assessment.

Table 6.2 **Summary of COC Concentrations (µg/m³) in Air PM$_{10}$ samples**

COI	COC	Min	Max	Mean[a]	95% UCLM
Coniston (n=61)	As	0.0019	0.0077	0.0022	0.0024
	Co	0.0006	0.0045	0.0007	0.0009
	Cu	0.0029	0.0509	0.0142	0.0162
	Ni	0.0009	0.0427	0.0086	0.0121
	Pb	0.0009	0.0424	0.0064	0.0080
	Se	0.0031	0.0100	0.0032	0.0034

COI	COC	Min	Max	Mean[a]	95% UCLM
Copper Cliff (n=60)	As	0.0019	0.0229	0.0031	0.0050
	Co	0.0006	0.0073	0.0017	0.0025
	Cu	0.0061	0.3426	0.0641	0.0809
	Ni	0.0020	0.2401	0.0476	0.0595
	Pb	0.0009	0.0924	0.0145	0.0220
	Se	0.0031	0.0301	0.0045	0.0055
Falconbridge (n=61)	As	0.0019	0.0058	0.0022	0.0024
	Co	0.0006	0.0097	0.0016	0.0025
	Cu	0.0029	0.0843	0.0227	0.0264
	Ni	0.0028	0.1027	0.0226	0.0280
	Pb	0.0009	0.0539	0.0072	0.0152
	Se	0.0031	0.0082	0.0032	0.0034
Sudbury Centre (n=116)	As	0.0018	0.0284	0.0041	0.0061
	Co	0.0006	0.0648	0.0040	0.0097
	Cu	0.0032	1.05	0.0771	0.17
	Ni	0.0009	0.87	0.0610	0.0947
	Pb	0.0009	0.13	0.0133	0.0254
	Se	0.00308	0.0808	0.00550	0.0092
Hanmer (n=56)	As	0.0019	0.0357	0.0029	0.0056
	Co	0.0006	0.0014	0.0006	0.0007
	Cu	0.0024	0.3242	0.0472	0.0992
	Ni	0.0009	0.0321	0.0057	0.0123
	Pb	0.0009	0.0271	0.0050	0.0098
	Se	0.0031	0.0135	0.0035	0.0040

n, number of samples analyzed.
95% UCLM, 95% upper confidence level on the arithmetic mean.
[a] The arithmetic mean was used for the current statistical presentation.

6.3 Sudbury Locally-Grown Food Consumption Survey

6.3.1 Purpose and Approach

It was brought to the attention of the study team early in the assessment that many Sudbury residents consumed locally grown or locally obtained food including vegetables from private gardens, wild berries, mushrooms, fish and game (i.e. moose, deer, partridge). There was also a preconceived notion that local foods contained elevated metal concentrations due to the historic smelter emissions. Therefore, a key set of information required for the HHRA was a profile of the various local foods that residents consumed on a seasonal and annual basis. To address this data gap, a food consumption survey was conducted with the results considered as part of the exposure assessment. The key research questions addressed by the survey were:

1. What types of local foods do residents consume?
2. What approximate quantities of local foods do residents consume?
3. What are the sources of local food consumed by residents?

The survey was designed to collect relatively detailed information on consumption patterns from population sub-groups considered to have higher levels of local food consumption (i.e. gardeners, hunters, First Nation residents) as well as to obtain broad information from the general public. Respondents were asked to recall consumption of local foods over the past year. The detailed information was collected through in-person interviews with representatives from the higher consumption groups. The broader information was collected via a telephone interview with a random sample of representatives from Sudbury households. The survey region for vegetables and fruit included those grown in the respondent's garden or a neighbor's garden, as well as local fruit and vegetables grown in the Greater Sudbury area (available at local markets and/or grocery stores). Local fish and game included species caught or hunted within a 100-km radius of the Sudbury city core.

In-person interviews were conducted with members of Whitefish Lake First Nations, local gardeners, and local hunters and anglers following the recruitment process. Residents of the Whitefish Lake First Nations reserve were notified of the survey via notices in a local newsletter and announcements at community meetings. Gardeners who had previously participated in the Vegetable Garden Survey (see Section 6.8) were contacted to determine whether they were willing to participate in the food consumption survey. This group was considered to be representative of a population that consumed a higher than average proportion of local vegetables and fruit. Finally, through interviews with representatives of the Sudbury Game and Fish Protective Association, the study team determined that in the winter months, one way to recruit local anglers and hunters was to focus on the ice-fishing community. The rationale presented by the representatives of the Association was that many people from the hunting and fishing community were involved in ice fishing. The interview team visited ice-fishing locations on local lakes to inform members of this sub-population of the food consumption survey.

Consumption statistics for the general population were gathered using telephone interviews. It is important to note that the survey used a self-reporting data collection methodology. While self-reporting methods are convenient for community-based surveys, some of the limitations include under- or over-reporting, difficulties with recall, and social desirability with respect to responses. Of particular importance in considering the limitations with this survey was the challenge involved in accurately reporting food consumption. Most respondents found it challenging to recall frequency of consumption, accurately estimate portion sizes, and few had accurate knowledge of where local fruits and vegetables were grown and harvested, if not from their own gardens. As a result, the data collected in this survey were not necessarily considered as representative of local diets. Rather, it was considered as suitable for providing estimate ranges required for the purposes of the HHRA. The data provided in this report were not validated using any other food consumption reporting techniques such as 24-h diaries, in-home monitoring, or secondary recall.

6.3.2 Survey Respondent Profiles

Typical Sudbury Residents

For the telephone survey of the general population, interviewers contacted 1470 households. Of the 1470 households contacted, 426 households (29%) agreed to participate. The interviews collected household-level data as well as individual-level data for 1226 individuals from the community. The following number of interviews were conducted according to community of interest: Sudbury and New Sudbury ($n=105$); Hanmer, Val Therese, Capreol, Val Caron ($n=107$); Falconbridge, Garson, Coniston ($n=107$); and Copper Cliff ($n=107$). Respondents were asked about household members' participation in hunting, fishing and gardening. Results of the survey indicated that 48% did not participate in any hunting or fishing, while 22% participated in both hunting and fishing activities. Approximately one-third of households (38%) reported that they plant a garden. When asked about their source of local drinking water, the majority of households (65.7%) reported to be on the municipal water supply. The second-most commonly reported water supply was bottled water (23%).

Whitefish Lake First Nations Residents

Food consumption interviews were conducted with 71 households (65%) of the 110 households from the Whitefish Lake First Nations reserve. The interviewers collected household-level data as well as individual-level data for 218 individuals. The study sampled a wide range of respondents, spanning all ranges of age. With respect to gender distribution, the study collected interview data from 105 male and 113 female respondents. Respondents were asked about household members' participation in hunting, fishing and gardening. Almost all (85%) of the households reported that they had not planted a garden in the past 12 months. A large (77%) majority of households reported to either fish or hunt. When asked about their source of local drinking water, the majority of households (78%) reported to be on the municipal water supply. Only a small minority (1%) of Whitefish Lake First Nations households reported well water as their primary source for drinking water.

Hunters and Anglers

Interviews were conducted with 29 households, representing 70 respondents. The interviews collected household data as well as individual-level data. The response group was comprised of 40 males and 30 females. Respondents were asked about household members' participation in hunting, fishing and gardening. A large majority of households (79%) reported that they both fish and hunt, while less than one-quarter (21%) reported that they only fish. More than half (55%) of the households reported that they had planted a garden in the last 12 months. When asked about their source of local drinking water, a large majority of households (72.4%) reported to be on the municipal water supply. Well and bottled water were equally reported as the second-most common water sources (10.3%) for the hunter and angler sub-group.

Gardeners

Interviews were conducted with 29 households, representing 65 respondents in the gardening sub-group. The interviews collected household data as well as individual-level data. The response group was comprised of 34 males and 29 females. Respondents were asked about household members' participation in hunting, fishing and gardening. Almost all households (92.7%) reported that they had planted a garden in the last 12 months. A large majority of households (75%) reported that they do not fish or hunt. When asked about their source of local drinking water, the majority of respondents (65.5%) reported to be on the municipal water supply. Ground well water was reported as the second most common water source (17.2%) for the gardening sub-group.

6.3.3 Survey Results

Data representing consumption rates for specific food groups reported by each respondent group were evaluated and compared to data presented in larger studies published in the scientific literature (as outlined in U.S. EPA 1997). Results of this comparison provided food group-specific consumption rates, appropriate for GSA residents, for use in the HHRA as discussed in Chapter 7.1.

6.4 COC in Private Drinking Water Supplies

Potential consumption of COC via potable drinking water was considered one of the primary exposure pathways for Sudbury residents. The majority of households (88%) in the study area were serviced by a municipal water supply. The municipal water supplies all undergo routine chemical monitoring to satisfy the requirements of the Ontario Safe Drinking Water Act which includes analysis for the COC in this study. Information on metal concentrations in the municipal water supply was obtained from the City of Greater Sudbury Public Works Department and is reported in Chapter 7.1. However, many people in the study area were not on municipal water supplies and obtained their drinking water from private wells or directly from surface water in a nearby lake. Recent existing data on the concentration of COC in these private water supplies were not available and this was considered an area of uncertainty in the HHRA. To address this concern, a drinking water survey was initiated in the fall of 2004.

The four possible drinking water sources and the approximate number of homes in the Sudbury area supplied by each of these sources were reported (Richards 2002) as:

- municipal supply, drawn from surface water (41,000 homes)
- municipal supply, drawn from groundwater (17,000 homes)
- private supply, drawn from surface water (1000 homes)
- private supply, drawn from groundwater (1300 homes with wells in bedrock and 5700 homes with wells in overburden).

An advertisement was placed in local Sudbury newspapers notifying residents that volunteers were required for a survey of metals in drinking water from private water supplies. Water samples were collected at the kitchen tap, as it is the primary site for drinking water. The cold water tap was turned on, and the water run for 2 min to ensure the sample represented water from the surface water or well source, and not water which had been in the homeowners' pipes. A 300 mL non-acidified high-density polyethylene (HDPE) plastic bottle was used to collect the sample. The bottle was placed in a cooler for storage until shipment to the laboratory for analysis. Water samples were not filtered, because the Ontario drinking water guidelines are based on unfiltered samples and filtering the sample would not be representative of what the homeowners are consuming.

All water samples were analyzed by Inductively Coupled Plasma Mass Spectrophotometry (ICP-MS) at SGS Lakefield Laboratories in Ontario for 20 inorganic parameters including the COC. The instrument minimum detection limits (MDLs) for the COC were As (2 µg/L), Co (0.3 µg/L), Cu (0.5 µg/L), Ni (1 µg/L), Pb (0.1 µg/L) and Se (3.0 µg/L). Triplicate water samples were collected at one in every 25 sites (4 triplicates in total). As an additional QA/QC measure, a certified reference material (CRM) was submitted for analysis. Drinking water samples were collected from 94 residential properties, including both private wells drawing water from groundwater and residences drawing surface water from lakes. Where applicable, the results of the analysis were compared to provincial drinking water standards set out in the Safe Drinking Water Act of 2002 (no provincial drinking water standards have been established for Co, Cu or Ni). In the case of Cu, the 1996 Canadian Drinking Water Quality Guideline was used for comparative purposes. Summary statistics for the concentration of COC in private water supplies are provided in Table 6.3. Results of the survey indicated that concentrations of all COC in the water supplies surveyed were below their respective drinking water guidelines (where available). These data were compared to the concentrations reported in the municipal water supply as part of the HHRA (see Chapter 7.1).

Table 6.3   **Private Drinking Water Survey Concentrations (µg/L)**

Potable water source	COC	Ontario Drinking Water Standard	Min	Max	Mean
Groundwater (n=76)	As	25	1.0	23.00	2.37
	Co	NA	0.15	8.70	0.56
	Cu	1000[b]	0.25	216	45
	Ni	NA	0.50	123	11.2
	Pb	10	0.05	8.00	0.70
	Se[a]	10	1.50	1.50	1.0
Lake water (n=18)	As[a]	25	1.00	1.00	1.0
	Co	NA	0.15	0.40	0.16
	Cu	1000[b]	20.9	302	98
	Ni	NA	9.9	126	56
	Pb	10	0.20	5.00	1.5
	Se[a]	10	1.50	1.50	1.5

NA, not available.
n, number of samples analyzed.
[a] All samples were below the minimum detection limit (MDL).
[b] 1996 Canadian Drinking Water Guideline based on aesthetic water quality.

6.5 Evaluation of COC in Indoor Dust

Potential exposure to the COC in indoor dust within Sudbury homes was considered one of the most important pathways in this HHRA. The Vale and Xstrata Nickel smelters release atmospheric emissions containing chemicals and particulate matter, including the COC. Gradually, wet and dry deposition causes the COC to settle onto local soils and other surfaces. Both the settled material and the airborne chemicals may be transferred into residential homes via human activity and local meteorological patterns. Outdoor yard soil can also be transported indoors on clothing or shoes of humans or by animals, and combines with other sources to form household dust (U.S. EPA Region VIII 2001). Studies have reported that between 20% and 30% of indoor contamination comes from outdoor soil sources (Rutz et al. 1997). Exposure to concentrations of COC present in indoor environments is an important pathway of exposure for human health, especially for children.

During the Problem Formulation phase of the HHRA, it was recognized that there was no information on the concentrations of the COC in indoor dust in Sudbury. Therefore, an indoor dust survey was developed to fill this significant data gap. The primary objectives of this survey were as follows:

- Measure concentrations of COC in indoor dust in the Greater Sudbury area (GSA).
- Measure concentrations of the COC in co-located outdoor soil samples to identify if a relationship existed between indoor dust and outdoor soil concentrations.
- If a relationship existed between COC concentrations in outdoor soil and indoor dust, use this relationship to predict indoor dust levels in indoor living spaces over the range of COC levels reported in the 2001 soil survey.
- Compare the Sudbury data with other information and relationships reported in the literature.
- Use these data to estimate human exposure to COC in indoor environments in the HHRA.

Settled dust typically collects on hard surfaces such as windowsills and is often sampled using a swipe method. However, swipe sampling is better suited for the measurement of COC loading (amount in a defined area) rather than COC concentrations in dust (Lanphear et al. 1998). Residential vacuum cleaner bag samples are commonly used to collect dust samples (Lioy et al. 1992) and have been used to collect dust in a number of other studies (Colt et al. 1998; Hysong et al. 2003; Hinwood et al. 2004). However, this method lacks the precision of systematic designed vacuum sampling methods and likely would not retain particles below 10 μm (Morawska and Salthammer 2004). Zhipeng et al. (2003) evaluated five methods of sampling Pb-contaminated dust on carpets including wipe, adhesive label, C18 sheet, vacuum, and hand rinse. The wipe and vacuum methods showed the best reproducibility and correlation with other methods. The authors concluded that surface wipe sampling was the best method to measure accessible Pb from carpets for exposure assessment, while vacuum sampling was most effective for providing information on total Pb accumulation (long-term concentrations). Based on a review of the literature and previous dust studies, it was determined that a soft-surface vacuum collection method would be most effective to collect COC concentrations in household dust for this study. The HVS3 (high-volume small surface sampler) provided the ability to collect dust samples of smaller particles and was selected as the most effective method for determining metal levels in indoor dust.

Dust was collected using a high volume surface vacuum sampler (HVS3) following a standard operating procedure (SOP) based on the ASTM Standard Practice for Collection of Floor Dust for Chemical Analysis (Designation: D 5438-00) (ASTM 2004). The process involved vacuum sampling a composite of at least three 1 m^2 carpeted quadrants in each home. Sample quadrants were selected from high-traffic areas and areas most frequented by children (e.g. floor areas in front of the main television, in a child's bedroom, in a playroom or family recreation room). Dust was collected directly into wide-mouth 250 mL HDPE (high-density polyethylene) sample bottles for analysis. Following each dust sample collection, the HVS3 sampler was disassembled and thoroughly cleaned.

Concurrent surface soil samples were also collected from the front yard of each house to examine the potential relationship between indoor dust and outdoor soils. Duplicate soil cores were collected in an 'X' pattern from the front yard. Each soil core was separated into three depths: humic material, 0–5 cm, and 5–10 cm, and combined into three separate containers (plastic bags). Following each soil sample collection, soil corers were decontaminated. Soil samples were collected in the same manner as samples in the 2001 soil survey as described in Chapter 3. In addition to the collection of dust and soil, a short questionnaire was administered at each household in an effort to examine possible confounding variables as part the data interpretation. The questionnaire was used to collect information on the following types of variables; age of the dwelling, primary source of heating (i.e. gas, oil, electric), renovation/redecoration history, number of occupants; adults, children, and pets, lifestyle factors including occupation and smoking habits, and vacuuming behavior (e.g. How often? Last time?).

Samples of dust and soil were delivered to accredited analytical laboratories for determination of metal content. The samples were diluted to 50.0 mL with 2% HNO_3. The diluted sample was tested directly by Inductively Coupled Plasma-Mass Spectrometry (ICP/MS). For every 10 samples or less, a blank and a control standard were used to verify the calibration standard and the level of instrument performance for blank detection. Every 20 samples or less had a replicated sample to verify precision.

Homes and schools from five regions throughout the GSA were selected for sampling. The five areas represent the primary Communities of Interest (COI) identified for the HHRA. Dust and soil samples from a total of 86 homes were used in this investigation, including: Copper Cliff (19), Coniston (19), Falconbridge (21), Sudbury Centre (17), and Hanmer (10). Summaries of mean indoor dust and outdoor soil concentrations are provided in Tables 6.4 and 6.5, respectively, from each community.

Table 6.4 Summary of Mean Indoor Dust Concentrations by Community of Interest

Community	Indoor dust (µg/g)					
	As	Co	Cu	Pb	Ni	Se[a]
Coniston (n=19)	19.6	32.2	916	202	768	2.6
Copper Cliff (n=19)	27.3	70	1308	379	1543	10.6
Falconbridge (n=21)	32.1	101	624	132	781	2.3
Hanmer (n=10)	15.6	17.8	375	94.2	297	1.7
Sudbury Centre (n=17)	14.8	29.7	662	108	428	4.1
Total residential dataset (n=86)	22.9	55.2	818	193	820	4.5

[a] Using half the minimum detection limit for all non-detect samples (<0.8 µg/g).

Results of the residential survey indicated that the concentrations of the COC in dust and soil differed between the five COI, which is consistent with the 2001 soil study results discussed in Chapter 4. For example, As levels in soil and dust were higher in the Town of Falconbridge relative to the other communities examined. In dust, the levels of Cu, Pb and Ni were higher in Copper Cliff, near the Vale smelter, compared with the other communities. In all cases, COC concentrations were lowest in soil and dust samples obtained from Hanmer. This finding was expected considering that Hamner is located furthest from point source (i.e. smelter) emissions of the COC and was used as a reference community.

Table 6.5 Summary of Mean Outdoor Yard Soil Concentrations by Community of Interest

Community	Yard soil (µg/g)					
	As	Co	Cu	Pb	Ni	Se[a]
Coniston (n=19)	7.26	11.6	166	37.8	213	0.7
Copper Cliff (n=18)	15.36	29	1047	88.1	610	5.7
Falconbridge (n=21)	100.05	61.2	1065	93.3	1130	3.4
Hanmer (n=11)	3.06	4.2	31.5	10.6	38.3	0.4[b]
Sudbury Centre (n=17)	6.54	8.9	141	27.2	121	1.05
Total residential dataset (n=86)	30.93	25.9	548	56.3	479	2.4

[a] Using half the minimum detection limit (MDL) for all non-detect samples (<0.8 µg/g).
[b] All samples were <MDL.

Indoor dust was also collected from eight elementary schools in the Rainbow District School Board, across the GSA: five in the core of the City of Greater Sudbury, one in Hanmer, one in Copper Cliff, and one in Garson, which is attended by children living in Falconbridge. The results indicate that COC levels in the schools were similar to or lower than in private household dust (Table 6.6).

Table 6.6 Concentrations of Metals in Indoor Dust from Elementary Schools

Parameter	As (µg/g)	Co (µg/g)	Cu (µg/g)	Pb (µg/g)	Ni (µg/g)	Se (µg/g)
All schools (n=8)						
Min	6.6	13.6	119	54.0	138	1.6
Max	17.4	45.1	600	100	700	8.4
Mean	10.9	28.8	391	78.3	464	4.8
Standard deviation	3.9	9.4	172	17.6	198	2.4

Two approaches were used to examine possible relationships between the concentration of COC in outdoor soil and indoor dust. First, the ratio of soil/dust concentration (S/D Ratio) was calculated, and secondly the linear relationship between soil and dust concentrations was examined. The S/D ratio was defined as the concentration of a specific metal observed in indoor dust (µg/g) divided by the concentration observed in co-located yard soil. The results are summarized in Table 6.7. With the exception of Cu, all median S/D ratios ($n=80$) were greater than 1.0. This indicated that indoor dust COC levels were 2.8 to 5.9 times higher than corresponding soil levels. However, these data also showed that the ratios did not remain constant over a large range of yard soil concentrations. For example, as the concentrations of COC in outdoor yard soil increased, the S/D ratios decreased.

Table 6.7 **Summary Statistics of Soil/Dust Concentration Values**

Variable	Mean	Std. Dev.	Std. Error	N	Minimum	Maximum	Median	Skewness
As	2.89	3.31	0.370	80	0.055	17.33	1.61	2.29
Co	3.27	2.50	0.279	80	0.318	10.61	2.56	1.30
Cu	0.28	0.27	0.030	80	0.021	1.18	0.186	1.43
Pb	5.95	6.40	0.713	80	0.326	42.76	4.68	3.10
Ni	4.26	5.60	0.626	80	1.50	32.41	2.22	3.18

Linear regression equations (based on the naturally log transformed data) were developed for each COC, describing indoor dust concentrations as a function of co-located outdoor yard soil concentrations. Table 6.8 shows the linear regression equations (i.e. *ln*-transformed) which provided the best-fit based upon the paired outdoor soil and indoor dust concentration sets obtained from this study. Visual examination of the residuals indicated that linear regressions using the raw data resulted in a violation of at least one of the classical assumptions; as a result, data were transformed using the natural logarithm. The R^2-value represents the proportion of variance observed in indoor dust levels that could be explained by co-located yard soil concentrations, the *P* model fit represents the significance level of the relationship (a *P*-value <0.05 was considered statistically significant) and *N* represents the number of samples included in each analysis (Table 6.8).

Table 6.8 **Summary of Best Fit Linear Regression Equations for Each COC**

COC	Equation (ln[indoor dust] = $\beta_0 \pm SE \times$ ln[soil] + $C \pm SE$)	R^2	P model fit	N
As	ln[indoor dust] = 0.22±0.06 × ln[soil] + 2.27±0.15	0.148	0.0004	79
Co	ln[indoor dust] = 0.57±0.07 × ln[soil] + 2.09±0.21	0.441	<0.0001	81
Cu	ln[indoor dust] = 0.21±0.05 × ln[soil] + 5.22±0.26	0.203	<0.0001	81
Pb	ln[indoor dust] = 0.26±0.06 × ln[soil] + 3.82±0.23	0.182	<0.0001	80
Ni	ln[indoor dust] = 0.36±0.06 × ln[soil] + 4.32±0.33	0.317	<0.0001	82

Statistically, outdoor soil could not account for a large percentage of the variance observed in indoor dust concentrations. The regression models presented in Table 6.8 suggested that outdoor soil concentrations were able to explain approximately 15% to 44% of the variation observed in indoor dust COC concentrations. All regression equations were statistically significant and considered appropriate for the development of Sudbury-specific dust-to-soil relationships. Additional regression analyses revealed that the addition of a second explanatory variable such as house age did not explain any additional variance in the indoor house dust concentration. Based upon the results of the indoor dust survey, statistically significant regression equations were used to characterize Sudbury-specific indoor dust-to-outdoor soil relationships for each of the COC. It was decided that the use of linear regression models was more appropriate than using the soil/dust concentration ratio since, at higher COC concentrations in soil, the ratios tended to grossly overestimate corresponding indoor dust levels relative to those predicted using the linear regression models.

6.6 Bioavailability/Bioaccessibility

6.6.1 Overview of Bioaccessibility

The ingestion of soils is often considered to be the major route of potential exposure to metals in humans (Sheppard et al. 1995; Paustenbach 2000). To effectively assess the dose of soil metals received by humans, the determination of bioavailability becomes an invaluable tool in risk assessment. The approach for oral bioavailability assessment of contaminants can typically be divided into four fundamental processes: i) the oral intake of soil/dust including metals; ii) bioaccessibility; iii) gastric-intestinal absorption; and iv) metabolism in the liver/intestines (Oomen et al. 2006; Sips et al. 2001). The inclusion of bioaccessibility testing as part of the assessment process allowed for a more realistic estimate of the systemic exposure to metals from soil and dust ingestion than using generic assumptions such as those employed to derive soil guideline values (EAUK 2005a, 2005b).

Oral bioaccessibility can be defined as the fraction of a substance that is released from the soil or dust matrix during digestion, thus making it soluble and available for absorption through the gastrointestinal tract (Defra and Environment Agency 2002). In effect, this fraction represents the upper limit of bioavailability. The bioaccessible fraction is the fraction of the substance of interest that is dissolved from soil into chyme, and represents the maximum fraction available for intestinal absorption (Ruby et al. 1999; Sips et al. 2001). The dissolved substance may be absorbed and transported across the intestinal wall into the blood or the lymphatic system. Once distributed into the systemic circulation from the intestines or the liver, substances can ultimately start to exert systemic toxicity (Sips et al. 2001). Thus, one can see the importance in assessing bioaccessibility as it will determine the amount of a soil- or dust-bound material that will actually become bioavailable to potentially exert effects in the body.

Toxicity data employed in most risk assessments (e.g. reference doses [RfDs] and cancer slope factors [CSFs]) are typically developed, in part, from toxicological studies using animals. These studies generally use a highly bioavailable chemical form (e.g. soluble inorganic salts) and delivery medium (e.g. food, water) to ensure a high dose reaches the target tissue. As such, RfDs and CSFs do not inherently address the availability of compounds in other environmental media, such as soils and dust. It is, therefore, important that the bioavailability of the compound present in soil or dust, relative to the bioavailability of the chemical species and delivery medium used by the critical toxicological study (i.e. the study used to develop either the RfD or CSF), be quantitatively supported.

Absolute bioavailability refers to the fraction or percentage of a compound that is ingested, inhaled or applied to the skin that is absorbed and reaches systemic circulation (Hrudey et al. 1996). Relative bioavailability, as it pertains to risk assessment, has been defined as *'the difference in absorption of a compound from the environmental medium of concern (e.g. food, soil and/or water) versus the absorption from the vehicle (or medium) used in the toxicological study from which the toxicity-based reference value is derived'* (Kelley et al. 2002). Traditionally, *in vivo* studies (animal studies) have been used to determine the relative bioavailability of metals; however, *in vivo* studies can have significant associated time and cost constraints (Ruby et al. 1999). Therefore, more rapid and inexpensive *in vitro* extraction studies (designed to simulate the human stomach and intestinal system) have been developed to provide a reasonable, yet conservative, approximation of true bioavailability by assuming relative bioavailability is equal to bioaccessibility. *In vitro* extraction studies have been designed to simulate the human gastrointestinal tract (e.g. pH, temperature, and chemical composition of solutions in both the stomach and small intestine) to assess the mobilization of compounds from soil during the digestion process. Given the importance of evaluating the potential toxicity of soil and dust-bound COC to Sudbury residents, *in-vitro* bioaccessibility analyses were conducted. The objective of these analyses was to estimate the bioaccessible fractions of As, Cu, Co, Pb, Ni and Se in Sudbury soil and dust samples. These results were then used to derive a relative absorption factor (RAF) for each COC.

A RAF based on a bioaccessibility evaluation is a simple quotient comparing the solubility of COC in soil and the exposure medium used to develop the RfD/CSF (i.e. spiked food) in simulated digestive fluids. The RAF makes no assumptions about digestive differences between humans and other mammalian species, and is calculated as follows:

$$RAF = \frac{Bioaccessibility\ of\ Chemical\ in\ Soil}{Bioaccessibility\ of\ Chemical\ in\ Exposure\ Medium\ used\ to\ Develop\ the\ RfD}$$

Many different *in-vitro* test methods were available to measure bioaccessibility of inorganic compounds in soil. Oomen et al. (2002) evaluated five different types of *in-vitro* digestion models for three different soil types, producing a wide range of bioaccessibility results. Although data on bioaccessibility of Pb and As in soil were available, limited data were available for other metals such as Ni, Cu, Zn, Cd and Cr (DEPA 2003). At the time of this study, no single *in-vitro* method was universally accepted (DEPA 2003). It is also important to note that oral bioaccessibility testing is only applicable to the soil and dust human exposure pathways, and not the food consumption pathways (EAUK 2005a). Furthermore, bioaccessibility test results have been reported to be significantly affected by various factors such as physical-chemical properties of the chemicals (Dieter et al. 1993; Freeman et al. 1996; Gasser et al. 1996; Ruby et al. 1996, 1999), soil characteristics (Ruby et al. 1993, 1996, 1999; Hamel et al. 1998, 1999), the composition of digestive fluids (Guyton 1991; Ruby et al. 1992; Oomen et al. 2000), and the presence of food constituents (Hack and Selenka 1996). Hence, it is not realistic to propose a single value to represent the bioaccessibility of a given metal; site-specific values must be developed on a case-by-case basis.

6.6.2 Approach for Bioaccessibility Testing in this Study

Following a review of the available literature (including Hamel et al. 1998, 1999; Drexler and Brattin 2007; Oomen et al. 2000; Rodriguez and Basta 1999; Rasmussen 2004; 2006; Yu et al. 2006; Ruby 2004; Ruby et al. 1992, 1993, 1996, 1999; U.S. EPA 2007), a two-phase bioaccessibility protocol was selected to estimate the bioaccessibility of each COC in Sudbury soils and indoor dust. The two-phase approach simulated both gastric and intestinal phases of absorption adapted from the standard operating procedure (SOP) developed by the Solubility/Bioavailability Research Consortium (SBRC) (Ruby et al. 1999). This study employed an *in-vitro* procedure to determine bioaccessibility of the COC in Sudbury soils and dust (Golder 2006). To our knowledge, published *in-vivo* validation of this method has only been conducted for Pb and As at the time of this study.

Soil and dust collected as part of the indoor dust study (Section 6.5) were utilized for the bioaccessibility study. In the first round of analyses, 87 soil samples and 10 dust samples were submitted to the laboratory for bioaccessibility testing. The initial results showed very low bioaccessibility values for all of the COC (less than 5% except for Se which showed 100% bioaccessibility) in dust (Golder 2006). Due to concerns with regard to these low values, a second round of analyses was conducted by submitting soil and dust samples to two independent laboratories for analysis (Golder Associates, Mississagua, Ontario, and Dr John Drexler, University of Colorado, USA). In total, 40 split soil samples were submitted to both laboratories and 10 split dust samples were also submitted. An additional 15 dust samples were submitted to the University of Colorado lab where there was insufficient material to split the samples.

It is important to note that both Golder and Drexler used similar *in-vitro* methods to establish bioaccessibility, with small changes in the overall methodology. One key difference is that Golder (2007) provided bioaccessibility results for both phases (i.e. gastric and gastric+intestinal), while Drexler only provided results for the gastric phase. Spiked rat chow was also subjected to the bioaccessibility assay and an overall bioaccessibility of 94.7% was measured. This was used in the RAF determination for Ni.

Sample Preparation

Soil and dust samples were dried to constant weight at 40°C. Soil was screened through a 250 μm sieve as the smaller particles are considered to be the more important size fraction for incidental soil ingestion. Dust samples were sieved through a 60 μm sieve to exclude fine material such as hair, clothing fibers and carpet particles from dust samples.

Gastric Fluid Preparation and Extraction

Gastric extraction fluid was prepared by adding 60.06 g of glycine to 1.9 L of distilled water. Concentrated trace metal-grade hydrochloric acid (HCl) was added to the solution until pH 1.5 was achieved. The solution was then filled to a final volume of 2.0 L with distilled water to yield a 0.4 M glycine solution. A 1 g sample of soil or dust was accurately weighed and placed in an 18 oz. Whirlpak bag. A 100 mL aliquot of heated gastric extraction fluid was added to the bag, and 1 g of pepsin was added for every 100 mL of extraction fluid. The bag was placed in a shaking incubator at 37°C for 2 h at a maximum setting of 25 strokes/min.

Intestinal Phase Extraction Technique

Upon completion of the gastric extraction phase, the WhirlPak bag was removed from the incubator and sodium hydroxide (NaOH) was added until the pH of the solution was between 7 and 8. Bile (0.35 g; Bovine Bile, Sigma) was added for every 100 mL of extraction fluid. Pancreatin (0.035 g; Sigma, pancreatin consisted primarily of lipase, amylase and proteinase) was added to every 100 mL of extraction fluid. The bag was placed in the shaking incubator at 37°C for 2 h at the same setting as previously. Following this stage, the WhirlPak was removed, and 15 mL of the leachate was drawn using a disposable pipette and placed into a disposable centrifuge tube. The sample was then centrifuged at 10,000g for 10 min. The supernatant was carefully decanted, stabilized with nitric acid and stored at 4°C until submission to the laboratory for analysis of metals using ICP-MS. Since bioavailability is variable and complex, the intended purpose of this study was to adjust for the difference in bioaccessibility between exposure medium (soil) relative to the exposure medium used in the key toxicological study (spiked rat chow for Ni).

6.6.3 Bioaccessibility Results

The results of the bioaccessibility analyses on soil and dust samples for both gastric and gastric+intestinal phases are provided in Table 6.9. The results indicated that, for most of the COC, there was little difference in bioaccessibility between the gastric and gastric+intestinal phases. The one exception to this was Pb, which showed considerably higher bioaccessibility if only the gastric phase of absorption is considered as representative of lead's bioaccessibility in the gastrointestinal tract. While the gastrointestinal absorption of Pb (and all chemicals for that matter) in humans occurs in two organs: the stomach (gastric phase), and the small intestine (intestinal phase), only the gastric phase has been validated for *in vitro* bioaccessibility testing for Pb and only the gastric phase bioaccessibility results for Pb are considered acceptable by many regulatory authorities (including Health Canada, U.S. EPA, and MOE). This is somewhat counter-intuitive as research has shown that, for metals, the actual absorption occurs in the small intestine, with very small to negligible amounts being absorbed in the stomach (Mushak 1991). While Pb and probably most other metals will certainly be solubilized at stomach pH, it is important to remember that solubilization does not necessarily equal uptake.

Table 6.9 **Summary of Bioaccessibility Results (%)**

COC	Gastric Phase				Gastric+Intestinal	
	Golder 2007		Drexler		Golder 2007	
	Mean	95UCLM	Mean	95UCLM	Mean	95UCLM
Soil Samples						
As	24	26	31	39	30	33
Co	26	28	26	28	23	25
Cu	50	54	69	74	61	65
Pb	62	66	69	78	14	16
Ni	35	39	40	44	35	38
Se	7	12	15	26	21	33
Dust Samples						
As	39	43	41	45	41	45
Co	28	32	28	30	32	38
Cu	44	50	46	49	58	67
Pb	79	83	83	95	18	21
Ni	32	36	29	31	37	43
Se	24	NC[1]	43	67	NC	

95UCLM, 95% upper confidence level on the arithmetic mean.

[1] NC = not calculated. Due to concentrations below detection limit, bioaccessibility values could not be calculated for Se.

The results from the Sudbury dust bioaccessibility analysis were similar to those reported by Rasmussen (2004) for Ni and Pb. The dust bioaccessibility reported by Rasmussen (2004) for Ni ranged from 30% to 44% for different size fractions, and from 55% to 74% for Pb. In another study, Yu et al. (2006) reported significantly lower bioaccessibility for intestinal extraction of Pb in vacuumed house dust for particle size fractions <75 μm versus larger particle size fractions up to 250 μm. Yu et al. found that, for the <75, 75 to 150, and 150 to 250 μm fractions, bioaccessibilities for Pb were as follows: 9.4 ± 4.2%, 17.1± 7.2%, and 16.2 ± 8.0%, respectively (mean and 95% confidence interval). The results for the <75 μm fraction were similar to those found for this study for the gastric and intestinal phase digestion. There was virtually no dust greater than 60 μm in the Sudbury dust samples. Differences in dust bioaccessibilities between studies may be due to site-specific differences in dust, including differences in the amount of organic matter, as well as differences in the bioaccessibility methods. Unfortunately, there is a paucity of data available with regard to the bioaccessibility of metals from house dust.

An issue that results from the regulatory policy on the use of bioaccessibility data in HHRA is that site-specific HHRA which utilize only gastric phase soil bioaccessibility data for Pb, and which are much higher than the two phase results, could result in remedial objectives that are lower than regulatory human health-based soil quality guidelines (SQGhh) for Pb. Thus, while agencies such as CCME, MOEE and U.S. EPA endorse SQGhh for lead that include 140, 200, 400, and 1000 mg Pb/kg soil, detailed site-specific HHRA that rely only on gastric phase soil bioaccessibility can generate remedial objectives for Pb in soil that are less than these generic SQGhh values. This situation is counter-intuitive as the more refined and site-specific one gets in an HHRA, the remedial objectives tend to be higher than generic SQGhh, as the generic values are designed to be highly conservative and protective of human health under the majority of common exposure conditions. Site-specific remedial objectives are almost always higher than generic SQGhh as they incorporate detailed site-specific information, and do not need to rely on the same degree of conservatism as the generic SQGhh.

6.6.4 Recommended Relative Absorption Factors

The results of the bioaccessibility study indicated that as much as 78% of the Pb present in Sudbury soils was solubilized (i.e. was available for absorption) in the gastric phase of the study. Similarly, 95% of the Pb present in dust was solubilized in the gastric phase of the bioaccessibility study. Drexler and Brattin (2007) related relative *in-vivo* bioavailability (RBA) and *in-vitro* bioaccessibility (IVBA) estimates from a large dataset of lead-contaminated soils and wastes. A highly significant correlation coefficient was found between the two sets of data and the following linear regression equation relating the two was developed: RBA = 0.878 * IVBA − 0.028

This equation allows an estimate of RBA when only IVBA is known. In the current study, the IVBA estimates for lead (78% for soil and 95% for dust) results in estimates for soil RBA of 66% and dust RBA of 83% (Table 6.10). As noted previously, a RAF (Relative Absorption Factor) based on a bioaccessibility evaluation is a simple quotient comparing the solubility of contaminants in soil and the exposure medium used to develop the RfD/CSF (i.e. spiked food) in simulated digestive fluids. Table 6.10 provides the recommended RAFs for each of the COC used in the current HHRA. It is important to keep the purpose of the bioaccessibility study in context. The purpose of the study was to estimate the relative difference in bioaccessibility between metals in soil and dust from the Sudbury area, and those used in the toxicological studies used to derive the Toxicity Reference Values (TRVs) utilized in the HHRA. The study was not intended to measure the absolute bioavailability of metals in soil and dust from the GSA. Since the results of the study are used in a relative manner, these uncertainties are not expected to affect the results or conclusions of the HHRA significantly.

Table 6.10 **Summary of Recommended Relative Absorption Factors (RAF)**

Chemical	Relative Absorption Factor (RAF)	
	Soil	Dust
As	39	45
Co	28	30
Cu	74	49
Pb[a]	66	83
Ni[b]	42	30
Se	26	67

[a] The RAF for Pb in soil and dust was adjusted based on the Drexler and Brattin (2007) regression equation.

[b] The bioaccessibility of Ni in soil and dust was adjusted by the bioaccessibility of Ni in spiked rat chow (media used in TRV development).

6.7 Speciation of the COC in Environmental Media

6.7.1 Introduction

The highly mineralized nature of the Sudbury area has significant implications on the form in which many of the COC are present and available for potential exposure. Potential exposures related to natural deposits versus those arising from smelting and processing activities are likely to be different in structure. For example, Ni may be present in the environment in a variety of forms including: metallic Ni; water soluble forms of Ni (like Ni sulfate); sulfidic Ni (including Ni subsulfide); and various oxides of Ni (including Ni oxide and complex oxides). Studies of emissions from metallurgical processes such as those employed by Vale and Xstrata Nickel have demonstrated that oxidic, sulfidic and soluble forms of Ni tend to be the predominant forms present, particularly when considering emissions from historic operations. The process of determining the actual form of metal COC present within a given sample matrix is typically referred to as *speciation*.

Within the last two decades, toxicologists, environmental chemists and scientists have increasingly realized that determining total concentrations of an element may not provide adequate information about the element's mobility, bioavailability, and potential toxicity on ecological systems or biological organisms (Michalke 2003; Peijnenburg and Jager 2003). Therefore, methods were developed for identifying and measuring the different forms of metals (or metalloids) in environmental matrices (soil, air, water, biological tissues). The metal species or form of a given metal or metalloid will influence its bioavailability and bioaccessibility in the environment as well as biological systems. The relative risk of trace metals and elements in the environment will depend upon the state of solubility or 'bioaccessability'. Different forms of the same metal can range from essential to innocuous to toxic (Caruso and Montes-Bayon 2003). Metals in particular interact as parts of macromolecules (proteins, enzymes, hormones, etc.) according to their oxidation state. Health risk research that focuses on speciation may eventually lead to regulatory criteria based on maximal element species concentrations rather than total element concentrations (Michalke 2003). Speciation analyses are required to perform adequate risk assessments for potential exposure to metals within a community. For example, inorganic As species are clearly toxic, while the innocuous organic form of As, arsenobetaine (commonly found in seafood), poses little risk and does not influence the outcome of a community-based health risk assessment. Like As, Cr can be either essential (i.e. Cr(III)) or harmful (i.e. Cr(IV)), depending on its oxidation state.

Reliable data for *both* the concentration and bioavailability of specific hazardous species contribute to the assessment of initiators of adverse health effects associated with the inhalation of airborne soils or PM (Huggins et al. 2004). It is, therefore, important to develop quantitative methods of speciating elements in the sample media to correlate the presence of specific chemical species with the potential for adverse effects on the human body, and to improve our understanding of their formation and reaction mechanisms. Direct determination of such species would both improve the quality of predictions of human health risks, and aid epidemiological studies by providing unambiguous data on specific, potentially toxic, inorganic substances. For example, the correlation between total Ni and health effects in a sample population is *unlikely* to be the same as the correlation with health effects of a minor, toxic species such as Ni sulfide (Huggins et al. 2004). A chemical species is defined as a specific form of a chemical element, such as molecular or complex structure or the oxidation state of a metal. Consequently, a speciation analysis is defined as the analytical activity of identifying and measuring species as necessary (Caruso et al. 2003).

The risk assessment team held many discussions with members of the TC (Technical Committee) and Scientific Advisors for developing a plan to assess speciation of the COC in the most effective manner. After considerable discussion and review of the available literature, it was agreed that only speciation of COCs in soil, dust and outside air particulate matter would be undertaken. With respect to food materials (i.e. those gathered as part of the vegetable garden survey), the available literature suggested that the COCs would already be in an organically bioavailable form within the media, therefore, speciation of food materials was not recommended for the current study. In addition, speciation of water samples was not considered necessary since any COCs present in water would be in a soluble form, and would be evaluated as such within the risk assessment. It was also agreed that:

- speciation of Ni in soil, air and dust samples was the priority for the Human Health Risk Assessment (HHRA)
- metal speciation was not necessary for the Ecological Risk Assessment (ERA)
- total metal concentrations would be used to assess health risks for COC other than Ni
- a weight-of-evidence approach to speciation and interpretation of the results would be employed.

Based upon recommendations stemming from these discussions, it was decided that speciation would be examined by two primary approaches: 1) chemical sequential leach extraction analysis, and 2) mineralogical analysis for trace minerals and bulk mineralogical analysis (i.e. using a scanning electron microscope (SEM)). These two approaches are described in more detail in the following sections.

6.7.2 Sample Selection

Soil Samples

A total of 84 soil samples were analyzed using the sequential leach technique from the following communities (sample size in parentheses): Copper Cliff (19), Falconbridge (21), Coniston (18), Sudbury Centre (16) and Hanmer (10). These samples were splits of the residential soil samples taken as part of the indoor dust survey. Of these 84 samples, 10 were selected for additional mineralogical analyses using the SEM (four from Falconbridge, three from Copper Cliff, and three from Coniston). These particular samples were selected for the additional analyses by SEM due to their locations in the three original smelting communities and the presence of elevated Ni concentrations in the samples.

Air Filters

A total of 10 air filters were selected for evaluation by both sequential leach and SEM analyses. Due to the very low concentrations of all metals, including Ni, detected in the air samples, to ensure sufficient material for the speciation analyses, air filters were selected from the day demonstrating the highest PM and Ni concentrations. A review of the monitoring data indicated that 8 June 2004 consistently demonstrated the highest concentrations across all monitoring locations. To evaluate potential differences in COC speciation between the key communities of interest, PM_{10} filters were selected from five locations on 8 June 2004. To evaluate the potential differences in COC speciation at different size fractions, $PM_{2.5}$ filters were also selected from three stations for the same date.

Metal concentrations detected at the Sudbury Centre West monitoring locations were consistently higher than those observed at other monitoring locations. Concentrations at this particular location appeared to be influenced by wind direction. Wind direction on 8 June was out of the south-west, blowing across the Vale Copper Cliff facility and nearby waste piles toward the Sudbury Centre West monitoring station. The location chosen for this particular monitoring site was intended to assess potential impacts on air quality from both of these sources. Five additional PM_{10} filters were submitted from the Sudbury Centre West location on different dates when the wind was blowing from different directions to examine whether wind direction influenced Ni speciation present in the air samples.

Indoor Dust Samples

A total of 25 indoor dust samples were selected for evaluation by both sequential leach and SEM analyses. These included four from Falconbridge, seven from Copper Cliff, four from Sudbury Centre, five from Coniston and three from Hanmer. These particular samples were selected for speciation analyses due to their locations in the various COI, and due to the presence of elevated COC concentrations, in particular, Ni in the samples.

6.7.3 Sequential Leach Analysis

6.7.3.1 Sequential Leach Methodology

Sequential leach analysis is a long-standing, documented analytical technique used to predict metal associations in soils. The chemical models that provide the rationale for these methods have been based on equilibrium reactions, or on empirical determinations from wet chemical methods that rely on the sequential extraction of various phases (Tessier et al. 1979; Tessier and Campbell 1988; Gaillard et al. 2001; Fernádez Espinosa et al. 2002). The determination of species of inorganic Ni in particulate matter (PM) through the application of a sequential dissolution method was described previously (Profumo et al. 2003; Vincent et al. 2001). Sequential extraction protocols are prone to artifacts (Tipping et al. 1985) and require careful evaluation and calibration before being used on a specific sample (Tessier and Campbell 1988; Profumo et al. 2003). For the current study, the sequential leach protocol referred to as a modified Tessier method was used (Tessier et al. 1979). This technique partitions the metals of interest into six fractions (water soluble, exchangeable, bound to carbonates, bound to Fe-Mg oxides, bound to organics, residual). The method and nominally defined speciation fractions are outlined in Table 6.11.

Table 6.11 **Tessier Leach Fractions and Methodology (from SGS 2005)**

Definition	Fraction sought	Method used
Fraction 1 Exchangeable	Metals bound by sorption/desorption processes. Readily bioavailable	1 M MgCl$_2$ shaken for 1 h at neutral pH
Fraction 2 Carbonate-hosted	COC bound to carbonate. Bioavailable after degradation/dissolution of carbonate	Residue from 1 leached with sodium acetate (NaOAc) adjusted to pH 5 with acetic acid (HOAc) to completion
Fraction 3 Reducible[a]	Bound to Fe-Mn oxides. Complete free Fe-oxide dissolution evaluated	Residue from 2 leached with 0.04 M NH$_2$OH.HCl in 25% (v/v) HOAc at 96°C
Fraction 4 Organic-bound or Oxidizable	Bound to organic matter	Residue from 3 leached with 30% v/v H$_2$O$_2$, 0.02 M HNO$_3$, 85°C. 3.2 M NH$_4$Ac (20% v/v HNO$_3$) added, shaken for 3 min
Fraction 5 Residual	Nitric-acid soluble species. Excludes silicate-bound and thus inert/stable/benign COC	Residue from 4 leached with 25% v/v HNO$_3$ heated to dryness. Then leached in 10% v/v HNO$_3$

[a] A combined leach, rather than two steps, usually separating an easily and moderately reducible fraction (e.g. easily reducible targets Mn-oxides).

6.7.3.2 Sequential Leach Results

The predominant two COC in each sample type were Ni and Cu (Table 6.12). Overall, Ni was most prominent in soils over dust and air samples, while Cu was highest in air filters. The next most abundant element was Pb which was similar in soil and air filters, about 6.8%, but was more than twice that (15.5%) in indoor dust. This might suggest some indoor sources of Pb. In the case of the soil samples, the results indicated that the speciation 'fingerprint' (i.e. the breakdown of percentages in each leaching step) was generally consistent across the samples taken from each COI. When evaluating the air filter samples, the results indicated that the speciation fingerprint did vary from sample to sample. This variability appeared to be particularly evident with respect to the fractions that were extracted in the exchangeable and organic phases. This was likely due to the presence of exchangeable metals which have been coated by an organic layer, preventing their quantification before the organic extraction step. Finally, the speciation fingerprints for the indoor dust samples were also very similar across the various COI.

Table 6.12 **Results of Sequential Leach Analyses in Soils (data from SGS 2005)**

Leach fractions	Soil mean (%) (n=84)	Soil range (%)	Air filter mean (%) (n=10)	Air filter range (%)	Indoor dust mean (%) (n=25)	Indoor dust range (%)
Arsenic	2.8	1.0–4.9	1.8	0–2.9	1.2	1.1–2.3
As Exchangeable	0.0	–	26.3	17.6–100	0.0	–
As Carbonate	0.0	0	0.0	–	0.0	–
As Reducible	32.2	24.0–54.3	14.5	0–21.2	18.6	10.7–33.3
As Organic	65.4	45.7–76.0	59.2	0–58.4	79.6	66.7–89.3
As Residual	–	–	0.0	–	1.8	0–3.8
Cobalt	3.1	1.8–8.5	2.6	1.6–4.4	2.2	1.3–2.4
Co Exchangeable	3.1	0–4.5	21.4	14.5–68.1	4.1	2.2–14.4
Co Carbonate	0.6	0–0.9	4.1	2.4–22.6	0.7	0–1.2
Co Reducible	31.7	21.6–36.1	13.1	0–25.0	16.8	12.0–25.8

Sampling and Analyses to Fill Data Gaps for the Human Health Risk Assessment

Leach fractions	Soil mean (%) (n=84)	Soil range (%)	Air filter mean (%) (n=10)	Air filter range (%)	Indoor dust mean (%) (n=25)	Indoor dust range (%)
Co Organic	34.5	24.3–40.9	44.2	0–52.8	52.9	36.8–59.9
Co Residual	30.1	23.2–47.8	17.2	0–22.0	25.5	18.7–27.0
Copper	**42.7**	**25.7–50.2**	**54.9**	**34.1–70.1**	**47.2**	**39.3–55.2**
Cu Exchangeable	0.9	0–1.2	34.6	25.2–73.9	8.3	3.8–20.3
Cu Carbonate	4.5	0.5–6.3	4.8	2.5–11.0	1.2	0.7–2.9
Cu Reducible	24.3	13.1–34.7	2.2	0.8–12.6	5.7	3.9–15.9
Cu Organic	62.9	52.8–74.0	41.7	6.3–54.2	81.6	58.6–86.6
Cu Residual	7.5	4.9–10.7	16.6	7.6–38.1	3.2	1.9–4.1
Lead	**6.7**	**4.2–20.9**	**6.9**	**6.2–15.2**	**15.5**	**8.2–21.7**
Pb Exchangeable	0.7	0–2.0	53.6	35.4–74.8	6.2	2.2–23.4
Pb Carbonate	5.0	3.5–11.6	13.7	10.3–18.9	16.9	6.6–19.7
Pb Reducible	34.8	30.8–48.6	9.4	5.8–18.3	40.7	37.3–41.9
Pb Organic	50.2	29.1–57.6	14.7	0.3–22.4	34.4	22.8–38.2
Pb Residual	9.3	6.9–25.2	8.6	3.7–16.8	1.8	1.5–2.8
Nickel	**44.7**	**39.2–50.1**	**29.6**	**7.3–33.7**	**33.7**	**26.2–35.3**
Ni Exchangeable	13.0	3.8–14.1	22.7	0–29.4	6.9	3.3–20.5
Ni Carbonate	3.8	0.3–4.2	4.3	3.9–36.8	1.5	1.3–2.7
Ni Reducible	30.0	26.4–38.7	4.4	0.9–7.3	13.8	11.5–20.5
Ni Organic	27.4	20.2–33.1	41.4	0–46.9	61.7	45.1–68.7
Ni Residual	25.9	18.2–39.4	27.3	9.9–61.7	16.1	13.5–17.6
Selenium	**0.03**	**0–0.1**	**4.2**	**1.3–27.4**	**0.2**	**0.03–0.7**
Se Exchangeable	0.0	–	17.1	7.1–26.8	0.0	–
Se Carbonate	0.0	–	14.4	0–20.0	0.0	–
Se Reducible	0.0	–	18.5	14.6–28.0	74.0	33.3–100
Se Organic	100.0	100	31.5	16.7–45.2	18.7	0–67.0
Se Residual	0.0	–	18.5	0–41.4	7.3	0–24.3

Note: Bolded and shaded values for the specific COC are the percentage of all COC represented by that particular COC (i.e. 2.8% of the metals detected in soil samples were As, while the remainder of the rows depict the percentage of that particular COC which was extracted at each leaching step.

'–' indicates that particulate COC was not detected in any of the samples.

6.7.4 Mineralogical Procedure

6.7.4.1 Mineralogical Methods Overview

Mineralogical analysis of soils is typically conducted in two phases: 1) trace mineral analysis, and 2) bulk mineral analysis. These vastly different objectives require different methodologies. The trace mineral analysis involves detailed, systematic, high magnification scanning of polished grain mounts prepared from soil size fractions, with the COC-bearing phases characterized by elemental composition, particle size and association (Stanley and Laflamme 1998). Bulk mineral analysis involves X-ray diffraction and scanning electron microscopy (SEM) to characterize mineral weight percent particle size, calculated chemistry and elemental/mineral associations (Jambor and Blowes 1998). Mineralogical analyses, while very useful, do have a number of methodological limitations. In this study, accurate assessment of Ni speciation in air samples posed a challenge, therefore, a number of different approaches were used at different laboratories to obtain results that could be reliably used in the risk assessment. This matter was considered very important since one form of Ni (Ni subsulfide or Ni_3S_2) is much more toxic via inhalation than other forms of Ni and, therefore, the proportion of Ni_3S_2, if present in air samples, would have a significant influence on the outcome of predicted risk to human health.

Mineralogical examination of the various sample types included analysis by Scanning Electron Microscope (SEM), X-Ray Absorption Near-Edge Structure (XANES) Spectroscopy, and Electron Microprobe Analysis (EMPA) at different laboratories in Canada and the United States. The first round of mineralogical analyses on samples of soil, air and dust were carried out by SEM using a Leo 440 SEM combined with energy dispersive X-ray spectrometry (EDS) and equipped with both a secondary electron and back-scattered electron (BSE) detector by SGS Lakefield Research Ltd. in Ontario (SGS 2005). Results of the initial SEM analyses indicated that some of the air filters and dust samples contained small amounts of Ni_3S_2. However, due to the small amounts of material present in the air filters, the small particle size, and limitations of the SEM equipment used in the speciation analyses (i.e. the SEM beam size was too coarse to properly identify the ultra-small particles on some of the air filter mounts), there was some question as to whether these Ni species were indeed Ni_3S_2 or a similar looking form (such as millerite (NiS), a natural form of Ni found in Sudbury ore). As a result, several air filter samples were mounted in epoxy impregnated mounts and the surface polished. This mounting technique eliminated geometric uncertainty providing more reliable X-ray analysis in the SEM.

Samples of air filters were also simultaneously submitted to other laboratories (EnviroAnalytix, Nova Scotia, Canada; National Synchrotron Light Sources (NSLS) at Brookhaven National Laboratories, Upton, NY, USA; and Synchrotron Radiation Centre (SRC) at the University of Wisconsin, USA) depending on availability of beam time for examination of Ni and sulfur compounds using XANES spectroscopy. X-ray absorption is capable of probing *in situ* a particular element in a complex sample in any physical state. Using XANES, it was possible to select the element to probe by tuning the X-ray energy to an appropriate absorption edge. The details of XANES have been described in the literature quantify S in environmental samples (Solomon et al. 2003) as well as Ni (Galbreath et al. 2000). In addition, five outdoor soil samples and nine indoor dust samples were submitted for Electron Microprobe Analysis (EMPA) to the Laboratory for Environmental and Geological Studies (LEGS) at the University of Colorado, Boulder (LEGS 2007) using an electron microprobe (JEOL 860).

6.7.4.2 Mineralogical Analyses Results

Scanning Electron Microscope (SEM)

Results of the SEM analyses indicated considerable variability between and within sample locations for all media types evaluated. Most of the COC species detected probably originated from an industrial source, such as smelting and refining.

Soil

The predominant form of Pb detected was anglesite (Pb sulfate) an emission from smelting/refining sources. With the exception of one sample taken in Coniston, galena (a natural Pb-bearing ore) represented little to none of the total Pb present. Very little As was present in any of the soil samples, even in the town of Falconbridge. In those samples that did have As, it was in the form of arsenopyrite or enargite, both forms naturally present in rock ores. The predominant forms of Cu in soil were chalcopyrite (a natural ore) and a Cu alloy. Interestingly, brass (Cu-Zn alloy) was detected in most of the soil samples, probably due to contamination from domestic or other industrial sources. The predominant forms of Ni present in all soils throughout Sudbury were oxides of Ni related to smelting and refining emissions. Pentlandite, a natural ore form of Ni, was also present in most samples, though at lesser amounts. Stainless steel was also observed in a number of the samples, probably due to contamination from domestic sources. No Ni subsulfide (Ni_3S_2) was detected in any of the soil samples.

Dust Samples

Similar to the soil and air filter samples, the predominant form of Pb in dust samples was anglesite. Arsenic was detected in less than half of the dust samples, with the highest number of detections in samples taken from residences in the Town of Falconbridge. The form of As was typically arsenopyrite, though one sample did identify As oxide (a smelter emission) as a major component. The predominant forms of copper present in the dust samples were Cu matte (Cu_2S) and Cu sulfate, both smelting/refining emission products. Not surprisingly, Cu metal and brass were also detected in most of the dust samples, probably arising from domestic sources around the home. Finally, a variety of different forms of Ni were detected in dusts from homes throughout Sudbury. While pentlandite and various oxides of Ni were the predominant forms identified, small amounts of Ni_3S_2, also called heazlewoodite, were identified in a number of dust samples taken from the various communities, including one sample from the reference community of Hanmer. This finding was unexpected given the relatively long distance between Hanmer and any of the three smelters, and may be indicative of historic impacts or alternate sources of the Ni_3S_2 material. Not surprisingly, stainless steel was detected in many of the dust samples, probably arising from domestic sources around the home.

Air Filters

During the initial round of SEM analyses, the predominant form of Pb detected in the air filters throughout Sudbury was anglesite. No obvious forms of As were detected in any of the air filter samples analyzed for metal speciation. Unlike soil, the predominant form of Cu observed in the air filters was chalcopyrite, with Cu matte (a smelting product) as a lesser form found in most of the samples. While the predominant forms of Ni found in soil samples were various oxides of Ni, interestingly, the predominant form of Ni found in all air samples (both PM_{10} and $PM_{2.5}$) was the natural ore form of pentlandite. The laboratory was able to confirm small amounts of Ni subsulfide in two of the PM_{10} air filters (Copper Cliff and Sudbury) and one of the $PM_{2.5}$ air filters (Copper Cliff).

One of the primary issues arising from the initial round of analytical work was whether wind direction played a role in the 'fingerprint' of the Ni species present in air filters surrounding the Sudbury Centre West monitoring station. To address that question, five new PM_{10} filters from the Sudbury Centre West monitoring station were submitted for further SEM speciation analyses using polished section mounts. These samples were collected at different times of the year, and corresponded to differing wind directions. The results in the normalized Ni ratios confirmed previous analytical results, showing that Ni oxide and pentlandite (Fe Ni sulfide – an important Ni ore) were the two predominant species of Ni found in these air filters, regardless of wind direction. However, the presence of Ni_3S_2 was only noted in three of the five air filters (Table 6.13). In fact, Ni_3S_2 was only identified in those filters taken on days when the wind blew from a westerly direction across the Vale facility to the Sudbury Centre West monitoring station. When the wind blew from an easterly direction, no Ni_3S_2 was detected (Table 6.13). These results also indicated that, where Ni_3S_2 was present, it only comprised between 7.9% and 11.5% of the total Ni species in the filters. This was important in later calculations of human exposure to Ni_3S_2.

Table 6.13 **Results of SEM Analyses of Ni Species at Sudbury Centre West Station Air Filters (SGS 2006)**

Nickel species	Particles (N)	Area (µm²)	Area %	Relative mass %	Contained % Ni	Normalized Ni ratios (%)
29 November 2003 (wind from west and north)						
Pentlandite	18	339	62.0	53.2	34.2	35.7
Ni-Subsulfide	2	44	8.0	8.0	73.3	11.5
Ni-Oxide	7	164	30.0	34.3	78.6	52.9
Ni-Sulfate	–	0	0.0	0.0	22.3	0.0
4 January 2004 (wind from the north)						
Pentlandite	–	0	0.0	0.0	34.2	0.0
Ni-Subsulfide	–	0	0.0	0.0	73.3	0.0
Ni-Oxide	–	0	0.0	0.0	78.6	0.0
Ni-Sulfate	30	178	100.0	35.4	22.3	100.0
10 March 2004 (wind from the south-southwest)						
Pentlandite	5	284	59.0	50.7	34.2	60.3
Ni-Subsulfide	3	15	3.1	3.1	73.3	7.9
Ni-Oxide	7	36	7.5	8.6	78.6	23.4
Ni-Sulfate	13	146	30.4	10.7	22.3	8.3
2 July 2004 (wind from the north and east)						
Pentlandite	11	122	48.0	41.3	34.2	61.2
Ni-Subsulfide	–	0	0.0	0.0	73.3	0.0
Ni-Oxide	4	15	5.9	6.8	78.6	23.1
Ni-Sulfate	13	117	46.1	16.3	22.3	15.8
30 September 2004 (wind from south-southwest)						
Pentlandite	2	8	0.9	0.8	34.2	0.3
Ni-Subsulfide	14	118.38	13.2	13.1	73.3	11.0
Ni-Oxide	5	770	85.9	98.4	78.6	88.7
Ni-Sulfate	–	0	0.0	0.0	22.3	0.0

Analyses Using XANES Techniques

The second set of analyses involved using a synchrotron light beam to determine all phases of Ni (including Ni_3S_2) present in select air filter samples. This technique was able to evaluate the K shell absorption spectrum of the sample to determine the various species of both Ni and S present. Results of the XANES analyses indicated the following:

- The majority of the S present in the air filters was in sulfate form.
- Only the TSP and PM_{10} filters from the Sudbury Centre West station showed the presence of sulfide.
- Analyses of the sulfide present in the Sudbury Centre West station samples (11% to 16% of total) indicated that it more closely resembled NiS than Ni_3S_2.
- Ni_3S_2 was not detected in the dust sample analyzed.

Electron Microprobe Analysis

Five outdoor soil samples and nine indoor dust samples were submitted for Electron Microprobe Analysis (EMPA) at the Laboratory for Environmental and Geological Studies (LEGS) at the University of Colorado, Boulder (LEGS, 2007). These analyses were conducted using an electron microprobe (i.e. JEOL 8600) equipped with four wavelength spectrometers, energy dispersive spectrometer (EDS), Backscattered Electron Image (BEI) detector and the Geller, dQuant data processing system. This round of speciation analysis focused primarily on As, Pb and Ni elements present within the soil and dust samples, and provided a detailed percentage breakdown of the specific species in relation to the overall mass of COC.

Results of the EMPA speciation indicated a similar pattern of results as reported using the other techniques. However, one observation from EMPA provided information potentially important for future risk management decisions. As noted previously, the primary form of Pb identified by SEM was in the form of anglesite (i.e. Pb sulfate), which is known to be emitted from smelting/refining sources. However, a major proportion of Pb present could not be accounted for mineralogically by SEM. The results of the EMPA speciation work indicated that a significant percentage of the Pb present in some of the dust samples was in the form of cerussite (Pb carbonate). This form of Pb was detected in most of the dust samples analysed (but none of the soil samples), and typically ranged between approximately 20% and 85% of the total Pb present in the sample. This was of considerable risk management significance because cerussite, or 'white lead', is a key ingredient in Pb-based paints and is not a product of high-temperature smelter emissions.

6.7.5 Discussion and Conclusions

Results of the speciation analyses conducted on the soil, air filter, and dust samples indicate that emissions from smelting and refining sources have impacted each of the sample media. The following are some key findings arising from of the speciation analysis:

- The speciation fingerprint noted in the Tessier leach analyses indicated that similar metal species were present in each of the COI. In particular, the As species were similar in each community including those found in Falconbridge.
- Ni and Cu were the two predominant COC detected in most of the samples.
- Oxidic Ni appears to be ubiquitous throughout each of the COI, in each of the sample media, and in soil and dust samples, in particular.
- A form of Pb carbonate called 'cerussite' was detected by EMPA speciation in a number of indoor residential dust samples analysed, but no soil samples. Cerussite, or 'white Pb', is known to be a key ingredient in Pb-based paints.
- The highest PM and COC concentrations were detected at the Sudbury Centre West monitoring location, probably resulting from fugitive dusts blowing off of the Copper Cliff facility property.
- The species present in dust samples appeared to be similar to those observed in air filters, indicating that the metals present within the dust probably originated from airborne emission sources, rather than being tracked in from outdoor soil sources.
- Ni_3S_2 was detected in a number of indoor dust samples. Follow-up analyses indicated that it appears to be limited to less than 2% of the total Ni present within a given dust sample.
- Ni_3S_2 was detected in a small number of air filters taken from the Sudbury Centre West monitoring station, but only on those days when the wind was blowing eastward across the Vale facilities. Based upon the results of the follow-up SEM and XANES analyses, it was concluded that a reasonable upper-bound estimate for the proportion of Ni_3S_2 present would be approximately 10% of the total Ni species in airborne particulates, under these conditions.

Air quality around the Sudbury Centre West monitoring station, and to a lesser extent the Copper Cliff monitoring station, appeared to be strongly influenced by fugitive dusts arising from the Vale Copper Cliff smelter facility. When the wind was blowing across the facility, one particular Ni species 'fingerprint' was evident, including the limited presence of nickel Ni_3S_2. However, when the wind was blowing from the opposite direction (i.e. not across the facility property), a different Ni species 'fingerprint' was present, without any Ni_3S_2. Conservative estimates of the two specific Ni species 'fingerprints' were established for use in the exposure assessment as summarized in Table 6.14.

Table 6.14 **Summary of Proposed Ni Species Fingerprints**

Ni species	Typical ambient fingerprint	Copper Cliff facility impacted fingerprint
Ni Oxide	80%	75%
Ni Sulfide	10%	10%
Ni Subsulfide	0%	10%
Ni Sulfate	10%	5%

6.8 Levels of COC in Locally Grown Vegetables

To address a key identified data gap in the HHRA, a study was conducted to measure total metal concentrations in locally grown fruits and vegetables from May to October 2003. The purpose of the vegetable garden survey was to obtain site-specific data on the range of concentrations of COC found in fruit and vegetables that comprise a portion of the dietary intake of Sudbury residents. Three sources of food crops were studied: residential home gardens, commercial farms and edible wild plant sites. A total of 89 sites were sampled, which included 64 residential properties, 15 commercial properties, and 10 natural sites. The sampling locations were chosen to reflect a variety of soil metal concentrations, site types and soil types.

At each residential and commercial site, a minimum of three and a maximum of five vegetable and/or fruit species were collected. Minimum collection at a site consisted of three species, to include an example of: a below-ground crop such as potatoes, carrots, beets, onions; a leafy vegetable such as lettuce, Swiss chard, spinach or cabbage and; an above-ground fruit vegetable such as tomatoes, cucumbers, beans, peas and zucchini. From the commercial sites, the predominant samples collected were potatoes, strawberries, cabbage, cucumbers, and squash. At each wild plant site, mushrooms and/or blueberries were collected. The produce was collected in a manner consistent with normal harvesting practices for that crop, and only the edible portions of each crop were collected. For above-ground leafy and fruit crops, stainless steel secateurs (clippers) were used to cut the edible portions of the crop from the non-edible portions; for berry crops the berries were gently removed by hand, and; for the below-ground crops, the edible portions were dug from the ground by hand. The plant samples were cleaned as per a realistic 'worst-case exposure scenario'. For instance, below-ground crops were scrubbed rather than peeled and blueberries were unwashed. The moisture content of all tissue was determined so that all measurements of total metal content could be presented on both a dry weight and wet weight basis. The produce samples were analyzed for total metal content by ICP-MS. It should be noted that the analysis of As in the produce samples was challenging in many of the samples. This was considered to be due to the high water content and relatively low As concentrations present in the samples. Several QA/QC measures including many replicate samples, and submission of Certified Reference Material (CRM) to the analytical laboratories were provided to increase confidence in the results.

Co-located soil samples were collected from gardens in private residences as well as from commercial grow operations. Soil samples were collected at all sites at depths of 0 to 15 cm and 15 to 30 cm. The soil samples were submitted for physical and chemical analysis. The range of COC concentrations and the pH in the 0 to 15 cm soil layer are presented in Table 6.15. The Ontario Ministry of the Environment (MOE) soil quality guidelines are also provided for comparative purposes. The concentrations of metals (particularly Cu and Ni) were generally higher in residential and natural soils compared to commercial soils. At some of the residential and natural sites, the concentration of metals in the soil exceeded the MOE criterion (Table 6.15). The metal levels at commercial sites were generally quite low with just one sample that had concentrations above guidelines. At the residential and commercial sites, there was little difference in the concentration of the COC with depth, suggesting that the soils were well mixed. At the natural sites, the concentration of metals was elevated in the upper 0 to 15 cm layer, reflecting that the soil layers were not naturally mixed and that atmospheric deposition was a likely source of the COC. The pH of the samples was variable with the mean pH at the residential sites higher (pH 6.7) than either the commercial (pH 5.6) or natural sites (pH 4.6).

Table 6.15 **Range of COC Concentrations (µg/g dry wt.) in Garden Soils (0–15 cm layer)**

MOE soil screening criteria		As	Co	Cu	Ni	Pb	Se
		20	40	225	150	200	10
Residential (n=70)[a]	pH 5.1–7.9	<dl–173	4–56	2–1170	31–1100	5.9–520	<dl–11
Commercial (n=24)[b]	pH 4.2–7	<dl–14.7	2.8–11	6–110	9–78	6.2–35	<dl–2.1
Wild plant sites (n=10)	pH 4–5.2	5.7–36.5	3–15	38–440	38–400	10–79	<dl–3.5

dl, detection limit.
[a] At some sites, more than one garden or field was sampled, so there are more soil samples than sites.

The concentration of COC in commercial vegetables tended to be lower than produce grown in residential gardens. This may be related to the fact that soil metal concentrations were generally lower than in residential gardens (Table 6.15) and the sites tended to be further from the smelter emissions. Data for above-ground vegetables from commercial farming operations are provided in Table 6.16.

Table 6.16 **Metal Concentrations (µg/g, wet wt.) in Above-Ground Commercial Vegetables (n=18)**

	As	Co	Cu	Ni	Pb	Se
Range	0.01–0.03	0.02–0.2	0.2–0.9	0.08–3.9	0.04–0.52	<dl
Mean	0.02	0.05	0.46	1.2	0.13	<dl
Standard deviation	0.010	0.05	0.24	1.01	0.14	na
95% Confidence on mean	0.04	0.02	0.12	0.5	0.06	na
Upper 95th percentile	0.02	0.05	0.75	2.3	0.25	na

dl, detection limit.
na, not applicable.
Approximate detection limits (µg/g dry wt.) As: 0.2, Co: 0.2, Cu: 0.5, Ni: 0.5, Pb: 0.5, Se: 0.2.

The concentrations of As, Co and Se were generally below detection limits in wild blueberries (Table 6.17). The maximum Cu and Ni concentrations were 0.9 and 1.0 µg/g, respectively, with a mean concentration of 0.5 µg/g for both elements. The metal concentrations in wild mushrooms were among some of the highest measured in the study, however, only three samples were analyzed. Arsenic ranged from 0.09 to 0.3 µg/g, where the mean was 0.18 µg/g. The maximum Co concentration was 0.09 µg/g, with a mean of 0.07 µg/g. The mean Cu and Ni concentrations in wild mushrooms were 3.4 µg/g and 1.4 µg/g, respectively. The mean Pb concentration was 0.16 µg/g, while Se concentrations ranged from 0.6 to 1.3 µg/g.

Table 6.17 **COC Concentrations in Fruits and Wild Blueberries (μg/g wet wt.)**

Food type	COC	Min	Max	Mean (arithmetic)	95% UCLM
Fruits (n=4)	As	0.01	0.02	0.02	0.02
	Co	0.01	0.02	0.02	0.02
	Cu	0.55	0.90	0.74	0.90[a]
	Ni	0.40	2.9	1.5	2.75
	Pb	0.02	0.05	0.03	0.05
	Se	0.01	0.06	0.03	0.06[a]
Local fruits (n=12)	As	0.01	0.02	0.01	0.02
	Co	0.01	0.06	0.02	0.04
	Cu	0.24	0.90	0.54	0.65
	Ni	0.03	2.9	1.0	1.5
	Pb	0.02	0.08	0.03	0.04
	Se	0.01	0.06	0.02	0.02
Wild blueberries (n=7)	As	0.01	0.02	0.02	0.02
	Co	0.01	0.02	0.02	0.02
	Cu	0.23	0.93	0.50	0.68
	Ni	0.26	1.03	0.52	0.71
	Pb	0.01	0.09	0.04	0.07
	Se	0.01	0.02	0.01	0.02

n, number of samples analyzed.
95% ULCM, 95% upper confidence level on the arithmetic mean
[a] Recommended 95% UCLM value is greater than the maximum value due to the small sample size. Maximum value was used as surrogate for 95% ULCM value.
Minimum Detection Limit (μg/g dry wt.): As=0.2, Co=0.2, Cu=0.5, Pb=0.5, Ni=0.5, Se=0.2.

Data from the residential produce survey are presented by community for above-ground and below-ground vegetables in Tables 6.18 and 6.19, respectively. The concentrations of As, Pb, Co and Se were either low or not detected in the majority of samples. The two dominant metals in produce from all three site types were Cu and Ni. The exception was wild mushrooms which had elevated concentrations of As, Se and Pb relative to the other sample types. The highest concentration of Cu was found in mushrooms from natural sites. The concentrations of Pb and Co tended to be slightly higher in above-ground vegetables compared to below-ground vegetables. The concentrations of Cu, Ni, Pb and Se appeared similar between the two groups. The concentrations of Cu and Ni tended to be higher in vegetables from the smelter communities compared to Hamner, the reference community. For example, the mean concentration of Cu in above-ground vegetables from Copper Cliff was 0.73 mg/kg, compared to 0.34 mg/kg in Hamner. The mean concentration of Ni in above-ground vegetables from Falconbridge was 1.6 mg/kg, compared to 0.18 mg/kg in Hamner. Concentrations of COC measured in this survey were used to establish exposure point concentrations (EPCs) for produce in each of the COI for use in the HHRA exposure assessment. The EPC chosen for the exposure assessment was the 95% upper confidence level on the arithmetic mean (95% UCLM).

Table 6.18 **COC Concentrations (µg/g wet wt.) in Above-Ground Vegetables by Community**

COI	COC	Min	Max	Mean (arithmetic)	95% UCLM
Coniston (n=31)	As	0.01	0.06	0.01	0.02
	Co	0.01	0.05	0.01	0.02
	Cu	0.19	1.23	0.46	0.54
	Ni	0.01	2.17	0.42	0.57
	Pb	0.01	0.42	0.06	0.09
	Se	0.01	0.10	0.02	0.03
Copper Cliff (n=34)	As	0.01	0.11	0.02	0.04
	Co	0.01	0.36	0.03	0.13
	Cu	0.24	2.39	0.73	0.92
	Ni	0.19	5.28	1.45	1.81
	Pb	0.01	0.63	0.08	0.13
	Se	0.01	1.61	0.13	0.68
Falconbridge (n=13)	As	0.01	0.14	0.02	0.13
	Co	0.02	0.21	0.07	0.11
	Cu	0.18	1.46	0.57	0.75
	Ni	0.46	2.96	1.6	2.04
	Pb	0.01	0.07	0.03	0.04
	Se	0.01	0.04	0.01	0.03
Sudbury Centre (n=61)	As	0.01	0.07	0.01	0.02
	Co	0.01	0.11	0.02	0.03
	Cu	0.15	2.54	0.64	0.75
	Ni	0.04	4.32	0.61	0.75
	Pb	0.01	0.73	0.07	0.09
	Se	0.01	0.21	0.02	0.06
Hanmer (n=8)	As	0.01	0.02	0.01	0.01
	Co	0.01	0.01	0.01	0.01
	Cu	0.16	0.73	0.34	0.46
	Ni	0.03	0.42	0.18	0.28
	Pb	0.01	0.13	0.03	0.09
	Se	0.01	0.01	0.01	0.01

n, number of samples analyzed.
95% UCLM, 95% upper confidence level on the arithmetic mean.
Minimum Detection Limit (µg/g dry wt.): As=0.2, Co=0.2, Cu=0.5, Pb=0.5, Ni=0.5, Se=0.2.

Table 6.19 **Summary of COC Concentrations (µg/g wet wt.) in Below-Ground Vegetables by Community**

COI	COC	Min	Max	Mean (arithmetic)	95% UCLM
Coniston (n=18)	As	0.01	0.02	0.02	0.02
	Co	0.01	0.07	0.02	0.02
	Cu	0.22	1.36	0.67	0.81
	Ni	0.12	1.22	0.43	0.56
	Pb	0.02	0.62	0.13	0.26
	Se	0.01	0.06	0.02	0.03
Copper Cliff (n=15)	As	0.01	0.04	0.02	0.02
	Co	0.01	0.04	0.01	0.02
	Cu	0.14	2.42	0.95	1.23
	Ni	0.24	2.51	1.38	1.69
	Pb	0.02	0.27	0.09	0.13
	Se	0.01	1.68	0.2	0.42
Falconbridge (n=6)	As	0.01	0.07	0.04	0.06
	Co	0.03	0.16	0.07	0.13
	Cu	0.46	1.40	0.90	1.18
	Ni	1.13	4.93	2.13	3.73
	Pb	0.02	0.28	0.14	0.23
	Se	0.01	0.02	0.01	0.02
Sudbury Centre (n=25)	As	0.01	0.03	0.02	0.02
	Co	0.01	0.02	0.01	0.02
	Cu	0.28	2.38	0.97	1.14
	Ni	0.02	1.69	0.55	0.79
	Pb	0.01	0.11	0.05	0.08
	Se	0.01	0.09	0.02	0.04
Hanmer[a] (n=2)	As	0.01	0.10	0.05	0.10
	Co	0.01	0.10	0.06	0.10
	Cu	0.87	1.09	0.98	1.09
	Ni	0.24	0.31	0.28	0.31
	Pb	0.04	0.25	0.14	0.25
	Se	0.03	0.10	0.07	0.10

n, number of samples analyzed.

95% UCLM, 95% upper confidence level on the arithmetic mean.

[a] 95% UCLM values were unable to be calculated due to the small sample size. Maximum values were used as surrogates for 95% ULCM value.

Minimum Detection Limit (µg/g dry wt.): As=0.2, Co=0.2, Cu=0.5, Pb=0.5, Ni=0.5, Se=0.2.

6.9 Levels of COC in Local Fish and Livestock

A key data gap identified during the problem formulation step of the HHRA was the lack of COC concentration data for local fish and livestock from the study area. To address these issues, a local fish survey and a local livestock survey were conducted to provide data for the HHRA. The following two sections provide a summary of each survey and the results utilized in the assessment.

6.9.1 Local Fish Survey

The local fish survey was intended to obtain site-specific data on the range of metal concentrations found in a variety of fish species typically caught and consumed by anglers in the study area. A total of eight lakes were selected and sampled by the Freshwater Co-op Unit at Laurentian University in Sudbury. The names of the lakes were: Ashigami, Crooked, Long, Massey, McFarlane, Ramsey, Vermillion and Whitson. These specific lakes were chosen based upon proximity to the smelters and urban populations and all of the eight lakes were known to have a moderate amount of recreational fishing activity. Some of the lakes are in close proximity to the smelters and represent lakes with elevated metal concentrations. Analytical results for three species of fish, lake herring (*Coregonus artedii*), yellow perch (*Perca flavescens*), and walleye (*Stizostedion vitreum*) are discussed here as they represent the species most likely to be eaten by anglers.

Sampling was conducted between 2 July and 30 October 2003. A total of 211 fish muscle tissue samples were submitted for metal analysis but only 145 of these represented fish species typically consumed by humans. Summary statistics for these species are provided in Table 6.20. For the purpose of the HHRA, mean and 95% UCLM values were used to represent the concentration of each COC within fish tissues consumed by Sudbury residents.

Table 6.20   **COC Concentrations in Fish Tissues (µg/g wet wt.)**

COC	Min	Max	Mean (arithmetic)	95% UCLM
As	0.01	0.66	0.07	0.11
Co	0.01	0.14	0.02	0.02
Cu	0.08	4.98	0.36	0.52
Pb	0.01	2.60	0.23	0.30
Ni	0.01	0.20	0.02	0.03
Se	0.30	4.47	1.64	1.96

95% UCLM, 95% upper confidence level on the arithmetic mean.
Note: The 95% ULCM values are based on $n = 145$.

6.9.2 Livestock

Tissues of 10 beef cattle raised in the Sudbury area were sampled to obtain site-specific data on the range of metal concentrations found in local beef cattle. These animals were raised and consumed locally, thereby comprising a portion of the dietary intake of Sudbury residents. The animals ranged in age from 9 months to 2 years. A 10 g sample was collected from each animal under the direction of Dr Glenn Parker, Laurentian University. Samples of kidney, liver and muscle were collected using stainless steel cutting instruments. A 1.0 to 1.7 g portion of each sample was first chopped and then blended. Samples were then prepared by microwave digestion. All collected samples were analyzed for metals and metalloids using ICP-MS. Results of analyses for the COC are presented in Table 6.21. In cases where concentrations were below laboratory detection, one-half of the detectable limit was substituted as the concentration for those samples for statistical purposes. Metal concentrations varied among tissues. For example, the concentration of Cu was markedly higher in liver followed by kidney, then muscle. In contrast, levels were generally higher in kidney tissue for As, Pb, Ni and Se. The levels of all elements were generally lowest in muscle, which represents the most significant tissue from a human consumption perspective.

Table 6.21 **Concentrations of COC in Tissue Samples (μg/g wet wt.) from Beef Cattle**

Statistics	As	Co	Cu	Pb	Ni	Se
Kidney (*n*=6)						
Mean	0.06	0.02	3.52	0.03	0.07	1.47
95% UCLM	0.07	0.03	4.01	0.04	0.09	1.71
Min	0.05	0.01	2.94	0.03	0.04	1.15
Max	0.08	0.04	4.42	0.05	0.11	1.89
Liver (*n*=8)						
Mean	0.04	0.08	43.2	0.02	0.04	0.26
95% UCLM	0.05	0.17	50.9	0.02	0.05	0.34
Min	0.01	0.03	24.0	0.01	0.01	0.15
Max	0.06	0.37	55.9	0.03	0.07	0.43
Muscle (*n*=10)						
Mean	0.04	0.01	1.42	0.01	0.06	0.17
95% UCLM	0.06	0.01	1.84	0.01	0.14	0.22
Min	0.00	0.00	0.53	0.01	0.01	0.06
Max	0.12	0.02	2.09	0.01	0.44	0.35

95% UCLM, 95% upper confidence level on the arithmetic mean.

6.10 Falconbridge Urinary Arsenic Study

6.10.1 Background to the Study

In response to community concerns over elevated levels of As in soil on some residential properties within the community of Falconbridge, Xstrata Nickel (then Falconbridge Ltd.) commissioned the risk assessment team to conduct an As exposure study. While not directly part of the Sudbury Soils Study, the results of this study provided a unique dataset for use in the HHRA. This section provides an overview of that study which was also reported by Do et al. (2011).

In the spring of 2003, the local Medical Officer of Health (MOH) for the Sudbury and District Health Unit (SDHU) notified the residents of the town of Falconbridge that elevated concentrations of As were present in some residential soils. Based on a sense of precaution, the MOH recommended that the town residents take precautions when working in gardens, handling of vegetables and other outdoor activities where people may come in contact with soil. As a result of these warnings, the residents became concerned about the health implications, particularly for young children, and for property values. Representatives from Xstrata Nickel and the research team worked with the Falconbridge Citizens' Committee to design a study that addressed their concerns, and provide results that would be useful to the community. The research team developed a study methodology to address two specific questions that were deemed to be most important by residents:

1. Do Falconbridge residents have higher urinary As levels than residents living in a comparison area with lower levels of As in their soil?
2. What health risks relative to other communities are associated with the urinary As levels of Falconbridge residents?

6.10.2 Arsenic Metabolism

Arsenic is a naturally occurring element in the earth's crust, and is found throughout the environment. It is usually found combined with other elements such as oxygen, chlorine, and sulfur. When As is combined with carbon containing compounds, it is usually referred to as organic arsenic. However, when combined with other elements, it is referred to as inorganic As. Inorganic forms (As III and As V species) are bioavailable and toxicologically significant. Inorganic As occurs in groundwater and foods and these forms are absorbed readily, in contrast to As in soil, whose absorption may vary considerably depending on many factors.

Absorbed inorganic arsenic is distributed throughout the body, excreted into sweat, skin, nails and urine. Absorbed As is cleared from the blood very quickly. The Agency for Toxic Substances and Disease Registry of the US (ATSDR) reports that the half-life of urinary As is 1–3 days and 24 h in blood. Hence, any evaluation of blood or urine reflects current exposure. Arsenic is classified as a human carcinogen by the International Agency for Research on Cancer (IARC), and the U.S. EPA. These classifications reflect the human and animal research evidence indicating that As should be treated as a human carcinogen for regulatory purposes.

6.10.3 Communities

The town of Falconbridge was the primary community of interest, while the nearby town of Hanmer was selected as the reference community. The community of Falconbridge is located approximately 20 km northeast of the Sudbury city core and was also a primary COI for the HHRA. Falconbridge was comprised of approximately 250 households with 700 residents. Situated on the eastern perimeter of the community are the smelting operations of Xstrata Nickel. Although the current population of the community has a variety of employment characteristics, the community initially developed as a residential site for mine and smelter employees and their families. The comparison community of Hanmer is located approximately 15 km northwest of Falconbridge. The municipality of Hanmer has a population of approximately 8000 residents with similar demographics as Falconbridge. The geology and physiography are also similar between the two towns, but there is no smelter or mining activities in Hanmer. The average soil As levels in Falconbridge were approximately 21 times higher than those measured in Hanmer. The arithmetic mean for As in soil in Falconbridge was 78.5 µg/g (range 2.5–620 µg/g) while in Hanmer, the average As soil level was 3.7 µg/g (range 2.5–25 µg/g).

6.10.4 Study Approach

Discussions with the residents of Falconbridge began in the summer and fall of 2003 and continued as the study was designed during the winter of 2004. Sampling took place in September and early October 2004 after the period of summer exposure to uncovered soils. All Falconbridge residents were invited to participate in the study. The research team also randomly recruited a similar number of families from the comparison community of Hanmer to participate in the study. Particular effort was directed at getting participation from children and toddlers. Generally, toddlers and children are considered most likely to experience exposure to chemicals from unsodded soil during warm months. Seasonal variations in exposure to metals in soil have been clearly shown for Pb, for example. Typically, soil exposures are higher for children than adults because of frequent direct contact with soils during play and other activities.

To address the primary questions, the research team developed a methodology that combined both the analysis of first morning void urine samples, and interviews that captured lifestyle information pertaining to potential As exposure. The study was comparative in nature, meaning that the main questions above were addressed by comparing Falconbridge with a similar community with lower soil As concentrations. The data collection process employed in Hanmer was identical to the one used in Falconbridge. Initially, potential participants were sent a letter indicating that a member of the study team would visit their house to provide a sample of the consent form, and to explain the study process. If they were willing to participate in the study, an appointment was scheduled. At the appointment time, the study team walked the participants through the consent/assent forms in detail, had the participants sign them, and then conducted an in-home interview with the adults of the household. At the conclusion of the interview, each family was left a urine sampling kit with instructions. The study team then picked up the sample the following morning. Sample collection and interviewing occurred between early September and mid-October 2004. Samples were processed and shipped to London Health Sciences Trace Elements Laboratory at the University of Western Ontario. All samples were analyzed for creatinine, total As and inorganic As and its major metabolites (i.e. monomethylarsonic acid (MMA) and dimethylarsinic acid (DMA)).

Screening As concentrations were determined before sampling to help identify elevated As levels in urine. The *a priori* levels chosen for 'normal' ranges were 0–20 µg/L for inorganic As and 0–100 µg/L for total As when adjusted for creatinine levels. These ranges did not refer to health effects, but rather were population distribution levels where 95% of a population would likely fall within these levels according to previous Canadian studies. Any participant with results outside the *a priori* 'normal range' for either inorganic or total As was contacted by the team physician and re-sampled. In addition, if the study team had a consent form allowing release of information to the family physician, the physician was also notified. For those participants who had an As level in the 'normal range', a letter was sent to the resident.

6.10.5 Participation Rate

A total of 273 households in the Town of Falconbridge were invited to participate in the study, of which 148 (54%) agreed. Overall, information was collected for 393 participants in the interview portion of the study and, of these, 369 participants provided a urine sample (Table 6.22). In Hanmer, 129 (36%) out of the 360 households approached agreed to participate in the study. Interviews captured information on 335 respondents and 321 participants provided urine samples. Approximately 70% of the participants were adults 18 years or older. Samples were also obtained from 195 children in three different age categories with very similar representation of age groups between the two communities.

Table 6.22 **Summary of Participants in the Urinary As Exposure Study**

Community	# Participating households	# Samples	Adults (>18 years)	Children (0–5 years)	Children (6–12 years)	Children (13–17 years)
Falconbridge	148	369	268 (73%)	18 (5%)	53 (14%)	29 (8%)
Hanmer	129	321	226 (70%)	17 (5%)	61 (19%)	17 (5%)
Total	277	690	494 (72%)	35 (5%)	114 (16%)	46 (7%)

6.10.6 Study Results

Results of the study indicated that Falconbridge residents' urinary As levels were very similar to those in the comparison community of Hanmer. With respect to inorganic As, the type of As most closely associated with health effects, the average levels in each community were nearly identical (Table 6.23). Falconbridge residents had a mean level of 7.1 μg/L and a median level of 6.0 μg/L in comparison with Hanmer residents who had a mean level of 7.2 μg/L and a median level of 6.0 μg/L. Approximately 80% of the urine samples in each community had an inorganic As level below 10 μg/L, and approximately 2% to 3% of samples in each community were at or above 20 μg/L. It was noted that urinary As generally decreased with increasing age, with children aged 6–12 years having the highest average values compared to the other age groups. In addition, males tended to have higher As levels, and fish consumption within the past 7 days was associated with higher As levels.

Total As concentrations (both organic and inorganic forms) were also similar between the two communities. The median level among Falconbridge residents was 8.9 μg/L compared to 9.7 μg/L for Hanmer residents (Table 6.23). Although the mean levels were higher, 21.2 μg/L for Falconbridge residents compared to 14.1 μg/L for Hanmer residents, there were two extreme outliers (of approximately 600 μg/L and 900 μg/L) measured in the Falconbridge community that strongly impacted the mean value, but limited impact on the median as a measure of central tendency. The distribution was positively skewed with over 80% of the samples having As levels below 20 μg/L. Approximately 2% to 3% of samples in each community were at or above 100 μg/L. Statistical comparisons (non-parametric Mann–Whitney U-test) that were less influenced by extreme outliers indicated that there was no statistically significant difference between the two communities. Similar to the inorganic levels, the total levels generally decreased with age, with children aged 6–12 years having the highest average values compared to the other age groups.

In summary, urinary As levels were almost identical between the two communities despite soil As concentrations being 20 times higher in Falconbridge compared to the reference community. Results of the survey also indicated that As intakes for Falconbridge and Hanmer residents on average were within the typical daily intake of As by Canadians, and therefore, residents were not at any increased risk from As exposure compared to other Canadians in general. The median levels in Falconbridge were within the lower portion of the range estimated for typical daily intake of arsenic by Canadians (Health Canada). Results of the survey were incorporated in the weight-of-evidence approach used to characterize overall health risks related to exposures of Sudbury residents to environmental concentrations of As.

Table 6.23 **Summary of Inorganic and Total Urinary As Concentrations (µg/L)**

	Mean	Minimum	Maximum	SD	Sample size
Inorganic As					
Falconbridge	7.11	1.5	32.9	4.5	369
Hanmer	7.19	1.5	67.4	5.6	321
Total	7.14	1.5	67.4	5.1	690
Total As					
Falconbridge	21.2	2.2	904	65.5	369
Hanmer	14.1	3.0	236	19.8	321
Total	17.9	2.2	904	65.5	690

6.11 Acknowledgements

Many people were involved with the various studies to collect Sudbury-specific exposure data that helped make the HHRA successful. The efforts and dedication of Monika Greenfield (formerly with RWDI) for designing and implementing the air monitoring study were greatly appreciated. Many staff and summer students participated in the vegetable garden study, indoor dust and potable water collection including Mary Kate Gilbertson, Devon Stanbury, Allison Merla, Robert Price and Janice Linquist. Fish for tissue analysis were collected by the Freshwater Co-op Unit of Laurentian University under the supervision of Dr John Gunn and Mr George Morgan. The logistical and analytical support of Testmark Laboratories in Garson was very helpful and greatly appreciated throughout the study. Soil and dust samples were also processed at the Stanford Synchrotron Radiation Laboratory (SSRL) at Stanford University, California, and at the National Synchrotron Light Source (NSLS) at Brookhaven National Laboratory, New York. The assistance of Mr Christopher Hamilton at SGS Lakefield Laboratories, Ontario, for analytical and speciation work was extremely helpful, as was the expertise and insight of Dr Ford and Dr Bruce Conard of Vale with respect to Ni speciation issues. The urinary As exposure study was expertly designed and supervised by Drs M T Do, L F Smith, and C L Pinsent, with field support from Gina Musca, Adam Safruk, Suzanne Goldacker and Josephine Archibold. Finally, we would like to express our deep appreciation to the many residents of Sudbury, Falconbridge and Hanmer who invited us into their homes for interviews, provided samples for analysis, and expressed interest in the study.

6.12 References

ASTM (2004) ASTM standard practice for collection of floor dust for chemical analysis (Designation: D 5438-00). ASTM International, West Conshohocken, PA

Caruso JA, Montes-Bayon M (2003) Elemental speciation studies – new directions for trace metal analysis. Ecotoxicol Environ Saf 56:148–163

Caruso JA, Klaue B, Michalke B, Rocke DM (2003) Group assessment: elemental speciation. Ecotoxicol Environ Saf 56:32–44

Colt JS, Zihm SH, Camann DE, Hart JA (1998) Comparison of pesticides and other components in carpet dust samples collected from used vacuum cleaner bags and from a high volume surface sampler. Environ Health Perspect 106:721–724

Defra and The Environment Agency (2002) Contamination in soil: Collation of toxicological data and intake values for humans. R&D Publication CLR 9. Environment Agency, Bristol

DEPA (2003) Danish Environmental Protection Agency. Technology programme for soil and groundwater contamination. Human bioaccessibility of heavy metals and PAH from soil. Environmental Project No. 840 2003. www.mst.dk/homepage/. Accessed 15, October 2007

Dieter MP, Matthews HB, Jeffcoat RA, Moseman RF (1993) Comparison of lead bioavailability in F344 rats fed lead acetate, lead oxide, lead sulfide, or lead ore concentrate from Skagway, Alaska. J Toxicol Environ Health 39:79–93

Do M, Smith L, Pinsent C (2011) Urinary inorganic arsenic in residents living in close proximity to a nickel and copper smelter in Ontario. Can J Public Health (submitted)

Drexler JW, Brattin WJ (2007) An in vitro procedure for estimation of lead relative bioavailability: with validation. Hum Ecol Risk Assess 13:383–401

EAUK (2005a) Bioaccessibility testing. Questions and answers. Environment Agency UK, January 2005. Available at: www.environmentagency.gov.uk/commondata/acrobat/bioaccessibility_faq_1284070.pdf. Accessed 8 February 2007

EAUK (2005b) Environment Agency's science update on the use of bioaccessibility testing in risk assessment of land contamination. February, 2005. Environment Agency UK. Available at: www.environmentagency.gov.uk/commondata/acrobat/science_update_1284046.pdf. Accessed 8 February 2007

Fernádez Espinosa AJ, Rodríguez MT, de la Rosa JB, Sánchez JCJ (2002) A chemical speciation of trace metals for fine urban particles. Atmos Environ 36:773–780

Freeman GB, Dill JA, Johnson JD, Kurtx PJ, Parham F, Matthews HB (1996) Comparative absorption of lead from contaminated soil and lead salts by weanling Fisher 344 rats. Fundam Appl Toxicol 33:109–119.

Gaillard JF, Webb SM, Quintana JPG (2001) Quick X-ray absorption spectroscopy for determining metal speciation in environmental samples. J Synchrotron Radiat 8:928–930

Galbreath KC, Toman DL, Zygarlicke CJ, Huggins FE, Huffman GP, Wong JL (2000) Nickel speciation of residual oil fly ash and ambient particulate matter using X-ray absorption spectroscopy. J Air Waste Manage Assoc 50(11):1876–1886

Gasser UG, Walker WJ, Dahlgren RA, Borch RS (1996) Lead release from smelter and mine waste impacted materials under simulated gastric conditions and relation to speciation. Environ Sci Technol 30(3):761–769

Golder (2006) Bioaccessibility testing of soil and house dust samples from the Sudbury soils study. Final Report. Golder Associates Ltd. 04-1112-069. June 2006. 412 pp

Golder (2007) Bioaccessibility testing of soil and house dust samples from the Sudbury soils study. Technical Memorandum. Report. Golder Associates Ltd. 04-1112-069 (6000). February 2007. 101 pp

Guyton AC (1991) Textbook of medical physiology. WB Saunders, Philadelphia, PA, USA

Hack A, Selenka F (1996) Mobilization of PAH and PCB from contaminated soil using a digestive tract model. Toxicol Lett 88:199–210

Hamel SC, Buckley FL, Lioy PJ (1998) Bioaccessibility of metals in soils for different liquid to solid ratios in synthetic gastric fluid. Environ Sci Technol 32:358–362

Hamel SC, Ellickson KM, Lioy PJ (1999) The estimation of the bioaccessibility of heavy metals in soils using artificial biofluids by two novel methods: Mass-balance and soil recapture. Sci Total Environ 243/244:273–283

Hinwood AL, Sim MR, Jolley D, et al. (2004) Exposure to inorganic arsenic in soil increases urinary inorganic arsenic concentrations of residents living in old mining areas. Environ Geochem Health 26:27–36

Hrudey SE, Chen W, Rousseaux CJ (1996) Exposure routes and bioavailability factors for selected contaminants. I. Arsenic and III. Chromium and chromium compounds. In: Hrudey SE (ed) Bioavailability in environmental risk assessment. CRC Press, Boca Raton, FL

Huggins FE, Huffman GP, Linak WP, Miller CA (2004) Quantifying hazardous species in particulate matter derived from fossil-fuel combustion. Environ Sci Technol 38(6):1836–1842

Hysong TA, Burgess JL, Cebrian Garcia ME, O'Rourke MK (2003) House dust and inorganic urinary arsenic in two Arizona mining towns. J Expo Anal Environ Epidemiol 13:211–218

Jambor JL, Blowes DW (1998) Theory and applications of mineralogy in environmental studies of sulfide-bearing mine wastes. In: Cabri LJ, Vaughan DJ (eds) Modern approaches to ore and environmental mineralogy. Mineralogical Association of Canada Short Course Series Volume 27. Mineralogical Association of Canada, Ottawa, Ontario

Kelley ME, Brauning SE, Schoof RA, Ruby MV (2002) Assessing oral bioavailability of metals in soil. ISBN 1-57477-123-X. Battelle Memorial Institute Press, Columbus, OH

Lanphear BP, Matte TD, Rogers J, et al. (1998) The contribution of lead-contaminated house dust and residential soil to children's blood lead levels – a pooled analysis of 12 epidemiological studies. Environ Res 79(1):528–532

LEGS (2007) Sudbury soils study data package. Laboratory for Environmental and Geological Studies (LEGS) University of Colorado. Unpublished report to the SARA Goup. Feb 12

Lioy PJ, Freeman NC, Wainman T, et al. (1992) Microenvironmental analysis of residential exposure to chromium-laden wastes in and around New Jersey homes. Risk Anal 2(2):287–299

Michalke B (2003) Element speciation definitions, analytical methodology, and some examples. Ecotoxicol Environ Saf 56:122–139

MOE (1999). Air Quality in Ontario 1999 Appendix. Available at: www.ene.gov.on.ca/envision/techdocs/4159eappendix.pdf. Accessed 15 June 2006

MOE (2000). Air Quality in Ontario 2000 Appendix. Available at: www.ene.gov.on.ca/envision/techdocs/4226e_appendix.pdf. Accessed 15 June 2006

MOE (2001). Air Quality in Ontario 2001 Appendix. Available at: www.ene.gov.on.ca/envision/techdocs/4521e_appendix.pdf. Accessed 15 June 2006

MOE (2002). Air Quality in Ontario 2002 Appendix. Available at: www.ene.gov.on.ca/envision/techdocs/4521e01_appendix.pdf. Accessed 15 June 2006

Morawska L, Salthammer T (2004) Introduction to sampling and measurement techniques. In: Morawska L, Salthammer T (eds) Indoor environment – Airborne particles and settled dust. Wiley-VCH

Mushak P (1991) Gastro-intestinal absorption of lead in children and adults: Overview of biological and biophysico-chemical aspects. Chem Species Bioavailab 3(3–4):87–104

Oomen AG, Sips AJA, Groten JP, Sijm DTH, Tolls J (2000) Mobilization of PCBs and lindane from soil during in vitro digestion and their distribution among bile slat micelles and proteins of human digestive fluid and the soil. Environ Sci Technol 34(2): 297–303

Oomen AG, Hack A, Minekus M, et al. (2002) Comparison of five in vitro digestion models to study the bioaccessibility of soil contaminants. Environ Sci Technol 36(15):3326–3334

Oomen AG, Brandon EFA, Swartjes FA, Sips AJA (2006) How can information on oral bioavailability improve human health risk assessment for lead-contaminated soils? Implementation and scientific basis. RIVM report 711701042/2006. RIVM, Bilthoven, The Netherlands

Paustenbach DJ (2000) The practice of exposure assessment: a state-of-the-art review. J Toxicol Environ Health Part B 3:179–291

Peijnenburg WJGM, Jager T (2003) Monitoring approaches to assess bioaccessibility and bioavailability of metals: Matrix issues. Ecotoxicol Environ Saf 56:63–77

Profumo A, Spini G, Cucca L, Pesavento M (2003) Determination of inorganic nickel compounds in the particulate matter of emissions and workplace air by selective sequential dissolutions. Talanta 61:465–472

Rasmussen PE (2004) Can metal concentrations in indoor dust be predicted from soil geochemistry? Can J Anal Sci Spectrosc 49(2):166–174

Richards PA (2002) Hydrogeology of the Sudbury area. In: Rousell DH, Jansons KJ (eds) The physical environment of the City of Greater Sudbury. Ontario Geological Survey, Special Volume 6, pp. 103–126. Ontario Geological Survey, Greater Sudbury, Ontario

Rodriguez RR, Basta NT (1999) An in vitro gastrointestinal method to estimate bioavailable arsenic in contaminated soils and solid media. Environ Sci Technol 33:642–649

Ruby MV (2004) Bioavailability of soil-borne chemicals: abiotic assessments. Hum Ecol Risk Assess 10:647–656

Ruby MV, Davis A, Kempton JH, Drexler JW, Bergstrom PD (1992) Lead bioavailability: Dissolution kinetics under simulated gastric conditions. Environ Sci Technol 26(6):1242–1248

Ruby MV, Davis A, Link TE, et al. (1993) Development of an in vitro screening test to evaluate the in vivo bioaccessibility of ingested mine-waste lead. Environ Sci Technol 27(13):2870–2877

Ruby MV, Dvais A, Schoof R, Eberle S, Sellstone CM (1996) Estimation of lead and arsenic bioavailability using a physiologically based extraction test. Environ Sci Technol 30(2):422–430

Ruby MV, Schoof R, Brattin W, et al. (1999) Advances in evaluating the oral bioavailability of inorganics in soil for use in human health risk assessment. Environ Sci Technol 33(21):3697–3705

Rutz E, Valentine J, Eckart R, Yu A (1997) Pilot study to determine levels of contamination in indoor dust resulting from contamination of soils. J Soil Contam 6(5):525–536

SGS (2005) A combined mineralogical and analytical study of speciation of chemicals of concern (COC) in soils, dusts and air filters. Prepared for the SARA Group. SGS Lakefield Research Ltd. LR 11007-001 – MI5001-AUG05. 18 August 2005

SGS (2006) A mineralogical study of speciation of Ni in two dust- and five air-filter samples: SARA Project. Prepared for the SARA Group. SGS Lakefield Research Ltd. LR 11060-004 – MI5001-MAR06. 17 March 2006

Sheppard SC, Evenden WG, Schwartx WJ (1995) Heavy metals in the environment. Ingested soil: bioavailability of sorbed lead, cadmium, cesium, iodine, and mercury. J Environ Qual 24:498–505

Sips AJA, Bruil MA, Dobbe CJB, et al. (2001) Bioaccessibility of contaminants from ingested soil in humans. Method development and research on the bioaccessibility of lead and benzo[a] pyrene. RIVM report 711701012/2001. RIVM, Bilthoven, The Netherlands

Solomon D, Lehmann D, Martinez CE (2003) Sulphur K-edge XANES spectroscopy as a tool for understanding sulfur dynamics in soil organic matter. Soil Sci Soc Am J 67:1721–1731

Stanley CJ, Laflamme JHG (1998) Preparation of specimens for advanced ore-mineral and environmental studies. In: Cabri LJ, Vaughan DJ (eds.) Modern approaches to ore and environmental mineralogy. Mineralogical Association of Canada Short Course Series Volume 27. Mineralogical Association of Canada, Ottawa, Ontario

Stern EA, Heald SM (1983) *In* Handbook on synchrotron Radiation, ed. E.E. Koch (North-Holland, Amsterdam, Vol. 1, chap. 10.

Tessier A, Campbell PGC (1988) Comments on the testing of the accuracy of an extraction procedure for determining the partitioning of trace metals in sediments. Anal Chem 60:1475–1476

Tessier A, Campbell PGC, Bisson M (1979) Sequential extraction procedure for the speciation of particulate trace metals. Anal Chem 51:844–851

Tipping E, Hetherington NB, Hilton J, Thompson DW, Howles E, Hamilton-Taylor J (1985) Artifacts in the use of selective chemical extraction to determine distributions of metals between oxides of manganese and iron. Anal Chem 57:1944–1946

U.S. EPA (1997) Exposure factors handbook. Volume II – Food ingestion factors. Office of Research and Development. EPA/600/P-95/002Fa. U.S. EPA, Washington, DC

U.S. EPA Region VIII (2001) Baseline human health risk assessment Vasquez Boulevard and I-70 superfund site, Denver, Co. U.S. Environmental Protection Agency, Region VIII. August 2001. Available at: www.epa.gov/region8/superfund/sites/VB-I70-Risk.pdf. Accessed: 20 Aug 2005

U.S. EPA (2007) Guidance for evaluating the bioavailability of metals in soils for use in human health risk assessment. Available at: www.cfpub.epa.gov/si/osp_sciencedisplay.cfm?dirEntryID=143311&ActType=project&kwords=Human%2520Health. Accessed 8 February 2007

Vincent JH, Ramachandran G, Kerr SM (2001) Particle size and chemical species 'fingerprinting' of aerosols in primary nickel production industry workplaces. J Environ Monit 3:565–574

Yu CH, Yiin LM, Lioy PJ (2006) The bioaccessibility of lead (Pb) from vacuumed house dust on carpets in urban residences. Risk Anal 26(1):125–134

Zhipeng B, Lih-Ming Y, Rich DQ, et al. (2003) Field evaluation and comparison of five methods of sampling lead dust on carpets. AIHA J 64:528–532

6.13 Appendix Chapter 6: Abbreviations

AAQC, ambient air quality criterion

ATSDR, Agency for Toxic Substances and Disease Registry of the US

BEI, backscattered electron image

BSE, backscattered electron

COC, chemical of concern

COI, community of interest

CRL, cancer risk level

CRM, certified reference material

CSF, cancer slope factor
DMA, dimethylarsinic acid
EDS, energy dispersive X-ray spectrometry
EMPA, electron microprobe analysis
EPC, exposure point concentration
ERA, ecological risk assessment
GSA, Greater Sudbury Area
HHRA, human health risk assessment
HVS3, high volume surface vacuum sampler
IARC, International Agency for Research on Cancer
IVBA, *in-vitro* bioaccessibility
LEGS, Laboratory for Environmental and Geological Studies
MDLs, minimum detection limits
MMA, monomethylarsonic acid
MOH, Medical Officer of Health
MOE, Ministry of the Environment (Ontario)
NAPS, National Air Pollution Surveillance Program
NSLS, National Synchrotron Light Sources at Brookhaven National Laboratories, Upton, NY, USA
PM, particulate matter
RAF, relative absorption factor
RBA, relative *in-vivo* bioavailability
RfD, reference dose
RME, reasonable maximum estimate
SBRC, Solubility/Bioavailability Research Consortium
SDHU, Sudbury and District Health Unit
SEM, scanning electron microscope
SOP, standard operating procedure
SQGhh, human health-based soil quality guidelines
SRC, Synchrotron Radiation Centre
TC, technical committee
TRV, toxicity reference value
TSP, total suspended particulate matter less than 44 microns in diameter
UCLM, upper confidence level on the arithmetic mean
USEPA, United States Environmental Protection Agency
XANES, X-ray absorption near-edge structure spectroscopy

DOI: 10.5645/b.1.7

7.0 Detailed Human Health Risk Assessment Methods

Authors

Glenn Ferguson, PhD
Intrinsik Environmental Sciences Inc.
6605 Hurontario Street, Suite 500, Mississauga, ON Canada L5T 0A3
Email: GFerguson@intrinsikscience.com

Elliot Sigal
Intrinsik Environmental Sciences Inc.
6605 Hurontario Street, Suite 500, Mississauga, ON Canada L5T 0A3
Email: Esigal@intrinsikscience.com

Chris Bacigalupo
Intrinsik Environmental Sciences Inc.
6605 Hurontario Street, Suite 500, Mississauga, ON Canada L5T 0A3
Email: Cbacigalupo@instrinsikscience.com

Table of Contents

7.1 Exposure Assessment 159
   7.1.1 Sudbury-Specific Media Concentrations and Exposure Estimates 159
   7.1.2 Background Exposure Assessment 167
   7.1.3 Market Basket Estimated Daily Intakes 169
   7.1.4 Summary of Exposure Data Used in the HHRA 174
7.2 Hazard Assessment 176
   7.2.1 Overview 176
   7.2.2 Exposure Limits Selected for the HHRA 178
   7.2.3 Toxicological Profiles for the COC 180
   7.2.4 Bioavailability/Bioaccessibility 190
7.3 Risk Characterization 191
   7.3.1 Overview 191
   7.3.2 Hazard Quotients 192
   7.3.3 Cancer Risk Level 193
7.4 Risk Management Recommendations 193
7.5 References 194
7.6 Appendix Chapter 7: Abbreviations 204

Tables

7.1 Summary of Surface Soil Concentrations (µg/g) by Community of Interest 160
7.2 Drinking Water Concentrations (µg/L) by Community of Interest 163
7.3 Local Fish Consumption Rate 166

7.4 Local Wild Game Consumption Rates ... 167
7.5 Typical Ontario Ambient Air Concentrations (µg/m³) (Toronto) ... 168
7.6 Typical Ontario Drinking Water Concentrations (µg/L) (MOE 2005) ... 168
7.7 Databases Selected for Use in the Development of the EDI_{MB} ... 170
7.8 Fraction of Inorganic As in Various Food Groups ... 171
7.9 Summary Values for All Exposure Point Concentrations (EPCs) Used in the HHRA ... 174
7.10 Summary of Toxicological Criteria Used in the Sudbury HHRA ... 179

The detailed human health risk assessment (HHRA) was conducted using the data collected from Phase 2 and followed the next three major steps of the HHRA framework: i) exposure assessment, ii) hazard assessment and, iii) risk characterization. The problem formulation step (Phase 1) was discussed in Chapter 5, while the additional sampling and analytical work conducted to fill identified data gaps were outlined in Chapter 6. This chapter describes the remaining three steps of the HHRA framework including the exposure assessment, hazard assessment and risk characterization.

7.1 Exposure Assessment

The exposure assessment evaluated data related to all chemicals, receptors and exposure pathways identified during the problem formulation using a multimedia approach. The multimedia approach took into account all potential exposure to the chemicals of concern (COC) from the different sources or media (i.e. soil, air, dust, water, food, etc.) which typical Sudbury residents could come in contact with as part of their daily activities. The primary objective of the exposure assessment was to predict, using a series of conservative assumptions, the rate of exposure (i.e. the quantity of chemical and the rate at which that quantity is received) of the selected receptors to the COC via the various exposure scenarios and pathways. The rate of exposure to chemicals from the various pathways was usually expressed as the amount of chemical taken in per unit body weight per unit time (e.g. μg chemical/kg body weight/day).

The degree of exposure of receptors to chemicals in the environment depends on the interactions of a number of parameters, including: the concentrations of chemicals in various environmental media; the physical-chemical characteristics of the COC and, the physiological and behavioral characteristics of the receptors (e.g. respiration rate, soil/dust intake, time spent at various activities and in different areas). The rate of exposure to the COC available to residents of the Greater Sudbury Area (GSA) was evaluated through the estimation of an exposure point concentration (EPC) for each media type. Based upon U.S. EPA (2004a) guidance, the 95% upper confidence limit of the mean (i.e. 95% UCLM) was used to estimate the reasonable maximum exposure (RME) point concentration. The arithmetic mean of the dataset was used to estimate the central tendency estimate (CTE) point concentration to represent an average exposure. This value was calculated for each COC using ProUCL software developed by the U.S. EPA (2004b). For the purpose of statistical analyses of the data, any negative concentration or zero value was set equal to half the detection limit. Any value measured at, or below, the detection limit was also set equal to half the detection limit.

For the CTE estimate, human receptor characteristics were defined in such a way as to reflect the central tendency within a given population. Most receptor characteristics (e.g. soil, water and food ingestion rates, etc.) were obtained using the 50th percentile values of the sample distribution. The arithmetic mean was selected to represent parameters such as body weight. Body weight is typically located in the denominator of most exposure calculations and therefore use of the 50th percentile versus the arithmetic mean (in the case of a lognormal distribution) would result in an inflated exposure estimate relative to the arithmetic mean body weight. The RME estimate typically employed the use of upper percentiles (typically 90 to 95th percentile) for most receptor characteristics, with the exception of food intake rates. To avoid unrealistic daily caloric intake diets (i.e. multiple 95th percentile ingestions of various food groups), food intake rates were based upon mean or median values. The receptor and scenario-specific input parameters were provided earlier in Chapter 5.

Exposure estimation was facilitated through the use of an integrated multi-pathway environmental risk assessment model. The model was spreadsheet-based (MS Excel). Models of this type have been used in hundreds of peer-reviewed human health risk assessments, including those conducted for contaminated sites, smelters, refineries, incinerators, landfills and a variety of other industrial facilities. The current version of this model incorporated the latest techniques and procedures for exposure modeling developed by various regulatory agencies (e.g. U.S. EPA, MOE, CCME, Cal/EPA, U.S. EPA Region VI, WHO, etc.) and published academic and scientific literature sources. The model integrated recent statistical and probabilistic techniques, and was capable of conducting complex modeling involving human receptors, and a myriad of exposure pathways.

Human health risks were calculated for individuals living in five Communities of Interest (COI) within the Greater Sudbury Area (GSA). For comparative purposes, a 'Typical Ontario Resident' (TOR) was also evaluated. Individuals were assumed to move in a random fashion within each COI and, over time, come into contact with the exposure point concentration (EPC) of the COC in a variety of environmental media.

7.1.1 Sudbury-Specific Media Concentrations and Exposure Estimates

7.1.1.1 Surface Soil Concentrations

The overall results of the comprehensive soil survey conducted in 2001 have been presented in Chapter 4. This comprehensive database of soil concentrations provided the foundation of the HHRA, and formed the basis of screening and selection of COC for the assessment. For the purposes of the HHRA, the soils database was screened using several criteria to evaluate exposure scenarios and soil concentrations most relevant to the exposure of residents within the study area. Samples taken from the surficial layer of soil at each sampling location were selected as these top layers of soils were most available for human exposure. Initially, more than 2100 surface soil samples were extracted from the database, which included samples for each COC in each COI. The data were also screened for the presence of statistical outliers which would not be representative of the overall dataset. In addition to the 2001 soil survey, several other surveys were conducted to collect Sudbury-specific media concentration data for filling data gaps and for use in the HHRA (see Chapter 6). For the HHRA, residential soil samples collected during the vegetable garden survey (n=55) and the indoor dust survey (n=86) were incorporated into the soil concentration dataset. Examination of the complete soils dataset indicated a notable proportion of As (35%) and Se (49%) levels detected at, or below, their minimum detection limits (MDLs) of 5.0 µg/g and 1.0 µg/g, respectively. The soils data were grouped and summarized according to each COI (Table 7.1). The 95th upper confidence level of the mean (UCLM) was calculated using ProUCL (U.S. EPA 2004a). In general, the data indicated that soil levels in each of the COI with smelters (historically or presently) tended to have higher concentrations than those communities without smelter operations, such as Sudbury Centre and Hanmer.

Table 7.1 **Summary of Surface Soil Concentrations (µg/g) by Community of Interest**

COI	COC	Min	Max	Mean (arithmetic)	95% UCLM	95th Percentile
Coniston (n=203)	As	2.50	55.7	9.48	12.2	29.3
	Co	3.46	66.8	14.8	18.5	40.2
	Cu	8.64	1200	215	315.5	710
	Ni	16.0	1800	290	433	1043
	Pb	2.00	309.8	45.0	52.0	139
	Se	0.25	5.00	1.04	1.31	3.0
Copper Cliff (n=197)	As	2.50	72.0	17.4	19.0	41.5
	Co	6.00	150	30.6	33.4	77.3
	Cu	31.4	5290	1240	1370	3042
	Ni	28.5	3260	886.7	976	2505
	Pb	3.00	582	88.4	98.0	251
	Se	0.50	42.0	6.80	7.51	16.9
Falconbridge (n=188)	As	2.50	400	69.4	78.7	205
	Co	4.47	159	46.0	56.5	106
	Cu	11.0	2900	733.9	1010	1774
	Ni	18.4	3390	780.3	1070	1990
	Pb	2.00	335	73.9	82.3	191
	Se	0.40	10.0	2.53	3.09	5.76
Sudbury Centre (n=597)	As	2.20	59.0	6.00	7.17	17.4
	Co	3.00	100	10.7	11.3	25.4
	Cu	6.20	1640	149.4	204.0	569
	Ni	11.0	3260	165.0	210.1	579
	Pb	1.00	310	26.8	35.9	109
	Se	0.25	12.5	1.07	1.30	3.46
Hanmer (n=80)	As	1.50	22.4	3.68	4.27	6.70
	Co	2.70	11.0	6.16	6.55	10.0
	Cu	9.30	330	42.7	67.0	74.1
	Ni	14.0	272	46.8	67.9	69.2
	Pb	2.00	78.5	12.2	19.2	43.1
	Se	0.25	3.00	0.59	0.68	0.77

n = number of samples.

7.1.1.2 Indoor Dust Concentrations

Sudbury-specific indoor dust concentrations, along with co-located outdoor soil concentrations, were measured as part of the indoor dust survey conducted in 2004. The survey consisted of the collection of indoor dust from 91 residential homes and eight schools. Co-located yard soil was collected concurrently from the front yards of residential homes to determine whether there was a relationship between COC levels in indoor dust and outdoor soil in the GSA. The results of the indoor dust survey were provided in Chapter 6. Initial review of the indoor dust and outdoor soil data indicated that indoor dust levels for each COC were 2.8 to 5.9 times higher than corresponding soil levels. It was also noted that as outdoor soil concentrations increased, soil appeared to become a more significant contributor to indoor dust concentrations of the COC. To most accurately describe this relationship, linear regression equations were developed for each COC to predict indoor dust concentrations as a function of outdoor soil concentrations. These relationships were used to generate dust exposure values for the HHRA.

Comparison with Other Studies

Based on a review of the literature, there seems to be adequate evidence to suggest that a relationship exists between levels of contaminants measured in indoor dust and the levels observed in nearby outdoor soil. However, there were varying degrees of certainty and strength to this relationship. A study by Rasmussen (2004) collected indoor dust and outdoor soil and dust from 48 homes across the city of Ottawa, Ontario. The results of the study indicated that the multi-element composition of indoor dust differed significantly from that of garden soil or street dust. For most elements, levels in household dust exceeded natural background concentrations for the region whereas most concentrations in garden soil and street dust were low in comparison with local background concentrations. The authors found no significant correlations for element concentrations in household dust versus street dust or household dust versus garden soil.

A study by Harrison (1979) found similar indoor/outdoor dust concentration ratios in a study of household and street dust in the Lancaster area of the UK. The mean levels of total metals in household dust were higher than in dust collected from rural roads, and to a lesser degree, urban roads. Hwang et al. (1997) examined the relationship between exterior soil and indoor dust in homes near a historic Cu smelter operation in Montana. Geometric mean As concentrations in five types of soil collected around the exterior of the homes ranged from 121 to 236 µg/g, with a total average soil concentration of 192 µg/g. An average As level of 75 µg/g was reported for indoor dust and there was a significant correlation between As concentration in indoor dust and each of the five soil types. Calabrese (unpublished, as reported in Walker and Griffin 1998) reported a similar indoor dust/outdoor soil concentration ratio of 0.387 for average As concentrations in samples collected from a smaller subset of the same homes in Anaconda, Montana. Calabrese used different sampling methods and reported significantly different As concentrations in soil (average=74.7 µg/g) and indoor dust (average=29.3 µg/g); however, the average indoor/outdoor concentration ratio was similar to the ratio observed by Hwang et al. (1997).

The Ontario Ministry of the Environment (MOE) reviewed the available literature pertaining to the relationship between concentrations of metals in outdoor soil and indoor dust for the Rodney Street risk assessment in Port Colborne, Ontario (MOE 2002). The review included an evaluation of a study conducted by PTI Environmental Services (PTI 1994) at a contaminated site in Bartlesville, Oklahoma. The study reported concentration ratios (CRs) of indoor dust to outdoor soil for As, Cd, Pb, and Zn of 0.20, 0.35, 0.50, and 0.36, respectively (MOE 2002). The MOE selected a CR value of 0.39, derived from the Hwang and Calabrese studies, to define the relationship between indoor dust and outdoor soil concentrations of Ni in Port Colborne.

Studies conducted in areas that have been impacted by neighboring industrial activities provide evidence that a significant relationship exists between metal concentrations in outdoor soil and levels found in household dust. Regions or cities with less industrial activity have not shown evidence to support this relationship; however, in most studies, higher levels of contaminants are reported in household dust than in surrounding outdoor soil, which tends to be similar to natural background levels. Regression analysis is often used to examine the relationship between metal concentrations in indoor dust and outdoor soil. The slope of a regression equation (dust concentration/soil concentration) for a dataset can be used to define the changes in dust concentration over a range of soil concentrations. However, it should be noted that the slope of a regression equation for the plot of indoor dust concentrations over outdoor soil concentrations and mean concentration ratios for the same dataset are not analogous and should not be used for comparison of different data.

Regression analysis conducted by Murgueytio et al. (1998) on indoor dust and outdoor soil data collected during an exposure study in the Big River Mine Tailings site, south of St. Louis, Missouri, found a significant correlation between indoor dust Pb concentrations and outdoor soil Pb levels (r^2=0.36; P < 0.001). The relationship between contaminants in indoor dust and outdoor soil in residential homes was also examined by the U.S.

Environmental Protection Agency (U.S. EPA) at several Superfund sites across the United States. Residential dust sampling at one Superfund site near Denver, Colorado, found only a weak correlation between the levels of As and Pb. However, regression analysis of the paired soil and dust data revealed statistically significant regression line slopes, for both arsenic and lead (U.S. EPA 2003).

7.1.1.3 Exposure Rates to Soil and Dust

Dermal Contact with Indoor and Outdoor Soil and Dust

A literature review was conducted to examine methods used to predict chemical exposures via dermal contact with impacted soils and dusts (including U.S. EPA 1997a, 2002b, 2004a; Richardson 1997; Burmaster 1998; Garlock et al. 1999). The overall approach used to evaluate exposures via direct dermal contact with soil and dust for this assessment was taken from U.S. EPA (2004a). The fraction of total surface area method was selected for use in the current assessment. Indoor and outdoor area-weighted soil adherence values were derived using data presented in U.S. EPA (1997a, 2002b, 2004a). Area-weighted adherence factors were derived using the percentage of the total surface area of each body part (hands, arms, legs and feet) in conjunction with body-part specific adherence values for a given activity. Indoor adherence/loading factors were developed based on children playing indoors on carpeted areas. Adherence factors for adults were based on indoor Tae Kwon Do activities or outdoor soccer activity.

For children, teenagers and adults, the percentage of the total surface area for each body part was selected from the U.S. EPA (1997a). The mean percentage of total body surface area reported by U.S. EPA (2002b) was used for infants and preschool children. The U.S. EPA (2002b) presents various clothing scenarios in which 10% to 25% of skin surface area is estimated to be exposed. A prorated, seasonally adjusted estimate of the area of exposed skin was developed by dividing the year into spring (61 days), summer (92 days), fall (91 days) and winter (121 days) with each season associated with a different fraction of exposed skin. During the spring and fall seasons, it was assumed that 15% of the total body surface area would be exposed, while during the summer months, exposure of 25% of the total body surface area was considered to be reasonable. The fraction of skin exposed during the winter season was considered to be much less at only 5% of the total body surface area. A seasonally-adjusted fraction of exposed skin was estimated to be 14.2% based on the duration of each season and the fraction of exposed skin within each season. The ability to come into direct contact with surface soil during the winter season was considered less likely due the additional clothing worn during these months and the fact that much of the ground is either frozen and/or covered with snow. A winter covering factor of 10% was applied to the outdoor soil ingestion pathways during the winter season only.

Incidental Soil and Dust Ingestion

A significant amount of both regulatory and scientific literature was reviewed with regard to the application of incidental soil intake rates of children for use during chronic exposure assessments, including Calabrese et al. (1997a,b), U.S. EPA (1997b, 1999c, 2002b, 2004a), Stanek and Calabrese (2000), Stanek et al. (2001a,b), and Health Canada (2004). The U.S. EPA (1997a, 2002b) recommended a mean soil intake rate of 100 mg/day and an upper conservative mean of 200 mg/day. The Ontario MOE also recommends the use of 100 mg/day, while Health Canada (2004a) recommended a soil ingestion rate of 80 mg/day for children and 20 mg/day for all other receptors. The 80 mg/day value was selected for use in this current assessment; however, due to the uncertainty with regard to this variable, evaluations of potential risk using either soil intake rate (100 or 80 mg/day) were completed as part of the sensitivity analyses with regard to potential Pb risk to children (see Chapter 8).

The U.S. EPA's Integrated Exposure Uptake Biokinetic (IEUBK) model for Pb in children employs central tendency 'total soil and dust' ingestion rates for five individual age classes of children ranging from 85 mg/day to 135 mg/day. The IEUBK model also uses a default 45:55 split which assumes that 45% of the total intake rate is applied to soil while 55% is applied to dust. The 45:55 ratio of soil to dust was used for the Sudbury HHRA.

7.1.1.4 Ambient Air Concentrations

A year-long air monitoring program was conducted to collect Sudbury-specific ambient air concentrations for all seasons under variable wind and climate conditions. A description of the air monitoring program was presented in Chapter 6 (Section 6.2). High and low volume samplers were used to collect three size fractions of particulate matter on quartz fiber filters: i) total suspended particulate matter (TSP); ii) particulate matter less than 10 microns (PM_{10}); and, iii) particulate matter less than 2.5 microns ($PM_{2.5}$). These size fractions were selected for sampling because they are most relevant to human exposure and can be retained in the nose (TSP), upper lung (PM_{10}) and lower lung ($PM_{2.5}$) of an individual. For this assessment, PM_{10} concentrations were used to calculate 95% UCLM values for each COC. This fraction conservatively represents the most toxicologically significant particle size for human exposure and toxicity (i.e. it contains both the PM$_{2.5}$ fraction and slightly larger particles which may cause impacts/irritation within the upper lungs). The 95% UCLM values for each COI for ambient air concentrations used in the HHRA exposure assessment were provided in Table 6.2 of Chapter 6. The direct air inhalation pathway utilized the basic exposure equations from U.S. EPA (2004a).

For this assessment, indoor air concentrations were assumed to be equal to measured outdoor air concentrations. This is considered a conservative assumption as a number of studies (Chao and Wong 2002; Komarnicki 2005; Molnar et al. 2005) demonstrated that outdoor concentrations of heavy metals can be significantly greater than measured indoor air concentrations. Lower indoor air concentrations appear to be a result of outdoor air filtration and dilution with the existing indoor aerosol. Although there may be some minor indoor sources of metals, their contribution does not appear to be significant compared to the contribution of outdoor air.

7.1.1.5 Drinking Water Concentrations

The majority of homes in the GSA (88%) were serviced by municipal drinking water in each of the five communities. Drinking water quality is monitored under the provincial Drinking Water Surveillance Program (DWSP), a voluntary monitoring program managed by the MOE in conjunction with municipalities across Ontario. For this study, data from 1995 to 2005 were obtained from water supply systems and water treatment plants that service each COI. The 95% UCLM was calculated for human exposure to each COC through consumption of residential drinking water. The values represented total metal concentrations in unfiltered samples. The drinking water supply system servicing the Town of Falconbridge was owned and operated by Xstrata Nickel and was not part of the DWSP. In the summer of 2005, the drinking water source supplying the Town of Falconbridge was switched from the original well to a new deeper well. Initial sampling data, including MOE audits, indicated that concentrations of Pb greatly decreased with the change to the new well. While both sources are displayed in Table 7.2, as the purpose of the HHRA was to evaluate current risks for the residents of Falconbridge and in the near future, exposures to drinking water from the new well were used in the current assessment. A voluntary drinking water survey for households with private water supplies was also conducted (see Chapter 6.4) as part of the assessment. In general, COC concentrations in private drinking water supplies in the study area were within the range of drinking water concentrations measured in the municipal water sources. Therefore, COC concentrations in the municipal drinking water supply were selected for estimating exposure of individuals through the consumption of tap water because these represented the majority of the population in the study area (88%) and the data were similar between the private and municipal sources. Table 7.2 provides a summary of COC concentrations in potable water used in the current assessment.

Table 7.2 **Drinking Water Concentrations (µg/L) by Community of Interest**

COI	COC (No. of samples)	Min	Max	Mean[a]	Standard Deviation	95% UCLM	95th Percentile
Coniston	As (62)	0.20	2.00	0.87	0.50	1.14	1.7
	Co (62)	0.01	2.20	0.16	0.29	0.20	0.44
	Cu (62)	1.50	212	36.2	37.2	44.6	96.2
	Ni (62)	6.80	120	35.3	31.7	52.8	97.2
	Pb (62)	0.03	1.67	0.26	0.26	0.31	0.73
	Se (62)	0.31	5.00	0.91	0.67	1.28	2
Copper Cliff	As (9)	0.25	4.50	1.23	1.37	2.53	3.5
	Co (4)	0.02	0.05	0.03	0.011	0.05	0.045
	Cu (7)	18.3	248	103	90.8	170	226
	Ni (4)	8.37	49.3	20.1	19.6	49.3	43.9
	Pb (9)	0.30	2.80	0.82	0.79	1.39	2.08
	Se (9)	0.50	3.00	1.66	1.12	3.00	2.8

COI	COC (No. of samples)	Min	Max	Mean[a]	Standard Deviation	95% UCLM	95th Percentile
Falconbridge	As (194)	0.40	5.70	2.46	0.90	2.57	4
	Co[b] (62)	0.01	2.20	0.16	0.29	0.20	0.44
	Cu (421)	1.00	500	18.9	37.7	30.4	100
	Ni (431)	2.00	160	30.4	16.4	31.7	60
	Pb Old well (10)	0.60	6.00	1.46	1.60	3.67	3.75
	Pb New well (7)	0.18	1.4	0.50	0.44	0.97	1.19
	Se (3)	2.50	2.50	NA	0	2.50	NA
Sudbury Centre	As (62)	0.20	2.00	0.87	0.50	1.14	1.7
	Co (62)	0.01	2.20	0.16	0.29	0.20	0.44
	Cu (62)	1.50	212	36.2	37.2	44.6	96.2
	Ni (62)	6.80	120	35.3	31.8	52.8	97.2
	Pb (62)	0.03	1.67	0.26	0.26	0.31	0.73
	Se (62)	0.31	5.00	0.91	0.67	1.28	2
Hanmer	As (18)	0.90	1.96	1.33	0.301	1.46	1.83
	Co (18)	0.01	0.14	0.04	0.031	0.06	0.081
	Cu (18)	0.50	115	53.1	29.4	65.2	96.0
	Ni (18)	0.10	2.10	0.29	0.47	0.80	0.66
	Pb (18)	0.05	1.14	0.34	0.31	0.49	1.05
	Se (18)	0.12	2.00	0.87	0.40	1.28	1.15

[a] Arithmetic mean.
[b] Cobalt concentrations from Wanapitei and David Street WTPs used as surrogate.
NA = not applicable.

Drinking Water Ingestion Rates

Receptor-specific water intake rates and body weights were used to estimate chemical-specific daily intake rates for all individuals (e.g. preschool children, children, teenagers and adults) and COI. The CTE estimate employed the 50th percentile water intake rates provided by Richardson (1997). The RME exposure estimate employed the 95% UCLM drinking water intake rates. Each exposure estimate (i.e. the CTE and RME) used age-specific mean body weights provided by Richardson (1997).

7.1.1.6 Locally Grown Produce and Wild Blueberries

Local Fruits and Vegetables

The daily intake of fruits and vegetables for area residents was supplemented by local produce from commercial farms and backyard gardens. A vegetable garden survey was conducted to collect Sudbury-specific COC concentration data to address concerns of the local citizens. The study involved the collection of vegetable, fruit and crop samples from residential and commercial gardens across the GSA, wild blueberry and wild mushroom samples from natural areas, as well as co-located soil samples from each sample site. A description of the vegetable garden survey was provided in Chapter 6.8. Sudbury-specific COC concentrations from residential gardens were used to calculate 95% UCLM concentration values for each COC in both above ground and below ground vegetables in each of the COI. Refer to Tables 6.16 through 6.21 in Chapter 6 for a complete summary of the vegetable and fruit concentrations used in the current exposure assessment.

Information gathered from the local food consumption survey (see Chapter 6.3) and Volume II of the U.S. EPA Exposure Factors Handbook (U.S. EPA 1997b) was used to approximate the amount of fruits and vegetables a Sudbury resident might consume from an individual home garden. The intake rate data provided in the local food consumption survey were highly skewed which was not unexpected for such a limited recall-based survey. Reported mean intake rates were often more than twice the reported median value, indicating a highly skewed distribution. Given these uncertainties, it was decided that data from the U.S. EPA (1997b) would be used to approximate local vegetable intake for this study. Food intake rates as a receptor characteristic were presented in Chapter 5. For locally grown fruits, the local food intake survey indicated that cultivated strawberries were the main source of locally derived fruit (wild blueberries were considered separately). The median intake for cultivated strawberries originating from the GSA was reported to be 10 cups/year, equivalent to approximately 0.1 g/kg/day or approximately 2.3 kg of local strawberries per person per year. The total daily food intake of an individual was kept constant by expressing local food intake rates as a proportion of the total daily intake for a specific food group. In other words, it was not assumed that an individual consuming local food was consuming more total food per day than a person not consuming local products, but rather, it was assumed that an individual would derive a certain proportion of their total food intake from local sources.

The gathering and consumption of local wild blueberries was common practice in the study area and, therefore, was considered as a separate exposure pathway. Data for metal concentrations in wild blueberry samples ($n=10$) were provided in Table 6.19. The Local Food Consumption Survey (Chapter 6.3) reported mean and median blueberry consumption rates of 173 and 12 cups/year, respectively. The median intake rate of 12 cups/year was used to form the central tendency estimate of the consumption rate of local wild blueberries. The upper 95th percentile consumption rate of 'other berries' was 1.28 g/kg/day (or 33 kg/year) (U.S. EPA 1997b). The 'other berries' food category includes all berries other than strawberries, including a wide range of commercially frozen and canned berry produces (e.g. pie fillings, cranberry sauces, juices, etc.).

For the purpose of the point estimate assessment, the CTE estimate employed a daily wild blueberry consumption rate equivalent to the reported median intake of 12 cups/year (or approximately 0.12 g/kg/day). For the RME estimate, it was assumed that an individual may consume up to twice the amount of blueberries as the CTE estimate or 0.24 g/kg/day (approximately 5.5 kg/year for a female adult).

7.1.1.7 Fish Tissue Concentrations and Consumption

Angling and fishing are significant recreational activities in the Sudbury area and represented an important exposure pathway for human health. Consumption of locally caught fish by freshwater anglers and fisherman in the GSA may result in higher exposure to one or all of the COC being evaluated. A survey of local fish species commonly caught and consumed was conducted as part of the assessment and the data used in the HHRA exposure assessment were provided in Table 6.22 of Chapter 6.

The consumption of local fish was considered for two distinct populations including the general population and an angler sub-population within the overall general population. The population of anglers would be expected to consume much larger quantities of local fish than the typical resident. The local food consumption survey provided self-reported consumer-only intake rates of anglers for the top four most commonly consumed fish species (walleye, trout, pike and perch). Daily consumption rates of freshwater fish were estimated using the consumption frequency data provided by the local food consumption survey and information from the Great Lakes Sport Fish Consumption Advisory Task Force (GLSFATF 1993). The GLSFATF (1993) suggests that a typical serving of fish is approximately 227 g. Combining the site-specific consumption frequency data with an assumed serving size of 227 g produced a mean (or CTE) intake rate of fish for the general population of 12.44 g/day. For the current assessment, the CTE and RME estimates (for the general population) employed local fish consumption rates equivalent to 12.44 g/day and 35.94 g/day, respectively (Table 7.3). For the angling population, CTE and RME estimates employed intake rates of 35.64 and 85.39 g/day. It should be noted that these intake rates were significantly greater than the 95th percentile intake rate for fresh water anglers of 25 g/day reported by the U.S. EPA (1997b).

Table 7.3 **Local Fish Consumption Rate**

Fish Species	General Sudbury Population		Angling Population	
	CTE (meals/year)	RME (meals/year)[a]	CTE (meals/year)	RME
Walleye	5.4	12.23	18.2	42.2
Trout	4.7	10.71	—	—
Pike	5.9	21.33	17.6	46.6
Perch	4	13.52	21.5	48.5
Sum (meals/year)	20.00	57.79	57.30	137.30
TOTAL (g/day)[b]	12.44	35.94	35.64	85.39

[a] RME intake rates were derived by adding one standard deviation to the reported mean intake frequency data.
[b] The number of meals/year was converted to g/day by assuming a serving size of 227 g (8 ounces) and a constant intake rate of fish over the entire year.

7.1.1.8 Wild Game Tissue Concentrations and Consumption

Limited monitoring data were available for metal concentrations in game meat in the study area; therefore, predicted concentrations were used for input to the human health exposure model to quantify potential exposures from this pathway. The moose was selected as an appropriate wildlife receptor for consumption by humans since: i) moose require large home ranges and consume large amounts of forage; ii) local residents stated that consumption of moose meat was common; iii) moose are herbivores like cows that were used to develop the empirical models to predict beef meat concentrations; and, iv) moose consume grasses in the study area that were measured by field investigations for metal concentrations.

Moose are large mammals (body weight = 325 + 59 kg) and COC concentrations in muscle tissues were calculated following the U.S. EPA OSW (1998) methodology for predicting metal concentrations in beef cattle. For the purpose of estimating game tissue residue levels, wildlife was assumed to be exposed to chemicals through consumption of soil and food derived from the rural portion of the study area. Only the game meat concentrations predicted in Zone 2 of the ERA (see Chapter 11) were used for input to the human exposure model. Measured COC concentrations in soils from Zone 2 were the highest for the wild land areas and would provide a conservative estimate of game meat concentrations from other areas where concentrations are expected to be lower. The predicted moose meat concentrations were assumed to represent exposures that might be received by humans from other game such as deer and upland birds (i.e. grouse). Water-to-algae bioconcentration factors (BCFs) provided by U.S. EPA (1999a) were used to estimate the potential distribution of metal concentrations in aquatic plants (described in more detail in Chapter 11). The final predicted game meat concentrations used in the HHRA are presented in Table 7.10 as wild game.

Local Wild Game Consumption Rates

According to the local food consumption survey, local hunters and anglers reported that they most commonly consumed five types of game including grouse, moose, deer, wild rabbit and ducks/geese. Grouse, moose and deer were identified as the three most commonly consumed game, by 65%, 62% and 48% of hunters, respectively. Less than 20% of all hunters interviewed indicated consuming wild rabbit while less than 5% of the general Sudbury population sampled indicated consuming rabbit. The method used to approximate local wild game intake on a g/day or g/kg/day basis was the same method used to approximate local fish intake rates. For the general Sudbury population, CTE and RME wild game consumption rates of 7.2 and 13.0 g/day were estimated, respectively. CTE and RME wild game intake rates for those individuals who reported hunting were approximately 29.8 and 60.3 g/day, respectively (Table 7.4). Again, these estimates were derived using an assumed 227 g serving size and the reported mean and upper percentile consumption frequencies provided by the local food consumption survey.

Table 7.4   **Local Wild Game Consumption Rates**

Species	General Sudbury Population		Hunting Population	
	CTE (meals/year)	RME (meals/year)[a]	CTE (meals/year)	RME (meals/year)[a]
Moose	4.5	6.74	24.6	48.6
Deer	4.3	8.46	10.8	20.8
Grouse	2.8	5.73	12.5	27.5
Sum (meals/year)	11.60	20.93	47.90	96.90
TOTAL (g/day)[b]	7.21	13.02	29.8	60.3

[a] RME intake rates were derived by adding one standard deviation to the reported mean intake frequency data.
[b] The number of meals/year was converted to g/day by assuming a serving size of 227 g (8 ounces) and a constant intake rate over the entire year (GLSFATF 1993).

7.1.2 Background Exposure Assessment

7.1.2.1 Rationale and Overview

The COC considered in this HHRA are naturally present within the environment, and/or have a number of anthropogenic sources, which are not associated with historic or ongoing emissions from the existing smelting operations. Therefore, an additional exposure and risk assessment was conducted to evaluate the degree of exposure of the human receptors to the COC without the contribution of the Vale and Xstrata Nickel smelters. Such an assessment provided an indication of the exposures experienced by a typical Ontario resident (TOR), based on ambient or background concentrations in water, air, soil, dust, and food sources. Predicted TOR exposures were then compared with the exposures attributed to smelting activity, to give an indication of total exposure to COC from all known sources. In addition to using background exposure to account for an individual's estimated total daily intake, background assessments can also be used as benchmarks of comparison that aid in determining the significance of the exposures from the study area relative to typical Ontario background exposures. Such relative contribution analysis can be useful in putting exposure and risk estimates into perspective, and guiding the development of risk management recommendations. If study area exposures and risks are estimated to be less than or similar to typical Ontario exposures, the need for risk management measures may be reduced or become unnecessary. Evaluation of typical background exposures also assists in the interpretation and validation of predictive modeling data, which increases stakeholder confidence in the overall results of the HHRA process. Background COC concentrations used in the current background exposure assessment were derived from monitoring programs in Ontario and across Canada. The background COC concentrations, described in more detail below, were used to calculate 95% UCLM values in outdoor air, soil and drinking water for a TOR scenario.

7.1.2.2 Data Used in Exposure Assessment for Typical Ontario Residents

By incorporating background sources of exposure into the assessment, total estimated exposure from all sources (including background) were compared to the reference exposure value (e.g. a tolerable daily intake (TDI) or reference dose (RfD)), without needing to make decisions about how the TDI or RfD value should be reduced to account for exposures that were not explicitly evaluated. The exposure equations used for the typical Ontario background assessment were the same as those used to calculate exposure rates of individuals in the GSA. However, it should be noted that the data used to evaluate TOR exposures were derived from a variety of generic sources, and the TOR receptor scenario, on its own, would not be appropriate for use in assessing potential health risks to the typical Ontario resident. It was simply included in the current assessment to provide a comparison with the evaluation of the Sudbury-specific health risk predictions, and to place them into an overall Ontario context.

Background Outdoor Air Concentrations

The National Air Pollution Surveillance (NAPS) monitors air quality at monitoring stations across Canada. The NAPS data provided a long-term archive of air quality data including the six COC at urban and rural locations in all regions of Canada. The NAPS dataset was selected as background outdoor air concentrations for the TOR scenario because it provided the advantage of consistency across chemicals, regions and time periods. The 2002 data from a monitoring station in a residential area of Toronto provided the most robust set of data and was selected to represent background air concentrations in Ontario. However, there were insufficient data on which to generate a 95% UCLM for each COC, as such the arithmetic mean for each COC was used in the exposure assessment (Table 7.5).

Table 7.5 **Typical Ontario Ambient Air Concentrations (µg/m³) (Toronto)**

COC	No. of samples	Arithmetic mean of PM$_{10}$ samples
As	53	0.001
Co	53	0.002
Cu	53	0.009
Pb	53	0.008
Ni	53	0.001
Se	53	0.002

Background Soil Concentrations

The Ontario Ministry of the Environment developed Ontario Typical Ranges (OTRs) that represented the expected range of contaminants in surface soil from areas in Ontario not subjected to the influence of known point sources of emissions (OMEE 1994). The OTR report presents ranges, means and the 98th percentile soil concentration (OTR$_{98}$) for each of the COC. The OTR$_{98}$ represents an upper limit of normal concentrations in Ontario. These values apply to the land use and soil type in Ontario for which they were developed. More recently, the MOE (2004) derived background values, based on the OTR values, to represent the upper limits of typical province-wide background concentrations that are not contaminated by point sources. These values were used as the background soil concentrations for the TOR scenario as follows: As (17 µg/g), Co (21 µg/g), Cu (85 µg), Pb (120 µg/g), Ni (43 µg/g) and Se (1.9 µg/g).

Background Drinking Water Concentrations

To estimate exposure of a TOR to COC in drinking water, monitoring data from over 170 water treatment plants and well supplies across Ontario were used to calculate 95% UCLM values for each COC. Drinking water monitoring data were provided by the DWSP system operated by the MOE in cooperation with municipalities across the province (MOE 2005). For the purpose of estimating exposure to COC concentrations in typical drinking water under a TOR scenario, only data from 1997 through 2002 were included in the calculation. These samples were unfiltered and were analyzed for total metal concentrations. A summary of the estimated 95% UCLM COC concentrations in drinking water for a TOR exposure scenario are presented in Table 7.6.

Table 7.6 **Typical Ontario Drinking Water Concentrations (µg/L) (MOE 2005)**

COC	No. of samples	Min	Max	Mean	95% UCLM
Arsenic	2296	0.03	15.0	0.56	0.64
Cobalt	2296	0.0023	2.85	0.08	0.09
Copper	2296	0.0025	0.00126	32.5	40.8
Lead	2301	0.0074	331	0.91	1.89
Nickel	2296	0.0006	113	1.41	2.18
Selenium	2291	0.0086	19.9	1.42	1.58

7.1.3 Market Basket Estimated Daily Intakes

Food represents a critical pathway of exposure to the COC for the residents of the GSA. Foods consumed and purchased from grocery stores, supermarkets, butchers, etc, are considered background sources of exposure and contribute to an individual's total level of exposure to COC. The exposure to COC through the consumption of store-bought foods was termed the 'market basket estimated daily intake' or EDI_{MB}. An EDI was defined as the estimated daily intake of a chemical that was unrelated to any specific contaminated site (i.e. normal 'background' exposure) (CCME 2005). It was characterized by an average Canadian's exposure to low levels of chemicals commonly found in air, water, food, soil, and consumer products (CCME 2005). A market basket EDI (EDI_{MB}) was defined as the estimated daily intake of a chemical that was related to food commonly purchased in the supermarket and other points of purchase, prepared, and consumed by urban Canadians. The purpose of the EDI_{MB} is to incorporate background exposure when characterizing an individual's exposure to COC. This is to ensure that a portion of a chemical's TDI is apportioned to background sources such that the total exposure to background levels, plus soil concentrations at the acceptable benchmark level do not exceed the TDI (CCME 2005). In the context of the HHRA, the purpose of the EDI_{MB} was twofold: i) to ensure that background sources were included in the exposure assessment of Sudbury residences; and ii) to ensure that background sources were accounted for when calculating a Sudbury-specific soil risk management.

A literature review was conducted to obtain published data on the concentrations of COC in store-bought foods (i.e. supermarket or market basket food items) which Sudbury residents may be consuming. The purpose of the literature review was to identify the most appropriate food data to characterize Sudbury area residents' background exposure to store-bought foods. In Canada, most supermarkets foods are from sources distributed across North America and are generally not specific to the location of the supermarket. Thus, food purchased in Sudbury should resemble the foods purchased in other cities in Canada, particularly those in Ontario. The purpose of the market basket review was to: i) identify the key food item categories making up the diet of Sudbury residents; ii) determine the estimated daily intake rates for each food category; and, iii) determine the range of COC concentrations in each food category. The information generated from this phase of the study was incorporated into the exposure pathway model of the HHRA as the EDI for each COC. The food concentrations used in the derivation of the EDI_{MB} were based on the most applicable data available for food purchased in a Canadian supermarket. The 95% UCLM was calculated from food concentration data for use in the HHRA. To determine the most appropriate data to use in the Sudbury HHRA, the following criteria were used:

- Priority was given to Canadian food concentration data (if Canadian data were unavailable, the literature search extended to international studies, preferably American).
- Food was from a supermarket or other public point-of-purchase (e.g. bakery, butcher, etc.).
- Food was prepared and/or cooked for normal consumption.
- Data were reported with adequate summary statistics (raw data, or at a minimum, the sample number, mean concentration and range).
- The minimum detection limits were adequately low to detect the metal in most of the food items.
- The quality of the study design and the comprehensiveness of the data collected were considered appropriate for use in this HHRA.

For the purposes of applying the food concentrations to the EDI_{MB}, the raw data were obtained for all of the datasets and the 95% UCLM of the food categories was calculated. A brief summary of each COC is provided in the following section. The databases selected for calculating the EDI_{MB} are summarized in Table 7.7.

Table 7.7 **Databases Selected for Use in the Development of the EDI_{MB}**

COC	Location	Date	Description	Reference
As	Six Canadian cities	1985 and 1988	Canadian Total Diet Study[a]: Total As analyzed in supermarket foods	Dabeka et al. 1993
Co	Eight Canadian cities	1993 to 1999; and 2000; 2002	Canadian Total Diet Study[a]: Total Co analyzed in supermarket foods, supplemented with green leafy vegetable data from Port Colborne	Health Canada 2004a; Dabeka and McKenzie 2005 pers. comm.; JWEL 2004a
Cu	Eight Canadian cities	1993 to 1999 and 2000	Canadian Total Diet Study[a]: Total Cu analyzed in supermarket foods	Health Canada 2004a; Dabeka and McKenzie 2005 pers. comm.
Ni	Port Colborne	2002	Total Ni analyzed in foods from local supermarkets, food outlets, butchers eateries, and markets[b]	JWEL 2004a
Pb	Canada	2000	Canadian Total Diet Study[a]: Total Pb analyzed in supermarket foods	Dabeka and McKenzie 2005 pers. comm.
Se	United States	1991 to 2002	U.S. FDA Total Diet Study[b]: Total Se analyzed in supermarket foods	U.S. FDA 2004

[a] All non-detected food concentrations were assumed by the authors to be the full detection limit.
[b] All non-detected food concentrations were assumed to be half of the detection limit.

7.1.3.1 Arsenic

There were a number of Canadian market basket surveys available for As (MOE 1987; Dabeka et al. 1993; JWEL 2004a). Some of the market basket studies analyzed total As (e.g. Dabeka et al. 1993; JWEL 2004a), while others analyzed both total and inorganic forms (MOE 1987). The database selected for use in the Sudbury HHRA was the Dabeka et al. (1993) Canadian Total Diet Study (CTDS) because it fulfilled all of the selection criteria and was found to be the most appropriate for As. In this survey, food was sampled from supermarkets in six Canadian cities and prepared as for normal consumption by Canadians (Dabeka et al. 1993). Unfortunately, As was not analyzed in the CTDS data for the period 1993 to 1999, and 2000 due to limited government resources (Dabeka and McKenzie 2005 pers. comm.). Therefore, the available data were greater than 10 years old. More recent data were available from another smelter related risk assessment in Port Colborne, Ontario (JWEL 2004a), but these data were not selected because they had inappropriately high detection limits for As (i.e. As was non-detectable in 97% of food samples; detection limit was ~50 ng/g dw vegetables), resulting in highly uncertain estimates of food concentrations.

Many studies that estimated the dietary intake of As traditionally were based on surveys of total As in food, including both organic and inorganic forms of As. According to Schoof et al. (1999), As concentrations in food were dominated by the relatively non-toxic organic forms of As found in seafood. Schoof et al. (1999) conducted a market basket survey of inorganic As in 40 different commodities which were anticipated to provide approximately 90% of the dietary intake of inorganic arsenic. The results provided by Schoof et al. (1999) were consistent with other studies, in that total As concentrations among seafood products were highest; however, inorganic As concentrations in seafood were not elevated and ranged between less than 1 ng/g and 2 ng/g. According to Schoof et al. (1999), raw rice was found to have the highest inorganic As content among all food commodities tested.

The As concentration data used to establish estimated daily intake rates of inorganic As from market basket and local foods were based on total As measurements (i.e. organic plus inorganic species). As a result, total As concentrations reported for various food groups were corrected by the fraction of total As present as inorganic species using the data from Schoof et al. (1999). The arithmetic mean ratio of different food groups was used to adjust the total As concentration of a particular food group to an inorganic As concentration. Table 7.8 provides data from Schoof et al. (1999) that were used to calculate the mean fraction of inorganic As in different food groups.

Table 7.8 **Fraction of Inorganic As in Various Food Groups**

Food group	Total As (ng/g)	Inorganic As (ng/g)	Fraction inorganic
Fats, oils, sweets, nuts			
Beet sugar	12.2	3.50	0.29
Cane sugar	23.8	4.40	0.18
Corn syrup	6.00	0.40	0.07
Butter	1.80	1.10	0.61
Soybean oil	1.80	1.10	0.61
Salt	4.80	0.80	0.17
Beer	2.70	1.80	0.67
Peanut butter	43.6	4.70	0.11
Mean value	12.1	2.23	0.34
Milk, yogurt, cheese			
Milk, skim (non-fat)	2.60	1.00	0.38
Milk, whole	1.80	1.00	0.56
Mean value	2.20	1.00	0.47
Meat, poultry, eggs			
Beef	51.5	0.40	0.01
Chicken	86.4	0.90	0.01
Pork	13.5	0.60	0.04
Eggs	19.9	1.00	0.05
Mean value	42.8	0.73	0.03
Vegetables			
Beans	2.10	1.20	0.57
Carrots	7.30	3.90	0.53
Corn	1.60	1.10	0.69
Cucumber	9.60	4.10	0.43
Onions	9.60	3.30	0.34
Potatoes	2.80	0.80	0.29
Tomato	9.90	0.90	0.09
Mean value	6.13	2.19	0.42
Fruit			
Apple, raw	4.80	1.80	0.38
Apple, juice	7.60	2.80	0.37
Banana	2.30	0.60	0.26
Grapes	10.2	3.60	0.35
Grape Juice	58.3	9.20	0.16

Food group	Total As (ng/g)	Inorganic As (ng/g)	Fraction inorganic
Orange Juice	4.80	1.00	0.21
Peaches	3.40	2.30	0.68
Watermelon	40.2	8.90	0.22
Mean value	16.5	3.78	0.33
Bread			
Corn (meal)	38.6	4.40	0.11
Flour	39.1	10.9	0.28
Rice	303	73.7	0.24
Mean value	127	29.7	0.21
Fish			
Saltwater finfish – mean ($n=4$)	2360	0.50	0.0002
Canned Tuna – mean ($n=4$)	512	1.00	0.002
Shrimp – mean ($n=4$)	1890	1.90	0.001
Freshwater finfish – mean ($n=4$)	160	1.00	0.006
Mean value	**1230**	**1.10**	**0.002**

Cobalt

A number of Canadian market basket studies were available for Co (Dabeka and McKenzie 1995; Health Canada 2004b; JWEL 2004a; Dabeka and McKenzie 2005 pers. comm.). The Co concentrations reported by these different studies were comparable; however, the databases did not include an analysis of green leafy vegetables. The datasets selected for use in the Sudbury HHRA were the consecutive years (1993 to 2000) of the CTDS (Health Canada 2004b; Dabeka and McKenzie 2005 pers. comm.) because they fulfilled all of the selection criteria and were the most appropriate for Co. The datasets were combined to increase the Canadian coverage (eight cities) and the statistical robustness of the data. The Canadian TDS results for 1986 to 1988 were not included because Co concentrations in approximately half of the samples were not detected. To include all important sources of Co, the results for green leafy vegetables provided in JWEL (2004a) were integrated into the database.

Copper

Canadian market basket data were available for Cu (Health Canada 2004b; JWEL 2004a; Dabeka and McKenzie 2005 pers. comm.). There was good agreement among the results for the CTDS (Health Canada 2004b; Dabeka and McKenzie 2005 pers. comm.). The Port Colborne results were lower than the other databases but within the same order of magnitude (JWEL 2004a). The Cu levels for organ meats were significantly higher than the rest of the meat and poultry samples for all three studies. For example, the mean Cu concentrations for the meat category with and without the organ meats for three different studies were: 10,911 and 1342 ng/g in the 2000 CTDS; 3496 and 1006 ng/g in the 1993 to 1999 CTDS; and 21,935 and 685 ng/g in the Port Colborne study. The databases selected for use in the Sudbury HHRA were the consecutive years (1993 to 2000) of the CTDS (Health Canada 2004b; Dabeka and McKenzie 2005 pers. comm.) because they fulfilled all of the selection criteria and were the most appropriate for Cu. The datasets were combined to increase the Canadian coverage (eight cities) and the statistical robustness of the data.

Lead

There were a number of Canadian datasets available for Pb, all conducted as part of the CTDS (Dabeka and McKenzie 1995; Health Canada 2004b; Dabeka and McKenzie 2005 pers. comm.). The databases selected for use in the Sudbury HHRA were Dabeka and McKenzie (2005 pers. comm.) because they fulfilled all of the selection criteria and were the most appropriate for Pb. The older Total Diet Study results were not used because Pb lead concentrations in environmental media and biological tissues/fluids were generally much higher in the 1970s and 1980s than today (ATSDR 2005).

Nickel

There were two Canadian market basket studies available for Ni (Dabeka and McKenzie 1995; JWEL 2004a). While food products were analyzed for Ni as part of the CTDS conducted in 2000, the data were accidentally contaminated during analyses (Dabeka and McKenzie 2005 pers. comm.). Therefore, the 2000 CTDS concentration data for Ni were not used for the current study. There was good agreement in Ni concentrations between the 1986–1988 Total Diet Study (Dabeka and McKenzie 1995) and Port Colborne market basket study (JWEL 2004a) for the categories that were uncooked (i.e. other vegetables; sugars and sweets; fats, nuts and oils; and, beverages). However, the Port Colborne mean Ni concentrations in the cooked food categories were approximately three times lower than those calculated and reported in the CTDS by Dabeka and McKenzie (1995).

Concern was expressed (JWEL 2004a) with the interpretation of the Ni concentrations in the cooked food analyzed in the 1986 to 1988 Canadian Total Food Study (i.e. Dabeka and McKenzie 1995). The food samples were prepared using new stainless steel frying and roasting pans. Food was analyzed before and after cooking and the results indicated that significant Ni contamination occurred, particularly by roasting some of the meat samples (Dabeka and McKenzie 1995). Jacques Whitford (JWEL 2004a) conducted an extensive literature review and a series of experiments to explore the role of cooking with stainless steel utensils on the leaching of Ni into food samples (some key papers include Christensen and Moller 1978; Kuligowski and Halperin 1992; Tupholme et al. 1993; Kumar et al. 1994). Their review revealed that significant Ni is leached during cooking; however, this contamination decreased to negligible amounts after the first few uses of the utensil (JWEL 2004a). They also conducted a screening-level cooking study with a well-used stainless steel frying pan and ceramic pan. This study demonstrated that the foods were not contaminated by Ni during normal preparation and cooking (use of 'well used' stainless steel pan) (JWEL 2004a). Thus, they concluded that contamination of the food items in the Dabeka and McKenzie (1995) study did not appropriately characterize the long-term contribution of Ni to the general public from cooking using stainless steel utensils.

The U.S. FDA (2004) also conducted an analysis for Ni in market basket foods. Approximately 320 different food items were sampled for the period 1991 to 2002, from over 36 cities across the United States. The foods were prepared as they would be consumed (table-ready), and three samples per food item were combined to form a single analytical composite for each food item. Details of the nature of the cooking of the samples were not available. Nickel was not detected in 23% of the 6459 samples evaluated. In the calculation of the mean values for each food item, U.S. FDA (2004) used a value of zero for samples with Ni levels below detection. The results of this study were also lower than the Dabeka and McKenzie (1995) analysis, but higher than the Port Colborne analysis (JWEL 2004a). There was good agreement between the JWEL and the U.S. FDA dataset for fish and shellfish, dairy products, root vegetables, other vegetables and fats and oils. For cereals and grains, the U.S. FDA data were in good agreement with the Dabeka and McKenzie database. All three databases agreed well for other vegetables.

Based on this review, the Port Colborne data were determined to be the most recent and reliable food dataset for a Canadian population. Therefore, the dataset selected for use in the Sudbury HHRA was market basket data sampled from the Port Colborne area (JWEL 2004a) because it fulfilled all of the selection criteria and was found to be the most appropriate for Ni. The Port Colborne data were gathered in 2002 with between 1 and 10 samples per food item (this number includes replicates and duplicates). The Port Colborne market basket study found 16.5% of food items were below the MDL (0.0091 mg/kg dw) (JWEL 2004a). Most samples with non-detectable concentrations of Ni were in the meat, poultry, eggs, milk and milk products food categories.

Selenium

No Canadian food data for Se were found in the published literature. A recent survey conducted by the U.S. FDA, which analyzed foods consumed in the United States during the period 1991 to 2004, detected Se in 5586 out of 10,026 food samples (U.S. FDA 2004). The Canadian Nutrient File (2001) contained data on the Se content of foods; however, the data were derived from American sources (i.e. United States Department of Agriculture) and were reported in a manner that was inconsistent with the purpose of the Sudbury HHRA (e.g. g/cup; g/8 nuts; g/sandwich). Thus, the FDA (2004) data were selected as the dataset to use in the Sudbury HHRA because of the robustness of the dataset (>10,000 food samples) and the lack of suitable Canadian alternatives. The mean Se values reported by the U.S. FDA assumed that any non-detectable values were equal to zero. For the purpose of this study, the recalculated UCLs on the mean assumed that non-detectable values were equal to half the detection limit.

Background Market Food Basket Exposure

Market basket exposures were defined as exposures resulting from the consumption of typical supermarket foods. Metal concentrations in market basket foods were considered representative of the typical levels observed in supermarket foods across Canada. For the current assessment, market basket exposures were classified as background exposures (i.e. exposures which are independent of the GSA). The 95% UCLM intake rates of specific food groups were used to determine market basket exposures for both the CTE and RME estimates. Food intake rates provided by Richardson (1997) were based on a 24-h recall study collected during the 1972 to 1973 National Food Consumption Survey (NFCS). The lognormal probability distributions representing food consumption provided by Richardson (1997) do not reflect long-term food consumption patterns of an individual, but rather the variability of reported consumption rates of many individuals over a 24-h recall period. It is not considered realistic (nor is it recommended) to use the Richardson (1997) probability density functions (describing variability in 24-h consumption rates) when characterizing long-term exposures.

7.1.4 Summary of Exposure Data Used in the HHRA

Table 7.9 provides a summary of the exposure point concentration (EPC) data outlined in the previous sections and Chapter 6, which were used in the Sudbury risk assessment.

Table 7.9 **Summary Values for All Exposure Point Concentrations (EPCs) Used in the HHRA**

Community of Interest	As[a]	Co	Cu	Pb	Ni	Se
Soil concentrations			µg/g			
Coniston	12	19	320	52	433	1.3
Copper Cliff	19	33	1370	98	976	7.5
Falconbridge	79	57	1010	82	1070	3.1
Hanmer	4.3	6.6	67	19	68	0.68
Sudbury Centre	7.2	11	204	36	210	1.3
Typical Ontario Resident	17	21	85	43	120	1.9
Dust concentrations (calculated)[b]			µg/g			
Coniston	87	98	204	127	221	49
Copper Cliff	98	113	298	150	273	77
Falconbridge	142	130	276	143	280	61
Hanmer	67	74	136	98	137	41
Sudbury Centre	76	85	182	116	183	49
Typical Ontario Resident	95	101	145	121	158	54
Air concentrations (outdoor and indoor)			$µg/m^3$			
Coniston	0.0024	0.00087	0.016	0.0080	0.012	0.0034
Copper Cliff	0.0050	0.0025	0.081	0.022	0.059	0.0055
Falconbridge	0.0024	0.0025	0.026	0.015	0.028	0.0034
Hanmer	0.0056	0.00066	0.099	0.0098	0.012	0.0040
Sudbury Centre						
Combined data (2 stations)	0.0061	0.0097	0.17	0.025	0.095	0.0092
Travers Street only	0.0090	0.018	0.20	0.031	0.26	0.014
Typical Ontario Resident	0.001	0.0019	0.0091	0.0080	0.0014	0.0019

Detailed Human Health Risk Assessment Methods

Community of Interest	As[a]	Co	Cu	Pb	Ni	Se
Drinking water			µg/L			
Coniston	1.1	0.2	45	0.31	53	1.3
Copper Cliff	2.5	0.05	170	1.4	49	3
Falconbridge	2.6	0.2	30	0.97	32	2.5
Hanmer	1.5	0.06	65	0.49	0.8	1.3
Sudbury Centre	1.1	0.2	45	0.31	53	1.3
Typical Ontario Resident	0.64	0.088	0.41	2.2	1.9	1.6
Home garden – below ground vegetables			µg/g wet weight			
Coniston	0.0069	0.024	0.81	0.26	0.56	0.029
Copper Cliff	0.0088	0.019	1.2	0.13	1.7	0.42
Falconbridge	0.025	0.13	1.2	0.23	3.7	0.016
Hanmer	0.042	0.10	1.1	0.25	0.31	0.10
Sudbury Centre	0.0075	0.017	1.1	0.075	0.79	0.040
Home garden – above ground vegetables			µg/g wet weight			
Coniston	**0.0069**	0.21	0.54	0.095	0.57	0.030
Copper Cliff	**0.016**	0.13	0.92	0.13	1.8	0.68
Falconbridge	**0.052**	0.11	0.75	0.038	2.0	0.02
Hanmer	**0.0046**	0.0074	0.46	0.089	0.28	0.0083
Sudbury Centre	**0.0067**	0.027	0.75	0.094	0.75	0.059
Home garden – fruits			µg/g wet weight			
All COI	**0.0063**	0.019	0.90	0.046	2.7	0.058
Wild berries			µg/g wet weight			
All COI	**0.0052**	0.016	0.68	0.074	0.71	0.016
Local commercial produce			µg/g wet weight			
Root vegetables	**0.0086**	0.037	1.0	0.11	0.91	0.13
Above ground vegetables	**0.0079**	0.038	0.71	0.078	1.1	0.10
Fruit	**0.0061**	0.035	0.65	0.042	1.5	0.024
Fish and wild game			µg/g wet weight			
Wild game	**0.00013**	0.040	0.68	0.0040	0.62	1.4
Fish	**0.00022**	0.019	0.52	0.30	0.032	2.0
Market basket foods – TEDIs			µg/g			
Infant formula	**7.2×10-6**	0.0046	0.90	0.0023	0.011	0.020
Dairy	**0.0032**	0.010	0.36	0.0060	0.015	0.072
Meat and eggs	**0.00046**	0.011	1.1	0.0066	0.022	0.25
Fish	**0.00041**	0.0093	1.3	0.0069	0.037	0.43
Root vegetables	**0.0043**	0.033	1.1	0.0073	0.075	0.014

Community of Interest	As[a]	Co	Cu	Pb	Ni	Se
Other vegetables	**0.0093**	0.013	1.2	0.0050	0.28	0.023
Fruits	**0.0022**	0.025	1.7	0.014	0.080	0.0092
Cereals and grain	**0.0059**	0.025	1.8	0.012	0.17	0.13
Sugar and sweets	**0.0077**	0.024	1.4	0.040	0.27	0.021
Fats and oils	**0.0091**	0.022	0.25	0.00038	0.057	0.025
Nuts and seeds	**0.0073**	0.063	14	0.014	2.0	0.32

[a] The As exposure point concentration (values in bold) for all food products was adjusted to represent only the inorganic As fraction content of the food (on which the TRV is based).

[b] Indoor dust concentrations calculated based upon regression equation developed from paired soil and indoor dust data collected during the Sudbury indoor dust survey.

7.2 Hazard Assessment

7.2.1 Overview

This phase of the study is the hazard assessment, which is also sometimes termed the 'toxicity assessment'. The primary objectives of the hazard assessment are to:

- Provide the reader with an understanding of the toxicological effects that have been reported to be associated with exposure to the COC by various routes.
- Identify whether each COC is considered to cause carcinogenic (non-threshold) or non-carcinogenic (threshold) effects.
- Identify the most appropriate and scientifically-defensible exposure limits against which exposures can be compared to provide estimates of potential health risks.

Toxicity refers to the potential for a chemical to produce any type of damage, permanent or temporary, to the structure or functioning of any part of the body. The toxicity of a chemical depends on the amount of chemical taken into the body (referred to as the 'dose') and the duration of exposure (i.e. the length of time the person is exposed to the chemical). For every chemical, there is a specific dose and duration of exposure necessary to produce a toxic effect in humans (this is referred to as the 'dose–response relationship' of a chemical). The toxic potency of a chemical (i.e. its ability to produce any type of damage to the structure or function of any part of the body), is dependent on the inherent properties of the chemical itself (i.e. its ability to cause a biochemical or physiological response at the site of action), as well as the ability of the chemical to be absorbed into the body (i.e. bioavailability), and then to reach the site of action. The dose–response principle is central to the HHRA methodology. There are two main types of dose–response relationships for chemicals:

Threshold Response Effects: For some chemicals, it is thought that there is a dose–response threshold below which no adverse effects would be expected to occur. This relationship is true for all chemicals that do not cause cancer by altering genetic material (e.g. most metals). Thresholds are generally assumed for non-carcinogens because, for these types of effects, it is generally believed that homeostatic, compensating, and adaptive mechanisms must be overcome before toxicity is manifested. Exposure limits derived for threshold-response chemicals are called reference doses (RfD), acceptable daily intakes (ADI), tolerable daily intakes (TDI) or permissible daily intakes (PDI) and are generally derived by regulatory agencies such as Health Canada and the U.S. Environmental Protection Agency (U.S. EPA). These values indicate doses of chemicals that individuals can receive on a daily basis without the occurrence of adverse health effects. Exposure limits derived for threshold-response chemicals are typically expressed as: g/kg body weight/day, and are typically based on experimentally-determined 'No Observed Adverse Effect Levels' (NOAELs), with the application of extrapolation factors that are often referred to as 'safety factors' or 'uncertainty factors' (U.S. FDA 1982; U.S. EPA 1989; Health Canada 1993). The magnitude of these factors is dependent on the level of confidence in the available toxicology database, and reflects differences in species, duration of exposure, sensitivity, and overall quality of available data (i.e. the weight-of-evidence of the supporting data).

Non-threshold Response Effects: For these chemicals, it is assumed that there is no dose–response threshold. This means that any exposure greater than zero is assumed to have a non-zero probability of causing some type of response or damage. This relationship is typically used for chemicals which can cause cancer by damaging genetic material. Under a 'no threshold' assumption, any exposure has some potential to cause damage, so it is necessary to define an 'acceptable' level of risk associated with these types of exposures. For the purposes of evaluating exposures to chemicals in the environment, the 'acceptable' level of risk is usually defined as a risk of one-in-one hundred thousand to one-in-one million. These numbers can be better explained as the daily dose that may cause an additional incidence of cancer (i.e. one cancer that would not be expected in the absence of the exposure) in a population of one hundred thousand (or a million) people exposed every day over their entire lifetime. The acceptable level of risk is a policy rather than a scientific decision, and is set by regulatory agencies, as opposed to risk assessors. For example, the MOE has indicated that an incremental lifetime cancer risk level less than one-in-one million would be considered a *de minimis* risk level; in other words, a risk which is considered so small, it is of little or no significance and is acceptable from a regulatory perspective (MOEE 1987). Exposure limits derived for non-threshold chemicals that are believed to be potential carcinogens are typically expressed as 'increased risk per unit of dose'. The potency estimates are called cancer slope factors (SF) or cancer potency factors (e.g. $[\mu g/kg$ body weight/day$]^{-1}$). These values are derived using a mathematical model unit risk estimation approach with the built-in assumption that the condition of 'zero increased risk of cancer' would only be observed when the dose is zero.

The health endpoint of concern for carcinogenic chemicals in the HHRA framework was considered to be incremental lifetime cancer risk. As such, the exposure period that was assessed was an assumed lifetime which is typically a period of 70 years (U.S. EPA 1989). However, for exposure periods that comprise less than 70 years (which is generally the case), the exposures must be amortized (or averaged) over the entire lifetime. Thus, if an individual is exposed to COC for 5 years, the exposure estimate would typically be multiplied by a factor of 5/70, to yield an amortized exposure estimate. For each exposure scenario assessed, all five receptor age classes were evaluated to provide an evaluation of lifetime cancer risks through the use of a composite receptor.

It must be recognized that the assumption of no dose–response threshold for carcinogens is an assumption which is not directly testable by experimentation. Thresholds may exist, even for assumed non-threshold chemicals and effects. The 'no threshold' assumption ignores a large number of factors, such as the ability of the body to repair damage to genetic material, that are known to be important responses of people to naturally occurring genotoxic carcinogens. Exposure to small concentrations of chemicals which have the potential to cause cancer happens on a daily basis to everyone in the world, because non-threshold chemicals (along with other chemicals which do not cause cancer) are present in soils, air, food and water, either from natural sources or as a result of human activities. The human body has many ways of handling these substances once they enter the body. In many cases, the body can repair damage that may be caused by exposures to low levels of carcinogenic chemicals; therefore, adverse effects do not necessarily occur.

The development of toxicological criteria or exposure limits for any given chemical must consider factors which affect the potential toxicity of that chemical. These factors may be scenario-specific, such as variation in duration or levels of exposure. Where possible, it is important that exposure limits be derived from 'realistic' exposure situations that are representative of those occurring under the conditions assessed in the HHRA. For many chemicals, the toxic endpoint is also dependent on the route of exposure, as exposure via different routes may impact different tissues, such as those at the site of entry. In such a case, different exposure limits may be identified or developed for the different routes of exposure. Toxic potency may be modified by species- or individual-specific factors such as the ability to resist, repair or adapt to the effects of chemical exposures. In these situations, separate exposure limits might be used to ensure protection of sensitive sub-populations.

Exposure limits for chemicals are based on scientific information, professional judgment and technical review by experienced scientists with expertise in a wide range of scientific disciplines. Exposure limits are derived based on the most sensitive endpoints in individuals (e.g. cancer, organ damage, neurological effects, reproductive effects, etc.). In many cases, large uncertainty factors (i.e. 100-fold or greater) were used in establishing exposure limits for chemicals causing effects that are expected to have thresholds. Thus, exceedance of the exposure limit does not necessarily mean that adverse effects will occur. Rather, this result would necessitate a more detailed evaluation of both exposure and the toxicity-based exposure limit to better understand the likelihood of adverse effects occurring. Exposure rates less than an exposure limit are usually considered unlikely to be associated with adverse health effects and are, therefore, less likely to be of concern. As the frequency or magnitude of exposures exceeding the exposure limit increase, the probability of adverse health effects in a human population is usually presumed to increase, subject to scientific judgment and critical evaluation of the exposure limit and the exposure estimate, as discussed above. However, it should not be categorically concluded that all exposures below an exposure limit will be unlikely to result in adverse health effects or that all exposures above such a limit are likely to result in adverse health effects.

7.2.2 Exposure Limits Selected for the HHRA

A detailed toxicological assessment was conducted for each COC, involving identification of mechanism of action and relevant toxic endpoints, and determination of receptor- and route-specific toxicological criteria. These profiles were not intended to provide comprehensive reviews of the available toxicological and epidemiological literature on each COC. Rather, the purpose of the toxicological profiles was to: i) summarize the most relevant toxicological and epidemiological information on the substances; ii) outline any recent information that may challenge previous findings; and iii), provide supporting rationale for the exposure limits selected for use in the risk assessment. The toxicological reviews were based primarily on secondary sources, such as ATSDR toxicological profiles and other detailed regulatory agency reviews, and were supplemented with recent scientific literature.

Exposure limits for the COC in the current HHRA were identified from regulatory agencies such as MOE, Health Canada, U.S. EPA, U.S. Agency for Toxic Substances and Disease Registry (ATSDR), California Environmental Protection Agency Office of Environmental Health Hazard Assessment (CalEPA OEHHA), U.S. Centers for Disease Control (CDC), the European Union (EU), and the World Health Organization (WHO). The exposure limits (or toxicological criteria) employed in the current assessment were obtained from a review of toxicological criteria from various regulatory agencies mentioned above. Review of the regulatory exposure limits (toxicological criteria) was supplemented by detailed toxicological assessments conducted for each COC, involving identification of mechanism of action and relevant toxic endpoints, and determination of receptor- and route-specific toxicological criteria. Together, this information was used to select toxicological criteria for each COC that were based on the best available science. In some instances, several regulatory agencies and/or authorities have recommended different exposure limit values for the same chemical. In this situation, a rationale was provided for using one regulatory criterion over another in this study.

The U.S. EPA derives exposure limits for both threshold and non-threshold effects when data are available. The reference dose (RfD) and reference concentration (RfC) are based on the assumption that a threshold exists for certain toxic non-carcinogenic effects. In general, the RfD (or RfC) is an estimate (with uncertainty spanning perhaps an order of magnitude) of a daily exposure to the human population (including sensitive subgroups) that is likely to be without an appreciable risk of deleterious effects during a lifetime (U.S. EPA 2005). For a number of chemicals, exposure limits were not always available for all exposure routes of concern. In these circumstances, exposure limits may be extrapolated from other routes. For example, it is common in human health risk assessments to assess the risks posed by dermal absorption of a chemical based on the exposure limit established for oral exposure (U.S. EPA 1989, 1992). The systemic dose absorbed dermally is scaled to the 'equivalent' oral dose by correcting for the bioavailability of the dermally applied chemical relative to an orally administered dose.

As of January 1991, IRIS and NCEA databases no longer presented RfDs or slope factors (SFs) for the inhalation route (U.S. EPA 2004c). These criteria were replaced with reference concentrations (RfC) for non-carcinogenic effects and unit risk factors (URF) for carcinogenic effects. However, for purposes of estimating risk and calculating risk-based concentrations, inhalation reference doses (RfDi) and inhalation slope factors (SFi) were preferred. This was not a problem for most chemicals because the inhalation toxicity criteria were easily converted. The toxicological criteria selected for use in the Sudbury HHRA are summarized in Table 7.10.

Table 7.10 **Summary of Toxicological Criteria Used in the Sudbury HHRA**

Chemical	Route	Toxicological criterion		Endpoint	Study	Regulatory Agency
Arsenic	Oral	RfD	0.3 µg/kg/day	Hyperpigmentation, keratosis, possible vascular complications (human)	Tseng et al. 1968; Tseng 1977	U.S. EPA 1993
		SF_O	0.0015 (µg/kg/day)$^{-1}$	Skin cancer, basal and squamous cell carcinoma (human)	Tseng et al. 1968; Tseng 1977	U.S. EPA 1998
	Inhalation	Chronic REL	0.03 µg/m³	Decreased fetal weight; increased incidences of intrauterine growth retardation and skeletal malformations in mice	Nagymajtényi et al. 1985	OEHHA 2000
		SF_i (IUR)	0.015 (µg/kg/day)$^{-1}$ [4.3×10^{-3} (µg/m³)$^{-1}$]	Lung cancer (human)	Enterline and Marsh 1982; Higgins 1982; Brown and Chu 1983a,b,c; Lee-Feldstein 1983	U.S. EPA 1998
Cobalt	Oral	RfD	10 mg/kg/day	Polycythemia	Davis and Fields 1958	ATSDR 2001
	Inhalation	RfC	0.5 mg/m³	Interstitial lung disease	Sprince et al. 1988	RIVM (Baar et al. 2001)
Copper	Oral	UL	140 µg/kg/day	Liver damage (human)	Pratt et al. 1985	IOM 2001; Health Canada 2005
	Inhalation	TCA	1 µg/m³	Subchronic NOAEC (respiratory and immunological effects) (rabbits)	Johansson et al. 1984	RIVM (Baars et al. 2001)
Lead	Oral, Inhalation, Dermal	IOC_{POP}	1.85 µg/kg/day	Subclinical neurobehavioral and developmental effects (child)	Various	MOE 1994; 1996a
Nickel	Oral	RfD	20 µg/kg/day	Decreased body and organ weight (rats)	Ambrose et al. 1976	U.S. EPA 1996
	Inhalation	RfC	0.02 µg/m³ (total nickel)	Respiratory effects (lung inflammation and lung fibrosis)	European Commission DG Environment 2001	OJEU 2005
Selenium	Oral	RfD/TRV	5.00 µg/kg/day	Selenosis, including hair loss and nail sloughing (human) Clinical selenosis	Yang and Zhou 1994 Yang et al. 1989a,b	IOM 2000; Health Canada 2005 U.S. EPA 1991a
	Inhalation	Chronic REL RfC	20 µg/m³	Hepatic, cardiovascular, neurological effects (human)	Yang et al. 1989a,b Dudley and Miller 1941	OEHHA 2001

Note: For chemicals with no identified inhalation toxicological criteria, it was assumed that inhalation bioavailability and toxic potency are equivalent to that which occurs via the oral exposure route.

7.2.3 Toxicological Profiles for the COC

The following section provides a very brief toxicological profile for each of the COC. Much more detailed and extensive profiles were prepared for the technical reports in the actual Sudbury HHRA, but an overview is provided here. Note that the literature was current to 2005, and more recent information may be available for each of these COC.

7.2.3.1 Inorganic Arsenic

Arsenic has not been demonstrated to be essential in humans (WHO-IPCS 2001). The following text relates to inorganic As species only, as all regulatory TRVs that exist for As have been developed for inorganic As exposure. The following organizations were consulted to select exposure limits for As: the U.S. EPA; MOE; ATSDR; Health Canada; the Dutch National Institute for Public Health and the Environment (RIVM); NRC; WHO; and OEHHA. Exposure limits derived by the U.S. EPA were selected for use in this assessment, with the exception of the inhalation RfC, for which the U.S. EPA has not derived a value. Thus, the chronic reference exposure level (REL) developed by OEHHA was used as a threshold inhalation exposure limit.

Other risk assessment studies in several Ontario communities [e.g. Deloro (Cantox Environmental 1999), Wawa (Goss Gilroy 2011), Falconbridge (Do et al. 2011), Port Colborne (JWEL 2004b)] revealed that As is a complex substance to evaluate in HHRAs. It is important to evaluate As exposures and risks using a weight-of-evidence approach that includes risk assessment, biomonitoring (urinary As), predictive modeling and medical surveillance to collectively and definitively address concerns related to As exposures at contaminated sites (Sigal et al. 2002a,b). This approach was successfully applied in other communities in Ontario where As was a concern (e.g. Deloro, Wawa and Port Colborne) and proved effective in ensuring public safety and satisfying the concerns of the local community and regulators.

The cancer potency of As continues to be a source of controversy in the risk assessment and management of As-contaminated sites. The use of U.S. EPA slope factors to estimate possible cancer risks to people through all pathways (air, water, food, soils) consistently results in risk values from natural (i.e. background) sources at or higher than *de minimis* risk levels. Several studies and reviews have questioned the relevance of the Taiwanese dataset for the North American population (Brown and Chen 1995; Lamm et al. 2004; U.S. EPA 2007). For example, Lamm et al. (2004) considered the relationship between As exposure through drinking water and bladder cancer mortality. County specific mortality ratios were considered for 133 counties across the U.S. where the primary source of drinking water was groundwater. No As-related increase in bladder cancer mortality was found over an exposure range of 3 to 60 µg/L. In Ontario, background As soil levels (17 µg/g) and the generic residential/parkland soil criterion (25 µg/g) were associated with predicted incremental lifetime cancer risk levels in the one-in-one-hundred thousand range. Cancer risk estimates well above the *de minimis* risk level were also routinely predicted for As exposures associated with typical North American diets, air quality and regulated North American drinking water supplies. These elevated As risk levels that result from typical and/or natural exposure conditions create challenges in communicating risk estimates for both incremental and total As exposures. In discussing As risk estimates in a HHRA, it is critical to provide additional perspective using information from a weight-of-evidence approach that includes a variety of 'tools' in addition to risk assessment, such as bio-monitoring, predictive modeling and medical surveillance. In combination, these tools can be helpful for regulators and other stakeholders in considering the real-world implications of hypothetical risk predictions based upon HHRA.

Arsenic Oral Exposure Limits

Non-Carcinogenic (Threshold) Effects

The U.S. EPA (1993) calculated an oral RfD of 0.3 µg As/kg body weight/day based on the epidemiological studies of chronic exposure to arsenic through drinking water (Tseng et al. 1968; Tseng 1977). Critical effects were hyperpigmentation, keratosis, and possible vascular complications at a lowest-observable-adverse-effects-level of 14 µg As/kg body weight/day. The RfD was based on a NOAEL of 0.8 µg As/kg body weight/day, with the application of an uncertainty factor of three to account for both lack of data on reproductive toxicity in humans, and for differences in individual sensitivity. New data that could possibly impact on the recommended RfD for As will be evaluated by the U.S. EPA Work Group as it becomes available. Confidence in the chosen principal study and the resulting oral RfD is considered medium. The MOE (1996b) adopted 0.3 µg As/kg body weight/day, based on information provided on IRIS in 1993. This conservative exposure limit was in use by the U.S. EPA (1998) and the value of 0.3 µg As/kg body weight/day was selected as the oral exposure limit for non-carcinogenic effects in this study.

Carcinogenic (Non-threshold) Effects

Arsenic exposure via the oral route was considered by the U.S. EPA to be carcinogenic to humans, based on the incidence of skin cancers in epidemiological studies examining human exposure through drinking water in Taiwan (Tseng et al. 1968; Tseng 1977). Based on the application of a linear-quadratic mathematical model to the data from these studies, the U.S. EPA (1998) calculated an oral slope factor of 0.0015 (μg As/kg body weight/day)$^{-1}$. The slope factor (SF) was based on the assumption that carcinogenic effects do not have a threshold (i.e. dose–response relationship is linear to zero exposure). It was assumed that the Taiwanese individuals had a constant exposure from birth. Recently, there has been concern on the part of regulators with regard to the applicability of the As cancer potency estimates for cancers at other sites (specifically bladder cancer) in setting exposure limits for As. The National Research Council (NRC) (1999, 2001) re-evaluated drinking water criteria for the United States, based on bladder cancer incidence data in the Taiwanese population as presented in Wu et al. (1989), Chen et al. (1992) and Smith et al. (1992). NRC (1999, 2001) emphasized that the evaluation of cancer potency factors for bladder cancer was limited by the amount and the quality of data available for use in the linear model. As the intended use of the cancer potency factor is in the estimation of risk to a particular population in comparison to a 'background' or 'typical' population, and risks for both will be assessed with the same methodologies and the same exposure limit, the use of the skin cancer potency factor was considered acceptable and conservative. The SF of 0.0015 (μg As/kg body weight/day)$^{-1}$, corresponding to an RsD of 0.00067 μg As/kg body weight/day for an acceptable risk level of one-in-one million, was adopted as the oral exposure limit for carcinogenic effects of As for this assessment.

Arsenic Inhalation Exposure Limits

Non-cancer (Threshold) Effects

At the time of this study, the U.S. EPA had not established an inhalation reference concentration or dose for As. Thus, the chronic REL developed by OEHHA (2000) was used. The OEHHA (2000) was based on the study by Nagymajtenyi et al. (1985) for deriving the chronic REL. This was accomplished by using the average experimental exposure for the lowest observed adverse effect level (LOAEL) group (determined to be 33 μg As/m^3) and applying a cumulative uncertainty factor of 1000 (10-fold each for use of a LOAEL, interspecies extrapolation and intraspecies differences in sensitivity) to yield a chronic REL of 0.03 μg As/m^3. According to the OEHHA (2000), route-to-route conversion of the LOAEL in the key study indicates that this chronic REL should also be protective of non-cancer adverse effects that have been observed in studies with oral exposures, either in food or drinking water.

Cancer (Non-threshold) Effects

The U.S. EPA (1998) considered As to be a non-threshold carcinogen. Based on this assumption, the U.S. EPA (1998) calculated an inhalation unit risk value of 0.0043 (μg As/m^3)$^{-1}$, based on studies by Brown and Chu (1983a,b,c), Lee-Feldstein (1983), Higgins (1982), and Enterline and Marsh (1982) which indicated increased lung cancer mortality of exposed populations. It was assumed that the increase in age-specific mortality rate of lung cancer was a function only of cumulative exposures. The unit risk was converted to a slope factor of 0.015 (μg As/kg body weight/day)$^{-1}$ assuming a 70 kg adult breathes 20 m^3/day. It should be noted that all of the studies used to derive the U.S. EPA unit risk value had a number of confounding factors and uncertainties. These included: concurrent exposure to airborne dusts, SO$_2$ and other chemicals; lack of measured air concentrations in some studies; failure to consider latent periods for lung cancer development; and, confounding by smoking.

Arsenic Dermal Exposure Limit

No dermal As exposure limits were developed by regulatory agencies at the time of this study. Route-to-route extrapolation was used to derive an appropriate limit for the current assessment.

7.2.3.2 Cobalt

Cobalt is an essential micronutrient in humans and most other organisms, as it is a required element in vitamin B12, and is also associated with the regulation of several cofactors and enzymes, and the production of erythropoietin (Lison et al. 2001). The Recommended Dietary Allowance (RDA) for vitamin B12 is 2.4 μg/day for adults, which corresponds to 0.1 μg/day of Co (ATSDR 2001). Due to its essentiality, Co occurs in many tissues of individuals with no known occupational or environmental exposure, with the highest concentrations occurring in the liver, where vitamin B12 is stored (ATSDR 2001). Adverse health effects will typically occur only at doses that exceed the daily nutritional requirements for Co.

Exposure Limits

ATSDR, U.S. EPA, MOE and RIVM were the regulatory agencies consulted to select exposure limits for Co. For the current assessment, Co was not considered a non-threshold carcinogen by the inhalation or oral routes of exposure. There was inadequate information available from oral studies to determine whether or not Co is carcinogenic via this route. IARC (2004) classified Co compounds as 'possibly carcinogenic to humans' and the American Conference of Governmental Industrial Hygienists (ACGIH) classifies Co in category A3 – confirmed animal carcinogen with unknown relevance to humans. Furthermore, under the old 1986 Guidelines for Carcinogen Risk Assessment (U.S. EPA 1986), Co was classified as group B1 (Probable Human Carcinogen), based on limited evidence of carcinogenicity in humans and sufficient evidence of carcinogenicity in animals, as evidenced by increased incidence of alveolar/bronchiolar tumors in both sexes of rats and mice (U.S. EPA 2002a). Under the U.S. EPA (1999b) cancer guidelines, Co was considered likely to be carcinogenic to humans (U.S. EPA 2002a). Health Canada had no TRVs for Co and had not classified Co compounds as to their carcinogenicity.

Oral Exposure Limits

Non-Carcinogenic (Threshold) Effects

The most sensitive indicators of the effects of Co following oral exposure appear to be related to an increase in hemoglobin in both humans and animals and the elicitation of dermatitis in sensitized individuals (ATSDR 2001). The U.S. EPA (2002a) reported an oral RfD for Co of 20 μg/kg/day. The exposure limit was based upon a study by Duckham and Lee (1976) which demonstrated an increased level of hemoglobin in anemic patients treated therapeutically at a level of 0.18 mg/kg/day. The oral RfD was calculated by dividing this LOAEL by 10 (three to account for the use of a LOAEL, three for deficiencies in the database, primarily the use of a sub-chronic study; uncertainly factor rounded to 10). In the derivation of the *Guideline for Use at Contaminated Sites in Ontario* (MOE 1996b), MOE utilized an oral RfD of 60 μg/kg/day for Co.

ATSDR (2001) derived an oral intermediate-duration MRL of 10 μg of Co/kg/day. The MRL was based on a LOAEL of 1 mg of Co/kg/day for polycythemia as reported in a study by Davis and Fields (1958). These authors exposed six male volunteers to 120 or 150 mg/day of CoCl (~1 mg of Co/kg/day) for up to 22 days. Exposure resulted in the development of polycythemia in all six patients, with 16% to 20% increases in red blood cell numbers above pre-treatment levels. Oral MRL values were not derived by ATSDR for acute or chronic exposure to Co. An acute MRL was not derived because the reported effects in animals were serious and occurred at levels above those reported in the few available human oral studies. No chronic oral studies were available for humans or animals. RIVM (Baars et al. 2001) derived a tolerable daily intake (TDI) of 1.4 μg/kg/day based on a LOAEL of 0.04 mg/kg/day for cardiomyopathy in humans after intermediate oral exposure (Morin et al. 1971). RIVM used an uncertainty factor of 30 (three for intra-human variation and 10 for extrapolation to a NOAEL) to yield the TDI. For the purposes of the current assessment, the ATSDR (2001) MRL (10 μg of Co/kg/day) was selected as the oral exposure limit for non-carcinogenic effects.

Carcinogenic (Non-threshold) Effects

Cobalt does not appear to cause cancer in humans via inhalation, oral, or dermal exposure routes (ATSDR 2001). No studies were located in the literature with regard to carcinogenic effects in animals after oral or dermal exposure to Co. In a review of genotoxicity and carcinogenicity studies published between 1991 and 2001, Lison et al. (2001) concluded there was no evidence for genotoxic or carcinogenic activity of Co in humans. However, a more recent study by Hengstler et al. (2003) reported that co-exposure to Cd, Co and Pb may cause genotoxic effects even at concentrations below current regulatory limits, and that the cancer hazard of Co exposure may be underestimated, especially when individuals are co-exposed to Cd or Pb. This hypothesis has not yet been substantiated by other studies identified in the scientific literature.

Inhalation Exposure Limits

Non-cancer (Threshold) Effects

The U.S. EPA (2002a) reported an inhalation RfD for Co of 5.7×10^{-6} mg/kg/day based on an RfC of 2.0×10^{-5} mg/m^3. The exposure limit was based upon an epidemiological study which showed a NOAEL of 0.0053 mg of Co/m^3 and a LOAEL of 0.015 mg of Co/m^3 for decreases in expiratory flow rates in diamond polishers (Nemery et al. 1992). ATSDR (2001) also used the Nemery et al. (1992) study to develop its inhalation MRL, but only applied a 10-fold safety factor to the time-adjusted NOAEL, resulting in the derivation of a less conservative limit of 1×10^{-4} mg of Co/m^3. WHO (2006) also established a tolerable concentration for inhaled Co of 1×10^{-4} mg/m^3, based on a NOAEL of 0.0053 mg of Co/m^3 in diamond polishers (Nemery et al. 1992). The Dutch Institute for Public Health

(RIVM) (Baars et al. 2001) derived a tolerable concentration in air (TCA) of 0.0005 mg/m³, based on a LOAEL of 0.05 mg/m³ for interstitial lung disease in humans (Sprince et al. 1988). An uncertainty factor of 100 (10 for extrapolation from a LOAEL and a factor of 10 for intrahuman variability) was applied to the LOAEL to yield the TCA. Medium reliability is suggested for this TCA by RIVM (Baar et al. 2001). This exposure limit was selected for use in the Sudbury assessment.

Cancer (Non-threshold) Effects

Overall, the weight-of-evidence indicates that Co does not cause cancer in humans by the inhalation, oral, or dermal exposure routes. The U.S. EPA (2002a) classified Co as a group B1 Probable Human Carcinogen based on limited evidence of carcinogenicity in humans and sufficient evidence in animals following inhalation exposure. The U.S. EPA (2002a) derived an inhalation unit risk for Co of 2.8×10^{-3} (µg Co/m³) based on tumorigenic effects (alveolar and bronchiolar) in rats and mice (NTP 1998; Bucher et al. 1999) which equates to a inhalation slope factor of 9.8 (mg/kg/day)$^{-1}$. No other identified regulatory agencies have derived exposure limits for Co based on carcinogenic endpoints. There appears to be a consistent increased risk of respiratory tract cancer in workers co-exposed to both Co and tungsten carbide (i.e. 'hard metal' workers). However, the exposure conditions experienced by 'hard metal' workers would not be expected to occur in the natural ambient environment.

Dermal Exposure Limit

No regulatory dermal exposure limits for Co were identified in the literature. Route-to-route extrapolation was used to derive an appropriate limit for the current assessment.

7.2.3.3 Copper

Copper is an essential trace element that is naturally present in all environmental media (air, water, soil, sediments), as well as all biota and all foods consumed by humans. The primary source of Cu in humans is the diet. It is estimated that the typical daily Cu intake from food is around 1.0 to 1.3 mg/day for adults (ATSDR 2004). The World Health Organization (WHO 1998) reports that total daily intake of Cu in adults ranges between 0.9 and 2.2 mg, with most studies indicating daily Cu intakes at the lower end of this range. WHO (1998) notes that intakes may occasionally exceed 5 mg/day. In some cases, drinking water may also make a substantial additional contribution to the total daily Cu intake, particularly if corrosive waters remain in Cu pipes for prolonged periods. Other common environmental routes of exposure, such as inhalation and dermal uptake, are insignificant relative to oral consumption of dietary items.

Since Cu is an essential element, its uptake, metabolism and excretion are physiologically regulated, and most tissues of the body have measurable amounts of Cu associated with them. It has been estimated that the whole human body contains 100 to 150 mg of Cu at any given time (WHO 1998). All mammals have metabolic mechanisms that maintain Cu homeostasis (a balance between metabolic requirements for copper and prevention against accumulation to toxic levels, such that Cu levels are generally maintained within a range that avoids both deficiency and excess). As with any substance, even essential trace elements, excessive exposures may result in toxicity. Toxicity is likely to occur only when such homoeostatic controls are overwhelmed and/or basic cellular defense or repair mechanisms are impaired. However, this has only been documented to occur in individuals with genetic Cu metabolism impairment (e.g. Wilson's disease, Indian childhood cirrhosis, idiopathic Cu toxicosis) or cases of intentional or accidental poisoning, where very large amounts of Cu were ingested (WHO 1998; ATSDR 2004).

Threshold levels for Cu toxicity in humans have not been firmly established. However, it appears that the main intracellular binding site for Cu, metallothionein, becomes saturated with Cu before toxicity occurs. Copper deficiency rarely occurs in humans since most diets have Cu in excess of what is required by the body (WHO 1998). Copper deficiency is more common in animals, particularly livestock species, and may lead to several different disorders such as anemia, bone, nerve and cardiovascular disorders, failure of keratinization and reproductive failure (Davis and Mertz 1987).

Exposure Limits

The following organizations were consulted to select exposure limits for copper: Health Canada; the U.S. EPA; ATSDR; WHO; MOE; JECFA; Health and Welfare Canada; RIVM; and the National Academy of Science.

Copper Oral Exposure Limits

Non-Carcinogenic (Threshold) Effects

The National Academy of Science (IOM 2001) derived an acceptable Upper Limit (UL) based on the NOAEL of 10 mg/day from Pratt et al. (1985). IOM (2001) considered this NOAEL to be protective of the general population and felt that no further uncertainty factor was warranted. This decision was supported by the large database of human information indicating no adverse effects with 10 to 12 mg/day and a paucity of observed liver effects from Cu exposure in humans with normal Cu homeostasis. This NOAEL results in an acceptable upper limit (UL) of approximately 140 µg/kg/day for adults (10 mg/day ÷ 70 kg). There were insufficient data to establish unique ULs for any other age group (similar sensitivity for all ages) (IOM 2001). Health Canada indicated that the agency will officially adopt ULs as toxicity reference values for all essential elements (Health Canada 2005 pers. comm.) for contaminated site HHRAs.

It is important to recognize that all available regulatory oral exposure limit values for Cu were similar in magnitude, and were based on either typical daily intakes, or intakes associated with gastrointestinal distress. Copper doses at, or below any of these values would not be expected to result in adverse health effects under conditions of continuous lifetime daily exposure. It is also important to recognize that all oral exposure limits, regardless of their basis, lie within the range of typical estimated daily dietary intakes, and/or recommended nutritional requirements when the body weight of various human age classes is taken into account (e.g. if a 70 kg adult is assumed, the Health Canada TDI of 0.03 mg/kg body weight/day equates to a daily intake of 2.1 mg of Cu/day).

RIVM (Baars et al. 2001) noted that Cu is an essential nutrient, with a minimum daily requirement of 0.02 to 0.08 mg/kg/day (as reported by WHO 1996). It was determined that a TDI for Cu cannot be lower than the levels required for nutrition essentiality. Thus, RIVM based a TDI on the typical daily intake of the population which was shown to be 0.02 to 0.03 mg/kg/day on average, with a range of 0.003 to 0.1 mg/kg/day and an upper limit of 0.14 mg/kg/day (Slooff et al. 1989). This latter upper limit daily intake value (0.14 mg/kg/day) was selected as the TDI by RIVM. For the purposes of this risk assessment, an oral RfD of 140 µg/kg/day was selected (IOM 2001, Health Canada 2005, pers. comm.).

Carcinogenic (Non-threshold) Effects

There was no evidence to suggest that Cu compounds are carcinogenic in humans or animals (WHO 1998; ATSDR 2004; U.S. EPA 2004e; TERA 2004). There were no data available on the genotoxicity of Cu in humans exposed via oral, inhalation or dermal routes. The existing genotoxicity database suggests that Cu is a clastogenic agent, and some studies have shown that exposure to Cu can result in DNA damage; however, point mutation assay results are mixed and inconclusive (ATSDR 2004). Overall, the database on mutagenicity and genotoxicity of Cu compounds was limited and equivocal, and considerably more research is required to determine whether or not Cu is mutagenic and/or genotoxic to mammals (including humans) *in vivo*.

Cu Inhalation Exposure Limits

Non-cancer (Threshold) Effects

RIVM derived a tolerable concentration in air (TCA) of 0.001 mg/m^3 based on a NOAEC of 0.6 mg/m^3 for lung and immune system effects in rabbits from a short-term toxicity study by Johansson et al. (1984). RIVM used an uncertainty factor of 100 (10 each for intra- and interspecies variability), and adjusted for continuous exposure (5/7×6/24) to yield the TCA. OEHHA (1999) derived an acute reference exposure level (REL) for a 1 h exposure of 0.1 mg/m^3. This acute REL was considered protective against mild adverse effects. The REL was derived based on studies by Gleason (1968), and Whitman (1957, 1962) which investigated metal fume fever in workers. A NOAEL of 1 mg/m^3 was identified from these studies. The NOAEL was mainly based on the report of Whitman (1957) indicating that exposure to copper dust was detectable by taste, but that no other symptoms occurred following exposure to 1 to 3 mg/m^3 for an unspecified short duration. Given that the exposure duration was not clearly stated in these studies, no extrapolation to a 1 h concentration could be conducted. Rather, the NOAEL was assumed to be applicable to a 1 h exposure. A cumulative uncertainty factor of 10 was applied to the NOAEL (for intraspecies uncertainty) to yield the acute REL. Given the limitations of the existing data, OEHHA suggested that re-evaluation of the acute REL for Cu be conducted when better methods or data are available and did not derive a chronic REL for inhalation exposure to Cu. For the purposes of this risk assessment, an inhalation TCA of 1 µg/m^3 derived by RIVM (Baars et al. 2001) was selected.

Cancer (Non-threshold) Effects

There was no evidence to suggest that Cu compounds are carcinogenic in humans or animals (WHO 1998; ATSDR 2004; U.S. EPA 2004e; TERA 2004). There were no data available on the genotoxicity of copper in humans exposed via oral, inhalation or dermal routes.

Dermal Exposure Limits

No regulatory dermal exposure limits for Cu compounds were identified in the literature reviewed for the current assessment. Route-to-route extrapolation was used to derive an appropriate limit for the current assessment.

7.2.3.4 Lead

Lead is not known to be an essential micronutrient in humans or other mammals.

Exposure Limits

The following organizations were consulted to select exposure limits for Pb: Health Canada, RIVM, MOE, ATSDR, U.S. EPA and OEHHA. Although the toxicological database for Pb was large, the majority of human effects data were expressed as a blood Pb concentration (PbB), rather than a dose or concentration in an environmental medium. In addition, there were inadequate empirical data for demonstrating a threshold for the health effects of Pb. In fact, many consider lead a non-threshold toxicant, indicating that any exposure to Pb may cause possible effects. Given these limitations, many regulatory agencies have not derived conventional exposure limits such as RfDs, TDIs or MRLs, and advocate that exposure to Pb should be minimized. In order to utilize the wealth of literature relating human PbB concentrations to health effects, such agencies (e.g. ATSDR, U.S. EPA) have developed models or other approaches to relate environmental Pb exposure to PbB levels. This is discussed further in Chapter 8. In addition, environmental quality guidelines for Pb have also been developed with a different approach than is used for most other chemicals. Instead of developing exposure limits based on no- or low-effects-levels observed in test organisms following controlled exposures, Pb guidelines are typically back calculated from a critical PbB concentration (usually 10 µg/dL, as recommended by CEOH 1994; CDC 2004, 2005; U.S. EPA 2004d).

Although recent scientific data indicate an association between intellectual performance in children and PbB levels < 10 µg/dL, it appeared that major agencies (CDC 1991, 2004, 2005; U.S. EPA 2003, 2007; ATSDR 2005; MOE 2007) acknowledged that a clear threshold for protection of neurological impacts in children had not yet been identified. In addition, derivation of acceptable exposure levels is complicated by numerous confounding factors that influence Pb toxicity, including socioeconomic status, pre-existing Pb body burdens, age, health status, nutritional status and lifestyle factors such as alcohol consumption and tobacco smoke (tobacco smoke has been associated with elevated PbB).

Lead Oral Exposure Limits

Non-Carcinogenic (Threshold) Effects

While the issue of whether or not a threshold exists for the cognitive effects of Pb in children continues to be debated, there was consistent information from the available Pb health effects literature indicating that PbB levels > 10 µg/dL are linked to decreased intelligence and impaired neurobehavioral development (WHO 1995; CDC 2004, 2005; U.S. EPA 2004d; Lanphear et al. 2003; ATSDR 2005) in children.

The MOE (1994) recommended an intake of concern for populations (IOC_{pop}) of 1.85 µg/kg/day to minimize the predicted number of children with individual blood Pb levels of concern. Subclinical neurobehavioral and developmental effects were the critical effects appearing at the lowest levels of exposure (MOE 1994). The intake of concern for individuals (IOC_{ind}) was based on a LOAEL in infants and young children of 10 µg/dL PbB divided by an intake/PbB slope factor of 0.21 µg Pb per dL PbB per µg/day. This resulted in an IOC_{ind} of 3.7 µg/kg/day for a 13 kg child (0.5–4 years). To derive the IOC_{pop} an uncertainty factor of 2 was applied to the IOC_{ind}, which resulted in a daily intake of 1.85 µg/kg/day (MOE 1994). This value was based on the same research as the other agencies' limits, as described below.

The U.S. EPA IRIS database did not recommend oral or inhalation reference doses (or concentrations) for Pb due to high levels of uncertainty, and because Pb is considered a non-threshold toxicant (U.S. EPA 2004d). The U.S. EPA believes that the effects of Pb exposure, particularly changes in blood enzyme levels, and children's neurodevelopment, may occur at blood levels so low as to be essentially without a threshold. Alternatively, the U.S. EPA has developed the Integrated Exposure Uptake Biokinetic Model (IEUBK) as a means of predicting the occurrence of blood Pb concentrations in children (U.S. EPA 1994). The IEUBK model predicts the geometric mean PbB concentration for a child exposed to Pb in various media (or a group of similarly exposed children). The model can also calculate the probability that the child's PbB exceeds 10 μg Pb/dL (P10), or some other value. Preliminary remediation goals (i.e. Soil Risk Management Levels) for Pb are generally determined with the model by adjusting the soil concentration term until P10 is below a 5% probability (U.S. EPA 2003, 2004c).

Health Canada (2003a,b) adopted the value of 3.6 μg/kg body weight/day as the provisional TDI for Pb, and CCME and Health Canada use this value as the basis for derivation of soil and drinking water guidelines that are protective of human health. In the Netherlands, RIVM (Baars et al. 2001) also derived a TDI of 3.57 μg/kg body weight/day, based on the PTWI of 25 μg/kg/week derived by the FAO/WHO (1993). For the purposes of this risk assessment, an oral, inhalation and dermal exposure limit of 1.85 μg/kg/day was selected for Pb (MOE 1994, 1996a).

Carcinogenic (Non-threshold) Effects

The U.S. EPA (2004d) classified Pb compounds as B2 – probable human carcinogen, based on sufficient animal evidence of kidney tumors, but inadequate human evidence. The U.S. EPA determined that an estimate of carcinogenic risk from oral exposure (such as a slope factor) using standard methods would not adequately describe the potential risk for lead compounds. IARC (2004) classified inorganic Pb compounds as probably carcinogenic to humans (Group 2A), based on limited evidence for carcinogenicity in humans and sufficient evidence for carcinogenicity in experimental animals. Health Canada has not formally classified Pb compounds with respect to their carcinogenic potential. The OEHHA (California Environmental Protection Agency Office of Environmental Health Hazard Assessment) considers Pb compounds to be human carcinogens and derived both oral and inhalation slope factors and unit risks for Pb. However, at this time, no other regulatory agencies, other than OEHHA, were known to have derived regulatory exposure limits for Pb that are based on carcinogenic effects.

Lead Inhalation Exposure Limits

Non-cancer (Threshold) Effects

There were no non-cancer inhalation exposure limits for Pb. However, the FAO/WHO TDI of 3.57 μg/kg/day accounts for Pb exposure from all sources and was considered protective of all humans, including infants and children. The MOE IOC$_{pop}$ of 1.85 μg/kg/day was also considered protective of multimedia Pb exposure (MOE 1994). For the purposes of this risk assessment, an oral, inhalation and dermal exposure limit of 1.85 μg/kg/day was selected for Pb (MOE 1994, 1996a).

Cancer (Non-threshold) Effects

Only one agency was identified as having developed quantitative toxicity estimates based on the carcinogenicity of Pb (OEHHA 2002). The U.S. EPA did not derive any exposure limits based on carcinogenic endpoints as its Carcinogen Assessment Group concluded that the uncertainties associated with Pb pharmacokinetics and factors affecting the absorption, release, and excretion of Pb preclude the development of a numerical estimate to predict carcinogenic risk (U.S. EPA 2004d). The OEHHA (2002) reported an inhalation unit risk factor of 1.2 E-5 (μg/m^3)$^{-1}$, an inhalation slope factor of 0.042 (mg/kg/day)$^{-1}$, and oral slope factor of 0.0085 (mg/kg/day)$^{-1}$. All values were originally calculated by OEHHA (1997) from rat kidney tumor incidence data (Koller et al. 1985). For the purpose of the current assessment, Pb was not considered a carcinogen for oral or inhalation exposures.

Dermal Exposure Limits

No regulatory dermal exposure limits for Pb compounds were identified in the literature. For the purposes of this risk assessment, an oral, inhalation and dermal exposure limit of 1.85 μg/kg/day was selected for Pb (MOE 1994, 1996a).

7.2.3.5 Nickel

Nickel is an essential trace element in animals, based on reports of Ni deficiency in several animal species (e.g. rats, chickens, cows, and goats) (ATSDR 2003). However, Ni deficiency has never been reported in humans as Ni intake generally exceeds dietary requirements (Anke et al. 1995; Denkhaus and Salnikow 2002). Nickel is widely considered to be a normal constituent of the diet, with daily intakes ranging from 100 to 300 μg/day (U.S. EPA 1991a). The functional importance of Ni has not been clearly demonstrated as no enzymes or cofactors that include Ni are known in humans (Denkhaus and Salnikow 2002). Therefore, the essentiality of Ni in humans has not been confirmed and Ni dietary recommendations have not been established for humans (Denkhaus and Salnikow 2002; ATSDR 2003).

Exposure Limits

TERA, ATSDR, OEHHA, Health Canada, and the U.S. EPA were consulted to select exposure limits for nickel. In addition, the inhalation exposure limits developed by Seilkop (2004) were also considered for this study given the current state of knowledge related to Ni compounds and their toxicological behavior.

Nickel Oral Exposure Limits

Non-Carcinogenic (Threshold) Effects

Health Canada (1996, 2003a) reported a TDI of 50 µg Ni/kg/day for Ni sulfate. This value was derived from a NOAEL of 5 mg/kg/day reported by Ambrose et al. (1976) for a 2-year dietary study in which rats were administered Ni sulfate hexahydrate. The critical effects included decreased body and organ weights. An overall uncertainty factor of 100 (10-fold for interspecies extrapolation and 10-fold for interspecies variation) was applied to the study NOAEL (considered the human-equivalent NOAEL) to yield the TDI. OEHHA (2003) used the same principal study (Ambrose et al. 1976) to derive a chronic oral reference exposure level (REL) of 50 µg/kg/day. The U.S. EPA (1991b) reported an oral RfD of 20 µg/kg/day for Ni which was also based on the chronic oral rat study by Ambrose et al. (1976). The NOAEL was then adjusted by an uncertainty factor of 300: 10 for interspecies extrapolation, 10 to protect sensitive populations, and an additional factor of three to account for inadequacies in the reproductive studies which was not included in the derivation of the Health Canada TDI. The 2-year feeding study in rats was supported by a subchronic gavage study in water (American Biogenics Corp 1986), which indicated the same NOAEL of 5 mg/kg/day.

In their review of the toxicology of soluble Ni compounds, TERA (2004) calculated an oral reference dose of 8 µg/kg/day Ni for ingested Ni-soluble salts. The most sensitive endpoint was determined to be increased albuminuria (indicating renal glomerular dysfunction) in male and female rats exposed to Ni in drinking water for 6 months (Vyskocil et al. 1994a,b). TERA (2004) noted that the Ni doses used in the principal study did not include the Ni present in the diet. Therefore, the RfD represents the dose of Ni in addition to the amount received in food. TERA (2004) considered this oral RfD to agree well with the U.S. EPA oral RfD of 20 µg/kg body weight/day for total Ni exposure, and is within the expected inherent uncertainty surrounding an RfD. The TERA RfD was not used in this study as reference values derived for all other COC were expressed on a total exposure basis, whereas this value was considered an incremental value. For consistency, the U.S. EPA value of 20 µg/kg/day was selected for the purposes of this risk assessment.

Carcinogenic (Non-threshold) Effects

There were no data available on the carcinogenicity of Ni in humans exposed via oral or dermal routes.

Nickel Inhalation Exposure Limits

Non-cancer (Threshold) Effects

Health Canada (1996) recommended various guidance values for inhalation exposure to different forms of Ni. A tolerable inhalation concentration (non-cancer effects) of 0.0035 µg/m^3 was recommended for Ni sulfate based on a study of lung and nasal lesions in rats and mice observed by Dunnick et al. (1989). Tolerable inhalation concentrations of 0.018 µg Ni/m^3 and 0.02 µg Ni/m^3 were recommended for metallic Ni and Ni oxide, respectively. OEHHA (2003) derived a chronic reference exposure level (REL) of 0.00005 mg/m^3 for Ni compounds (apart from Ni oxide). The principal study was NTP (1994a) and the critical effects were pathological changes in lung, lymph nodes, and nasal epithelium. The study NOAEL was 0.03 mg/m^3 using Ni sulfate hexahydrate.

OEHHA (2003) derived a chronic REL specifically for Ni oxide of 0.0001 mg/m^3. The principal study was NTP (1994b), and the critical effects considered were pathological changes in lung and lymph nodes. This study identified a LOAEL of 0.5 mg/m^3. The study NOAEL was adjusted for continuous exposure (multiplied by $6/24 \times 5/7$) and then converted to a NOAEL$_{HEC}$ by multiplying against an regional deposited dose ration (RDDR) of 0.29. Following this, the NOAEL was divided by a cumulative uncertainty factor of 300 (i.e. 10 for use of a LOAEL, three for interspecies uncertainty, and 10 for intraspecies uncertainty) to yield the chronic REL. The exposure limits derived by OEHHA (2003) were selected for use in this study.

Cancer (Non-threshold) Effects

Oller (2002) suggested that, in isolation, water soluble Ni compounds are not complete carcinogens. However, they may enhance the carcinogenic risks associated with other compounds when inhaled, if concentrations are large enough to induce chronic lung inflammation. By keeping exposure below levels resulting in chronic respiratory toxicity, Oller (2002) suggested that possible tumor-enhancing effects would be avoided. Seilkop

(2004) derived an inhalation unit risk of 1.9×10^{-4} $(\mu g/m^3)^{-1}$ for Ni sulfate when exposure is in the presence of a carcinogen. The basis of this unit risk is the incidence of an inflammatory response in the exposed animals. While there was unquestionably a link between inflammation and cancer promotion, the use of this endpoint in the derivation of a unit risk is uncertain. As such, this unit risk was not utilized in this assessment, as clear evidence exists to indicate that Ni sulfate acts via a non-mutagenic mechanism. Seilkop and Oller (2003) estimated safety limits for workers from fitted animal dose–response curves after accounting for interspecies differences in deposition and clearance, differences in particle size distributions, and human work activity patterns. Using a 10^{-4} risk level (which they deemed an acceptable occupational lifetime cancer risk level), they derived an occupational exposure limit concentration of 0.002 to 0.01 mg of inhalable Ni subsulfide/m³. Subsequently, Seilkop (2004) derived an inhalation unit risk (IUR) of 6.3×10^{-4} $(\mu g/m^3)^{-1}$ for Ni subsulfide (Ni_3S_2). This IUR was independent of any specific risk level and can be used to establish 10^{-5} or 10^{-6} acceptable risk levels as appropriate.

Seilkop and Oller (2003) also estimated safety limits for workers from fitted animal dose–response curves. Using a 10^{-4} risk level (which they deemed an acceptable occupational lifetime cancer risk level), they derived an occupational exposure limit concentration of 0.5 to 1.1 mg of inhalable Ni oxide/m³. The authors report that although the animal data for Ni oxide suggest a threshold response for lung cancer, this cannot be concluded with certainty as sampling uncertainty in data make the non-threshold response equally as plausible. They report that the non-linearity of the observed dose–response of Ni oxide is well represented by benchmark dose models (excluding the high dose–response). Subsequently, Seilkop (2004) derived an inhalation unit risk of 2.3×10^{-5} $(\mu g/m^3)^{-1}$ for Ni oxide.

Interestingly, when the IURs for Ni oxide and Ni sulfate (Seilkop 2004) were converted to a risk-specific concentration (RsC) assuming a target risk level of one-in-one hundred thousand, the RsCs were very similar to the OEHHA (2005a,b) chronic RELs for both Ni compounds apart from Ni oxide. The RsC that corresponds to the Ni sulfate IUR was 0.05 µg/m³ (the OEHHA REL was also 0.05 µg/m³), while the RsC that corresponds to the Ni oxide IUR was 0.4 µg/m³ (the OEHHA REL was 0.1 µg/m³). Thus, it can be extrapolated from this comparison that, although the OEHHA chronic RELs were not derived from a cancer endpoint, they would appear to be protective of both non-cancer and potential cancer effects of inhaled soluble Ni and Ni oxide. However, the OEHHA chronic RELs are not protective of the potential carcinogenic effects of Ni_3S_2 based on converting the Seilkop (2004) IUR for this substance to an RsC at a one-in-one hundred thousand target risk level.

Following a detailed evaluation of three different mechanistic approaches, an EU working group proposed a limit value range of 0.01 to 0.05 µg Ni/m³ (as an annual mean), based upon non-cancer effects. This working group also believed that a limit value in this range can be judged compatible with the objective of limiting excess lifetime cancer risks to not more than one-in-a-million. The majority of the working group proposed a limit value at the lower end of this range, to represent an annual mean of total airborne nickel (European Commission DG Environment 2001). Based upon this work, in 2004, the European Parliament adopted a target value for airborne Ni of 20 ng Ni/m³, or 0.02 µg Ni/m³, considered protective of both cancer and non-cancer health endpoints (OJEU 2005). This value was selected as the primary exposure limit for nickel inhalation used in this risk assessment.

Health Canada (1996, 2003a) considered oxidic, sulfidic and soluble Ni to be carcinogenic to humans. A Tumorigenic Concentration 05 (TC_{05}) of 70 µg/m³ was developed by Health Canada (1996) for soluble Ni (primarily Ni sulfate and Ni chloride) based on lung cancer mortality observed in a cohort in Norway (Doll et al. 1990). A TC_{05} of 40 µg/m³ was recommended for combined oxidic, sulfidic and soluble nickel. Assuming that a 70 kg person breathes at a rate of 20 m³/day, the inhalation slope factors for soluble Ni and combined oxidic, sulfidic and soluble Ni were estimated to be 0.0025 $(\mu g/kg/day)^{-1}$ and 0.0044 $(\mu g/kg/day)^{-1}$, respectively.

The World Health Organization (WHO 2000) developed an incremental unit risk of 0.0004 $(\mu g/m^3)^{-1}$ for Ni subsulfide based on epidemiological lung cancer data for Ni refinery workers (Doll 1977; Chovil et al. 1981; Magnus et al. 1982; U.S. EPA 1991c). WHO (2000) reassessed this unit risk based on updated epidemiology data for lung cancer in refinery workers, including follow-up studies of a cohort examined in Kristiansand, Norway used in the 1987 assessment (Andersen 1992; Andersen et al. 1996). Based on the estimated risk for this cohort of 1.9 and a lifetime exposure estimate of 155 µg/m³, the WHO (2000) calculated an incremental life-time unit risk of 0.00038 $(\mu g/m^3)^{-1}$ for inhalation of Ni. This unit risk of 0.00038 $(\mu g/m^3)^{-1}$ was then converted to an inhalation cancer slope value of 0.00133 $(\mu g/kg/day)^{-1}$ based on an adult body weight of 70 kg and a breathing rate of 20 m³/day. Based on these studies, the U.S. EPA derived slope factors for Ni refinery dust and Ni_3S_2 of 2.4×10^{-4} $(\mu g/m^3)^{-1}$ and 4.8×10^{-4} $(\mu g/m^3)^{-1}$, respectively. These slope factors were not directly attributed to the Sudbury environment given the concomitant presence of multiple Ni species in proportions that will differ from those found within refineries. However, these slope factors were used in a weight of evidence approach most notably for consideration of the potential presence of Ni sulfide in the ambient environment.

Dermal Exposure Limits

No regulatory dermal exposure limits for Ni compounds were identified in the literature reviewed for this study. Route-to-route extrapolation was used to derive an appropriate limit for the current assessment.

7.2.3.6 Selenium

There is widespread scientific consensus that Se is an essential trace element in both animal and human nutrition (Bennett 1982; Levander 1982, 1991; Robinson 1982; WHO 1986; Levander et al. 1987; Foster and Sumar 1997; NAS 2000). Selenium deficiency in isolation seldom causes overt illness; however it leads to biochemical changes that predispose Se-deficient individuals to illness associated with other stresses (NAS 2000). A deficiency of Se in the human diet is associated with Keshan disease (juvenile cardiomyopathy endemic to certain areas of China) and Kashin-Beck disease (osteoarthropathy endemic to Eastern Siberia) as well as numerous other diseases, conditions and effects (Chen et al. 1980; Whanger 1983; Sokoloff 1985; Wu et al. 1989).

Selenium is a rather unique element in that there is a small margin of safety (ranging from a factor of approximately 5 to 18 between levels of Se compounds that constitute dietary deficiency and those that result in toxicity; Lemly 1997). The U.S. National Academy of Sciences (NAS 2000) recommended a safe and adequate daily intake ranging from 20 to 70 µg per person per day for adults. The RDAs for Se of 70 µg/day for adult men and 70 µg/day for adult women were based on a daily dose 0.87 µg/kg bw/day derived from a series of depletion studies carried out in Chinese males (Yang et al.1989a,b; Levander 1991). RDAs for children and infants were extrapolated from the adult RDAs on the basis of body weight. The dietary requirement below which adverse human health effects resulting from deficiency may occur has been tentatively estimated to range from 2 to 120 µg/day (Stewart et al. 1978). Whanger et al. (1996) suggest that an intake of less than 40 µg/day will likely result in deficiency. The minimum dose to cause toxicity in humans is not well defined, but the threshold appears to lie in the range of 400 to 900 µg/day (Allegrini et al. 1985; Yang et al. 1989a,b; Longnecker et al. 1991; Whanger et al. 1996).

Selenium is also believed to have a protective function against certain types of cancers (Foster and Sumar 1997). Levander et al. (1997) hypothesized that the 'anti-cancer' protective effects of Se are due to its roles in alleviating oxidative damage, altering carcinogen metabolism, and selective toxicity against rapidly dividing tumor cells. It should be noted that there is conflicting evidence with respect to this function of Se. Nonetheless, relatively high levels of Se have been used successfully to protect against both chemically-induced and spontaneously occurring tumors in laboratory animals (Whanger 1983; Ip and Ganther 1992). Selenium supplementation has also been shown to significantly inhibit tumors induced by viruses, or ultraviolet radiation (ATSDR 2003). Methylated forms of selenium appear to be the most important with respect to cancer prevention.

Exposure Limits

The following organizations were consulted to select exposure limits for selenium: the U.S. EPA; ATSDR; Health Canada; National Academy of Science; MOE; and OEHHA.

Selenium Oral Exposure Limits

Non-Cancer (Threshold) Effects

In determining the oral RfD for Se compounds, the U.S. EPA selected the epidemiology study by Yang et al. (1989a,b) as the principal and supporting study. The study NOAEL of 15 µg/kg bw/day was used to calculate an oral RfD of 5 µg/kg bw/day; a threefold uncertainty factor was applied to account for sensitive individuals (U.S. EPA 1991a). The results of Longnecker et al. (1991) strongly corroborate the NOAEL identified by Yang et al. (1989a,b). In addition, numerous other epidemiological studies and animal studies also support the findings of Yang et al. (1989) (U.S. EPA 1991a). The ATSDR (2003) chronic MRL for Se compounds was also 5 µg/kg bw/day, and was based on the same endpoint (selenosis) and utilizes the same magnitude of uncertainty factor as the U.S. EPA oral RfD. The National Academy of Science (IOM 2000) derived an acceptable Upper Limit (UL) based on the NOAEL of 800 µg/day based on Yang and Zhou (1994). Application of a twofold uncertainty factor results in an upper limit of approximately 5 µg/kg/day for adults. In the derivation of the Guideline for Use at Contaminated Sites in Ontario (MOE 1996b), MOE utilized an oral RfD of 5 µg/kg/day for Se. For the purposes of this risk assessment, an oral RfD/TRV of 5 µg/kg/day was selected for Se (IOM 2000; Health Canada 2005).

Cancer (Non-threshold) Effects

Selenium compounds (with the exception of Se sulfide) are widely considered non-carcinogenic, therefore, no regulatory agencies were identified that developed health-based exposure limits based on carcinogenic endpoints. Since Se sulfide is typically not present in soils, foods or other environmental media to a significant extent (ATSDR 2003), human environmental exposure to Se sulfide would likely be negligible, relative to other forms of Se.

Inhalation Exposure Limits
Non-cancer (Threshold) Effects

The U.S. EPA, ATSDR, and Health Canada had not developed inhalation exposure limits for Se compounds. The Ontario MOE provided two health-based limits for Se in air, a point-of-impingement limit and a 24-h ambient air quality criterion of 20 and 10 µg/m^3, respectively (MOE 2001). The California Environmental Protection Agency developed a chronic Reference Exposure Limit of 20 µg/m^3, for effects on the alimentary, cardiovascular and nervous systems, based on route-to-route extrapolation from the Yang et al. (1989a,b) study (OEHHA 2001). The REL was derived by multiplying the U.S. EPA oral RfD of 5 µg/kg/day by an inhalation extrapolation factor of 3.5 µg/m^3 per mg/kg-day. For the purposes of this risk assessment, a chronic REL RfC of 20 µg/m^3 was selected (OEHHA 2001).

Cancer (Non-threshold) Effects

Since Se compounds (with the exception of Se sulfide) are widely considered non-carcinogenic, no regulatory agencies were identified that developed health-based exposure limits based on carcinogenic endpoints. Selenium sulfide is typically not present in soils, foods or other environmental media to any significant extent, so human environmental exposure to Se sulfide would likely be negligible (ATSDR 2003).

Dermal Exposure Limits

No regulatory dermal exposure limits for selenium compounds were identified in the literature. Route-to-route extrapolation was used to derive an appropriate limit for the current assessment.

7.2.4 Bioavailability/Bioaccessibility

One of the most important factors in determining exposure of target tissues to a substance, and the body's ultimate response, is *bioavailability*. Bioavailability is the fraction of the total amount of a substance to which an organism has been exposed that successfully enters the bloodstream. The bioavailability of a substance is dependent on the chemical form, the environmental medium, the route of exposure, the physiological characteristics of the organism at the time of exposure (e.g. ingested substances may be absorbed to different extents depending on whether the stomach is full or empty) as well as the tissues/organs with which the substance must interact as it passes from the point of entry to target tissues.

Since most exposure limits are based on administered doses, it is not appropriate to consider absolute bioavailability (fraction or percentage of an external dose which reaches the systemic circulation) in the assessment of exposures in most instances. A better measure may be that of relative bioavailability which can be determined by comparing the extent of absorption among several routes of exposure, forms of the same substance, or vehicles of administration (such as food, soil, and water). Systemic absorption of substances will differ according to whether the dose was received via dermal contact, ingestion or by inhalation. Also, systemic absorption will differ depending on whether the substance is delivered in a solvent vehicle (water, soil, food, etc.). For some substances, exposure limits were not available for all exposure routes of concern. In those cases, it may be necessary to extrapolate an exposure limit from one route to another. For example, it is common in human health risk assessment to assess the risks posed by dermal absorption of a substance based on the exposure limit established for oral exposure. The systemic dose absorbed dermally is scaled to the 'equivalent' oral dose by correcting for the bioavailability of the dermally applied chemical relative to an orally administered dose. The oral bioavailability of a substance is typically determined from absorption or excretion studies. The bioavailability, expressed as a percentage, is generally assumed to be 100% minus the percent of the ingested chemical excreted unchanged in the feces. In cases where only the fraction of chemical in the urine is reported, this fraction is selected as the minimum oral bioavailability with the maximum being 100%. In the absence of relevant data, this approach was considered to be reasonable, and to reflect the uncertainty in the oral bioavailability of the chemical.

Typically, adjustments of exposure limits for bioavailability are considered for systemic effects [i.e. following entry into and distribution by the bloodstream, as opposed to occurring at the site of entry (e.g. lungs, skin, gut)] when:

- the exposure limit is based on a different route of exposure (i.e. when the criterion is based on ingestion and the exposure routes of interest are inhalation or dermal exposure);
- the medium of administration in the study used to develop the exposure limit results in a different bioavailability than the exposure medium of interest (e.g. ingestion in drinking water versus ingestion in soil); or
- the bioavailability of the chemical, based on the particular study animal/receptor, is different from that of the receptor upon which the exposure limit is based (e.g. the exposure limit is based on a study using mice, the species of interest is human, and there are reported bioavailabilities for both mice and humans).

When evaluating the health risks related to exposures to metals, an important aspect of a substance's bioavailability is the *bioaccessibility* exhibited by that substance. Bioaccessibility is the mass fraction of a substance that is converted to a soluble form under conditions of the external part of the membrane of interest. If one is evaluating bioaccessibility via the oral route, it is the fraction of substance that becomes solubilized within the gastrointestinal tract (i.e. stomach and small intestine). In the case of dermal exposures, it is the fraction solubilized on the outside of the skin (i.e. in sweat). To better characterize this fraction, a detailed site-specific *in vitro* bioaccessibility study was conducted to estimate the bioaccessibility of each of the COC present in soil and indoor dust media collected as part of the overall study. These site-specific oral bioaccessibility studies were used to help address the differences in oral bioavailability observed in these media versus the medium used in the study from which the toxicological criterion was derived.

In addition to performing the site-specific bioaccessibility studies described in Chapter 6, the scientific and regulatory literature was reviewed to identify bioavailabilities for each route of exposure evaluated in the HHRA, and where possible, values specific to humans and to the environmental media of concern (e.g. soil, dust). To use the bioaccessibility results effectively in the risk assessment, the relative absorption factor (RAF) should be used, where appropriate. The RAF corrects for the differential media/matrix to which the samples for the bioaccessibility evaluations are conducted (i.e. soil) and the media used in the study that was used to derive the RfD. For example, the originating study for the RfD of Ni soluble salts was based on the ingestion of rat chow by mice. Therefore, a correction for the bioaccessibility of Ni in rat chow should be applied to all bioaccessibility data gathered for soil samples. The RAF was determined and applied to the exposure estimates for potential human exposure scenarios. Spiked rat chow was subjected to the bioaccessibility assay and an overall bioaccessibility of 94.7% was observed. This was then used in the RAF determination for Ni. For other COC, study absorption factors of 100% were assumed. The RAF values used in this HHRA were provided in Table 6.10 in Chapter 6.

7.3 Risk Characterization

7.3.1 Overview

The final phase of the risk assessment was risk characterization. The risk characterization step integrates the exposure and hazard assessments to provide a conservative estimate of human health risk for the receptors assessed in the various exposure scenarios. Potential risk is characterized through a comparison of the estimated or predicted exposures from all pathways (from the exposure assessment) with the identified exposure limits (from the hazard assessment) for each of the COC. For the COC which are thought to be non-carcinogens, this comparison is typically called the hazard quotient (HQ). Risk characterization for chemicals with non-threshold-type dose responses (i.e. carcinogens) consists of a calculation of the Cancer Risk Level (CRL). The approaches to calculating and interpreting these values are presented below in Sections 7.3.2 and 7.3.3, respectively.

The evaluation and interpretation of HQs and CRLs can be applied with greatest confidence to situations where comparisons are made between the HQs/CRLs of two or more independent exposure scenarios. From such comparisons, the incremental difference in the potential for occurrence of adverse health effects between the two or more different scenarios (e.g. study area versus typical Ontario) can be assessed with reasonable confidence since the same exposure and hazard assessment methodologies are used in addressing each situation. Most of the uncertainties in such comparative assessments are related to the ability to accurately estimate COC concentrations in the various environmental media that determine the different exposure pathways, and in the estimation of the toxicological criteria that exposure estimates are compared against.

HQs and CRLs were used to express the potential adverse health effects from exposures to the selected chemicals for several reasons:

- to allow comparisons of potential adverse effects on health between chemicals and different exposure scenarios (e.g. Typical Ontario versus site-specific conditions)
- to estimate potential adverse effects on health from exposures to mixtures of chemicals that act on similar biological systems (e.g. all chemicals that cause liver toxicity, or kidney toxicity, or respiratory tract cancers), and
- to simplify the presentation of the HHRA results so that the reader may have a clear understanding of these results, and an appreciation of their significance.

When predicted risks are substantially greater than the acceptable level (i.e. more than 10-fold), the potential for adverse effects in sensitive individuals or in some of the exposure scenarios is suggested. Again, however, the reevaluation of such HQs/ILCRs is extremely important since both the exposure estimation procedures and the toxicological criteria are based on a series of conservative assumptions that tend to overestimate exposures and risks. Often, a sensitivity analysis is conducted which facilitates the reevaluation by focusing on the proportional contribution of various parameters to the final HQ/ILCR value. Once the major contributing model parameters have been identified, they can be reevaluated to determine their impact on the resulting risk estimates and whether health risks have been underestimated or overestimated. Most often, the sensitivity analysis indicates that exposures and risks were overestimated. This occurs because a certain amount of overestimation of risk is inherently built into the risk assessment process. For example, in cases where there is considerable uncertainty in the data such as the determination of toxicological criteria for cancer causing chemicals (e.g. arsenic), a conservative dose–response extrapolation model is used to derive the toxicological criterion to ensure the protection of human health.

7.3.2 Hazard Quotients

As stated above, for the COC which are thought to be non-carcinogens, the comparison of exposure to the relative hazard is typically called the Hazard Quotient (HQ), where HQ is calculated by dividing the predicted exposure level by the exposure limit:

$$\text{Hazard Quotient (HQ)} = \frac{\text{Estimated Exposure (µg/kg/day)}}{\text{Exposure Limit (µg/kg/day)}}$$

Once HQ values have been determined for threshold chemicals (non-carcinogens), they are compared to a benchmark indicator of 'safety', which is sometimes called the Critical Hazard Quotient (CHQ). In general, if the total chemical exposure from all pathways is equal to, or less than the exposure limit, then the HQ would be 1.0 or less, and no adverse health effects would be expected. Therefore, the benchmark of safety would be 1.0, assuming that estimates of exposure from all relevant exposure pathways are included.

However, if risk assessments generally evaluate single or a few sources of contamination and a limited number of exposure pathways, the selection of a CHQ value of 1.0 for threshold chemicals is not always appropriate. In an attempt to address this issue, the CCME (1996) considered that a substance has the potential to be present in all media, and assumes an allocation of 20% of the residual tolerable daily intake for each of the five major media (i.e. air, water, soil, food, consumer products). Similarly, the MOE recommends apportioning 20% of the total exposure to any one pathway (MOEE 1987), in the absence of information to the contrary. This means that the overall CHQ (i.e. 1.0) must also be apportioned for the single source (e.g. a contaminated site) under consideration. This yields a value of 0.20, which can be considered as the CHQ representing a situation in which no adverse health effects are likely to be associated with the estimated level of exposure for a given pathway. Therefore, if threshold chemicals are determined to have HQ values less than 0.20, exposure rates are considered to be less than 20% of the exposure limit (toxicological criterion), and no adverse health effects would be expected to occur in the receptors and scenarios evaluated in the risk assessment. If HQ values are greater than 0.2, the estimated exposure rates are considered to exceed 20% of the exposure limits, indicating the potential for adverse effects in sensitive individuals or in some of the exposure scenarios considered.

In the Sudbury HHRA, all significant sources of exposure were accounted for, with the exception of consumer products. The SARA Group conducted a detailed literature search and was unable to locate any information that indicated that consumer products would be a significant source of inorganic exposures to the COC in the current study. The risk assessment team recommended an HQ value of 1.0 to represent an 'acceptable level' of exposure, while recognizing there is some uncertainty with respect to the potential contribution of consumer products to an individual's estimated daily intake. Where the HQ value is < 1.0, risk was considered negligible and no further action was deemed necessary. When HQ is > 1.0, risk was not screened out. If HQ > 1.0, it does not, in and of itself, indicate that adverse effects will take place, or even that the risk of adverse effects is unacceptable (Health Canada 2004a).

7.3.3 Cancer Risk Level

Risk characterization for chemicals with non-threshold-type dose responses (i.e. carcinogens) consists of a calculation of the Cancer Risk Level (CRL), which is defined as the predicted upper bound risk of an individual in a population of a given size developing cancer over a lifetime. The CRL is expressed as the prediction that one person per n people would develop cancer, where the magnitude of n reflects the risks to that population; for example, if the CRL is one person per 10, the predicted risks of any individual developing cancer would be higher than if the CRL is one per 1000. The following equation provides the method whereby the CRL was calculated:

$$\text{Cancer Risk Level (CRL)} = \text{Estimated Lifetime Exposure} \times \text{Cancer Slope Factor } (q_1^*)$$

An incremental lifetime cancer risk (ILCR) refers to the contribution that a facility or site makes to the total risk. In Ontario, the MOE specifies an acceptable ILCR of one-in-one million (1×10^{-6}). In other jurisdictions, negligible or *de minimis* cancer risk levels are generally considered to be in the range 1×10^{-4} to 1×10^{-7} (Health Canada 2004a). The selection of an acceptable risk level is predominantly a policy-based, rather than a science-based, decision. In situations such as Sudbury, it is difficult to tease out the actual incremental contributions that the facilities have made. In this case, Sudbury specific risks were calculated and as such, an alternate acceptable risk level may be appropriate.

Risk estimates that are substantially less than the acceptable level are not considered to require further evaluation. In situations where risks are predicted to be within the same order of magnitude as the acceptable level, reevaluation of certain model parameters (e.g. chemical concentration estimates, exposure parameters, and toxicological criteria) is conducted before the potential risks to health are fully characterized. In these situations, consideration must be given to the possibility of adverse health effects, but a slight exceedance of the acceptable risk benchmarks do not typically indicate a high potential for risk. The methods and assumptions used in this HHRA were designed to be conservative (i.e. health protective), and have a built-in tendency to overestimate, rather than underestimate, potential health risks. Thus, risk estimates that are within an order of magnitude of the acceptable risk benchmarks may reflect overestimation through the use of overly conservative assumptions and parameters. In these cases, interpretation of the risk estimates may indicate that, given the conservatism of the assessment, no adverse health effects would be expected despite the exceedance of the acceptable risk level or, that further assessment (i.e. progression to a more detailed and specific risk assessment that could involve further data collection or probabilistic exposure analysis), or mitigative measures are warranted.

7.4 Risk Management Recommendations

If, after careful review and consideration of the factors described previously, the results of the risk characterization indicate that there may be unacceptable risks posed to some receptors of concern, then preliminary recommendations toward mitigation of those risks can be made. Risk management recommendations may suggest possible ways in which exposure pathways contributing significantly to overall exposure and risk can be limited or eliminated. For example, if contact with surface soils is driving risk, depending on the current and future uses of the land, it may be appropriate to simply put a layer of asphalt or clean fill over the contaminated soil, thereby preventing soil contact and mitigating the risk. Soil amendments, such as liming, can also be used to mitigate risks, in that they can modify the availability of chemicals in the soil. In some cases, it may be necessary to remove contaminated media to mitigate risk. In cases where it is determined that risk management is necessary, soil risk management levels (SRMLs) can be used to guide potential remediation activities. SRMLs may also be referred to as risk management criteria (RMC) for intervention levels or preliminary remediation goals (PRGs) or site-specific remediation targets by different agencies.

The need to recommend a SRML is based on a number of key considerations including:

1. the nature, extent and duration of the risk and the uncertainties in how risks are estimated
2. evidence or lack of evidence of actual harm to health in the community, and
3. outcomes of risk assessments in other communities with similar or higher levels of exposure.

There may also be legal, financial, political, and community concern-based issues that play a role in the establishment of suitable SRMLs and subsequent action that may be taken. The SRML can be defined as the average concentration within an exposure unit (EU) that corresponds to an acceptable level of risk (U.S. EPA 2001). It is noted that, in this study, the EPCs used to facilitate the long-term (or chronic) exposure assessment and subsequent approximation of hazard and risk were defined as the upper 95% confidence limit on the arithmetic mean (95% UCLM) from a specific or community of interest. Risks were based on a conservative approximation of the true (or population) mean of community-specific environmental media, and in essence assume that individuals move in a random fashion with their residential community. In reality, individuals do not move in a random fashion within their residential community, but rather exhibit some type of predictable spatial pattern in their movements. For example, many individuals will tend to spend the majority of their time between home, work and/or school. If the SRML is defined as the EPC in soil within a given community which yields an acceptable level of risk, then some residential properties will exceed the EPC. Depending on how the soil concentration data are distributed, it is plausible that the remediation of a number of highly impacted soils within the community could bring the overall EPC for that community below the SRML. If the property or site of concern was a single residential lot, it would be reasonable to assume that an individual would move in random fashion within his or her own residential property. The removal of a number of highly impacted zones to facilitate the reduction in the EPC for this single property may be a reasonable approach. However, because the exposure units of interest represent entire communities, in which individuals do not spatially move in a random fashion, the remediation of locally impacted zones to reduce the overall EPC for the community is not valid. The SRML values should be applied to individual residential properties, not necessarily the community as a whole. The result is that on a community wide basis, no unacceptable risk may be predicted while on a site specific basis, some properties may exceed the SRML for that community.

A variety of SRMLs can be derived for each of the COC depending on the statistic (mean, UCLM, RME, CTE, percentile value of a probabilistic distribution of risk) deemed appropriate for the protection of human health in the Sudbury area. For the current assessment, the risk predictions from the RME receptor exposure scenario (i.e. reasonable upper bound) were used to generate the SRML for Pb (see Chapter 8 for a further discussion of this issue).

7.5 References

Allegrini M, Lanzola E, Gallorine M (1985) Dietary selenium intake in a coronary heart disease study in nortern taly. utr. Res. Suppl. I:398

Ambrose AM, Larson PS, Borzelleca JF, Hennigar GR (1976) Long term toxicologic assessment of nickel in rats and dogs. J Food Sci Technol 13:181–187

American Biogenics Corp (1986) Ninety day gavage study in albino rats using nickel. Draft Final Report submitted to Research Triangle Institute, Research Triangle Park, NC

Andersen A (1992) Recent follow-up of nickel refinery workers in Norway and respiratory cancer. In: Nieboer E, Nriagu JO (eds) Nickel and human health: Current perspectives, pp. 621–628. Wiley, New York

Andersen A, Berge SR, Engeland A, Norseth T (1996) Exposure to nickel compounds and smoking in relation to incidence of lung and nasal cancer among nickel refinery workers. Occup Environ Med 53:708–713

Anke M, Angelow L, Glei M, Muller M, Illing H (1995) The importance of nickel in the food chain. Fresenius J Anal Chem 352:92–96

ATSDR (2001) Toxicological profile for cobalt. Draft for public comment. U.S. Department of Health and Human Services, Public Health Service, Atlanta, GA www.atsdr.cdc.gov/toxprofiles/tp33.html. Accessed 15 Sepember 2005

ATSDR (2003) Toxicological profile for Selenium. U.S. Department of Health and Human Services, Public Health Service, Agency for Toxic Substances and Disease Registry, Atlanta, GA. September 2001. www.atsdr.cdc.gov/toxprofiles/tp92.html. Accessed August 2005

ATSDR (2004) Toxicological profile for copper – draft for public comment. U.S. Department of Health and Human Services, Public Health Service, Agency for Toxic Substances and Disease Registry Atlanta, GA. www.atsdr.cdc.gov/toxprofiles/tp132.html. Accessed 30 August 2005

ATSDR (2005) Draft toxicological profile for lead. U.S. Department of Health and Human Services, Public Health Service, Agency for Toxic Substances and Disease Registry Atlanta, GA. September 2005. www.atsdr.cdc.gov/toxpro2.html. Accessed June 2006

Baars AJ, Theelen RMC, Janssen PJCM et al. (2001) Re-evaluation of human-toxicological maximum permissible risk levels. RIVM report no. 711701025, pp. 62–65. National Institute of Public Health and the Environment, Bilthoven, The Netherlands, March 2001, www.rivm.nl/bibliotheek/rapporten/711701025.pdf. Accessed 21 August 2005

Bennett BG (1982) Exposure commitment assessments of environmental pollutants. Vol. 2. Summary exposure assessments for PCBs, selenium, chromium. Monitoring and Assessment Research Centre (MARC), Chelsea College, University of London, UK

Brown CC, Chu KC (1983a) Approaches to epidemiologic analysis of prospective and retrospective studies: Example of lung cancer and exposure to arsenic. In: Environmental epidemiology: Risk assessment. Proceedings of SIMS Conference, Alta, UT, 28 June 28–2 July 1982. SIAM Publications

Brown CC, Chu KC (1983b) Implications of the multistage theory of carcinogenesis applied to occupational arsenic exposure. J Natl Cancer Inst 70:455–463

Brown CC, Chu KC (1983c) A new method for the analysis of cohort studies: Implications of the multistage theory of carcinogenesis applied to occupational arsenic exposure. Environ Health Perspect 50:293–308

Brown KG, Chen CJ (1995) Significance of exposure assessment to analysis of cancer risk from inorganic arsenic in drinking water in Taiwan. Risk Anal 15(4):475–484

Bucher JR, Hailey JR, Roycroft JR et al. (1999) Inhalation toxicity and carcinogenicity studies of cobalt sulfate. Toxicol Sci 49:56–67

Burmaster DE (1998) Lognormal distributions of skin area as a function of body weight. Risk Anal 18(1):27–32

Calabrese EJ, Stanek III EJ, Barnes RM, Pekow P (1997a) Soil ingestion in adults – results of a second pilot study. Ecotox Environ Saf (36):249–257

Calabrese EJ, Stanek EJ, Pekow P, Barnes RM (1997b) Soil ingestion estimates for children residing on a Superfund site. Ecotox Environ Saf 36:258–268

Canadian Nutrient File (2001) Online: www.hc-sc.gc.ca/food-aliment/ns-sc/nr-rn/surveillance/cnf-fcen/e_index.html. Last updated March, 2003. Accessed April, 2005

Cantox Environmental (1999) Deloro Village exposure assessment and health risk characterization for arsenic and other metals. Final Report. Prepared for Ontario Ministry of the Environment, Toronto, Ontario. Cantox Environmental, Waterloo, Ontario

CCME (1996) A protocol for the derivation of environmental and human health soil quality guidelines. Canadian Council of Ministers of the Environment. March 1996

CCME (2005) A protocol for the derivation of environmental and human health soil quality guidelines. Draft. The National Contaminated Sites Remediation Program. ISBN 1-896997-45-7. Canadian Council of Ministers of the Environment

CDC (1991) Preventing Lead Poisoning in Young Children. Chapter 2: Background. Childhood Lead Poisoning Prevention Program. Centre for Disease Control and Prevention, Atlanta, GA. www.cdc.gov/nceh/lead/publications/books/plypc/chapter2.htm. Accessed 12 August 2005

CDC (2004) Why not change the blood lead level of concern at this time? Childhood Lead Poisoning Prevention Program, National Center for Environmental Health. Center for Disease Control and Prevention. www.cdc.gov/nceh/lead/spotLights/changePbB.htm. Accessed 10 January 2005

CDC (2005) Preventing lead poisoning in young children. Centers for Disease Control and Prevention, Atlanta, GA. www.cdc.gov/nceh/lead/publications/PrevLeadPoisoning.pdf. Accessed 15 Januar 2008

CEOH (1994) Update of evidence for low-level effects of lead and blood lead intervention levels and strategies – Final Report of the Working Group. Federal-Provincial Committee on Environmental and Occupation Health, Health Canada

Chao CY, Wong KK (2002) Residential indoor PM10 and PM2.5 in Hong Kong and the elemental composition. Atmos Environ 36:265–277

Chen X, Yang G, Chen J, Chen X, Wen Z, Ge K (1980) Studies on the relations of selenium and Keshan disease. Biol Trace Elem Res 2:91

Chen CJ, Chen JW, Wu M, Kuo T (1992) Carcinogenic potential in liver, lung, bladder and kidney due to ingested inorganic arsenic in drinking water. Br J Cancer 66:888–892

Chovil A, Sutherland RB, Halliday M (1981) Respiratory cancer in a cohort of nickel sinter plant workers. Br J Ind Med 38:327–333

Christensen OB Moller H (1978) Release of nickel from cooking utensils. Contact Dermatitis 4:343–346

Dabeka R, McKenzie AD (1995) Survey of lead, cadmium, fluoride, nickel, and cobalt in food composites and estimation of dietary intakes of these elements by Canadians in 1986-1988. J AOAC Int 78(4):897–909

Dabeka RW, Mckenzie AD (2005) Personal Communication. Food Research Division, Bureau of Chemical Safety, Food Directorate, Health Products and Food Branch, Health Canada, Ottawa. Canadian Total Diet Study results for 2000 and 1993 to 1999 to Cantox Environmental Inc. May 2005

Dabeka RW, McKenzie AD, LaCroix GMA (1993) Survey of arsenic in total diet food composites and estimation of the dietary intake of arsenic by Canadian adults and children. J AOAC Int 76(1):14–25

Davis JE, Fields JP (1958) Experimental production of polycythemia in humans by administration of cobalt chloride. Proc Soc Exp Biol Med 37:96–99

Davis KD, Mertz, W (1987) Copper. In: Trace elements in human and animal nutrition, 5th edition, Vol 1. Academic Press Inc

Denkhaus E, Salnikow K (2002) Nickel essentiality, toxicity and carcinogenicity. Crit Rev Oncol Hematol 42:35–56

Do MT, Smith LF, Pinsent CL (2011) Urinary inorganic arsenic in residents living in close proximity to a nickel and copper smelter in Ontario, Canada. Can J Public Health (submitted)

Doll R, Matthews JD, Morgan LG (1977) Cancers of the lung and nasal sinuses in nickel workers: A reassessment of the period of risk. Br J Ind Med 34:102–105

Doll R, Anderson A, Copper WC et al. (1990) Report of the international committee on nickel carcinogenesis in man. Scand J Work Environ Health 16:1–82

Duckham JM, Lee HA (1976) The treatment of refractory anemia of chronic renal failure with cobalt chloride. Q J Med 178:277–294

Dudley HC, Miller JW (1941) Toxicology of selenium. VI. Effects of subacute exposure to hydrogen selenide. J Ind Hygiene Toxicol 23:470–477

Dunnick JK, Elwell MR, Benson JM et al. (1989) Lung toxicity after 13 week inhalation exposure to nickel oxide, nickel subsulfide, or nickel sulfate hexahydrate in F344/N rats and B6C3F1 mice. Fundam Appl Toxicol 12(3):584–594

Enterline PE, Marsh GM (1982) Mortality among workers in a nickel refinery and alloy manufacturing plant in West Virginia. J Natl Cancer Inst 68(6):925–933

European Commission DG Environment (2001) Ambient air pollution by As, Cd and Ni compounds. Position paper. 14 KH-41-01-349-EN-N. Office for Official Publications of the European Communities, Luxembourg

FAO/WHO (1993) Evaluation of certain food additives and contaminants. 41st Meeting of the Joint FAO/WHO Expert Committee on Food Additives. World Health Organization Technical Report Series No. 837. WHO, Geneva

Foster LH, Sumar S (1997) Selenium in health and disease. A review. Crit Rev Food Sci Nutr 37(3):211–228

Garlock TJ, Shirai JH, Kissel JC (1999) Adult responses to a survey of soil contact related behaviors. J Expo Anal Environ Epidemiol 9:134–142

Gleason RP (1968) Exposure to copper dust. Am Ind Hyg Assoc J 29:461–462

GLSFATF (1993) Protocol for a uniform Great Lakes sport fish consumption advisory. Great Lakes Sport Fish Advisory Task Force (GLSFATF) Protocol Drafting Committee, September 1993

Goss Gilroy Inc (2001) Survey of urinary arsenic for residents of Wawa, Ontario. www.downloads.ene.gov.on.ca/envision/sudbury/survey_of_arsenic/01_summary.pdf Accessed November 2004

Harrison RM (1979) Toxic metals in street and household dusts. Sci Total Environ 11: 89-97

Health Canada (1993) Health risk determination. The challenge of health protection. ISBN 0-662-20842-0. Health Canada, Ottawa, ON

Health Canada (1996) Health-based tolerable daily intakes/concentrations and tumorigenic doses/concentrations for priority substances. ISBN 0-662-24858-9. Health Canada, Ottawa, ON

Health Canada (2003a) Federal contaminated site risk assessment in Canada. Part II: Health Canada toxicological reference values (TRVs). Version 1.0, 3 October 2003. Health Canada, Ottawa, ON

Health Canada (2003b) Summary of guidelines for Canadian drinking water quality. Prepared by the Federal-Provincial-Territorial Committee on Drinking Water of the Federal-Provincial-Territorial Committee on Environmental and Occupational Health. April 2003. www.hc-sc.gc.ca/hecs-sesc/water/dwgsup.htm.

Health Canada (2004a) Contaminated sites program. Federal contaminated site risk assessment in Canada. Part I: Guidance on human health preliminary quantitative risk assessment (PQRA). Health Canada, Ottawa, ON

Health Canada (2004b) Canadian total diet study: Concentrations of contaminants and other chemicals in food chemicals. Food Program. Health Canada. www.hc-sc.gc.ca/food-aliment/cs-ipc/fr-ra/e_tds_concentration.html.

Health Canada (2005) Personal Communication N. Roest, and S. Petrovic, Safe Environments Program, Health Canada

Hengstler JG, Bolm-Audorff U, Faldum A et al. (2003) Occupational exposure to heavy metals: DNA damage induction and DNA repair inhibition prove co-exposures to cadmium, cobalt and lead as more dangerous than hitherto expected. Carcinogenesis 24(1):63–73

Higgins I (1982) Arsenic and respiratory cancer among a sample of Anaconda smelter workers. Report submitted to the Occupational Safety and Health Administration in the comments of the Kennecott Minerals Company on the inorganic arsenic rulemaking (Exhibit 203-5)

Hwang YH, Bornschein RL, Grote J, Menrath W, Roda S (1997) Environmental arsenic exposure of children around a former copper smelter site. Environ Res 72(1):72–81

IARC (2004) Monographs on the evaluation of carcinogenic risks to humans. Inorganic and organic lead compounds. International Agency for Research on Cancer. Vol. 87, 10–17 February 2004. www.monographs.iarc.fr/htdocs/announcements/vol87.htm. Accessed November 2004

IOM (2000) Dietary reference intakes for: vitamin C, vitamin E, selenium, and carotenoids. A Report of the Panel on Dietary Antioxidants and Related Compounds, Subcommittees on Upper Reference Levels of Nutrients and Interpretation and Uses of Dietary Reference Intakes, and the Standing Committee on the Scientific Evaluation of Dietary Reference Intakes, Food and Nutrition Board, Institute of Medicine. National Academy Press, Washington, DC

IOM (2001) Dietary reference intakes for: vitamin A, vitamin K, arsenic, boron, chromium, copper, iodine, iron, manganese, molybdenum, nickel, silicon, vanadium, and zinc. A Report of the Panel on Micronutrients, Subcommittees on Upper Reference Levels of Nutrients and of Interpretation and Uses of Dietary Reference Intakes, and the Standing Committee on the Scientific Evaluation of Dietary Reference Intakes, Food and Nutrition Board, Institute of Medicine. National Academy Press, Washington, DC

Ip C, Ganther HE (1992) Relationship between the chemical form of selenium and anticarcinogenic activity. In: Wattenberg L et al. (eds) Cancer chemoprevention, pp. 479–488. CRC Press

Johansson A, Curstedt T, Robertson B, Camner P (1984) Lung morphology and phospholipids after experimental inhalation of soluble cadmium, copper, and cobalt. Environ Res 34:295–309

JWEL (2004a) Appendix 18: Local supermarket food basket. Port Colborne Community Based Risk Assessment. Human Health Risk Assessment – Volume V: Appendices 13 to 21. Jacques Whitford Environmental Ltd., May 2004

JWEL (2004b) Appendix 5: Toxicity assessment. Port Colborne Community Based Risk Assessment. Human Health Risk Assessment – Volume IV: Appendices 5 to 7. Jacques Whitford Environmental Ltd., May 2004

Koller LD, Kerkvliet NI, Exon JH (1985) Neoplasia induced in male rats fed lead acetate, ethyl urea and sodium nitrite. Toxicol Pathol 13(1):50–57

Komarnicki GJK (2005) Lead and cadmium in indoor air and the urban environment. Environ Pollut 136:47–61

Kuligowski J, Halperin KM (1992) Stainless steel cookware as a significant source of nickel, chromium, and iron. Arch Environ Contam Toxicol 23:211–215

Kumar R, Srivastava PK, Srivastava SP (1994) Leaching of heavy metals (Cr, Fe, and Ni) from stainless steel utensils in food simulants and food materials. Bull Environ Contam Toxicol 53:259–266

Lamm SH, Engel A, Kruse MB et al. (2004) Arsenic in drinking water and bladder cancer mortality in the United States: An analysis based on 133 U.S. counties and 30 years of observation. J Occup Environ Med 46:298–306

Lanphear BP, Succop P, Roda S, Henningsen G (2003) The effect of soil abatement on blood lead levels in children living near a former smelting and mining operation. Public Health Rep 118:83–91

Lee-Feldstein A (1983) Arsenic and respiratory cancer in man: Follow-up of an occupational study. In: Lederer W, Fensterheim R (eds) Arsenic: Industrial, biomedical and environmental perspectives. Van Nostrand Reinhold, New York

Lemly AD (1997) Environmental implication of excessive selenium: a review. Biomed Environ Sci 10:415–435

Levander OA (1982) Selenium: biochemical actions, interactions, and some human health implications. In: Prasad AS (ed) Clinical, biochemical and nutritional aspects of trace elements, pp. 345–368. Alan R. Liss, New York

Levander OA (1991) Scientific rationale for the 1989 recommended dietary allowance for selenium. J Am Diet Assoc 91(12):1572–1576

Levander OA, Moser PB, Morris VC (1987) Dietary selenium intake and selenium concentrations of plasma, erythrocytes, and breast milk in pregnant and postpartum lactating and nonlactating women. Am J Clin Nutr 46(4):694–698

Lison D, De Boeck M, Verougstraete V, Kirsch-Volders M (2001) Update on the genotoxicity and carcinogenicity of cobalt compounds. Occup Environ Med 58(10):619–625

Longnecker MP, Taylor PR, Levander OA et al. (1991) Selenium in diet, blood, and toenails in relation to human health in a seleniferous area. Am J Clin Nutr 53(5):1288–1294

Magnus K, Andersen A, Hogetvett AC (1982) Cancer of respiratory organs among workers at a nickel refinery in Norway. Int J Cancer 30:681–685

Murgueytio AM, Evans RG, Sterling DA et al. (1998) Relationship between lead mining and blood lead levels in children. Arch Environ Health 53(6): 414-423

MOE (1987) Organic vs. inorganic arsenic in selected food samples. Report No.87-48-45000-057. Hazardous Contaminants Coordination Branch, Ontario Ministry of the Environment, Toronto, Ontario, Canada

MOE (1994) Soil, drinking water, and air quality criteria for lead: Recommendations to the Minister of the Environment and Energy. ACES Report No. 94-02. Advisory Committee on Environmental Standards (ACES), Ontario Ministry of the Environment and Energy (MOEE), Toronto, Ontario, Canada

MOE (1996a) Guidance on site specific risk assessment for use at contaminated sites in Ontario. Appendix B: MOEE human health based toxicity values. Standards Development Branch, Ontario Ministry of Environment and Energy (MOEE), Toronto, Ontario, Canada

MOE (1996b) Rationale for the development and application of generic soil, groundwater and sediment criteria for use at contaminated sites in Ontario. Standards Development Branch, Ontario Ministry of Environment and Energy (MOEE), Toronto, Ontario, Canada

MOE (2001) Summary of point of impingement standards, point of impingement guidelines, and ambient air quality criteria (AAQCs). Standards Development Branch, Ontario Ministry of the Environment (MOE), Toronto, Ontario, Canada

MOE (2002) Soil investigation and human health risk assessment for the Rodney Street Community, Port Colborne. Ontario Ministry of the Environment. www.ene.gov.on.ca/envision/portcolborne/4255e.htm. Accessed 17 May 2006

MOE (2004) Reg. 153/04 Record of site condition. Soil, ground water and sediment standards for use under Part XV.I of the Environmental Protection Act – Table 1. Ontario Ministry of the Environment. www.ene.gov.on.ca/envision/gp/4697e.pdf. Accessed August 2005

MOE (2005) Drinking water surveillance program reports. Ontario Ministry of the Environment. www.ene.gov.on.ca/water.htm. Accessed May 2005

MOE (2007) Rationale for the development of Ontario air standards for lead and lead compounds. Standards Development Branch, Ontario Ministry of the Environment. www.ene.gov.on.ca/envision/env_reg/er/documents/2006/PA06E0006.pdf. Accessed 10 March 2008

MOEE (1987) Organic vs. inorganic arsenic in selected food samples. Report No. 87-48-45000-057. Hazardous Contaminants Coordination Branch, Ontario Ministry of Environment and Energy, Toronto, Ontario, Canada

Molnar P, Gustafson P, Johannesson S, Boman J, Baregard L, Sallsten G (2005) Domestic wood burning and PM2.5 trace elements: personal exposures, indoor and outdoor levels. Atmos Environ 39:2643–2653

Morin Y, Tetu A, Mercier G (1971) Cobalt cardiomyopathy: Clinical aspects. Br Heart J 33:175–178

Nagymajtényi L, Selypes A, Berencsi G (1985) Chromosomal aberrations and fetotoxic effects of atmospheric arsenic exposure in mice. J Appl Toxicol 5:61–63

NAS (2000) Selenium. In: Dietary reference intakes for vitamin C, vitamin E, selenium, and carotenoids, pp. 284–324. National Academies Press, Washington, DC

Nemery B, Casier P, Roosels D, Lahaye D, Demedts M (1992) Survey of cobalt exposure and respiratory health in diamond polishers. Am Rev Respir Dis 145:610–616

NRC (1999) Arsenic in drinking water. National Research Council. National Academies Press, Washington, DC

NRC (2001) Arsenic in drinking water, 2001 update. National Research Council, Subcommittee to update the 1999 arsenic in drinking water report. National Academies Press, Washington, DC

NTP (1994a) NTP Technical Report on the Toxicology and Carcinogenesis Studies of Nickel Subsulfide in F344/N Rats and B6C3F1 Mice. NTP TR 453, NIH Publication No. 94-3369. National Toxicology Program, U.S. Department of Health and Human Services

NTP (1994b) NTP Technical Report on the Toxicology and Carcinogenesis Studies of Nickel Sulfate Hexahydrate in F344/N Rats and B6C3F1 Mice. NTP TR 454, NIH Publication No. 94-3370. U.S. Department of Health and Human Services. National Toxicology Program

NTP (1998) Toxicology and carcinogenicity studies of cobalt sulfate heptahydrate (CAS No. 10026-24-1) in F344/N rats and B6C3F1 mice (inhalation studies). National Toxicology Program. NTP Technical Report Series, No. 471. U.S. Dept. of Health and Human Services, Public Health Service, National Institutes of Health

OEHHA (1999) Determination of acute reference exposure levels for airborne toxicants. Acute toxicity summary. Metallic copper and copper compounds. Office of Environmental Health Hazard Assessment, California Environmental Protection Agency. www.oehha.org/air/acute_rels/pdf/CusA.pdf. Accessed 2 August 2005

OEHHA (2000) Arsenic and arsenic compounds – chronic toxicity summary. Determination of noncancer chronic reference exposure levels batch 2B – December 2000. Office of Environmental Health Hazard Assessment, California Environmental Protection Agency, California. www.oehha.ca.gov/air/chronic_rels/AllChrels.html. Accessed 12 August 2005

OEHHA (2001) Selenium and selenium compounds (other than hydrogen selenide) – chronic toxicity summary. Determination of noncancer chronic reference exposure levels batch 2B – December 2000. Office of Environmental Health Hazard Assessment, California Environmental Protection Agency. www.oehha.org/air/chronic_rels/pdf/selenium.pdf. Accessed 15 August 2005

OEHHA (2002) Technical support document for describing available cancer potency factors. December 2002. Office of Environmental Health Hazard Assessment, Air Toxicology and Epidemiology Section, California Environmental Protection Agency. www.oehha.org/air/hot_spots/pdf/TSDNov2002.pdf. Accessed 15 August 2005

OEHHA (2003) Nickel and nickel compounds. Office of Environmental Human Health Hazard Assessment, California Environmental Protection Agency. www.oehha.org/ecotox.html. Accessed 22 July 2006

OEHHA (2005a) Nickel and nickel compounds – chronic toxicity summary. Office of Environmental Health Hazard Assessment, California Environmental Protection Agency, California. www.oehha.ca.gov/air/chronic_rels/AllChrels.html. Accessed 21 August 2005

OEHHA (2005b) Nickel oxide – chronic toxicity summary. Office of Environmental Health Hazard Assessment, California Environmental Protection Agency, California. www.oehha.ca.gov/air/chronic_rels/AllChrels.html. Accessed 24 August 2005

OJEU (2005) Directive 2004/107/EC of the European Parliament and of the Council of 15 December 2004 relating to arsenic, cadmium, mercury, nickel and polycyclic aromatic hydrocarbons in ambient air. Official Journal of the European Union – 23/3, 26 January 2005

Oller AR (2002) Respiratory carcinogenicity assessment of soluble nickel compounds. Environ Health Perspect 110 (Suppl 5):841–844

OMEE (1994) Ontario typical range of chemical parameters in soil, vegetation, moss bags and snow. Version 1.0a. PIBS 2792. Phytotoxicology Section Standards Development Branch, Ontario Ministry of Environment and Energy, Toronto, Ontario, Canada

Pratt WB, Omdahl JL, Sorenson JRJ (1985) Lack of effects of copper gluconate supplementation. Am J Clin Nutr 42:681–682

PTI (1994) Remedial Investigation Report. National Zinc Site. PTI Environmental Services. September 1994

Rasmussen PE (2004) Can metal concentrations in indoor dust be predicted from soil geochemistry? Can J Anal Sci Spectrosc 49(2):166–174

Richardson GM (1997) Compendium of Canadian human exposure factors for risk assessment. O'Connor Associates Environmental Inc., Ottawa, Ontario, Canada

Robinson MF (1982) Clinical effects on selenium deficiency and excess. In: Prasad AS, Clinical, Biochemical and Nutritional Aspects of Trace Elements. Alan R. Liss, New York. p. 325-343

Schoof RA, Yost LJ, Eickhoff J et al. (1999) A market basket survey of inorganic arsenic in food. Food Chem Toxicol 37:839–846

Seilkop SK (2004). Estimation of respiratory cancer risks associated with exposure to small airborne concentrations of nickel-containing substances. Presentation to the Ontario Ministry of the Environment and INCO, 10 February 2004

Seilkop SK, Oller AR (2003) Respiratory cancer risks associated with low-level nickel exposure: an integrated assessment based on animal, epidemiological and mechanistic data. Regul Toxicol Pharmacol 37:173–190

Sigal E, Bacigalupo C, Moore R, Ferguson G, Fleming S (2002a) A case study in arsenic risk assessment: Deloro Village, Ontario. Presented at the 2002 Annual Society for Risk Analysis Meeting, New Orleans, USA

Sigal E, Bacigalupo C, Moore R (2002b) The use of a weight of evidence approach to assess health risks from arsenic exposure. Toxicologist 66(1-S):103

Slooff W, Cleven RFM, Janus JA (1989) Integrated criteria document copper. Bilthoven, The Netherlands, National Institute of Public Health and the Environment (RIVM), Report No. 758474009

Smith AH, Hopenhayn-Rich C, Bates MN et al. (1992) Cancer risks from arsenic in drinking water. Environ Health Perspect 97:259–267

Sokoloff I (1985) Endemic form of osteoarthritis. Clin Rheum Dis 11:187–202

Sprince NL, Oliver LC, Eisen EA, Greene RE, Chamberlin RI (1988) Cobalt exposure and lung disease in tungsten carbide production. A cross-sectional study of current workers. Am Rev Respir Dis 138(5):1220–1226

Stanek III EJ, Calabrese EJ (2000) Daily soil ingestion estimates for children at a Superfund site. Risk Anal 20(5):627–635

Stanek III EJ, Calabrese EJ, Zorn M (2001a) Soil ingestion distributions for Monte Carlo risk assessment in children. Hum Ecol Risk Assess 7(2):357–368

Stanek III EJ, Calabrese EJ, Zorn M (2001b) Biasing factors for simple soil ingestion estimates in mass balance studies of soil ingestion. Hum Ecol Risk Assess 7(2):329–355

Stewart RDH, Griffiths NM, Thomson CD, Robinson MF (1978) Quantitative selenium metabolism in normal New Zealand women. Br J Nutr 40:45–54

TERA (2004) Toxicological review of soluble nickel salts. Toxicology Excellence for Risk Assessment. www.tera.org/vera/Nickel%20Doc%20page.htm. Accessed Aug 2005

Tseng WP (1977) Effects and dose-response relationships of skin cancer and blackfoot disease with arsenic. Environ Health Perspect 19:109–119

Tseng WP, Chu HM, How SW, Fong JM, Lin CS, Yeh S (1968) Prevalence of skin cancer in an endemic area of chronic arsenicism in Taiwan. J Natl Cancer Inst 40(3):453–463

Tupholme KW, Dulieu D, Wilkinson J, Ward NB (1993) Stainless steels for the food industries. Proceedings of Innovation Stainless Steel, Vol. 1, p. 49, Florence, Italy, 11–14 October 1993. Associazione Italiana di Metallurgia, Milan, Italy

U.S. EPA (1986) Guidelines for carcinogen risk assessment. Published on 24 September 1986. EPA/630/R-00/004. Federal Register 51(185):33992-34003. Risk Assessment Forum, U.S. Environmental Protection Agency, Washington, DC

U.S. EPA (1989) Risk Assessment Guidance for Superfund. EPA/540/01. U.S. Environmental Protection Agency, Washington, DC

U.S. EPA (1991a) Selenium and compounds (CASRN 7782-49-2). Integrated Risk Information System (IRIS). U.S. Environmental Protection Agency. www.epa.gov/iris/subst/0472.htm. Accessed August 2005

U.S. EPA (1991b) Nickel soluble salts. Integrated Risk Information System. U.S. Environmental Protection Agency. www.epa.gov/iris/subst/0271.htm. Accessed August 2005

U.S. EPA (1991c) Nickel subsulphide/refinery dust. Integrated Risk Information System. U.S. Environmental Protection Agency. www.epa.gov/iris/subst/0272.htm and www.epa.gov/iris/subst/0273.htm. Accessed 20 August 2005

U.S. EPA (1992) Dermal exposure assessment: Principles and applications. Interim Report. Exposure Assessment Group Office of Health and Environmental Assessment, U.S. Environmental Protection Agency, Washington, DC

U.S. EPA (1993) Arsenic, inorganic; CASRN 7440-38-2. Integrated Risk Information System (IRIS). On line database. www.epa.gov.iris. Date of last major revision for oral reference dose assessment

U.S. EPA (1994) Guidance manual for the integrated exposure uptake biokinetic model for lead in children. EPA/540/R-93/081. U.S. Environmental Protection Agency, Washington, DC

U.S. EPA (1996) Nickel, soluble salts; CASRN various. Integrated Risk Information System (IRIS). On line database www.epa.gov.iris. Date of last major revision for oral RfD assessment

U.S. EPA (1997a) Exposure factors handbook. Volume I – General factors. EPA/600/P-95/002Fa. Office of Research and Development. U.S. Environmental Protection Agency, Washington, DC

U.S. EPA (1997b) Exposure factors handbook. Volume II – Food ingestion factors. EPA/600/P-95/002Fa. Office of Research and Development. U.S. Environmental Protection Agency, Washington, DC

U.S. EPA (1998) Arsenic, inorganic; CASRN 7440-38-2. Integrated Risk Information System (IRIS). On line database. www.epa.gov.iris. Date of last major revision for lifetime carcinogenicity assessment

U.S. EPA (1999a) Screening level ecological risk assessment protocol for hazardous waste combustion facilities. Volume III – Appendix C. EPA530-D-99-001A. U.S. Environmental Protection Agency, Washington, DC

U.S. EPA (1999b) Risk Assessment guidance for Superfund: Volume 3 – (Part A, Process for conducting probabilistic risk assessment). Office of Solid Waste and Emergency Response, U.S. Environmental Protection Agency, Washington, DC

U.S. EPA (1999c) Short sheet: IEUBK model soil/dust ingestion rates. EPA#540-F-00-007. Office of Solid Waste and Emergency Response, United States Environmental Protection Agency, Washington, DC

U.S. EPA (2001) Risk assessment guidance for superfund. Volume I: Human health evaluation manual (Part E, Supplemental guidance for dermal risk assessment). Interim review draft – for public comment. EPA/540/99/005. Office of Emergency and Remedial Response, U.S. Environmental Protection Agency. www.epa.gov/superfund/programs/risk/ragse/index.htm. Accessed Oct 2003

U.S. EPA (2002a) Provisional peer reviewed toxicity values (PPRTV). Derivation support document for cobalt and compounds (CASRN 7440-48-4). EPA 00-122/1-15-02. Superfund Health Risk Technical Support Centre (STSC), U.S. Environmental Protection Agency, Washington, DC

U.S. EPA (2002b) Child-specific exposure factors handbook. EPA-600-P-00-002B. National Center for Environmental Assessment, U.S. Environmental Protection Agency, Washington, DC

U.S. EPA (2003) Superfund lead-contaminated residential sites handbook. Final. August 2003. OSWER 9285.7-50. Prepared by the U.S. Environmental Protection Agency Lead Sites Workgroup (LSW). U.S. Environmental Protection Agency, Office of Emergency and Remedial Response, Washington, DC

U.S. EPA (2004a) Exposure scenarios. EPA/600/R-03/036. National Center for Environmental Assessment, U.S. Environmental Protection Agency, Washington, DC

U.S. EPA (2004b) ProUCL version 3.0 user guide. U.S. Environmental Protection Agency. www.epa.gov/nerlesd1/tsc/download.htm. Accessed 10 October 2006

U.S. EPA (2004c) Users' guide and background technical document for USEPA Region 9's preliminary remediation goals (RMSL) table. U.S. Environmental Protection Agency Region IX. www.epa.gov/region09/waste/sfund/RMSL/files/04usersguide.pdf. Accessed 10 October 2006

U.S. EPA (2004d) Integrated risk information system. IRIS database on-line search profile for lead. US Environmental Protection Agency, Cincinnati, OH. Oral RfD Assessment updated 2004. Carcinogenicity Assessment updated 1993 www.epa.gov/iris/subst/0193.htm Accessed November 2004

U.S. EPA (2004e) Framework for inorganic metals risk assessment. Peer review draft. EPA/630/P-04/068B. U.S. Environmental Protection Agency, Washington, DC

U.S. EPA (2005) Science issue paper: Mode of carcinogenic action for cacodylic acid (dimethylarsinic acid, dmav) and recommendations for dose response extrapolation. July 26, Integrated Risk Information System. Glossary of IRIS terms. Updated December 2005. Health Effects Division, Office of Pesticide Programs, U.S. Environmental Protection Agency. www.epa.gov/iris/gloss8.htm. Accessed 15 September 2006

U.S. EPA (2007) Advisory on EPA's assessments of carcinogenic effects of organic and inorganic arsenic: A report of the US EPA Science Advisory Board. EPA-SAB-07-008. EPA Science Advisory Board Arsenic Review Panel, U.S. Environmental Protection Agency, Washington, DC

U.S. EPA OSW (1998) Human health risk assessment protocol for hazardous waste combustion facilities, Vol. I Peer review draft. U.S. Environmental Protection Agency Region 6. Multimedia Planning and Permitting Division. Center for Combustion Science and Engineering. Office of Solid Waste, U.S. Environmental Protection Agency, Washington, DC

U.S. FDA (1982) Toxicological principles for the safety assessment of direct food additives and color additives used in food. U.S. Food and Drug Administration, Bureau of Foods, Washington, DC

U.S. FDA (2004) Total diet study statistics on element results. Revision 2, 1991–2002, July 6, 2004. U.S. Food and Drug Administration. Total diet study market baskets 1991–3 through 2002–4. www.cfsan.fda.gov/~comm/tds-res.html. Accessed 15 Nov 2005

Vyskocil A, Senft V, Viau C, Cizkova M, Kohout J (1994a) Biochemical renal changes in workers exposed to soluble nickel compounds. Hum Exp Toxicol 13:257–261

Vyskocil A, Viau C, Cizkova M (1994b) Chronic nephrotoxicity of soluble nickel in rats. Hum Exp Toxicol 13:689–693

Walker S, Griffin S (1998) Site-specific data confirm arsenic exposure predicted by the U.S. Environmental Protection Agency. Environ Health Perspect 106(3):133–139

Whanger PD (1983) Selenium interactions with carcinogens. Fundam Appl Toxicol 3(5):424–430

Whanger PD, Vendeland S, Park YC, Xia Y (1996) Metabolism of subtoxic levels of selenium in animals and humans. Ann Clin Lab Sci 26(2):99–113

Whitman NE (1957) Letter to TLV Committee from Industrial Health Engineering. Bethlehem Steel Co, Bethlehem, PA, 12 March 1957

Whitman NE (1962) Letter to TLV Committee from Industrial Health Engineering. Bethlehem Steel Co, Bethlehem, PA, April 1962

WHO (1986) Selenium. Environmental health criteria #58. International Program on Chemical Safety, a joint programme of the International Labour Organization (ILO) and the World Health Organization, Geneva, Switzerland

WHO (1995) International Programme on Chemical Safety. Environmental Health Criteria 165. Inorganic Lead. World Health Organization, Geneva. www.inchem.org/documents/ehc/ehc/ehc165.htm. Accessed 21 August 2005

WHO (1996) Guidelines for drinking-water quality, 2nd ed. Volume 2, Health criteria and other supporting information. World Health Organization, Geneva, Switzerland

WHO (1998) Environmental health criteria #200: Copper. Published jointly by the United Nations Environment Programme, the International Labour Organisation, and the World Health Organization, Geneva, Switzerland

WHO (2000) Air quality guidelines. World Health Organization www.who.int/en/. Accessed 17 April 2006

WHO (2006) Cobalt and inorganic cobalt compounds. Concise international chemical assessment document 69. World Health Organization International Programme on Chemical Safety. www.inchem.org/documents/cicads/cicads/cicad69.htm#11.0. Accessed 17 April 2007

WHO-IPCS (2001) Environmental health criteria #224. Arsenic and arsenic compounds. International Programme on Chemical Safety. United Nations Environment Programme (UNEP), the International Labour Organization (ILO) and the World Health Organization, Geneva, Switzerland

Wu MM, Kuo TL, Hwang YH, Chen CJ (1989) Dose-response relation between arsenic concentration in well water and mortality from cancers and vascular diseases. Am J Epidemiol 130(6):1123–1132

Yang GQ, Zhou RH (1994) Further observations on the human maximum safe dietary selenium intake in a seleniferous area of China. J Trace Elem Electrolytes Health Dis 8:159–165

Yang G, Zhou R, Yin S et al. (1989a) Studies of safe maximal daily dietary selenium intake in a seleniferous area in China. I. Selenium intake and tissue selenium levels of the inhabitants. J Trace Elem Electrolytes Health Dis. 3(2):77–87

Yang G, Yin S, Zhou R et al. (1989b) Studies of safe maximal daily dietary Se-intake in a seleniferous area in China. Part II: Relation between Se-intake and the manifestation of clinical signs and certain biochemical alterations in blood and urine. J Trace Elem Electrolytes Health Dis 3(3):123–130. Erratum in: J Trace Elem Electrolytes Health Dis 3(4):250

7.6 Appendix Chapter 7: Abbreviations

ACGIH, American Conference of Governmental Industrial Hygienists
ADI, acceptable daily intake
ATSDR, Agency for Toxic Substances and Disease Registry
BBL, blood lead level
BCF, bioconcentration factor
CGS, City of Greater Sudbury
CHQ, critical hazard quotient
COC, chemical of concern
COI, community of interest
CR, concentration ratio
CRL, cancer risk level
CSF, cancer slope factor
CTDS, Canadian Total Diet Study
CTE, central tendency estimate
DWSP, Drinking Water Surveillance Program
EDI, estimated daily intake
EDI$_{MB}$, estimated daily market basket intake
EDTI, estimated daily total intake
EPC, exposure point concentration
EU, exposure unit or European Union
GLSFATF, Great Lakes Sport Fish Consumption Advisory Task Force
GSA, Greater Sudbury Area
HHRA, human health risk assessment
HQ, hazard quotient
IARC, International Agency for Research on Cancer
IEUBK, Integrated Exposure Uptake Biokinetic
ILCR, incremental lifetime cancer risk
IOC$_{POP}$, intake of concern (population)
IRIS, integrated risk information system
IUR, inhalation unit risk
IVBA, *in vitro* bioaccessibility
LOAEL, lowest observed adverse effect level
MDL, minimum detection limit
MOE, Ministry of the Environment (Ontario)
MOEE, Ministry of the Environment and Energy
MRL, minimal risk level
NAPS, National Air Pollution Surveillance
NFCS, National Food Consumption Survey
NOAEC, no observable adverse effect concentration
NOAEL, no observed adverse effect level
NRC, National Research Council
OEHHA, Office of Environmental Health Hazard Assessment (California)

OJEU, Official Journal of the European Union
OTR, Ontario Typical Range
OTR$_{98}$, 98th percentile soil concentration Ontario Typical Range
PbB, blood lead
PDI, permissible daily intake
PDF, probability distribution function
PRGs, preliminary remediation goals
RAF, relative absorption factor
RBA, relative *in vivo* bioavailability
RDA, recommended dietary allowance
RDDR, regional deposited dose ratio
REL, reference exposure level
RfC, reference concentration
RfD, reference dose
RIVM, Dutch National Institute for Public Health and the Environment
RMC, risk management criteria
RME, reasonable maximum exposure
SARA, Sudbury Area Risk Assessment
SF, slope factor
SFi, inhalation slope factor
SFo, oral slope factor
SRM, standard reference material
SRML, soil risk management level
TCA, tolerable concentration in air
TDI, tolerable daily intake
TERA, Toxicology Excellence for Risk Assessment
TOR, Typical Ontario Resident
TRV, toxicity reference value
UCL, upper confidence level
UCLM, upper confidence level of the mean
UL, upper limit
URF, unit risk factors
U.S. CDC, U.S. Centers for Disease Control
USDA, United States Department of Agriculture
USEPA, United States Environmental Protection Agency
USWA, United Steel Workers of America
WHO, World Health Organization
WHO – IPCS, World Health Organization – International Programme on Chemical Safety

8.0 Human Health Risk Assessment Results

DOI: 10.5645/b.1.8

Authors

Glenn Ferguson, PhD
Intrinsik Environmental Sciences
6605 Hurontario Street, Suite 500, Mississauga, ON Canada L5T 0A3
Email: Gferguson@intrinsikscience.com

Elliot Sigal
Intrinsik Environmental Sciences
6605 Hurontario Street, Suite 500, Mississauga, ON Canada L5T 0A3
Email: Esigal@intrinsikscience.com

Chris Bacigalupo
Intrinsik Environmental Sciences
6605 Hurontario Street, Suite 500, Mississauga, ON Canada L5T 0A3
Email: Cbacigalupo@instrinsikscience.com

Table of Contents

8.1 Overview of Approach	209
8.2 Results	213
8.2.1 Arsenic	213
8.2.2 Cobalt	216
8.2.3 Copper	217
8.2.4 Lead	218
8.2.5 Nickel	218
8.2.6 Selenium	223
8.2.7 Summary of HHRA Risk Characterization Results	224
8.3 Development of Soil Risk Management Level (SRML) for Lead	225
8.3.1 Weight-of-Evidence Evaluation for Pb	225
8.3.2 Supplemental Model Results for Selecting SRML	231
8.3.3 Recommended SRML for Lead	232
8.4 Other Risk Assessment Issues	233
8.4.1 Considerations for Children's Exposure and Toxicity	233
8.4.2 Sulfur Dioxide (SO_2)	234
8.4.3 Occupational Exposures	235
8.4.4 Chemical Mixtures	235
8.4.5 Soil Ingestion Rates in Children	236
8.4.6 Dermal Sensitization to Nickel	237
8.4.7 The Elderly as a Sensitive Subpopulation	237
8.4.8 COC Lifetime Body Burden	238

8.5 Uncertainties in the Human Health Risk Assessment ... 238
   8.5.1 General Sources of Uncertainty in Risk Assessment ... 238
   8.5.2 Uncertainty Analysis ... 240
   8.5.3 Uncertainties in the Sudbury HHRA ... 240
   8.5.4 Sensitivity Analysis ... 244
8.6 References ... 248
8.7 Appendix Chapter 8: Abbreviations ... 252

Tables
8.1 Comparison of Average and Maximum Exposure Scenarios for the Female Toddler – Sudbury Centre ... 211
8.2 Example Risk Characterization – RME Female Receptors – Falconbridge ... 215
8.3 Summary of Non-Cancer Ni Inhalation Risk Characterization ... 220
8.4 Summary of Proposed Ni Species Fingerprints ... 220
8.5 Summary of Ni Species-Specific Inhalation Unit Risks ... 221
8.6 Risk Estimates for the Typical Ni Species Fingerprint (no Ni_3S_2 exposure) ... 221
8.7 Risk Estimates Based on Exposure to Ni Species Fingerprint ... 222
8.8 Risk Estimates Based on Ambient Air Concentrations at Different Monitoring Stations ... 222
8.9 Different Soil Risk Management Levels (SRML) for Pb (mg/kg) ... 226
8.10 Sensitivity Analysis of Calculated SRML Values for Pb to the Female Child; Copper Cliff ... 227
8.11 Relationship between Pb in Soil (mg/kg) and Blood Pb Levels (µg/dL) in North American Populations ... 229
8.12 Relationship of Soil Pb to Blood Pb in Children Living Near Mine Waste and Urban Areas ... 230
8.13 SARA and IEUBK Pb Model Results at the Selected SRML ... 231
8.14 Relative Impact to Predicted Risk from Changing Key HHRA Assumptions for Lead in Copper Cliff ... 246
8.15 Relative Impact to Predicted Risk from Changing Key HHRA Assumptions for Nickel in Copper Cliff ... 247

Figures
8.1 HQavg Estimates for Ni (RME) – Sudbury Centre ... 211
8.2 Oral HQ Estimates in the General Population Versus Anglers and Hunters ... 212
8.3 As Exposure by Pathway for the Female Toddler ... 213
8.4 Estimated Pathway Contribution to Co for Female Preschool Child ... 217
8.5 Estimated Pathway Contribution to Cu for Female Preschool Child ... 217
8.6 Ni Hazard Quotients for a Female Preschool Child ... 218
8.7 Estimated Ni Exposure Pathway Contribution in Copper Cliff ... 219
8.8 Se Hazard Quotients for the Female Preschool Child ... 224

8.1 Overview of Approach

This chapter provides a summary of the results of the detailed human health risk assessment (HHRA) as well as a discussion of the implications of these results for residents of the Greater Sudbury Area. It represents a summary of the results since the detailed HHRA considered a variety of data, chemicals, communities, individuals, exposure pathways and assumptions including:

- six chemicals of concern (COC) (As Co, Cu, Pb, Ni and Se)
- five communities of interest (COI) (Copper Cliff, Falconbridge, Coniston, Sudbury Centre, Hanmer [as a background community], and the Typical Ontario Resident [TOR])
- five receptor age classes (i.e. infant, preschool child, child, adolescent and adult) and composite lifetime individuals for each of the two genders
- consideration of the general population and a special receptor category of avid anglers and hunters (including First Nation receptors)
- receptor characteristics characterized by: (i) average or *Central Tendency Exposure* (CTE); and, (ii) upper-bound or *Reasonable Maximum Exposure* (RME) estimates
- inhalation, oral and dermal exposure pathways, and
- a large database of site-specific media concentrations characterized by average (95% upper confidence limit on the mean or UCL) or upper-bound (maximum) statistics.

These combinations of COC, communities, receptors and exposure scenarios resulted in over 300 separate calculations of risk. Since it is clearly not practical, nor of general interest, to present the results of all these combinations, representative examples are provided in the following text and all key results are discussed. Potential risk is characterized by comparing predicted exposures from all pathways with the exposure limits or toxicity reference values (TRVs). For non-carcinogenic COC, this comparison is typically referred to as the *Hazard Quotient* (HQ) and is calculated by dividing the predicted exposure level by the exposure limit. If the total chemical exposure from all pathways is equal to or less than the exposure limit, then the HQ would be 1.0 or less, and no adverse health effects would be expected (refer to Chapter 7 for a more detailed discussion of risk characterization).

For chemicals with non-threshold-type dose responses (i.e. carcinogens), the comparison is referred to as the *Incremental Lifetime Cancer Risk Level* (ILCR) or more simply a *cancer risk level* (CRL) and is defined as the incremental risk of an individual in a population of a given size developing cancer over a lifetime. The ILCR is calculated by multiplying the predicted exposure by the slope factor or unit risk value. The ILCR is expressed as the prediction that one person per *n* people would develop cancer, where the magnitude of *n* reflects the risks to that population. In the case of carcinogens, the acceptable risk level in Ontario is considered to be an incremental increase in cancer risk of one-in-one million (i.e. one additional cancer per million people). Typically, incremental lifetime cancer risks are calculated by multiplying a chemical- and route-specific cancer slope factor by facility related exposures. For the current assessment, an evaluation of all potential exposure sources and pathways was completed (regardless of source).

As discussed previously, the 95% upper confidence limit on the arithmetic sample mean (95% UCLM) was utilized to characterize the exposure point concentration (EPC) of a given exposure unit (U.S. EPA 2001a). The sample mean was based on a collection of samples from the exposure unit and, therefore, uncertainty exists as to whether the sample mean is a true reflection of the population mean. As a result, the 95% UCLM can be considered an estimate of the true population mean for a given exposure unit. The underlying assumption was that individuals would move randomly within each community and, therefore, over time, come into contact with the average soil concentration within a given community (or exposure unit).

If the property or site of concern were a single residential lot, it would be reasonable to assume that an individual would move in a random fashion within his or her own property. In reality, individuals do not move in a random fashion within their community, but rather exhibit predictable spatial patterns in their movements. For example, many individuals will tend to spend the majority of their time between home and work or school. Therefore, the evaluation of risks on the basis of average EPCs (assuming random movement) in an area-wide risk assessment may underestimate exposure and associated risks for some receptors. As a result, in area-wide assessments, where data permit, it is also necessary to consider upper bound concentration estimates, in addition to averages for the EPC. For this reason, two statistics were used to characterize COC concentrations in soils within each community of interest: (i) the average soil concentration (based on the 95% UCLM); and, (ii) the maximum soil concentration. For all other media, data were not sufficient for a site-by-site evaluation, and as such, other media (i.e. air, indoor dust) were only considered on an area-wide basis, using average (95% UCLM) concentrations. The resultant risk estimates utilizing the average and maximum soil concentrations were referred to as the average hazard quotient (HQ_{avg}) and the maximum hazard quotient (HQ_{max}), respectively.

In addition to a benchmark comparison, both the strength-of-evidence and weight-of-evidence must be evaluated when considering the results of an HHRA, including consideration of non-site related exposures (i.e. a comparison of site conditions to background) and the consideration of additional pieces of information that may be available (e.g. biological monitoring results, public health information, etc.). As a result, the calculation and interpretation of estimated human health risk estimates was a multi-step process which involved the following key elements:

1. Risks estimated using both CTE and RME receptor characteristics and average (95% UCLM) concentrations for all of the exposure media, to obtain a general picture of overall risk for each COI. Non-cancer HQ estimates were calculated for all individual receptor groups and presented for the female preschool child as this receptor was considered the most highly exposed individual. Incremental lifetime cancer risks were also considered. CRL estimates were generated using the female lifetime or composite receptor which includes all five age categories. (i.e. infant, preschool child, child, adolescent and adult). The resultant values were referred to as HQ_{avg} and CRL_{avg}, and provide risk estimates for the *average* resident of that community (note: risk predictions presented as HQ or CRL are equivalent to HQ_{avg} and CRL_{avg}, respectively).

2. For RME receptor characteristics, risks were estimated using both the average (95% UCLM) and maximum soil concentrations. The resultant risk estimates are referred to as the HQ_{avg} and HQ_{max}, and CRL_{avg} and CRL_{max}, respectively. This provided an estimate of risk for receptors exposed to the maximum (i.e. worst-case) soil concentration in each COI.

3. If no unacceptable health risks were predicted using average (95% UCLM) and maximum soil concentrations, then no further assessment was considered necessary. If unacceptable risks were not screened out, further evaluation of the exposure assessment was undertaken through a weight-of-evidence approach.

4. If unacceptable risks were predicted, site-specific risk management goals for soil (termed a *Soil Risk Management Level* or $SRML_{soil}$) were developed.

5. With the exception of As and Pb, the results of the risk assessment were presented separately for oral/dermal and inhalation exposures. Arsenic has been found to act via a similar toxicological mechanism following either inhalation, oral or dermal exposure. The TRV established by the Ontario Ministry of the Environment (MOE) for Pb is for all pathways and, as such, the risk characterization includes oral/dermal and inhalation.

6. The relative contribution of each exposure pathway (i.e. soil, dust, drinking water, local foods, market basket, and air) was also calculated. This was useful for identifying circumstances in which meaningful pathway-specific interventions may be undertaken, if necessary.

Focus on the Female Toddler

For the assessment of non-cancer health risks, it is common to consider HQ (risk) estimates for the most sensitive or highly exposed life stage. Average lifetime risks were considered when the toxicological endpoint of concern is a result of lifetime exposures (as for Ni). Figure 8.1 illustrates the comparison of risk estimates (HQ values) for Ni oral/dermal exposure to five different life stages and a lifetime composite receptor in the Sudbury Centre. This is one example, but it clearly demonstrates that consideration of the preschool child (toddler) provides the most conservative evaluation (highest HQ value) of non-cancer risks relative to other receptor age classes. Lifetime risk estimates were significantly less for the other age groups, and as such, consideration of the toddler for risk assessment and risk management purposes was considered conservative. Although both male and female receptors were evaluated, for the purposes of discussing the results, the female toddler was used due to longer predicted life expectancy. The detailed results for all receptor age classes, COC and COI for both male and female receptors, general population and anglers and hunters occupied over 325 pages in one appendix of the original report (SARA 2008) and are not repeated here.

Receptor age class

[Bar chart showing Oral/dermal HQ estimate by age class:
- Infant: ~0.15
- Toddler: ~0.66
- Child: ~0.42
- Adolescent: ~0.27
- Adult: ~0.21
- Lifetime: ~0.27]

Oral/dermal HQ estimate

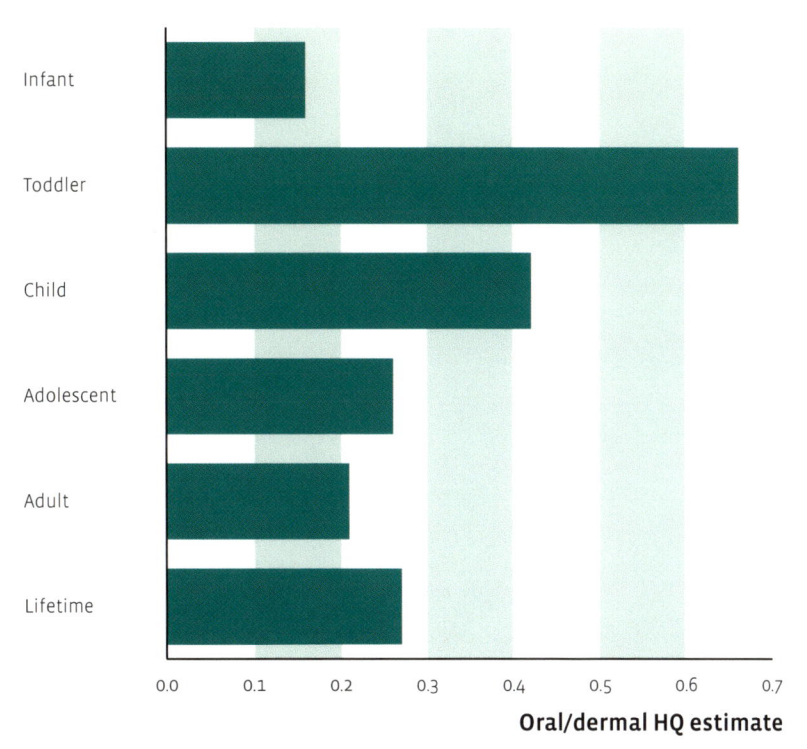

Figure 8.1   **HQavg Estimates for Ni (RME) – Sudbury Centre**

Comparing Risk from Average and Maximum Soil Concentrations

As discussed earlier, when considering community-wide risks that employ an estimate of the average (i.e. 95% UCLM) soil concentration for an entire COI, it is prudent also to consider smaller, more localized areas, which may be associated with COC soil concentrations in excess of the community-wide average. To demonstrate this point, Table 8.1 provides a comparison of the RME results for female preschool children (oral/dermal exposures) in Sudbury Centre using average (HQ_{avg}), and maximum (HQ_{max}) soil concentrations. Similar trends were observed with CTE results for all COI, age groups, and receptors, including hunters/anglers. Risk estimates with HQ > 1.0 are presented in bold type. On the basis of these results, risks for As and Se were not ruled out (HQ > 1.0) and these elements required further consideration. Although the HQ_{max} estimate of 1.1 for Pb marginally exceeded the target value of 1.0, due to public and regulator concerns and consideration of model uncertainties and sensitivities, Pb was also carried forward into the weight-of-evidence approach.

Table 8.1   **Comparison of Average and Maximum Exposure Scenarios for the Female Toddler – Sudbury Centre**

COC	Average exposure scenario – 95% UCLM soil concentration (µg/g)	HQ_{avg}	Maximum exposure scenario – Maximum soil concentration (µg/g)	HQ_{max}
As	7.2	**1.3**	59.0	**1.4**
Co	11.3	0.16	100	0.17
Cu	204	0.62	1640	0.63
Pb	36	0.83	310	**1.1**
Ni	210	0.66	3260	0.79
Se	1.30	**1.6**	12.5	**1.6**

HQ > 1.0 presented in bold type.

Hunting and Fishing Populations within the Greater Sudbury Area (GSA)

In addition to the general Sudbury population, individuals within the GSA who participate in hunting and fishing activities were considered separately when predicting HQ estimates. Members of the general population were assumed to consume local wild game and fish; however, members of the hunting/fishing sub-population (including First Nation members) were considered to consume significantly more local wild game and fish relative to those in the general population. Refer to Chapter 7 for a discussion with regard to wild-game and fish intake rates. Predicted HQ (female preschool child) and CRL (female composite receptor) values for the hunting and angling sub-populations of the GSA were compared to those of the general Sudbury population. Figure 8.2 provides a comparison of HQ_{avg} estimates for the general GSA population of female preschool children versus those of female preschool children in an avid hunting/fishing sub-population.

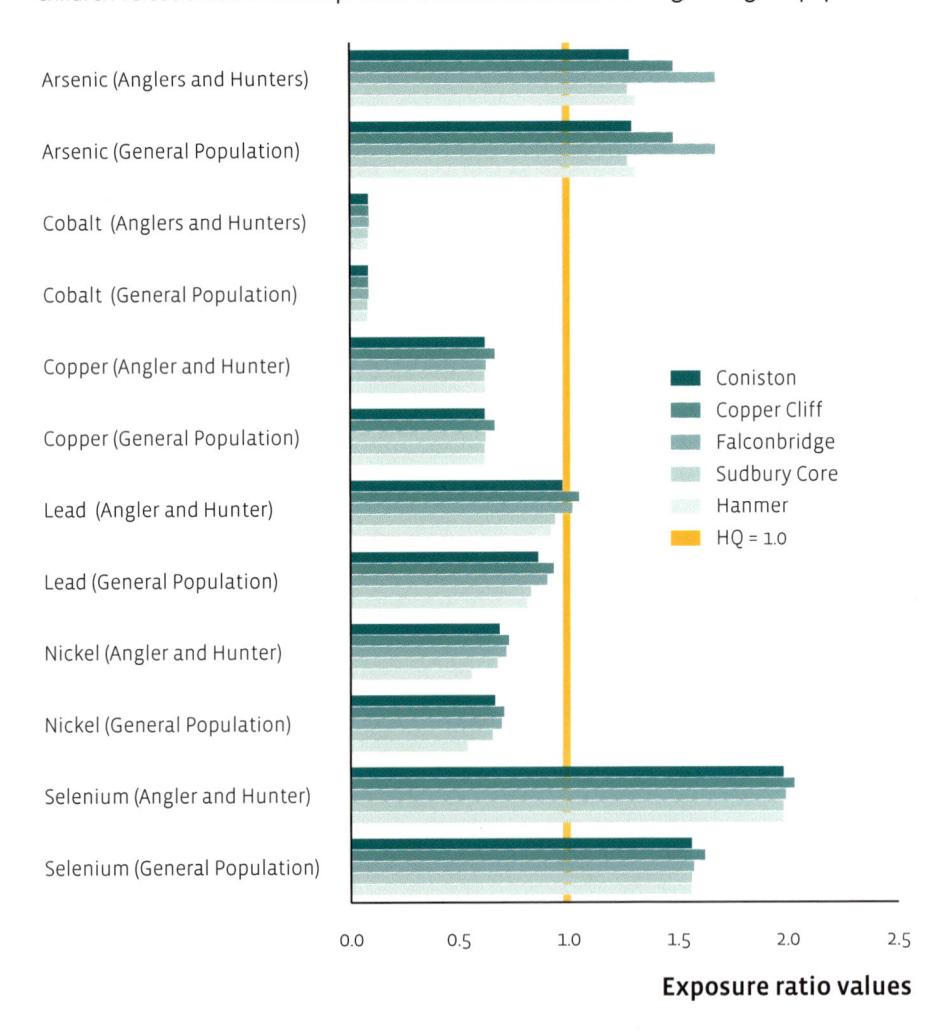

Figure 8.2   Oral HQ Estimates in the General Population Versus Anglers and Hunters

Little difference, if any, was observed in HQ_{avg} estimates (under a RME scenario) between female preschool children from the general GSA population and those from an avid hunting and angling population. The greatest differences in HQ_{avg} estimates between the two populations were observed for Se and Pb. Selenium HQ estimates of 1.6 and 2.0 were predicted for the female preschool child of the general and hunting/angling sub-populations, respectively, living in Sudbury Centre. Slight differences in HQ_{avg} estimates predicted for Pb were also observed between the two populations.

8.2 Results

8.2.1 Arsenic

Health-based As soil standards in Ontario (i.e. soil concentrations resulting in ILCRs of less than one-in-one million) result in the derivation of impractical soil remediation standards that are typically lower than the levels of As found to occur naturally in many soil environments. According to the *Guideline for Use at Contaminated Sites in Ontario* (MOE 1996) and the Risk Assessment Procedures document (MOE 2005), site-specific health-based soil standards for non-threshold (i.e. carcinogenic) compounds should be developed using an ILCR of 1×10^{-6} per exposure medium. A health-based soil standard based on direct contact pathways only and an ILCR of one-in-one million can produce health-based soil standards between 1.0 and 2.0 mg of As/kg soil. The use of site-specific relative absorption factor (RAF) adjustments to account for differences in accessibility between different environmental media (i.e. water versus soil) could potentially increase (depending on-site conditions) the health-based portion of the standard by 30 to 60. The MOE currently has an As soil standard for residential/parkland and agricultural land uses of 20 mg/kg (25 mg/kg for medium/fine textured soils). These standards are not health-based but rather a reflection of the 98th percentile background concentration of 14 and 17 mg/kg for rural and urban parkland soils, respectively, in Ontario.

Given the above information, risk assessments involving As are likely to produce quantitative health risk estimates (including both cancer and non-cancer effects) in excess of the acceptable level of risk set by a regional regulator. It is apparent that a variety of 'tools' such as risk assessment, community health status and/or urinary As studies must be used to capture the true context of arsenic-related health risks (see the subsection *Weight-of-Evidence Approach for As*, below). Figure 8.3 illustrates inorganic As exposures via different pathways for three communities, Falconbridge, Hanmer (reference community) and the Typical Ontario Resident (TOR). The figure shows that water and incidental soil/dust ingestion in Falconbridge were greater than those experienced by typical Ontario and Hanmer female composite receptors. The 'market basket' risks for typical Ontario were slightly lower than those of Sudbury-specific COI because a proportion of an individual's diet (living within a Sudbury COI) was apportioned to locally derived foods.

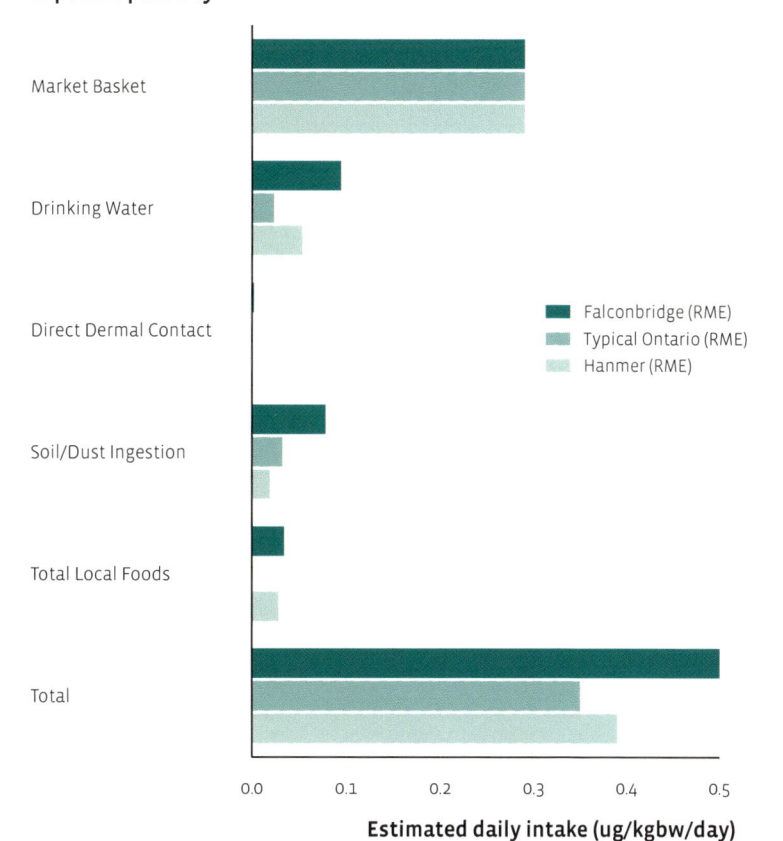

Figure 8.3 **As Exposure by Pathway for the Female Toddler**

The calculated lifetime CRLs associated with Sudbury-specific inorganic As exposure ranged from 9.6×10^{-5} to 2.6×10^{-4} (i.e. between approximately one- and three-in-ten thousand) for all exposure scenarios. Lifetime cancer risk estimates for a female composite receptor, under typical Ontario conditions, were between five- and-six-in-one hundred thousand (i.e. 5.5×10^{-5} and 6.3×10^{-5} for CTE and RME scenarios, respectively).

Although lifetime cancer risk is typically the endpoint of interest when assessing the human health implications of inorganic As, non-cancer endpoints also exist and the U.S. EPA has developed a reference dose (RfD). Therefore, non-cancer health risk estimates associated with exposures to inorganic As were also calculated. The HQ estimates for the female preschool child living in Falconbridge were between 1.4 and 1.7 for the CTE and RME scenarios, respectively. HQ estimates in Falconbridge were slightly higher than those predicted using typical Ontario assumptions for the CTE and RME scenarios.

If the EPC for As in Falconbridge soils (i.e. the 95% UCL on the arithmetic mean of 188 samples) was reduced from 79 mg/kg to 17 mg/kg (background for non-agricultural sites in Ontario), the reduction in lifetime cancer estimates was approximately 12% (from 2.6×10^{-4} to 2.3×10^{-4}). However, these were only rough estimations of overall improvement to health risks related to potential soil remediation activities, and did not account for any potential improvements observed in some secondary media impacted in some fashion by soil itself (e.g. home garden produce, water, indoor dust, etc.).

Table 8.2 provides an example of summing exposure from different routes for different receptor classes using As in the community of Falconbridge for the female receptor. The EPC values were reasonable maximum or 95% UCL values. Although this is just one example using one COC and one COI, a similar process was followed for all combinations of receptors and exposures. The exposure data were summed to calculate the estimated total daily intake (ETDI). For the female preschool child, the ETDI was 0.51 μg/kg bw/day. Of this total, inhalation accounted for about 0.3% of the total, direct soil contact (outdoor and indoor) accounted for about 11.8% (0.083 μg/kg/day), drinking water was about 29% of the total (0.099 μg/kg/day) and market basket ingestion accounted for 51.7% (0.29 μg/kg/day).

The next step in the non-cancer risk characterization was to compare the ETDI with the toxicity reference value (TRV). In this example, the oral TRV was an oral reference dose of 0.3 μg As/kg bw/day (as presented in Table 7.13, Chapter 7). This step produced a HQ value > 1.0 for the preschool child and the female lifetime receptor. Thus, non-cancer risk for As was not ruled out for these receptors in the community of Falconbridge and further evaluation of As risk was undertaken using a weight-of-evidence approach. For the calculation of cancer risk, the oral slope factor (SF_o) was 0.0015 (μg/kg/day)$^{-1}$ and the inhalation slope factor (SF_i) was 0.015 (μg/kg/day)$^{-1}$. These values and their sources were also presented previously in Table 7.13. The total incremental lifetime cancer risk (ILCR) was calculated to be 2.6×10^{-4}, or approximately 2.6 in 10,000, which clearly exceeds the Ontario government target ICLR of 1×10^{-6} or one in a million.

Table 8.2 Example Risk Characterization – RME Female Receptors – Falconbridge

Exposure pathway	Environmental media concentrations Value	Units	Average percent of total ETDI Percent	Female infant	Female preschool child	Female child	Female adolescent	Female adult	Female lifetime
Inhalation of fine particulate – Outdoors	0.0024	µg/m³	0.0	0.000039	0.000084	0.000061	0.000038	0.000036	0.000076
Inhalation of fine particulate – Indoors	0.0024	µg/m³	0.3	0.00058	0.0013	0.00090	0.00056	0.00054	0.0011
Dermal contact – Outdoors	79	µg/g	0.5	0.0018	0.0017	0.0013	0.0012	0.00036	0.0015
Dermal contact – Indoors	25	µg/g	0.0	0.00014	0.00013	0.000097	0.000090	0.000040	0.00012
Soil ingestion	79	µg/g	6.8	0.024	0.049	0.0055	0.0034	0.0031	0.016
Indoor dust ingestion	25	µg/g	4.4	0.015	0.032	0.0036	0.0022	0.0020	0.010
Home produced fruits and vegetables	0.025	µg/g fw	3.3	0	0.015	0.011	0.0080	0.0066	0.014
Local fruits and vegetables	0.028	µg/g fw	2.6	0	0.013	0.0089	0.0059	0.0046	0.011
Local wild blueberries	0.0052	µg/g fw	1.0	0	0.0066	0.0032	0.0016	0.0013	0.0037
Local wild game	0.00013	µg/g fw	0.0	0	0.000023	0.000015	0.000010	8.4×10^{-06}	0.000018
Local fish	0.00022	µg/g fw	0.0	0	0.000097	0.00015	0.000083	0.00010	0.00016
Drinking water	2.6	µg/L	29.1	0.095	0.099	0.059	0.046	0.062	0.10
Market basket contribution	NA	µg/g	51.7	0.00053	0.29	0.18	0.10	0.062	0.19
Summary									
Estimated total daily intake (µg/kg/day)			100.0	0.14	0.51	0.28	0.17	0.14	0.35
Oral TRV (µg/kg/day)				0.3	0.3	0.3	0.3	0.3	0.3
Hazard Quotient – oral			Unitless	0.46	**1.66**	0.93	0.56	0.47	**1.16**

Weight-of-Evidence Approach for As

Risk assessment of As-contaminated sites is a complex and problematic exercise which has been a source of controversy and complication when managing these sites. The issue of the cancer potency of As, and the interpretation of and response to predicted risks in excess of the traditional *de minimis* or negligible risk levels of one-in-one-million have complicated issues surrounding the risk assessment and management of sites with even relatively low concentrations of As. Use of the U.S. EPA slope factors to estimate possible risks from As exposures to people through all pathways (air, water, food, soils) results in elevated risk values from background (natural) sources. In Ontario, consideration of background soil levels (17 µg/g) and generic soil criteria (25 µg/g for residential), reveals risks in the one-in-one-hundred thousand range for the average Ontario resident living in an uncontaminated area. This immediately results in problems understanding and explaining what such risk estimates mean. Alternatively, a weight-of-evidence approach has been successfully used at several other sites across Canada and the U.S. – Port Hope (MOE 1991); Deloro (MOE 1999); Wawa (MOE 2001a); Anaconda, Montana (Calabrese unpublished data; Hwang et al. 1997; Walker and Griffin 1998); Balmerton (Gradient 1985). Risk assessments involving multi-pathway exposure assessment and use of the U.S. EPA slope factors revealed risk levels in the one-in-one thousand range for many of these sites. In fact, the results predicted for As in the Sudbury HHRA were consistent with those obtained at other similar sites.

In the case of As, risks well above the *de minimis* level are routinely predicted for exposures associated with typical North American diets, and high-quality, regulated North American drinking water supplies. Further investigation into the risk assessment results for communities within the GSA revealed the following: (i) market basket foods and drinking water were the main contributors to As related risks; (ii) generic criteria in Ontario (25 mg/kg) result in elevated risk levels (greater than one-in-one hundred thousand); (iii) the contribution of soil to overall As related risks was small, and all other pathways were less significant; (iv) health-based intervention levels (remediation goals), as determined by the risk assessment, typically are economically and technologically impossible; and, (v) removal of all soil above the generic criteria would only result in a small overall risk reduction, with the assessment still predicting risks at the generic criteria level.

It was clear that additional information, beyond that typically contained within a risk assessment, was needed to complete a *weight-of-evidence* approach. The ionic species of As (typically found in soils) forms insoluble salts with a number of cations and is adsorbed by organic matter, Fe and Al oxides. Thus, As becomes tightly bound to the soil and very difficult to liberate for biological uptake. Therefore, relatively high levels of As in soil may pose little risk if they are indeed highly insoluble and, therefore, not available for absorption if ingested. In fact, the measured bioaccessibility of As in the Sudbury soils was approximately 40% (see Chapter 7 for a discussion of soil bioaccessibility results).

The Falconbridge As Exposure Study (Chapter 7) compared the urinary As levels of several hundred individuals from an impacted community (Falconbridge) to those from a reference community (Hanmer). The results indicated that urinary As measurements from the two communities were similar despite the significantly higher soil concentrations present in the Town of Falconbridge. The results of speciation analysis clearly showed that the forms of As in the soil, dust and air were consistent between the various communities within the GSA. As such, the results of the Falconbridge As exposure study indicated that As exposure for all resident of the GSA were similar to those in other communities with significantly lower As soil concentrations. The exposure analysis also revealed that market basket foods, not soils, were the main contributor to As related risks for both the typical Ontario resident and the typical Sudbury resident.

Each of these elements provided strong, complementary lines of evidence to assist in the realistic evaluation of health risks associated with exposure to As. Only after consideration of all pieces of evidence (i.e. the risk assessment, a review of the scientific literature, urinary As study, and speciation study), and the relative strength-of-evidence associated with each of these elements, was it possible to conclude that there were no unsafe exposures or increased health effects associated with the observed As levels in the study area.

8.2.2 Cobalt

The HQ_{avg} estimates for the female preschool child associated with Co exposure (via the oral and dermal routes) were less than 1.0 and did not differ between COI or the typical Ontario scenario. Predicted exposures of female preschool children to Co under maximum exposure receptor scenarios were less than 10% of the recommended RfD of 10 µg/kg/day (HQ_{avg} estimates for both Copper Cliff and typical Ontario were less than 0.2). Figure 8.4 shows that approximately 91% of the ETDI of Co (for a female preschool child residing in Sudbury Centre) was the result of consuming market basket foods while 6% was attributed to soils and local fruit/vegetable consumption. All non-cancer inhalation HQ estimates for female preschool children were less than 1.0 for Co at all COI.

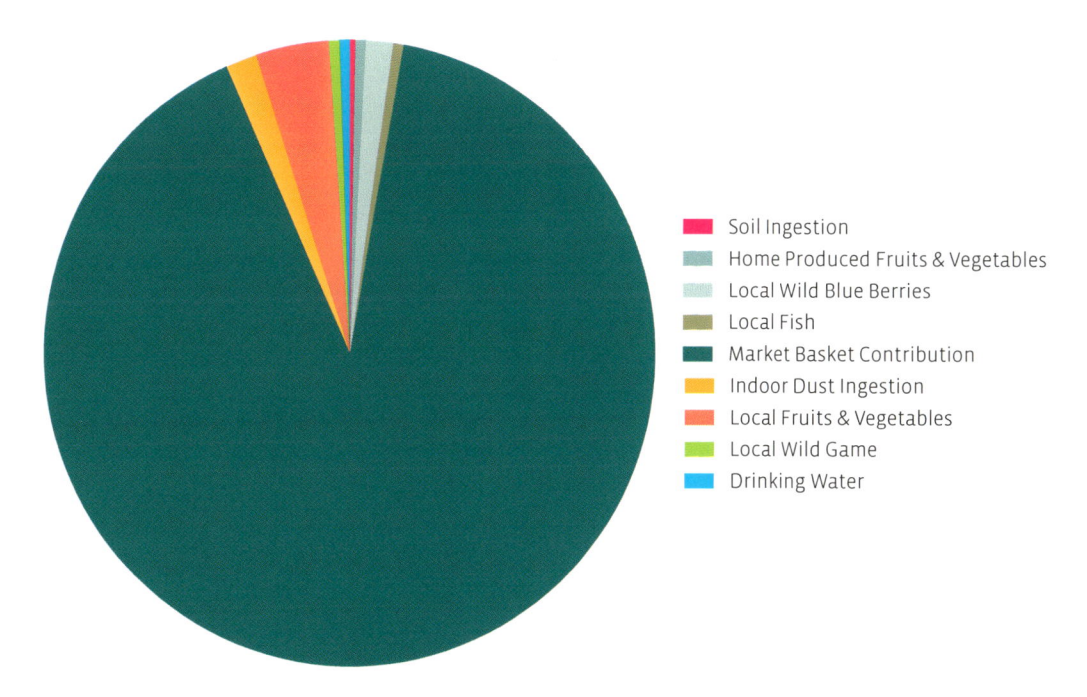

Figure 8.4   **Estimated Pathway Contribution to Co for Female Preschool Child**

8.2.3 Copper

Female preschool child HQ estimates associated with copper exposures were less than 1.0 under the RME scenario for all COI and typical Ontario assumptions. Under the RME receptor exposure scenario, HQ estimates of 0.67 and 0.62 were predicted for the female preschool child living in Copper Cliff and typical Ontario, respectively. Central tendency (CTE) HQ estimates for Copper Cliff and typical Ontario female preschool children were 0.56 and 0.53, respectively. According to Figure 8.5, approximately 86% of the ETDI of Cu (for a female preschool child) was the result of consumption of market basket foods; an additional 7% was associated with local drinking water, while 2% of the ETDI was a result of incidental soil ingestion.

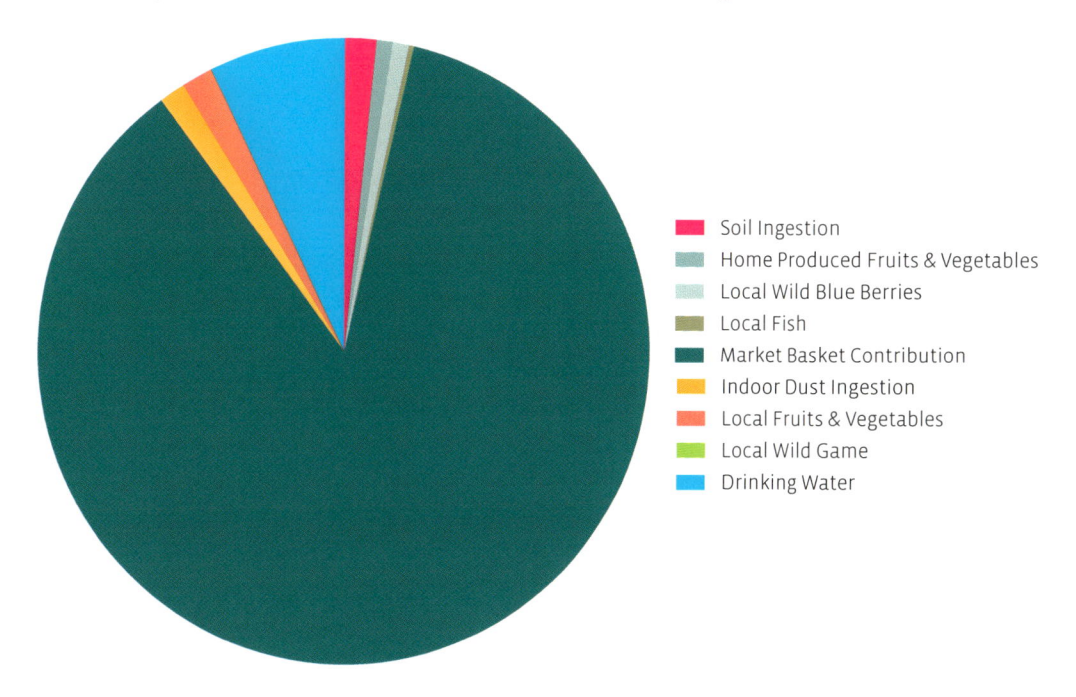

Figure 8.5   **Estimated Pathway Contribution to Cu for Female Preschool Child**

8.2.4 Lead

All lead HQavg estimates for the general population of female preschool children were below a value of 1.0 using average soil Pb concentrations (95% UCLM soil values). The highest HQavg estimate for Pb was 0.94 for a female preschool child living in Copper Cliff under a RME receptor exposure scenario. The EPCs for Pb in soil, based upon the estimated 95% UCLM concentration, at all Sudbury COI were less than the Ontario Typical Range (OTR) soil value of 120 mg/kg. When evaluating community-wide risks, it was thought prudent to consider exposures of individuals who may reside in areas that were associated with soil concentrations greater than the community average. When maximum soils concentrations of Pb in each COI were evaluated, the calculated HQ exceeded 1.0 for Copper Cliff (HQ=1.3), Falconbridge (HQ=1.1), Coniston (HQ=1.1) and Sudbury Centre (HQ=1.1). Based on these results, it was considered appropriate to derive a soil risk management level (SRML) for Pb to identify localized areas where risk management may be required. Development of the SRML for Pb is presented later in Section 8.3 of this chapter.

8.2.5 Nickel

8.2.5.1 Oral Exposure to Nickel

All HQ_{avg} estimates for female preschool children (under both the RME and CTE scenarios) were less than 1.0. The highest HQ_{avg} estimate of 0.70 was observed for a female preschool child living in Copper Cliff. Figure 8.6 provides a relative comparison of HQ values between Copper Cliff, typical Ontario resident and Hanmer (the regional background site) using the angling and hunting sub-population as an example. Figure 8.6 indicates that Ni HQ estimates predicted at Copper Cliff (0.70) were approximately twice as high as those predicted for a female preschool child living under typical Ontario conditions (0.35). Exposures via drinking water, incidental soil ingestion and diet all appear to be greater in Copper Cliff than in a typical Ontario resident or Hanmer resident.

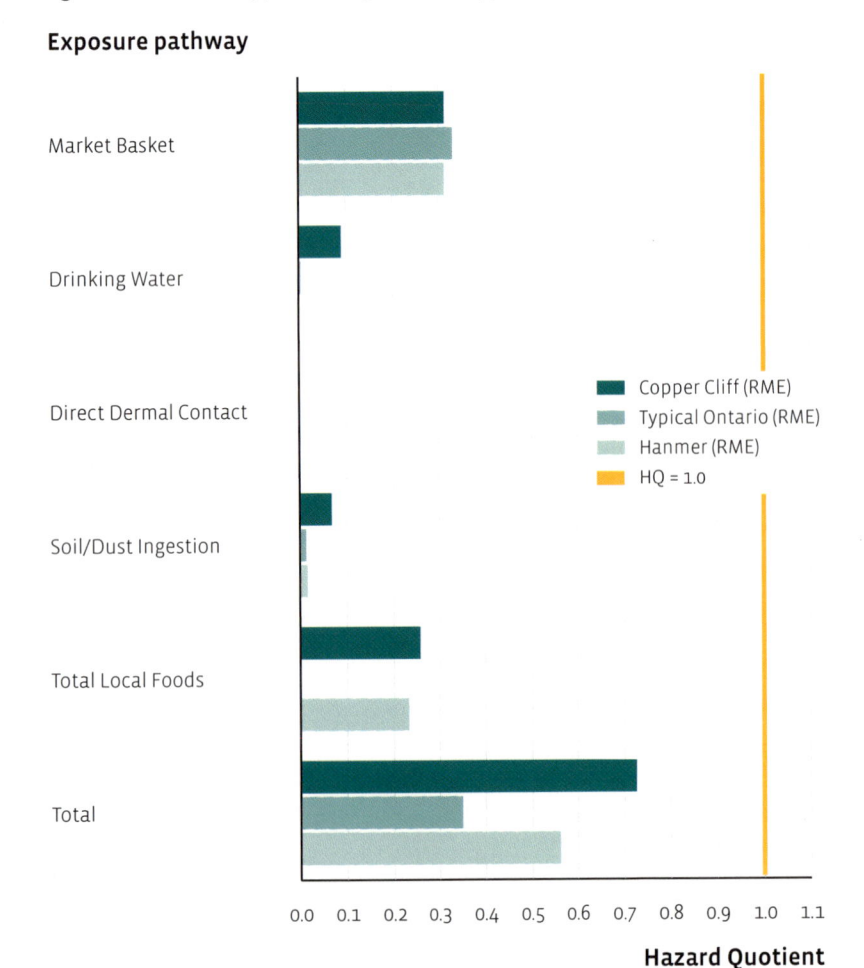

Figure 8.6   Ni Hazard Quotients for a Female Preschool Child

As discussed in Chapter 7, 'total local foods' included home garden vegetables, local fruits and vegetables, wild blueberries, wild game and fish caught and/or raised within the study area. Figure 8.7 provides a complete exposure pathway breakdown, reporting the percent contribution each pathway makes to the estimated total daily intake (ETDI) for a female preschool child. Market basket foods and the consumption of local drinking water represented 56% of the ETDI. The consumption of home produced/local fruits and vegetables represented an additional 25% of the ETDI. Together these exposure pathways comprised over 80% of the ETDI to Ni for a female preschool child. Under the current scenario, reducing the EPC of Ni in soils in Copper Cliff (976 µg/g) to typical Ontario background levels would only result in an 8.5% reduction in exposure and estimated HQ values for a female preschool child living within Copper Cliff.

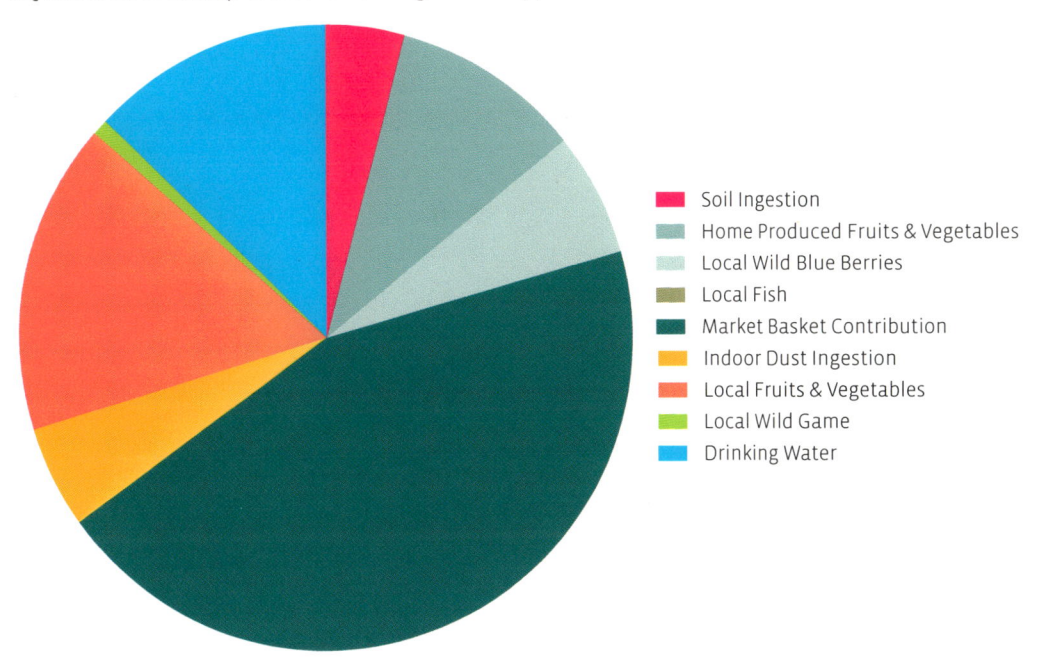

Figure 8.7   **Estimated Ni Exposure Pathway Contribution in Copper Cliff**

8.2.5.2 Inhalation Exposure to Ni

Potential risks from inhaling airborne COC were assessed by comparing annual average air concentrations in the study area with the TRVs from various agencies and sources. For Ni, these comparisons were conducted for both non-cancer and cancer endpoints. While a variety of valid TRVs for both cancer and non-cancer endpoints were evaluated as part of a weight-of-evidence approach (Seilkop and Oller 2003; MOE 2004b), the inhalation TRV established by the European Union (OJEU 2005) was ultimately selected as the primary benchmark to evaluate risks related to Ni inhalation in this study. The remainder of the TRVs included in the overall weight-of-evidence approach are provided for comparative purposes, and demonstrate potential risks for a variety of different endpoints and potential Ni species (many of which were not directly relevant to the current assessment). In addition, the weight-of-evidence discussion provides risk estimates based on the various chemical species of nickel present in the Sudbury air, under a variety of wind conditions.

Primary Risk Evaluation

The European Union (OJEU 2005) established a Ni TRV of 0.02 µg/m³ primarily based upon non-cancer data on respiratory effects (specifically lung inflammation and fibrosis). It is important to note that this TRV is based upon exposure to total Ni in ambient air, and not one particular Ni species. Based upon the available data, the European Union (EU) working group also believed that this value was compatible with the objective of limiting excess lifetime cancer risks to not more than one-in-a-million. Table 8.3 summarizes the predicted non-cancer health risks from comparison of total Ni concentrations measured in ambient air at each key monitoring station to this TRV. Note that Sudbury Centre was represented by two monitoring sites, Sudbury Centre South and Sudbury Centre West.

Table 8.3 **Summary of Non-Cancer Ni Inhalation Risk Characterization**

Monitoring station	Annual average air concentration µg/m³	OJEU (2005) RfC µg/m³	HQ
Coniston	0.012		0.6
Copper Cliff	0.059		**2.95**
Falconbridge	0.028		**1.4**
Hanmer	0.012	0.02	0.6
Sudbury Centre South	0.017		0.85
Sudbury Centre West	0.26		**13**
TOR	0.0014		0.07

Based upon this approach, potential health risks were noted at the Copper Cliff, Falconbridge, and Sudbury Centre West monitoring stations. These results indicated that further attention should be given to airborne Ni concentrations in the areas surrounding the Copper Cliff and Sudbury Centre West monitoring stations. The potential risks around the Falconbridge monitoring station were considered to be negligible given the degree of safety built into the assessment, and no further evaluation of that result was considered necessary.

8.2.5.3 Assessments of Alternative TRVs

As part of an overall weight-of-evidence approach, potential airborne risks were also evaluated based upon the specific Ni species identified in the year-long air monitoring program carried out in Sudbury as part of this study. As discussed in Chapter 6, the form of airborne Ni in the study area was very site-specific and dependent on a number of factors, such as proximity to various sources, wind direction, and other meteorological conditions (wind speed, precipitation, snow cover, etc.). Results of the speciation analysis conducted on the air filters from the air monitoring program demonstrated a fairly consistent Ni species 'fingerprint' across the study area, with the exception of one specific area. The speciation analysis revealed that when the wind was blowing across the Vale Copper Cliff facility, fugitive dusts appeared to give a unique Ni species fingerprint which included the presence of a small amount of Ni subsulfide (Ni_3S_2) at the Sudbury Centre West and Copper Cliff monitoring stations. As a result, the risk assessment team, in consultation with the TC, developed two specific Ni speciation fingerprints to assist in the evaluation of potential risks. These two speciation fingerprints were presented in Chapter 6 and are reiterated in Table 8.4.

Table 8.4 **Summary of Proposed Ni Species Fingerprints**

Nickel species	Typical ambient fingerprint	Copper Cliff facility impacted fingerprint
Nickel oxide (NiO)	80%	75%
Nickel sulfide	10%	10%
Nickel sub-sulfide (Ni_3S_2)	0%	10%
Nickel sulfate	10%	5%

Wind in the study area, on average, originated from a westerly direction 39% of the time and from an easterly direction 61% of the time. When originating from a westerly directly, ambient air monitors located at the Sudbury Centre West station showed a speciation fingerprint impacted by fugitive dusts from the Vale Copper Cliff facility. When wind was blowing from the opposite direction, the air monitors were not affected by fugitive dusts from the Copper Cliff facility, and the typical Ni species fingerprint was observed (i.e. absence of Ni_3S_2).

A summary of some alternative inhalation Ni TRVs available is provided in Table 8.5 for the evaluation of species-specific related risks. It should be noted that the IURs (inhalation unit risks) from U.S. EPA, WHO and Health Canada considered similar occupational data sets in the derivation of their unit risk values. The range in unit risk values and subsequent cancer risk estimates was a consequence of how each agency interpreted the cancer mortality data, as well as the mathematical models used to conduct low-dose extrapolation of the dose–response information. The occupational cohorts utilized by the U.S. EPA, WHO and Health Canada to derive their unit risk factors were developed for Ni refinery dust which contains varying percentages of oxidic, sulfidic, and soluble forms of Ni, as well as concurrent exposures to a myriad of other chemicals. Conversely, the Seilkop (2004) unit risk values were based on controlled animal studies which subjected groups of rats and mice to varying levels of oxidic or sulfidic forms of Ni.

Table 8.5 **Summary of Ni Species-Specific Inhalation Unit Risks**

Type of TRV	Source of TRV	IUR
IUR	Seilkop (2004) – NiO	$2.3 \times 10^{-5}\ (\mu g/m^3)^{-1}$
	Seilkop (2004) – Ni_3S_2	$6.3 \times 10^{-4}\ (\mu g/m^3)^{-1}$
	U.S. EPA (refinery dust)	$2.4 \times 10^{-4}\ (\mu g/m^3)^{-1}$
	U.S. EPA (subsulfide)	$4.8 \times 10^{-4}\ (\mu g/m^3)^{-1}$
	WHO	$3.8 \times 10^{-4}\ (\mu g/m^3)^{-1}$
	Health Canada	$1.3 \times 10^{-3}\ (\mu g/m^3)^{-1}$
REL[1]	OEHHA (NiO)	$0.1\ \mu g/m^3$
	OEHHA (non-NiO)	$0.05\ \mu g/m^3$

[1] Reference Exposure Level

The potential risks related to exposures to the typical Ni species fingerprint (i.e. no Ni_3S_2) are summarized in Table 8.6. This scenario existed year-round in all of the communities except at the Sudbury Centre West (61% of the time) and Copper Cliff (39% of time) monitoring stations. Those estimates were based on the following assumptions:

- Only Seilkop (2004) provided an IUR specific for NiO, therefore, cancer risks were estimated using this IUR only.
- HQ estimates were based on the OEHHA (2003) derived RELs.
- Risks were predicted separately for both Sudbury Centre (south and west) monitoring locations.
- Total Ni air concentrations were proportioned based on the typical fingerprint for all COI with the exception of typical Ontario resident which assumed 100% NiO.
- Air concentrations were pro-rated based on wind direction for the Copper Cliff and Sudbury Centre West monitoring locations.

Table 8.6 **Risk Estimates for the Typical Ni Species Fingerprint (no Ni_3S_2 exposure)**

Monitoring station	Nickel air concentration (µg/m³) NiO (90%)[a]	Non-NiO (10%)	Total	Occurrence frequency	HQ estimates NiO	Non-NiO	Total	Cancer risks NiO
Coniston	0.011	0.0012	0.012	100%	0.11	0.024	0.13	2.5×10^{-7}
Copper Cliff	0.053	0.0059	0.059	39%	0.21	0.046	0.25	4.8×10^{-7}
Falconbridge	0.025	0.0028	0.028	100%	0.25	0.056	0.31	5.8×10^{-7}
Hanmer	0.011	0.0012	0.012	100%	0.11	0.024	0.13	2.5×10^{-7}
Sudbury Centre South	0.015	0.0017	0.017	100%	0.15	0.034	0.19	3.5×10^{-7}
Sudbury Centre West	0.23	0.026	0.26	61%	1.4	0.32	1.7	3.3×10^{-6}
TOR	0.0014	–	0.0014	100%	0.014	–	0.014	3.2×10^{-8}

[a] Nickel sulfides which are not Ni_3S_2 were conservatively included in the NiO grouping.

Only the Copper Cliff and Sudbury Centre West monitoring locations experienced Ni fingerprints containing Ni$_3$S$_2$, and only when wind direction was such that the locations were influenced by the Vale Copper Cliff facility. The risk estimates are summarized in Table 8.7 for the periods when this fingerprint was expected.

Table 8.7 **Risk Estimates Based on Exposure to Ni Species Fingerprint**

Monitoring station	Nickel air concentration (µg/m³) NiO (85%)	Ni$_3$S$_2$ (10%)	Ni Sulfate (5%)	Total	Occurrence frequency	HQ estimates (OEHHA 2005) NiO	Non-NiO	Total	Cancer risks (Seilkop 2004) NiO	Ni$_3$S$_2$	Total[a]
Copper Cliff	0.0050	0.00059	0.0030	0.0059	61%[b]	0.31	0.11	0.41	7.0×10^{-7}	2.3×10^{-6}	3.0×10^{-6}
Sudbury Centre West	0.22	0.026	0.013	0.26	39%[c]	0.86	0.30	1.2	2.0×10^{-6}	6.4×10^{-6}	8.4×10^{-6}

[a] Cancer risks from NiO and Ni$_3$S$_2$ are assumed to be additive due to similar mechanisms.
[b] Wind blowing from east to west, across the Vale Inco facility, impacting the Copper Cliff station 61% of the time.
[c] Wind blowing from west to east, across the Vale Inco facility, impacting the Sudbury Centre West station 39% of the time.

When the risks from the two potential wind-directional Ni species fingerprints were combined, the overall risk estimates for each monitoring station were predicted. These results are summarized in Table 8.8. The shaded rows indicate situations where the risk HQ is > 1, or the cancer risk exceeds one-in-one million. The assessment indicated that all HQ estimates were less than 1.0, with the exception of the Sudbury Centre West monitoring station where the summed non-cancer risks were estimated to be approximately three-fold higher (HQ = 2.9) than the target HQ. Furthermore, all cancer risk estimates were less than one-in-one million, with the exception of the Sudbury Centre West and Copper Cliff monitoring stations. At the Sudbury Centre West station, cancer risk estimates are approximately 12 per million, while cancer risk estimates at the Copper Cliff station are approximately 3.5 per million. Therefore, a predicted risk greater than one in a million related to inhalation of Ni in ambient air was restricted to the area surrounding the Vale Copper Cliff facility.

Table 8.8 **Risk Estimates Based on Ambient Air Concentrations at Different Monitoring Stations**

Monitoring station	HQ estimates NiO	Non-NiO	Total	Cancer risks NiO	Ni$_3$S$_2$	Total
Coniston	0.11	0.024	0.13	2.5×10^{-7}	–	2.5×10^{-7}
Copper Cliff	0.51	0.15	0.67	1.2×10^{-6}	2.3×10^{-6}	3.4×10^{-6}
Falconbridge	0.25	0.056	0.31	5.8×10^{-7}	–	5.8×10^{-7}
Hanmer	0.11	0.024	0.13	2.5×10^{-7}	–	2.5×10^{-7}
Sudbury Centre South	0.15	0.034	0.19	3.5×10^{-7}	–	3.5×10^{-7}
Sudbury Centre West	2.3	0.62	2.9	5.3×10^{-7}	6.4×10^{-6}	1.2×10^{-5}
TOR	0.014	–	0.014	3.2×10^{-8}	–	3.2×10^{-8}

Note: one in one million = 1 × 10^{-6}

Summary of Airborne Ni Assessment

The results indicate that airborne Ni concentrations exceed the air quality regulatory benchmark selected (the EU TRV) which resulted in HQ values > 1.0 at three of the monitoring sites – Copper Cliff, Sudbury Centre West, and Falconbridge (see Table 8.3). While the predicted risks at the Copper Cliff and Sudbury Centre West stations were considered of potential concern, it was the opinion of the risk assessment team that the potential risks around the Falconbridge monitoring station were marginal given the degree of safety built into the assessment. The assessment of various alternative TRVs for Ni inhalation also resulted in risk predictions exceeding one in a million in the areas surrounding the Vale facility. Although there is some uncertainty in evaluating risks based upon different TRVs and IURs available from different regulatory agencies, the results were generally consistent. The weight-of-evidence evaluation indicated that the calculated risk to airborne Ni exceeded regulatory benchmarks for both cancer and non-cancer health effects in two communities. This information, as well as other elements of the weight-of-evidence evaluation, was used as a basis for making informed risk management decisions on addressing potential health risks related to airborne Ni in the study area.

8.2.6 Selenium

The predicted risk characterization from Se exposure to female preschool children in three communities for the angler and hunter sub-population is illustrated in Figure 8.8. The figure shows that the exposure exceeded the recommended RfD under the RME scenarios, resulting in HQ estimates greater than 1.0. Figure 8.8 also provides an exposure pathway analysis and a relative comparison between Falconbridge, Hanmer (regional background) and typical Ontario conditions for the angler and hunter sub-population group. The predicted Se risk was similar in all of the Sudbury COI, and was slightly higher in the angler and hunter group compared to the general population due to exposure to local foods. For example, the calculated HQ was approximately 2.0 for the angler and hunter sub-population in Falconbridge and Hanmer. In contrast, the predicted HQ under the same scenario for the TOR was about 1.3.

Figure 8.8 demonstrates that a significant proportion (approximately 75%) of the ETDI of Se (and hence risk) for the female preschool child resulted from consuming general market basket (or supermarket) foods, even when considering the angler and hunter subgroup. The ETDI from market basket foods alone exceeded the recommended selenium RfD under the RME scenario. Market basket foods are not specific to the study area and represent the best estimate of daily Se intake as a result of consuming general supermarket foods in Ontario. It is noted that the 'total local foods' exposure pathway applies only to Sudbury-specific COI. The 'Market Basket' HQ estimates for Sudbury-specific COI were slightly higher than those for typical Ontario because a proportion of an individual's diet (living in a COI) was applied to locally derived foods. The consumption of local foods (including local berries, vegetables, wild game and fish) represented the most significant site-specific pathways for Se exposure, accounting for more than 20% of the total HQ estimate for the female preschool child in Falconbridge.

Reducing Se soil concentrations in Sudbury soil to levels similar to that of Ontario background would result in only a marginal decrease in the overall Se HQ estimate. Since the target HQ was also exceeded in the typical Ontario resident, it was considered that health risks to Sudbury residents associated with exposure to Se were no different from those observed in other parts of Ontario or the rest of Canada, and no further action or evaluation of Se was considered necessary

Exposure pathway

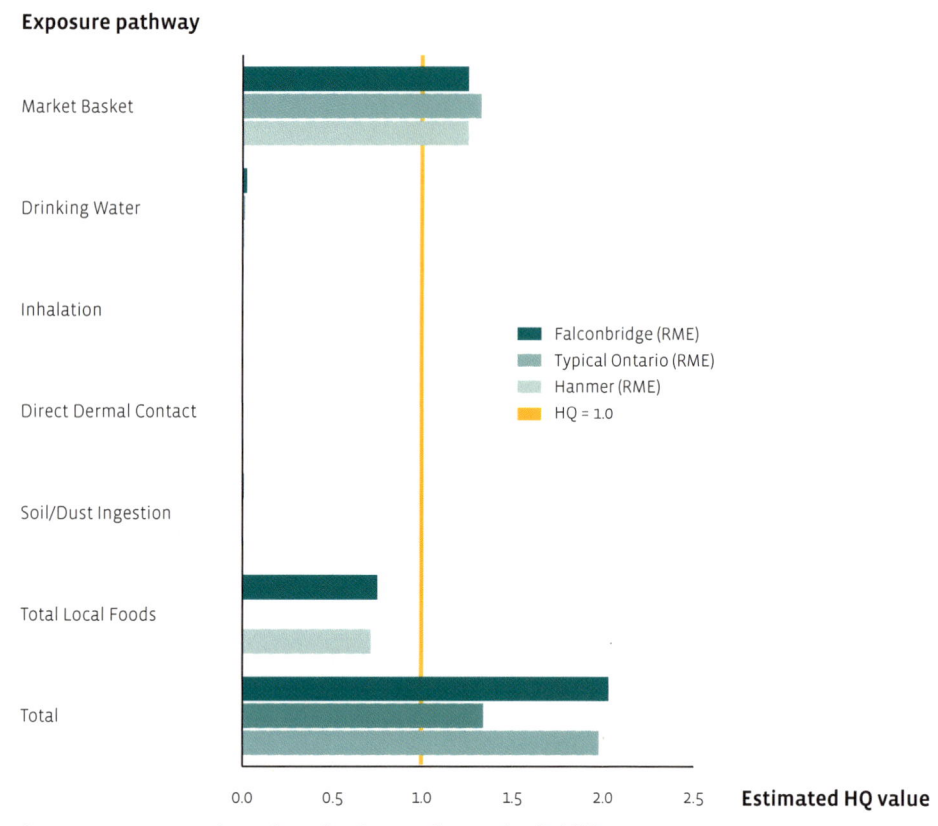

Figure 8.8 Se Hazard Quotients for the Female Preschool Child

8.2.7 Summary of HHRA Risk Characterization Results

The wide range of combinations of COCs, communities, receptor groups and exposure scenarios made simple interpretation of all of the results difficult. However, for the sake of communicating the results to the public, it was necessary to simplify the findings of the risk assessment in terms and language that all stakeholders understood and supported. The following five statements provided an overall summary of the risk assessment results as they were presented to the general public:

1. Based on current conditions in the Sudbury area, the study predicted little risk of health effects on Sudbury area residents associated with metals in the environment.

2. There were no unacceptable health risks predicted for exposure to four of the six Chemicals of Concern studied: As, Cu, Co and Se.

3. The risks calculated for typical exposures to Pb in the environment throughout the Greater Sudbury area are within acceptable benchmarks for protection of human health. However, levels of Pb in some soil samples indicate a potential risk of health effects for young children in Copper Cliff, Coniston, Falconbridge and Sudbury Centre.
   - Pb levels in soils and dust in the Sudbury area are similar to levels in other older urban communities in Ontario.

4. The study calculated a risk of respiratory inflammation from lifetime exposures (70 years) to airborne Ni in two areas: Copper Cliff and the western portion of Sudbury Centre.
   - Respiratory inflammation has been linked to the promotion of cancer caused by other agents.
   - Based on the conservative assumptions and approaches used in this risk assessment, it is unlikely that any additional respiratory cancers will result from nickel exposure over the 70-year lifespan considered in the risk assessment.
   - Health risks related to Ni inhalation were not identified in the other communities of interest.

5. Anglers, hunters and First Nation people who may consume more local and wild game are at no greater risk of health effects due to metals in the environment than the general population.

These statements were approved and supported by the TC and were also reviewed by the Public Advisory Committee (PAC) before being released to the public. Further discussion on the public communications component of the risk assessment is presented in Chapter 13.

8.3 Development of Soil Risk Management Level (SRML) for Lead

As discussed previously, the predicted HQ_{max} for Pb exceeded 1.0 when maximum soil Pb concentrations were used in the risk characterization for a number of the communities. For Coniston, Falconbridge, and Sudbury Centre, the maximum concentrations of Pb in soil were 310, 335 and 310 mg/kg, respectively. The use of the maximum concentration of Pb in soil (582 mg/kg) at Copper Cliff resulted in an HQ_{max} estimate of 1.3 for the general population of female preschool children. While these predicted risks were only marginally above the established HQ benchmark of 1.0, it was considered appropriate to derive a Sudbury-specific soil risk management level (SRML) for Pb to ensure the protection of receptors in locally impacted zones.

The term SRML was developed for use in this assessment. The team deliberately wanted to avoid using a term such as 'guideline' or 'soil target'. Some jurisdictions, particularly the USA, use the term 'preliminary remediation goal' (PRG). However, it was felt that this term implied remediation, or clean-up, if the PRG was exceeded, while risk management options other than remediation may be available. The PRG or SRML term, was defined as the average COC soil concentration within an exposure unit that corresponded to an acceptable level of risk (U.S. EPA 2001a). In other words, the SRML is the exposure point concentration (EPC) in soil within a given EU (i.e. a community of interest) which would yield an acceptable level of risk, or HQ < 1.0. Given the available data, the risk assessment team considered it appropriate to use a weight-of-evidence approach in the evaluation of health risk estimates and the development of SRML for Pb.

8.3.1 Weight-of-Evidence Evaluation for Pb

The following section provides an overview of the lines of evidence used to establish the recommended SRML value for lead in the study area. As part of an overall weight-of-evidence approach, the following lines of evidence were reviewed and evaluated to aid in the development of an appropriate SRML for Pb:

- Risk predictions from the Sudbury Exposure Model for each of the COI.
- Information with regard to the uncertainties in the model derived values.
- The empirical relationship between Pb in soil and blood lead level (BLL) reported in the literature and how this information formed the basis for SRML values derived at other sites.
- An evaluation of the selected SRML in both the Sudbury Exposure Model and the U.S. EPA IEUBK model to determine the level of estimated risk posed by soil concentrations at the SRML.

Sudbury Exposure Model Results

As the first line of evidence, the Sudbury exposure model was used to estimate potential site-specific Pb SRML values which would be protective of human health. For this study, the iterative forward calculation method as recommended by the U.S. EPA (2001a) was used to generate SRML with non-linear parameters. This method involved collecting data from multiple model runs where each run used a different EPC in soil. The Solver tool included in MS Excel was used to complete this calculation. The calculation was conducted until the EPC corresponding to an HQ value of 1.0 was determined. This EPC corresponded to the SRML as it indicated the soil Pb level within a specific community which corresponded to an acceptable level of risk (HQ < 1.0).

If the SRML is defined as the EPC (i.e. the 95% UCLM) in soil within a given community which yielded an acceptable level of risk, then it is possible that some residential properties exceed the EPC. Depending on how the soil concentration data are distributed, it is plausible that the remediation of a number of highly impacted soils within the community could bring the overall EPC for that community below the SRML. If the property or site of concern was a single residential lot, it would be reasonable to assume that an individual would move in random fashion within his or her own residential property. Under these circumstances, a reasonable approach may be to remove highly impacted soil to facilitate the reduction in the EPC of the single property. However, because the exposure units in this study represent entire communities, in which individuals do not move in a random fashion, the remediation of locally impacted zones to reduce the overall EPC for the community is not valid. Hence, the SRML values should be applied to individual residential properties, not necessarily the community as a whole.

It was deemed appropriate to calculate SRML for lead in all COI to ensure protection of locally impacted zones. Table 8.9 provides the community-specific SRML mathematically derived for Pb using Sudbury-specific data. Also included in the table for comparison are soil Pb criteria previously established by the U.S. EPA (2001b) and the MOE (2001c). Examination of the results provided in Table 8.9 indicates that the SRML derived from the Sudbury model appear very conservative (i.e. low) relative to published guidelines. The Ontario Ministry of the Environment recently provided a generic soil standard of 200 mg of Pb/kg soil (Ontario Regulation 153/04; MOE 2004a). Again, that standard is not considered an intervention level but rather is used for the purposes of screening.

Table 8.9 **Different Soil Risk Management Levels (SRML) for Pb (mg/kg)**

COI	Model derived SRML	US EPA SRML		MOE (2001c) SRML	
		Play area	Bare soil remainder	Bare play area	Elsewhere on property
Coniston	190	400	1200	400	1000
Copper Cliff	170				
Falconbridge	180				
Sudbury Centre	200				

Section 403 of the Toxic Substances Control Act (TSCA) established standards for bare residential soil (400 mg/kg in play areas and an average of 1200 mg/kg in bare soil in the remainder of the yard) (U.S. EPA 2001b). The U.S. EPA utilized a weight-of-evidence evaluation in the derivation of the criteria. A blood Pb level of 10 µg/dL was considered as the blood-level of concern while the environmental level of concern was established based on a 1% to 5% probability of an individual child exceeding the blood Pb level of concern.

Model Sensitivity Analyses

In risk assessment, it is useful to conduct a sensitivity analysis to identify how variation in the output of a model (e.g. SRML) is influenced by changes in the model input variables. Table 8.10 contains a number of model scenarios that demonstrate how changing input parameters impacted the model-derived SRML. This example used data from the community of Copper Cliff for the preschool female child. The key input parameters that were changed include soil ingestion rate, soil-to-dust ratio, food consumption rate, bioavailability of Pb in soil and dust, and the TRV used for comparison. By altering one or more of these parameters, the SRML for Pb in Copper Cliff varied substantially (−41% to +2200%). The range of SRML values demonstrates the sensitivity of the model to the input parameters. The community of Copper Cliff was used for example purposes only and was not intended to single out this community. It is also important to note that the selection process behind the choice of many of these input parameters is largely based on policy, rather than science as the input values in Table 8.10 all have scientific merit.

In Table 8.10, the projected SRML (bare soil remainder) was calculated by multiplying the calculated (play area bare soil) SRML by three-fold to account for the three-fold difference inherent in the two Pb soil action levels proposed by the U.S. EPA (400 versus 1200 mg/kg). The Food Consumption Database showed the effect on SRML between selection of food consumption data from the older Health Canada/Nutrition Canada database and the more recent USDA data from the Northeastern U.S. Adjustment of the soil-to-dust relationship had some impact on the calculation of the SRML because most of the allowable daily intake (ADI) was taken up by non-soil/dust related sources (such as market basket foods). The 95 UCLM of soil Pb concentration in Copper Cliff was 98 mg/kg. If the regression equation is used, developed using data from the indoor dust survey, a dust concentration of 150 mg/kg is calculated, resulting in an estimated HQ of 0.93. However, if the IEUBK soil-to-dust relationship of 0.7 is used, a dust concentration of 69 mg/kg is calculated, resulting in an estimated HQ of 0.84. As such, a more than doubling of the dust concentration changed the HQ by only 0.09. Lastly, the two-phase bioaccessibility (gastric plus intestinal) was originally utilized in this study as it inherently makes physiological sense. Independent peer review questioned the validity and validation of these values and, as such, the study relied on the one-phase (gastric only) results. The impact of utilizing two-phase results is provided for comparative purposes only.

Table 8.10 **Sensitivity Analysis of Calculated SRML Values for Pb to the Female Child; Copper Cliff**

Scenario	MOE soil consumption rate	Previously reported scenario	IEUBK soil-to-dust concentration ratio	USDA food consumption + MOE soil consumption rate	Bioavailability adjustment	USDA food consumption	USDA food consumption + two-phase bioaccessibility	Two-phase bioaccessibility	Using Health Canada TRV	USDA food consumption + two-phase bioaccessibility + Health Canada TRV
SRML (play area bare soils)	100	**170**	220	250	350	380	930	1100	1700	3900
Percent change	59%	–	129%	147%	206%	224%	547%	647%	1000%	2294%
Projected SRML (bare soil remainder)	300	**510**	660	750	1100	1100	2800	3300	5100	12000
Oral TRV (µg/kg bw/day)	1.85 (MOE)	**1.85 (MOE)**	1.85 (MOE)	1.85 (MOE)	1.85 (MOE)	1.85 (MOE)	1.85 (MOE)	1.85 (MOE)	3.57 (HC)	3.57 (HC)
Soil/dust consumption rate (mg/day)	100	**80**	80	100	80	80	80	80	80	80
Food consumption database	HC	**HC**	HC	USDA	HC	USDA	USDA	HC	HC	USDA
Soil-to-dust relationship	Regression equation	**Regression equation**	IEUBK soil-to-dust concentration ratio (0.7)	Regression equation	Regression equation	Regression equation	Regression equation	Regression equation	Regression equation	Regression equation
Bioaccessibility	Soil=66% Dust=83% (one-phase)	**Soil=66% Dust=83% (one-phase)**	Soil=66% Dust=83% (one-phase)	Soil=66% Dust=83% (one-phase)	Soil=66% Dust=83% (one-phase)	Soil=100% Dust=100%	Soil=66% Dust=83% (one-phase)	Soil=38% Dust=43% (two-phase)	Soil=38% Dust=43% (two-phase)	Soil=66% Dust=83% (one-phase)	Soil=38% Dust=43% (two-phase)
Bioavailability	100%	**100%**	100%	100%	50% (IEUBK)	100%	100%	100%	100%	100%
Percent change	-41%	–	29%	47%	105%	124%	447%	547%	900%	2200%

Note: numbers in bold represent base case scenario for model purposes

Conservatism of Selected Model Bioaccessibility Parameter

As mentioned above, the parameters and assumptions used in the exposure model had a significant impact on the calculated risk estimates, as well as the derived SRML values. A parameter that had particular impact on model estimates, and was believed to largely drive the conservatism inherent within the exposure model calculations, was the Pb bioaccessibility values for both soil and dust. van de Wiele et al. (2007) compared five *in vitro* digestion models to established *in vivo* experimental results as they applied to Pb bioaccessibility in the human gastrointestinal tract. In this multi-laboratory comparison study, the oral bioaccessible lead fraction was significantly different between the *in vitro* methods and, for the simulated fasted conditions, ranged from 2% to 33% and for the fed conditions, from 7% to 29%.

The soil and dust bioaccessibility values selected for the current Sudbury assessment were considerably higher (66% and 83%, respectively) than any of those estimated by models reported in the van de Wiele et al. (2007) study, as well as the *in vivo* bioavailability data from the literature for Pb exposures. It should be noted that the *in vitro* bioaccessibility results used in the current study were based upon gastric absorption only (i.e. one phase), and that the two-phase bioaccessibility (i.e. gastric+intestinal) results (38% for soil and 43% for dust) were more consistent with the range of *in vitro* study results summarized in the van de Wiele et al. (2007) paper, as well as the *in vivo* bioavailability data from the literature.

Relationship between Pb in Soil and Blood Pb Levels

Recent scientific literature suggests that exposure to Pb may cause adverse neurological changes in children at blood Pb concentrations lower than 10 µg/dL. Soil and dust are major exposure pathways for Pb. Therefore, understanding the relationship between soil and dust levels with corresponding blood Pb levels in children is important for development of environmental standards. The primary literature was reviewed to identify studies in which an empirical approach was used to investigate this relationship. The empirical approach generates a slope factor (µg of Pb/dL blood/mg of Pb/kg soil) based on the correlation between measured soil Pb concentrations and the blood Pb concentrations in children assumed to be exposed to the soil (Stern 1994). Empirical slopes reflect site-specific and study-specific exposure scenarios; therefore, these slope factors may not be generalized unless the factors that mediate soil Pb levels and blood Pb levels are taken into consideration (Stern 1994). In addition, the relationship between Pb intake and blood Pb level is sublinear for higher intake levels (U.S. EPA 1986); therefore, linear slopes derived from sites with high soil Pb levels will underestimate the relationship (Stern 1994).

The empirical results in the primary literature show that a blood Pb level of approximately 5 µg/dL results from exposure to soil containing Pb concentrations ranging from 500 to 1500 mg/kg (Angle et al. 1984; Steele et al. 1990; Stern 1994; Lewin et al. 1999; Johnson and Bretsch 2002; Mielke et al. 2007) (Table 8.11). Lewin et al. (1999) examined the relationship between the concentrations of Pb in soil and blood Pb levels in children residing near four Superfund sites in the U.S. by calculating a slope factor for the dose–response curve of children. Overall, concentrations of Pb in soil ranging from 500 to 1500 mg/kg resulted in blood concentrations ranging from 4.1 to 9.8 µg/dL. The high-risk population encompassed children who were male, and who lived in households with low income and education levels, without air conditioning, and that contained a smoker. For this population, soil concentrations of 500 mg/kg and 1500 mg/kg resulted in predicted blood lead levels of 8.4 µg/dL and 9.8 µg/dL, respectively. The low-risk population was defined as children who were female, and who lived in households with high income and education levels, with air conditioning, and with non-smokers. In this population, soil concentrations of 500 µg/g and 1500 mg/kg resulted in predicted blood lead levels of 4.1 µg/dL and 4.9 µg/dL, respectively.

Steele et al. (1990) examined 13 epidemiological studies that investigated the relationship between soil/dust Pb and blood Pb concentrations in children residing in urban and smelter areas, and in regions near mine wastes from inactive smelter sites in the U.S. Overall, slope factors ranged between 0.76 and 8.1 µg/dL per 1000 mg/kg of soil Pb (Table 8.11). The soil concentrations resulting in blood Pb levels of 5 µg/dL and 10 µg/dL ranged from 620 to 660 mg/kg. Stern (1994) employed a mechanistic model which estimated the total change in blood Pb concentration from ingestion exposure to soil and soil-derived dust under steady-state conditions, with various input parameters. A slope factor of 10 µg/dL per 1000 mg/kg of soil Pb was reported, which results from a blood level of 2 µg/dL in children with soil Pb concentrations of 200 mg/kg (Stern 1994). Using this slope factor, soil concentrations of 500 mg/kg and 1000 mg/kg resulted in blood Pb concentrations of 5 µg/dL and 10 µg/dL, respectively.

von Lindern et al. (2003) utilized a linear regression model to derive a slope factor of approximately 4 µg of Pb/dL blood per 1000 mg of Pb/kg soil. That study encompassed data from the Bunker Hill Superfund Site in Idaho near an abandoned Pb/Zn smelting complex. Using the slope factor, a blood Pb level of 5 µg/dL resulted in a calculated soil concentration of 1250 mg/kg. Mielke et al. (2007) derived a relationship between pooled soil Pb and child blood Pb data from census tracts of residential communities within metropolitan areas of New Orleans between 2000 and 2005. There was a highly significant curvilinear association between the soil and child blood Pb data. Based on the curvilinear model, a median blood Pb level of 5.9 µg/dL was associated with exposure to a median concentration of 500 mg/kg soil Pb. A median soil level of 300 mg/kg was associated with a predicted median blood level of 5 µg/dL. At higher Pb soil concentrations (1000–1500 mg/kg), median blood levels ranging from 7.5 to 8.7 µg/dL were reported. Due to the non-linear nature of the relationship between blood and soil Pb, a single slope factor was not reported. It was noted that, below 100 mg/kg of Pb in soil, blood levels increased 1.4 µg/dL per 100 mg/kg, and above 300 mg/kg of Pb in soil, blood levels increased 0.32 µg/dL per 100 mg/kg (Mielke et al. 2007).

Ren et al. (2006) also found that concentrations of Pb in soil less than 500 mg/kg resulted in blood Pb level concentrations of approximately 5 µg/dL in children living in urban areas (Table 8.12). The study measured Pb in children's blood and soils at 10 kindergartens in Shenyang, China where Pb pollution resulted primarily from automobile exhaust and industrial emissions. Concentrations of Pb in the soil at the kindergartens ranged from 53 to 350 mg/kg, and blood Pb levels in children aged 3–5 years ranged from approximately 1 to 5 µg/dL. Additional studies describing blood Pb levels in children living in urban populations in the 1970s and 1980s or living near actively emitting Pb smelters were excluded from this review. Urban studies were conducted during the 1970s and 1980s when Pb additives were commonly used in gasoline and measured soil Pb concentrations were extremely low indicating other sources of exposure. In addition, populations near active Pb smelters were also omitted as elevated Pb concentrations in air invalidate the relationship between soil Pb levels and blood Pb concentrations in children.

Table 8.11 **Relationship between Pb in Soil (mg/kg) and Blood Pb Levels (µg/dL) in North American Populations**

Slope factor (µg Pb/dL blood/1000 mg Pb/kg soil)	Receptor	Description	Blood lead concentration (µg/dL)	Soil concentration (mg/kg)	Reference
10	Children	Based on residential soil concentrations in the U.S. The calculated slope factor was assumed to be linear	5	500[d]	Stern 1994
			10	1000[d]	
	High-risk children[a]	Four concurrent investigations of populations residing near four National Priorities List sites in the U.S. A natural logarithm regression was assumed	8.4	500	Lewin et al. 1999
			9.2	1000	
			9.8	1500	
	No covariate children[b]	Four concurrent investigations of populations residing near four National Priorities List sites in the U.S. A natural logarithm regression was assumed	6.0	500	Lewin et al. 1999
			7.1	1000	
			7.6	1500	
	Low-risk children[c]	Four concurrent investigations of populations residing near four National Priorities List sites in the U.S. A natural logarithm regression was assumed	4.1	500	Lewin et al. 1999
			4.6	1000	
			4.9	1500	
	Children	Census tract data from residential communities in New Orleans, LA, 2000–2005. A curvilinear model was utilized	5	300	Mielke et al. 2007
			5.9	500	
			7.5	1000	
			8.7	1500	
6.8	Children	Based on urban/suburban soil concentrations in Omaha. The calculated slope factor was assumed to be linear	5	735[d]	Angle et al. 1984
			10	1470[d]	
4	Children	Based on paired blood lead/soil samples from the Bunker Hill Superfund Site in Idaho. The calculated slope factor is assumed to be linear	5	1250[d]	Von Lindern et al. 2003
			10	2500[d]	
	Children	Aggregation of geo-referenced datasets from Syracuse, New York. A logarithmic regression model was assumed	4–10[e]	50–350	Johnson and Bretsch 2002
0.76–8.1	Children	Thirteen epidemiological investigations of populations residing in urban towns and towns with operating smelters in the U.S. The calculated slope factors were assumed to be linear	5	620–6600[d]	Steele et al. 1990
			10	1200–13,000[d]	

[a] High risk population; defined as children who did not have air conditioning, who lived with a smoker, were male and were from low income households.

[b] A simple no-covariate regression model was used.

[c] Low risk population; defined as children who had air conditioning, lived with non-smokers, were female and were from high income households.

[d] The soil concentrations resulting in a blood Pb concentration of 5 and 10 µg/dL were calculated using the slope factor.

[e] Levels approximated from graphical representation.

Table 8.12 **Relationship of Soil Pb to Blood Pb in Children Living Near Mine Waste and Urban Areas**

Site (year)	Age group	Blood lead level (µg/dL)	Soil lead concentration (mg/kg)	Reference
Telluride, CO (1987)	0–71 months	6.2	178	Jin et al. 1997
Clear Creek/Central City, CO (1990)	0–71 months	5.9	201	
Socorro, New Mexico (1990)	All ages	5.8	317	
Granite City, IL (1991)	6–14 years	5	338	
Montreal, PQ (1990)	6–71 months	5.6	430	
Granite City, IL (1991)	6–71 months	6.9	449	
Smuggler Mt., Aspen, CO (1990)	6–14 years	1.8	544	
Smuggler Mt., Aspen, CO (1990)	6–71 months	2.6	641	
Palmerton, PA (1991)	6–71 months	6.5	691	
Leadville, CO (1987)	6–71 months	8.7	1034	
Tikkurila, Finland (1996–1999)	0–6 years	2.7 (<2.1–5)[a]	242	Louekari et al. 2004
		2.1 (<2.1–4.1)[a]	40	
		<2 (<2.1–2.5)[a]	20	
Midvale, UT (1989)	6–72 months	5.6	542	Lanphear et al. 2003
Shenyang, China (2003)	3–5 years	1–5[b]	135 (53–350)[a]	Ren et al. 2006

[a] Average, minimum and maximum values are presented.
[b] Levels approximated from graphical representation.

A blood Pb screening study was commissioned by the Regional Niagara Public Health Department in 2001 to determine exposure to and potential health impacts of Pb in Port Colborne, Ontario (Decou et al. 2001). In particular, the Eastside Community had elevated Pb concentrations in soils (arithmetic mean of 203 mg/kg and a maximum of 1350 mg/kg). Blood Pb was measured in 1065 individuals, with approximately one-third of all participants from the Eastside Community. The geometric means and their confidence intervals for the blood Pb concentrations for all participants were well below the 10 µg/dL screening benchmark, with the geometric mean for the Eastside Community reported as 2 µg/dL. No statistical relationship was apparent between soil Pb and blood Pb levels (Decou et al. 2001). The researchers concluded that children and pregnant women in the Eastside Community were not at an increased risk of Pb exposure compared to other communities in Ontario, even considering the localized elevated soil Pb levels. Furthermore, all studied children who lived on properties with surficial soil Pb concentrations in excess of 400 mg/kg had blood Pb concentrations less than 10 µg/dL. While the results of the survey indicated that no immediate intervention was required regarding Pb in soil in the Eastside Community, the Regional Niagara Public Health Department continued to recommend limiting exposure to soil containing known contaminants such as Pb (Decou et al. 2001).

The literature review indicated that a blood Pb level of approximately 5 µg/dL resulted from exposure to soil containing Pb concentrations ranging from 500 to 1500 mg/kg. Concentrations of Pb in soil less than 500 mg/kg were found to result in blood Pb concentrations of approximately 5 µg/dL in children living in urban areas and near mine waste or inactive smelter sites. The acceptable blood Pb level in most jurisdictions was generally considered to be 10 µg/dL in children. However, it is important to note that recent literature suggests that a level approaching 5 µg/dL may be more appropriate. In fact, the Ontario MOE now uses a probability of a 5% exceedance of 5 µg/dL blood Pb level as the policy basis of their new Pb air standard (MOE 2007).

8.3.2 Supplemental Model Results for Selecting SRML

The following section provides supplemental model results from both the SARA model and the U.S. EPA IEUBK lead model that evaluated the applicability of 400 mg of Pb/kg soil as an appropriate soil intervention level for the Sudbury area. As noted previously, a change in key assumptions and parameters within the Sudbury exposure model had a significant influence on the model-calculated SRML for Pb. To examine the sensitivity of the exposure model, estimated health risks for exposure to the SMRL soil concentrations of 400 mg of Pb/kg soil were evaluated by the Sudbury exposure model using a selected set of assumptions and parameters (Table 8.13). For comparative purposes, Table 8.13 also contains HQ_{avg} (based on average soil concentrations in each of the five COI) and HQ_{max} (based on maximum soil concentrations in each of the five COI) model results. Also provided are soil and dust Pb concentrations corresponding to an HQ of 1.0 (this soil concentration corresponded to the model calculated SRML).

The IEUBK model, developed by the U.S. EPA, was used to predict childhood Pb exposure and risk in this study to provide another line of evidence. The IEUBK model was also utilized to evaluate several scenarios related to the proposed SRML of 400 mg of Pb/kg soil. Table 8.13 provides blood Pb levels (geometric mean BLL, 95th percentile BLL) and the probability of exceeding both BLLs of 5 and 10 µg/dL, for the selected SMRL, HQ_{avg}, HQ_{max} and HQ = 1 scenarios. The results of this evaluation revealed that default IEUBK assumptions corresponded to estimated average blood Pb levels less than 5 µg/dL, for most scenarios. Upper bound (95th percentile) blood Pb level estimates ranged between 3.5 µg/dL and 5.6 µg/dL at average measured soil concentrations in the five COI and between 5.0 µg/dL and 12.1 µg/dL at maximum measured soil concentrations in the five COI. Upper bound Pb lead level estimates ranged between 9.7 µg/dL and 10.2 µg/dL at the SRML (400 mg/kg).

Table 8.13 **SARA and IEUBK Pb Model Results at the Selected SRML**

Parameters	Coniston	Copper Cliff	Falconbridge	Hanmer	Sudbury Centre
HQ_{avg}					
Soil concentration	52	98	82	19	36
Dust concentration	127	150	144	98	116
HQ	0.9	0.9	0.9	0.8	0.8
Geometric mean BLL	2.0	2.5	2.6	1.6	1.8
95th percentile BLL	4.3	5.4	5.6	3.5	3.9
Probability of exceeding a BLL of 5 µg/dL	2.5%	7.2%	7.7%	0.72%	1.5%
Probability of exceeding a BLL of 10 µg/dL	0.030%	0.17%	0.19%	0.004%	0.014%
HQ = 1					
Soil concentration	190	167	182	187	201
Dust concentration	178	173	176	178	181
HQ	1	1	1	1	1
Geometric mean BLL	3.1	3.1	3.3	3.1	3.2
95th percentile BLL	6.7	6.7	7.1	6.7	6.9
Probability of exceeding a BLL of 5 µg/dL	16.2%	15.0%	20.0%	16.2%	17.7%
Probability of exceeding a BLL of 10 µg/dL	0.70%	0.60%	0.98%	0.69%	0.82%

Parameters	Coniston	Copper Cliff	Falconbridge	Hanmer	Sudbury Centre
HQ$_{max}$					
Soil concentration	310	582	335	78.5	309.8
Dust concentration	203	239	207	142	203
HQ	1.1	1.3	1.1	0.9	1.1
Geometric mean BLL	4.0	5.6	4.4	2.3	4.0
95th percentile BLL	8.7	12.1	9.5	5.0	8.7
Probability of exceeding a BLL of 5 µg/dL	30.9%	59.7%	38.4%	4.70%	31.0%
Probability of exceeding a BLL of 10 µg/dL	2.4%	11.0%	3.8%	0.082%	2.4%
SRML					
Soil concentration	400	400	400	400	400
Dust concentration	217	217	217	217	217
HQ	1.2	1.2	1.2	1.2	1.2
Geometric mean BLL	4.5	4.6	4.7	4.5	4.5
95th percentile BLL	9.7	10.0	10.2	9.7	9.7
Probability of exceeding a BLL of 5 µg/dL	41.3%	42.7%	45.6%	41.5%	41.4%
Probability of exceeding a BLL of 10 µg/dL	4.5%	4.9%	5.6%%	4.6%	4.5%

The IEUBK model was also used to consider the soil concentrations which corresponded to a 5% probability of exceeding a blood Pb level of 5 µg/dL or 10 µg/dL. The following text provides an example of these calculations using the community of Copper Cliff for illustration purposes. A soil concentration of 405 mg/kg (and a corresponding indoor dust concentration of 217 mg/kg) was associated with a 4.98% probability of exceeding a blood Pb level of 10 µg/dL for the community of Copper Cliff assuming homogeneous concentrations of Pb in environmental media and diet. This also assumed soil and dust bioavailabilities of 33% and 40%, respectively. The geometric mean for blood Pb concentrations, when exposed to this soil concentration of Pb, was 4.6 µg/dL, with the 95th percentile at 10 µg/dL. Thus, a soil Pb concentration of 405 mg/kg would be considered protective of a 5% exceedance of a blood Pb level of concern of 10 µg/dL.

If the target blood Pb level was reduced to 5 µg/dL, the corresponding soil concentration was lowered to 75 mg/kg (with a corresponding indoor dust concentration of 140 mg/kg) to achieve a 5.041% probability of exceeding the blood Pb level of 5 µg/dL for the community of Copper Cliff. This assumed homogeneous concentrations of Pb in environmental media and diet and incorporated a geometric standard deviation of 1.6. This also used soil and dust bioavailabilities of 33% and 40%, respectively. Thus, a soil Pb concentration of 75 mg/kg would be considered protective of a 5% exceedance of a blood lead level of concern of 5 µg/dL.

8.3.3 Recommended SRML for Lead

Based on the weight-of-evidence including the conservative nature of the risk assessment, the strong indication provided in the literature that 500 mg of Pb/kg soil was a safe level for residential properties, and the previously established regulatory SRML for children's play areas of 400 mg of Pb/kg soil (U.S. EPA 2001b; MOE 2001c), and the relative strength-of-evidence associated with each of these elements, it was concluded that an SRML of 400 mg of Pb/kg soil was appropriate for the Sudbury study area. As the U.S. EPA indicated in the derivation of their Pb criteria, an SMRL of 400 mg of Pb/kg soil provides a sufficient level of protection to minimize the likelihood of harm to human health. This takes into consideration the uncertainty of the scientific evidence with regard to environmental Pb levels at which health effects would result. Lastly, blood Pb is the true marker of exposure, eliminating many of the assumptions and uncertainties inherent in a risk assessment and model predictions. Collection of blood Pb data in the future would aid in minimizing many of the uncertainties inherent in the risk assessment, and provide further confidence in the selection of an appropriate SRML.

8.4 Other Risk Assessment Issues

While conducting the Sudbury HHRA, a number of important issues were raised which required consideration during the risk assessment process. These included:

- Special considerations in assessing the exposure of children to the Sudbury COC and the implications of the inherent toxicity of these substances to this sensitive life stage.
- Implications of sulfur dioxide (SO_2) and acid precipitation on the mobility and toxicity of the COC, in an area historically impacted by SO_2.
- Potential impacts of occupational exposures for area residents.
- A discussion of the implications of metal–metal interactions, given that the COC can be present as complex mixtures in the environment.
- A brief review of soil ingestion rates in children and recommendations to address long-term 'pica' behavior within the risk assessment.
- Implications of dermal sensitization to nickel for GSA residents.
- A discussion of lifetime exposure, and how the elderly are addressed within the HHRA as a potentially sensitive life stage.
- Whether Sudbury residents are at an increased risk due to COC concentrations potentially accumulating within their bodies, leading to an elevated lifetime body burden.

These issues, in the context of the Sudbury HHRA, are discussed in the following sections.

8.4.1 Considerations for Children's Exposure and Toxicity

For the assessment of risks from exposures to environmental chemicals, children cannot be considered as small adults. Throughout childhood, children are growing and developing, and may be more susceptible to adverse effects from chemicals in the environment. Therefore, it was vital that the Sudbury HHRA took into account the potential sensitivity of this subpopulation. Children have heightened vulnerability to chemicals for the following reasons: children have disproportionately heavy exposures to many environmental agents; children's metabolic pathways, especially in fetal life and in the first months after birth, are immature; developmental processes are easily disrupted during rapid growth and development before and after birth, and children have more years of future life and thus more time to develop diseases initiated by early exposures (NAS 1993).

Children's Exposure

Children's exposure to environmental chemicals is different from that of adults because their bodies and their behaviors are different. Children consume more food and water for their body mass, have higher inhalation rates for their body mass, and have higher surface area to volume ratios than adults (NAS 1993; U.S. EPA 2002a). Children also have several unique exposure routes such as *in utero* and via breast milk (Landrigan et al. 2004). In addition to these exposure routes, children play outside, play close to the ground, touch and taste objects more than adults do, all of which may increase their exposure to environmental chemicals relative to that of adults. Children may also ingest non-food items, sometimes to an extreme (i.e. pica children), and may have a more limited diet than many adults due to life stage requirements or simply preference.

The exposure patterns of children may be addressed in risk assessments by estimating exposures for multiple age groups. However, the risk assessment community has not yet agreed on the most appropriate age groups for the assessment of children's exposure and/or risk (Ginsberg et al. 2004; U.S. EPA 2002a). The *Child-Specific Exposure Factors Handbook* (U.S. EPA 2002a) provides data on exposure factors that can be used to assess doses from oral, dermal and inhalation exposures among children. The handbook provides data in several areas including: breast milk ingestion, food ingestion, drinking water ingestion, soil ingestion, rates of hand-to-mouth and object-to-mouth activity.

Toxicological Susceptibility

Children are developing and constantly changing, and they may experience different susceptibilities to chemical perturbation during organ development. A risk assessor should ask a series of questions in the problem formulation stage to determine whether it is likely that children have a particular vulnerability to a chemical: Does the chemical cause known organ-specific toxicity; what organs are affected; how are these organs potentially differentially susceptible during development; and, what are the specific time periods of concern? (Daston et al. 2004). If the chemical is known to affect particular organ systems, or particular processes, then critical windows of vulnerability can be identified when the organs are developing or the processes are active (Daston et al. 2004).

There is also evidence in humans for the development of cancer in adults resulting from childhood exposures (U.S. EPA 2005). There are also examples from animal studies of transplacental carcinogens and suggestions that altered development can affect later susceptibility to cancer induced by chemical exposures in adult life (U.S. EPA 2005). U.S. EPA (2005) identified several factors that potentially lead to increased childhood susceptibility to carcinogenic agents relative to adults including: more frequent cell division during development can result in enhanced fixation of mutations due to the reduced time available for repair of DNA lesions; some embryonic cells (e.g. brain cells) lack key DNA repair enzymes, and some components of the immune system are not fully functional during development.

Risk Characterization

The most appropriate way to characterize children's risk is to compare age-specific exposure limits to exposure estimates derived for the same age group. Exposure limits developed specifically for a particular age group address both the differences in metabolism between children and adults, and the increased sensitivity of developing tissues. U.S. EPA (2005) recommends that age-specific exposure limits be used where available. Unfortunately, these limits are often not available, and other methods of addressing these issues must be used.

The U.S. EPA (2002b) described how they apply the child-protective uncertainty factor (UF) in risk assessments of pesticides conducted under the Food Quality Protection Act. The 10-fold child-protective UF (or a part thereof) was applied only where the adequacy and appropriateness of the toxicity assessment or the exposure assessment were judged to be insufficient. Ginsberg et al. (2004) examined the 10-fold uncertainty factor to accommodate variation within the human population. For the purpose of the current assessment, uncertainty adjustments for toxicological susceptibility as recommended by the U.S. EPA were conservatively applied to the evaluation of potential carcinogenic/mutagenic risks to each of the relevant modeled age stages, as follows:

- Infant (0 to <0.5 years)   10-fold UF
- Preschool Child (0.5 to <5 years) 10-fold UF
- Child (5 to 12 years)   3-fold UF
- Adolescent (12 to 19 years)   3-fold UF

It is important to remember that these UF values were intended to protect against carcinogens which have a mechanism of action relevant to the sensitive early life stages of the developing child, and are not relevant to late acting carcinogens. However, due to the uncertainty present for the mechanism of action for the current COC during these early life stages, these UFs were conservatively applied for the relevant life stages in the lifetime assessment of carcinogenic risk for As, Co and Ni in the Sudbury assessment.

8.4.2 Sulfur Dioxide (SO_2)

While SO_2 was specifically excluded as a COC for the current assessment, it was important to consider the potential effects it may pose as a modifying factor to the existing COC. Considerable study has historically been conducted into the impacts of SO_2 as a major precursor of acid precipitation. The phenomenon of 'acid rain' occurs because sulfuric acid (H_2SO_4) may be formed from SO_2 on contact with water, either in the atmosphere or on the surface. There may be direct human health effects of atmospheric SO_2 and acid precipitation which are probably related to the hydrogen ion (i.e. to the acidity) (Goyer et al. 1985). However, these direct effects were not addressed in this HHRA. The indirect, or post-deposition, effects of SO_2 and acid precipitation were of interest in the Sudbury area because acidification of soil and water can affect the speciation, mobility and solubility of metals.

There is no evidence that once deposited, H_2SO_4 and acid-forming S species represent a direct threat to human health; however, acidification of soil and water may mobilize metals from generally fixed sites and increase total human exposure to these COC (Goyer et al. 1985). Cations of various elements in the soil can be replaced by hydrogen ions (or various other ions) to cause their solubilization in water (Smith 1992). Some metals of toxicological significance that are affected by pH are Al, As, Cd, Cu Pb Mn, Hg and Se (Smith 1992; Gerhardsson et al. 1994). The solubility, and hence the availability and mobility of many metals is increased at lower pH values.

Regardless, any alterations in the concentrations of Cu, Pb and Se (three of the six COC that are known to be affected by acid precipitation) caused by acid rain were captured in the extensive monitoring conducted for the Sudbury Soils Study which included soils and drinking water. The effect of acid precipitation is to possibly alter exposures through existing pathways. Thus any incremental risk associated with the effects of acid precipitation on the COC was included as part of the total risk.

8.4.3 Occupational Exposures

One concern raised during the HHRA process was whether the study would consider the risks of occupational exposures on the health of community members, and whether this would make workers a particularly sensitive subpopulation. When discussing this issue, it is important to understand that the Sudbury HHRA was not designed or intended to examine occupational exposure to metals in the workplace. There were several reasons for this. Occupational exposure was a matter addressed by the Joint Health and Safety Committees that were comprised of company and union representatives, as well as Ministry of Labour personnel, among others. Both companies, Vale and Xstrata Nickel, have programs in place that examine and measure a worker's exposure to the COC addressed by the risk assessment. Most importantly, different levels of 'acceptable' risk are assumed for employees in the workplace compared to a resident of the general Sudbury population. The level of 'acceptable' risk to the resident is much lower, therefore, the standards being applied in the risk assessment were more rigorous than would be applied in an occupational setting. Additionally, occupational concerns lie with the worker, typically a healthy male adult, while risk assessments, by definition, protect sensitive individuals within the population (i.e. children, pregnant women, the elderly, and those with compromised health).

8.4.4 Chemical Mixtures

The Sudbury HHRA evaluated health risks related to individual exposures to each of the COC. However, the issue of potential metal–metal interactions as a result of chemical mixtures was frequently raised by members of the public and regulatory agencies. Under typical ambient environmental exposure conditions, humans are exposed to complex mixtures of metals (and various non-metallic substances), rather than individual compounds. Clearly, exposure to such complex mixtures can produce a broad range of health effects (U.S. EPA 1986). There can be a variety of types of interactions between metals in environmental or dietary mixtures that can alter the overall absorption, toxicokinetics, toxicodynamics and toxicity of metals in humans and animals (Newman et al. 2004). The potential for such interactions is an important consideration in the human health risk assessment of metals, as the nature of the interactions may increase or decrease the bioavailability and toxicity of metals present within the mixture.

Many of the studies available in the literature have methodological limitations that make it difficult to clearly ascertain the potential for interactions, and/or have produced conflicting results. Thus, there was little information available that was helpful in extrapolating available interaction data to the situations of low-level chronic exposure to complex chemical mixtures that are usually the focus of human health risk assessments (ATSDR 2004a; Krishnan and Brodeur 1994). Even for metals where reliable interaction data exist from laboratory studies, the data usually are not adequate for predicting the likely magnitude of the interaction's impact on toxicity (U.S. EPA 2004).

The vast majority of existing health criteria, guidelines, TRVs, exposure limits, and other health-based benchmarks for metals are derived for either elemental forms of individual metals, or a few types of single metal compounds (salts, oxides, sulfides, etc.). Because of this inherent limitation of the available toxicology database, regulatory agencies typically recommend that human health risk assessments of metals evaluate the individual components of the metals mixture, and then determine whether the exposures or risks for the individual metals in the mixture could reasonably be considered additive, based on the health effects associated with each metal. There is a very large body of literature that addresses these issues, and recent publications and guidance produced by ATSDR in 2004 (see www.atsdr.cdc.gov/iphome.html) have compiled and summarized a substantial amount of the available information on this complex topic.

ATSDR (2004a) defined toxicological interactions based on deviations from the results that are expected on the basis of additivity. Interaction is said to occur when the effect of a mixture is different from additivity based on the dose–response relationships of the individual components (ATSDR 2004a). Thus, interactions are sorted into three broad categories:

- Greater-than-additive (i.e. synergism, potentiation)
- Additive (additivity, no apparent influence)
- Less-than-additive (i.e., inhibition, masking).

Additivity is generally recognized as the most plausible type of interaction that may occur in situations of chemical exposure in the ambient environment. However, it requires that the chemicals act through the same or similar mechanisms of action and/or affect the same target tissue(s). In HHRAs where the COC act via different mechanisms of toxic action, and affect different target tissues, it is typically assumed that no potential toxicological interactions warrant consideration and the estimated exposures and risks for the COC are considered separately. Goyer et al. (2004) identified three main classes of interactions that occur between metals: i) interactions between essential metals; ii) interactions between non-essential metals; and iii) interactions between essential and non-essential metals. Based on the classification system of Goyer et al. (2004), Co, Cu and Se are considered as nutritionally essential metals, As and Ni have possible beneficial effects, whereas Pb has no known beneficial effect.

Dose-additivity is commonly applied in risk assessment by the calculation of a hazard index for those chemicals that produce the same or similar effects in the same organs by the same or similar modes of action. U.S. EPA (1986, 1989, 1990, 2000) guidance leans heavily towards the dose-additive approach and states that a strong case is required to indicate that two chemicals that produce adverse effects on the same organ system, even though by different mechanisms, should not be treated as dose additive. However, it should be recognized that, like the response addition approach, this can lead to toxicologically inappropriate summing of exposures and risks in some cases that may substantially overestimate risk.

Each of the COC in this study, with the exception of inhaled As and Ni, produce different critical effects on different organ systems. Although ATSDR (2004c) has indicated that the interaction of Pb and As may produce a neurological effect that is greater than additive, the toxicological criteria for these chemicals are not based on the same critical effects, nor do they target the same biological system. This is also true for Cu and Pb, which, when interacting, are suggested to produce a neurological effect that is sub-additive with no significant hepatic interaction (ATSDR 2004b). Based on these considerations, the overall limited nature of the metal–metal interactions literature, and consideration of the information presented in the previous sections, the potential risks from exposure to each of the COC was evaluated on an individual basis for all exposure routes assessed in the HHRA. However, despite the uncertainties involved with this approach, given the generous uncertainty factors built into the development of each of the COC-specific toxicological reference values, it was not expected that health risks would be underestimated even under worst case scenarios.

8.4.5 Soil Ingestion Rates in Children

Ingestion of soil by children may result in significant exposure to toxic substances at contaminated sites. The available literature was reviewed in some detail to address the issue of long-term soil intake rates in children, including those considered to display 'pica' behavior (the intentional ingestion of soil). The potential for exposure to contaminants via ingestion of soil is greater for children because they are likely to ingest more soil than do adults as a result of behavioral patterns present during childhood. Pica behavior is considered to be relatively uncommon and was estimated to be present in about 1% to 2% of the population (Calabrese et al. 1989, 1990). Other studies reported earth eating and pica 'dirt' eating to vary from 3% to 19% for children in different communities (Bruhn and Pangborn 1979; Clausing et al. 1987; Calabrese et al. 1989; Sedman and Mahmood 1994).

For children under 6 years of age, the U.S. EPA guidance recommends using a mean acute soil ingestion rate of 100 mg/day, and a conservative mean estimate of 200 mg/day (U.S. EPA 2002a). The U.S. EPA (2002a) determined that the 95th percentile values for soil ingestion, based on several key studies, ranged from 106 mg/day to 1432 mg/day, with an average of 358 mg/day. As a result, they have recommended a 95th percentile value for an acute soil ingestion rate in children of 400 mg/day (U.S. EPA 2002a).

Soil pica behavior is much less prevalent than normal, inadvertent soil ingestion, thus the available data on soil ingestion rates for pica children are limited. Calabrese et al. (1989, 1991) estimated that upper range soil ingestion values may range from approximately 5000 to 7000 mg/day. Based on a review of the key tracer studies, the U.S. EPA (2002a) proposed an ingestion rate of 10,000 mg/day for use in acute exposure assessment. ATSDR (2007) applied the soil ingestion rate of 5000 mg per day for the entire duration of acute (<14 days), intermediate (14 to 365 days), and chronic exposures (>365 days) to develop screening levels. One expert noted that extrapolations of short-term analytical studies to long-term exposure scenarios may be inappropriate, as few children likely ingest 5000 mg of soil a day throughout a year (Stanek and Calabrese 1995). In the end, experts agreed that there were limited data to support this approach. There are limited data for quantifying amounts of soil ingested by children, particularly by pica children. The increase in ingestion rates would range between 25- and 50-fold compared to soil ingestion by 'normal' children. Since the risk assessment results were used to drive soil risk management levels, using soil ingestion rates for pica children, who represent a very small proportion of the population, would likely not provide feasible and appropriate remediation goals.

In summary, it was decided that it would not be appropriate to use the short-term, acute soil consumption values associated with pica children for a long-term, chronic assessment of potential health risks related to soil contamination. Therefore, a long-term soil ingestion distribution was developed based upon information presented in Stanek et al. (2001). Based upon this distribution, a 95th percentile total soil and dust ingestion rate for preschool children and children was calculated as 202 mg/day (91 mg/day soil, 111 mg/day indoor dust), and a 50th percentile (or CTE) total soil and dust intake rate of 53 mg/day (24 mg/day soil and 29 mg/day dust).

8.4.6 Dermal Sensitization to Nickel

Nickel dermatitis (also called contact or allergic dermatitis) is the most commonly observed adverse effect of Ni in the general population (ATSDR 2003a). It is a form of allergic contact dermatitis where an inflammatory reaction is produced in the skin by contact with Ni in people who have acquired a hypersensitivity to Ni as a result of a previous exposure (Dorland 2000; Keczkes et al. 1982). Research has observed a relationship between specific human lymphocyte antigens and Ni sensitivity (Mozzanica et al. 1990). In sensitized individuals, exposure to Ni results in a red, itchy rash at the site of exposure (although the rash may spread). Sensitized individuals are of interest in risk assessment because they may experience adverse effects at lower exposures than non-sensitized individuals. Once an individual has been sensitized to Ni, subsequent inhalation, oral or dermal exposures to low levels of Ni may cause reactivation of the dermatitis (Keczkes et al. 1982). It is accepted that subsequent exposures via either the oral or dermal routes can reactivate Ni dermatitis; however, there is no evidence that airborne Ni nickel causes allergic reactions in the general population (ATSDR 2003a).

Under non-occupational exposure conditions, sensitization to Ni typically occurs primarily as a result of prolonged skin contact with Ni-containing metal objects (e.g. jewelry, coins, dental braces, stainless steel and metal fastenings on cloths) or when metal objects are inserted into body parts (e.g. ear piercing, orthodontics and orthopedic devices) (Menné et al. 1989; Menné 1994; Wilkinson and Wilkinson 1989; Dotterud and Falk 1994). ATSDR (2003a) reported that approximately 10–15% of the population is sensitized to Ni. Andreassi et al. (1998) reported that approximately 10–15% of women and 1–3% of men living in industrialized countries are sensitized to Ni. Although Ni is classified as an allergen of moderate potency (Kligman 1966), there is a high risk of developing Ni allergic hypersensitivity occupationally and in the general public due to the ubiquitous occurrence of Ni in all aspects of daily life (Hostynek 2002). Due to chemical and experimental variables in addition to individual variables (e.g. differences in susceptibility to Ni, age, gender, integrity of skin), Hostynek (2002) reported that a threshold value for Ni inducing sensitization cannot be developed at this time.

Without clear data on the exposure threshold for re-activation of Ni dermatitis, it was not possible to conclude whether environmental exposures to Ni would reactivate Ni dermatitis in sensitive individuals, or would have a desensitizing effect. Due to the confounding issues outlined above and the absence of obvious health concerns related to this form of sensitization in the Sudbury community, Ni dermatitis was noted as a potential uncertainty within the assessment, but was not evaluated further as part of the Sudbury HHRA.

8.4.7 The Elderly as a Sensitive Subpopulation

Children as a sensitive subpopulation in risk assessment have been the subject of intensive research and methodological development in recent years, while less focus has been given to the elderly as another potentially sensitive group. When discussing the fact that a subpopulation may be considered sensitive, it is important to note the distinction between those who may be considered 'sensitive' because they are more 'highly exposed' than other portions of the overall populace versus those who are specifically sensitive from a biological or toxicological point-of-view (e.g. asthmatics). Although many risk assessment paradigms describe the elderly as a sensitive subpopulation, specific methodologies for the assessment of risks to the elderly have not been developed. That child-specific risk assessment paradigms have been developed is, in part, a reflection of the general protective attitude of society toward the young.

In the elderly, liver and kidney function is impaired with age, limiting the body's ability to detoxify chemicals (Iyaniwura 2004). In addition to the physical factors influencing the vulnerability of the elderly to chemical toxicity, the mental, social, psychological and economic changes associated with aging may also increase vulnerability to chemical toxicity (Iyaniwura 2004). Two studies suggested that the elderly may be more sensitive to the effects of Pb (Muldoon et al. 1996; Payton et al. 1998). However, a review of the most recent ATSDR toxicological profiles did not locate any studies with regard to unusual susceptibility of any human subpopulation, including the elderly, to As, Co, Ni, Cu, Pb or Se (ATSDR (2000, 2003a,b, 2004b,d). In summary, no evidence was identified to indicate that the elderly may be more vulnerable to any of the COC than a young child. As such, the female preschool child was selected as the most sensitive receptor life stage for evaluation of non-carcinogenic risk for the current assessment. A lifespan of 70 years was conservatively selected for the evaluation of lifetime cancer risks for the current assessment.

8.4.8 COC Lifetime Body Burden

Some concern was raised by members of the community that long-term exposures to the COC in Sudbury over an individual's lifetime, could result in an accumulation of these COC leading to increasing health risk with age. While this could be a concern for certain organic compounds (e.g. PCBs, dioxins and furans, methyl mercury) which can bioaccumulate in the body's tissue, this is not the case for the COC under study in the current human health risk assessment. This is largely because the COC in question do not bioaccumulate, resulting in very little body burden over time.

Three of the COC are essential elements (Co, Cu and Se), meaning that a certain body burden must be maintained to prevent deficiencies and to maintain good health. Homeostatic mechanisms ensure that levels of the element are adequate for the body's needs, but do not reach toxic levels. These mechanisms are generally effective, but may be impaired or missing (which can lead to chronic poisoning), or they can be overwhelmed by high doses (i.e. acute poisoning). For the essential elements, uptake, distribution, storage and elimination are all strongly dependent on the nutritional status of the individual as the body seeks to maintain ideal concentrations. The remaining three COC (As, Pb and Ni) have no known functions in the body (though there is some evidence that As may be beneficial at very low doses). While the potential for accumulation of certain environmental contaminants is a concern for many risk assessments, none of the COC being evaluated in the Sudbury HHRA are prone to significant accumulation within the body over an individual's lifespan. In fact, most of the COC have a very short half-life within the body, and in some cases are essential nutrients for good health. As such, even long-term exposure to the COC in question would not have any additional risk other than that which is already evaluated using the selected toxicological limits.

8.5 Uncertainties in the Human Health Risk Assessment

8.5.1 General Sources of Uncertainty in Risk Assessment

There is no prescribed 'off the shelf' model or single approach to conduct a comprehensive human health risk assessment such as the current assessment. As such, many decisions are made along the way that can influence the outcome of the assessment. Seemingly simple, yet critical, decisions such as, which COC to evaluate, which communities of interest, which receptor groups, etc., all have considerable influence on how the HHRA will progress. In addition, the quantitative, or numerical, risk assessment requires the input of large amounts of data and numerical variables. Some of these input variables can be obtained from the general published literature, while other information must be Sudbury-specific and were obtained from the various surveys as described in Chapter 6. It must be realized that the goal of quantitative exposure assessment is to produce a conservative model to ensure that risks are never underestimated.

Each of the decisions and input variables contain some element of variability and uncertainty and can affect the outcome of the assessment to some degree. This leads to some amount of 'uncertainty' with the final results and conclusions. Risk managers need to know the uncertainties surrounding the study conclusions so they can make recommendations accordingly (for example, ask for more experimentation or monitoring, hedge decisions away from large losses). An uncertainty analysis can pinpoint the priorities for obtaining new information, so that uncertainty can be reduced and the decision-maker can have increased confidence in the decision ultimately taken. The traditional approach to dealing with uncertainties is to make the risk assessment conservative through the use of extreme assumptions and point estimates, and large uncertainty factors. There are, however, costs to this approach (Moore and Elliott 1996). In regulatory programs in which worst case assumptions are the norm, expensive risk mitigation measures may be enacted for chemicals that pose little threat to human health or the environment. Conversely, in programs that rely on best-guess values or so-called reasonable conservative values, chemicals having low likelihoods of causing effects may be ignored. In some cases, this results in the need to consider information beyond those generated by the quantitative risk assessment. The 'weight-of-evidence' evaluation was utilized for As and Pb in the Sudbury HHRA.

Generating a list of the various sources of uncertainties that affect a human health risk assessment is the first step en route to conducting a successful uncertainty analysis. Such a list will help structure the analysis and ensure that major sources of uncertainty are either quantified or explicitly excluded from the study (Finkel 1990). Uncertainty can be classified in many ways (Finkel 1990; Hoffman and Hammonds 1994; Rowe 1994; Hora 1996; Paté-Cornell 2002). There are many sources or components of uncertainty in a typical risk assessment. We may be uncertain about the identity of the sub-population at highest risk of exposure, possible routes of exposure, the appropriate multimedia exposure model, ingestion rates, chemical concentration in different media, sensitivity of different age groups to the chemical of interest, or the importance of modifying factors (e.g. diet, health). Despite the long list of possible sources of uncertainty, they all belong to one or several of four general types of uncertainty: i) variability; ii) incertitude arising from lack of knowledge about parameter values; iii) model structure; and, iv) decision rules.

Variability refers to the observed differences in a population or parameter attributable to true heterogeneity (Warren-Hicks and Moore 1998; U.S. EPA 1999). Examples include variation between individuals in size (e.g. height, weight), physiology (e.g. metabolic rate, food intake rate), and between environments (e.g. soil type, climate, chemical concentration).

Parameter uncertainty refers to our incertitude about the true values of the parameters or variables in a model (Warren-Hicks and Moore 1998; U.S. EPA 1999). Parameters are often estimated from laboratory, field or other studies. This type of uncertainty is introduced because the estimated value typically relies on insufficient, unreliable or partially relevant information for the parameter of interest. Measurement error often arises from the imprecision of analytical devices used, for example, to quantify chemical levels in different media or measure levels of detoxifying enzymes in humans. Errors in measurement, however, are not necessarily restricted to analytical hardware.

Random error or sampling error is a common source of incertitude in HHRA and arises when one tries to draw an inference about a quantity from a limited number of observations. For sample means, one can examine the importance of sampling error by calculating the standard deviation of sample means (Sokal and Rohlf 1981). Sample means based on 3000 observations will have a standard deviation only one-tenth that of means based on 30 observations. Systematic error occurs when the errors in the data are not truly random, such as might occur when the sample population is not representative of the entire population (e.g. when sampling is biased towards more contaminated areas). Systematic error, unlike random error, does not decrease with more observations and is not accounted for when calculating sample statistics (e.g. mean, standard deviation). When systematic error is pervasive, sample statistics such as 95% confidence intervals can be quite misleading. For example, nearly half of the 27 measures of the speed of light measured between the years 1875 and 1958 had 95% or 99% confidence intervals that did not bracket the most accurate value available today ($c = 299,792.458$ km/s) (Henrion and Fischhoff 1986).

Decision rule uncertainty comes into play during risk management (i.e. after a risk estimate has been generated). This type of uncertainty arises when social objectives, economic costs, value judgments, etc. are part of the decision-making process for deciding on what actions to take to remediate a problem. Individual decision makers are likely to be uncertain about how to best represent the complex preferences of their constituents. Even with the availability of formal analytical tools, controversial judgments remain about how to value life, distribute costs, benefits and risks among individuals and groups, and deciding whether to reduce risks now or sometime in the future (Finkel 1990).

This chapter discusses the topic of uncertainty analysis, and the related issue of sensitivity analysis. Uncertainty and sensitivity analysis both focus on the output of a model and are, therefore, closely related. The purposes of the two types of analyses, however, are different. An uncertainty analysis assesses the uncertainty in model outputs that derives from uncertainty (and variability) in the inputs. A sensitivity analysis assesses the contributions of the inputs to the total uncertainty in the output, and can evaluate the 'leverage' a given variable may have on the overall assessment results. The general concept of uncertainty analysis is described first in this chapter, followed by a discussion of specific areas of uncertainty attached to the Sudbury HHRA.

8.5.2 Uncertainty Analysis

Uncertainty is a widely recognized aspect of HHRA, but it is often ignored in regulatory applications. In decision-making for contaminated sites, there are compelling reasons to characterize uncertainties as part of the risk assessment to avoid the mistaken impression that model results are precise and well understood (Reckhow 1994; Finkel 1994). A balanced discussion of conclusions and uncertainties enhances the overall credibility of the assessment. Chao et al. (1994) provide an excellent example of how consideration of uncertainties about the consequences of ground level ozone can lead to a more cost-effective decision-making process. Thus, uncertainty analysis helps to discriminate among management options, identifies critical information needs, and, 'can spur on the iterative search for new decision options that may outperform any of the initial ones offered' (Finkel 1994).

Uncertainty analysis makes clear what is known and what is not, a huge advantage over the use of conservative assumptions and uncertainty factors. Thus, uncertainty analysis provides an objective and transparent means of comparing assumptions, models, and data put forth by stakeholders in an environmental dispute. Use of conservative assumptions and uncertainty factors in an analysis has the effect of blurring the distinction between science and decision-making, although a significant amount of work has been done to incorporate science into the establishment of uncertainty factors. Ultimately, the task of assessors is to come up with estimates of what is likely to happen, what might happen, and what is not likely to happen, to identify possible risk management options, but not to make decisions for society. The risk assessment approach does not negate a conservative approach, but rather moves it to the more appropriate risk management stage.

8.5.3 Uncertainties in the Sudbury HHRA

Assumptions were made during the risk assessment process, either because of data gaps or knowledge gaps, and consequently, each assumption results in some degree of uncertainty in the overall conclusions of the assessment. To understand the uncertainties within the HHRA and to ensure that the impact of these uncertainties is understood, it was important to document and characterize each of these. To ensure that the risk assessment did not underestimate the potential for the occurrence of adverse effects, it was necessary to make assumptions which were conservative. The following sections describe areas of uncertainty within the Sudbury risk assessment, and discusses the potential impacts of these uncertainties on the conclusions drawn from the assessment. Given the tendency for the assumptions used in this HHRA to overestimate both exposure and toxicity, it was considered extremely unlikely that the overall risk characterization underestimated potential health risks.

2001 Soil Survey Data

Soils data collected in 2001 formed the basis of the risk assessment. The soil survey was conducted before the risk assessors (SARA Group) were involved in the risk assessment. The study design was rigorous and encompassed over 8000 soil samples, and involved a significant degree of QA/QC by numerous stakeholders in the study. The parameters measured were chosen by scientists from the MOE, Laurentian University and other study stakeholders. Given the large sample size, there was high certainty that the data supplied for the risk assessment were representative of metal concentrations in Sudbury. However, since only approximately 10% of the residential properties were sampled, it was possible that local areas of higher soil COC concentrations were not captured in the 2001 survey.

Chemical concentrations in media used in the exposure modeling were assumed to remain unchanged over time

No attempt was made to predict future levels of COC in environmental media based on current emission rates. Smelter emissions in Sudbury are continuing to decline as a result of ongoing efforts by the mining companies. Therefore, media concentrations of the COC within the study area should continue to decrease over time, and the use of existing conditions would be conservative and may actually overestimate future exposure. It was, therefore, assumed that local soil metal levels had reached steady-state and would not increase in the future.

Chemical concentrations reported at 'below detection level'

There was some uncertainty with regard to the 'actual' concentration of a chemical for which laboratory analysis indicates a concentration below detection. Theoretically, the value of that concentration could be any value between zero and the detection limit. While the most conservative value to use in the assessment would be the detection limit itself, it was considered to be realistically conservative to employ a value of one-half the detection limit when the exposure concentration was listed as below the level of detection. Depending on the 'true' value, this may actually over- or underestimate the evaluated chemical concentration.

Use of outdoor air concentrations to represent indoor levels

The ambient air monitoring program provided a robust data set collected over a 1-year period. As part of the exposure modeling, it was assumed that indoor air concentrations were equivalent to outdoor PM_{10} concentrations. This was considered very conservative based on other studies which typically indicate lower indoor air levels compared to outdoor levels. Therefore, this approach would tend to over-predict exposure and potential risk.

Selection of appropriate soil ingestion rate (SIR)

There was some uncertainty and minor controversy with regard to the selection of an appropriate soil ingestion rate (SIR) for children in the current assessment. The SARA Group selected a value of 80 mg/day for children recommended in the Federal guidance on human health risk assessment by Health Canada (2004). However, the Ontario MOE (and the U.S. EPA) recommended the use of an SIR of 100 mg/day for children. Both values are rooted in a similar dataset of soil tracer studies in children, and largely differ due to differing statistical analyses and methodologies used in the development of the SIR. Neither value is incorrect and both involve appropriate interpretations of the underlying scientific data. The SARA Group ultimately selected the Health Canada regulatory value as it was based upon a more recent evaluation of the scientific literature. However, it is important to note that the conclusions and recommendations of the current assessment would not have changed had the slightly more conservative SIR of 100 mg/day been used in the Sudbury HHRA.

Treatment of replicate samples

Where duplicate soil, or other media, samples were analyzed, the geometric mean of the two samples was used, as opposed to the arithmetic mean or the highest value.

Potential impacts of Pb leaching from water distribution pipes

Use of provincial drinking water data based upon distribution system samples could potentially underestimate the concentrations of Pb in community drinking water (i.e. the potential for additional Pb to leach out of older distribution pipes while en route to Sudbury residences, as well as fixtures and fittings within the home itself). This could be addressed through the use of a tap water survey (similar to that conducted for the well water and lake survey). However, results of the exposure assessment modeling for Pb indicated that drinking water was not a significant pathway for exposure, compared to other dominant pathways such as market basket exposures.

Use of a non-linear model to calculate soil-to-dust regression relationships

A non-linear model, rather than alternative linear models, was used to calculate the soil-to-dust regression relationships, as part of the indoor dust survey. Visual examination of residuals indicated that linear regressions using the raw data resulted in a violation of at least one of the classical assumptions. A transformation of the raw data is commonplace and can, in some circumstances, help correct non-normality, non-linearity and/or lack of homeostatic variances. It was therefore decided to conduct the linear regression analysis on the natural-log transformed data. This approach was considered consistent with previous work conducted by the various regional U.S. EPA districts, and provided a more robust approach over the use of a soil/dust concentration ratio approach.

Food consumption patterns of Sudbury residents

It was assumed, for the most part, that typical Sudbury residents had similar eating habitats to other Canadians. To test, and support, this assumption, a food consumption survey of Sudbury residents was undertaken. In addition, separate exposure model runs were conducted for known eating preferences (i.e. high blueberry consumption, consumption of wild game and fish by anglers, hunters and members of the First Nation communities).

Levels of COC in mother's milk

No published methodology for consideration of mother's milk exposures to inorganic compounds was available, although maternal transfer exposures were considered by the IEUBK model for Pb. As the difference between formula exposures and those from mother's milk would be very small, the potential contribution of COC in mother's milk was not considered. Similarly, transplacental transfer of COC was not considered in the assessment. While it is likely that some *in utero* exposure does occur, no method of assessment for this exposure was identified in the literature. As such, this pathway was not considered.

Levels of COC in home grown produce

The concentrations of COC were measured in produce obtained from approximately 70 residential gardens, and 10 local commercial operations. A variety of produce types were analyzed from gardens with a wide range of soil conditions and COC concentrations. As a result, there was a great deal of confidence that the data were representative of Sudbury produce which helped to reduce the uncertainty with the exposure assessment. One issue that did arise was related to analytical reliability of the As concentrations measured in the garden vegetables. The QA/QC program which included submission of replicate samples as well as standard reference material (SRM) revealed that many of the As concentrations being reported were inconsistent between replicates, and higher than the certified SRM values. It was determined that the laboratory routine analytical method was having difficulty with samples with low As and high water content, such as lettuce. As a result, some of the As levels used in risk assessment were likely higher than what was actually present in the sample. In conclusion, although there was some uncertainty in the vegetable As data, the higher exposure concentrations used in the assessment would be conservative and overestimate risk.

Levels of COC in local fish tissue

Due to the prevalence of local lakes and the popularity of sport fishing, the consumption of local fish was considered a potentially important pathway for the HHRA. Therefore, COC levels were measured in common sport fish species from eight local lakes. A variety of lakes were sampled, with robust sample sizes of each species for analysis. Some of the lakes sampled were very close to the active smelters and sources of emissions. Therefore, the study team is confident that the data reflect some of the 'worst case' exposure conditions, resulting in conservative estimates of risk.

Level of COC in local wildlife (game)

COC concentrations within local wildlife were based on modeling COC levels in moose meat and not on measured concentrations. Concentrations were predicted for moose meat only, as moose had higher predicted body burdens than other types of wildlife (primarily due to the moose's higher consumption rate for forage and aquatic plants). Concentrations were modeled for ERA Zone 2, which encompassed much of the urban region within the study area, and would have higher soil metal concentrations than more remote areas. This was considered conservative since most hunting occurs in more remote locations, which are typically further removed from the emission sources.

Level of COC in consumer products

Background concentrations of the COC in consumer products were not evaluated in the current assessment. A detailed literature review was conducted to determine whether this potential route of exposure would be significant. The review failed to provide a quantitative value of the contribution of consumer products to total daily exposure for the study COC. While some of the COC are found in several consumer products (e.g. Pb in some hair dye and cosmetics; Co and Ni in cleaning products and cosmetics, such as eye shadow), the relative concentrations are considered minor compared to exposure contributions arising from other pathways, such as oral ingestion of food and water.

Level of COC in cigarettes

Nicotiana tabacum (tobacco) is known to be used effectively in biotechnology for the removal of metals from contaminated soils (Bernhard et al. 2005). It has been reported that As, Cu, Ni, Pb and Se are found in either tobacco, cigarette paper, filters and/or cigarette smoke (Bernhard et al. 2005; Arista 2003). The daily contribution of metals from cigarettes was further evaluated in an Austrian study conducted by Wolfsperger et al. (1994). That study showed that higher levels of Co, Pb, and Ni were found in the hair of cigarette smokers when compared to their non-smoking counterparts. Health Canada recently issued a statement that smoking cigarettes may contribute to an additional 0.01 to 0.04 µg/kg bw/day of As exposure (Health Canada 2006). Similarly, the

Nickel Institute (1997) also concluded that tobacco smoking may be a source of Ni exposure. They cite a study that suggested smoking a pack of 20 cigarettes per day can contribute up to 0.004 mg of Ni/day (Grandjean 1984). Other researchers have shown that 0.04–0.58 µg of Ni is released with the mainstream smoke of one cigarette (WHO 1991). While the data indicated that smoking cigarettes is a potential additional source of some of the COC, the degree of contribution would be highly dependent on the number of cigarettes smoked per day, and the conditions under which they are consumed. As such, the potential contribution from cigarette smoke to COC body burden could not be accurately quantified in the current HHRA, but does add an additional degree of uncertainty for those individuals who are smokers (or are routinely exposed to second-hand smoke).

The resuspended dust pathway was not considered

U.S. EPA recommended that inhalation of resuspended dust be evaluated only if site-specific exposure setting characteristics indicate that this is a potentially significant pathway. This could potentially be the case in areas of tailing or slag piles. However, the air monitoring program was designed to capture dust-borne contaminants originating from the tailing and or slag piles, and these data were used for the evaluation of indoor air exposures.

A site-specific bioaccessibility study was conducted as part of the HHRA

The use of bioaccessibility studies within HHRA is an emerging area that introduces some uncertainty into the assessment. There is no one accepted method for conducting a study of this nature and, as such, professional judgment was used in the development of the methods and the interpretation of results. Methodological changes are emerging in the literature on an ongoing basis, and the methods have not been validated for all COC. The purpose of the study was to estimate the relative difference in bioaccessibility between metals in soil and dust from the study area, and those used in the toxicological studies used to derive the TRVs utilized in the HHRA. Despite general methodological limitations, it was considered that the use of site-specific bioaccessibility data introduced less uncertainty than assuming 100% bioaccessibility or using non-site-specific literature-based values.

Ni species-specific fingerprints in ambient air

A variety of metal speciation techniques were used to assist in the development of species-specific 'fingerprints' to represent ambient exposures to the various forms of Ni in air, as well as for instances where air quality was impacted by fugitive dusts from the Vale facility. A weight-of-evidence approach was used to develop a fingerprint incorporating relative percentages of the various Ni species. While there was potential variability in particulate sources which may have implications on the corresponding Ni species observed, an effort was made to err on the side of conservatism in selecting an upper-end estimate percentage value for the more toxicologically relevant Ni species (i.e. Ni_3S_5 and Ni oxide).

Geographical extent of fugitive dust impacts from the Vale Inco Copper Cliff Facility

Fugitive dusts from the Vale facility were detected at the Sudbury Centre West monitoring station. This station is located very close to the Vale facility and did not provide a detailed insight into the geographical extent to which fugitive dusts may be dispersed into other Sudbury neighborhoods. Given the nature of the particulate, it is likely that this was a localized impact but there was some uncertainty surrounding the extent of impact from fugitive dust emissions. However, as a result of the HHRA, Vale undertook an aggressive program to reduce fugitive dust and emissions from their property so it would be expected that future airborne concentrations would be lower in the near future (see also Chapter 12).

The use of biomonitoring as part of the HHRA

The Sudbury HHRA was not a health study in which biomonitoring was used to predict potential health risks within the study population. Biomonitoring is often viewed as the gold standard for exposure assessment, and ultimately risk management. However, there are many pros and cons to the use of biomonitoring in health studies. One of the largest challenges is how the results are interpreted. Accurate benchmarks are generally lacking for comparison to measured biomonitoring data. Some biomonitoring data are easier to collect than others (e.g. breast milk, urine, and hair versus blood or adipose tissue). Due to ethical reasons, one cannot typically conduct a biomonitoring study without a clear demonstration of potential risk within the community under study. The Sudbury HHRA benefited from the urinary As data collected in the Town of Falconbridge. These results were used in the weight-of-evidence evaluation of As risks throughout the study area. Potential risks related to exposures to Pb were also evaluated with the U.S. EPA IEUBK model, which uses pharmacokinetics to predict blood Pb concentrations in children. This model has been validated using biomonitoring data, and provided an additional line of evidence for the evaluation of Pb risks to Sudbury residents.

Uncertainties in the Hazard Assessment

Animal models are used as surrogates for humans in the development of TRVs, thereby introducing uncertainties into the risk factors due to the interspecies variability in sensitivity. For genotoxic carcinogens, it was assumed that no repair of genetic lesions occurs, and therefore, no threshold can exist for chemicals that produce self-replicating lesions. However, the existence of enzymes that routinely repair damage to DNA is well documented in the scientific literature, and the potential adverse effects arising from damage to DNA are only observed if the ability of these repair enzymes to 'fix' the damage is exceeded.

In the derivation of limits by regulatory agencies, large uncertainty factors (i.e. 100-fold or greater) were used in the estimation of the reference dose (RfD) for threshold-type chemicals. These uncertainty factors were applied to exposure levels from studies where no adverse effects were observed (i.e. to the no observed adverse effect level (NOAEL)). Thus, exceeding the toxicological criterion does not mean that adverse effects would occur. Exposures greater than the calculated toxicological criterion may also be without risk (i.e. below the threshold for adverse effects in humans), but this could not be, or was not, determined by the agency which derived the toxicological criterion. Humans were assumed to be the most sensitive species with respect to toxic effects of chemicals. However, for obvious reasons, toxicity assays are not generally conducted on humans, so toxicological data from the most sensitive laboratory species were used in the estimation of toxicological criteria for humans.

Different age and gender categories were used as part of the exposure and hazard assessment components of the risk assessment to permit the evaluation of potential risks to sensitive subcategories (such as the female preschool child). As specific toxicity data were typically not available for specific life stages or genders, this added an additional layer of uncertainty to the results. In fact, the results of the assessment may distinguish a difference between genders or life stages which cannot be validated based upon existing toxicity data for most chemicals. However, it was considered a conservative approach to use chronic lifetime risk reference values with less-than-lifetime exposures.

The most sensitive toxicological endpoint (for example, decreased growth, body weight loss/gain, reproductive effects) was selected for each chemical to represent the exposure limit. TRVs, because of their inherent conservatism, are widely considered protective of sensitive subgroups and life stages. However, risk assessment, and TRVs and environmental quality guidelines for that matter, can only protect most of the people, most of the time. There can always be those individuals that are hypersensitive, and those situations require special consideration. But, risk assessments do not investigate these situations unless there is clear evidence that such a situation exists in the study area. There was no such evidence of this in the Greater Sudbury Area.

Only Seilkop (2004) provided species-specific cancer Inhalation Unit Risks (IURs) for airborne Ni. As such, these IURs were utilized in this assessment as part of the weight-of-evidence approach to evaluate Ni inhalation risks. These values have not been derived or endorsed by regulatory agencies, although MOE has acknowledged that these IURs were under consideration and have been the topic of discussion on several instances. The other IURs for airborne Ni used in the current assessment were derived by regulatory agencies such as U.S. EPA, Health Canada and WHO. It is important to note that these IURs were based on occupational cohorts that were exposed to refinery dust and, as such, the applicability of these IURs to the ambient environment in Sudbury was uncertain. The TRV relating to oral exposures to Ni was based upon the U.S. EPA RfD of 20 µg/kg bodyweight/day.

Thus, in our opinion, the level of effort and detail that went into preparation of the toxicological profiles for each COC was appropriate and adequate for this risk assessment. TRVs were selected based on a detailed review of several of the most well-known and well-regarded regulatory agencies in the world. A number of considerations went into selecting the TRVs, including the scientific basis, the underlying science policies and the date of last major revision.

8.5.4 Sensitivity Analysis

The purpose of the sensitivity analysis was to identify how variation in the output of a model (e.g. total daily intake of a chemical) was influenced by uncertainty in the input variables. If the output variance precludes effective decision making, sensitivity analysis may be used to identify the input variables that contributed the most to the observed output variance. Subsequently, research efforts may be initiated to reduce uncertainty in those particular input variables. Sensitivity analysis can also be used to simplify model structure by identifying those input variables that contribute little to the output (for example, a minor route of exposure) and thus can be removed from the analysis.

Sensitivity analysis methods may be classified into three groups: i) screening methods; ii) methods for local sensitivity analysis; and iii) methods for global sensitivity analysis. Screening methods are generally used to separate influential input variables from non-influential ones, rather than quantify the impact that an input variable has on the output of the model. Screening methods are useful for models with large numbers of input variables. They are able to identify important input variables with little computational effort, but at a cost of losing quantitative information on the importance of the input variables. In contrast, local and global sensitivity measures provide quantitative estimates of the importance of each input variable. The difference between them is that the former focuses on estimating the impact of small changes in input variable values on model output, while the latter addresses the contribution to model output variance over the entire range of each input variable distribution.

To investigate the relative sensitivity of risk predictions as part of the deterministic risk assessment, the impact of key input variables on the calculated health risk related to exposures of the female preschool child living in Copper Cliff to Pb and Ni were examined. The key variables evaluated included the following:

- The selected oral TRV for Pb (i.e. the MOE recommended value of 1.85 versus the U.S. EPA recommended value of 3.7 µg/kg bw/day) and Ni (i.e. the U.S. EPA recommended value of 20 µg/kg bw/day versus the OEHHA recommended value of 11 µg/kg bw/day).
- The soil/dust consumption rate (i.e. the MOE recommended value of 100 mg/day versus the Health Canada recommended value of 80 mg/day).
- The selected food consumption database (i.e. the older Health Canada/Nutrition Canada database versus the more recent USDA data from the Northeastern U.S.).
- In the case of Ni, the effect of evaluating risk to the female preschool child versus a hypothetical individual exposed for an entire lifetime.
- In the case of Pb, the relationship selected to estimate indoor dust COC concentrations from paired outdoor residential soil COC concentrations versus the IEUBK default soil-to-dust concentration ratio of 0.7.
- The bioaccessibility of Pb and Ni in soil and dust media (i.e. one-phase results from bioaccessibility testing versus two-phase results from bioaccessibility test versus assuming 100% bioaccessibility, but using the IEUBK bioavailability default value of 50%, for Pb).

The result of altering these variables on the risk characterization is presented in Table 8.14 and 8.15 for Pb and Ni, respectively. As would be expected, the outputs demonstrate that the variable with the largest impact on the estimated risk is the TRV value. For example, the calculated risk for Pb is reduced by 49% if the TRV is increased from 1.85 to 3.7 µg/kg bw/day. While the other variables had lesser impacts, they still played a significant role in the calculation of overall risk. In the case of Pb, the vast majority of the acceptable daily intake as dictated by the TRV was used up by background sources unrelated to Sudbury (i.e. the market basket).

Ultimately, the choice of assumptions used in the assessment scenario must be based upon an evaluation of the best available science and considers the implications of regulatory policy as it pertains to these assumptions. In the case of the Sudbury HHRA, these parameters were selected based upon the expert judgment of the SARA Group in consultation with the TC and the external scientific advisors.

Table 8.14 Relative Impact to Predicted Risk from Changing Key HHRA Assumptions for Lead in Copper Cliff

Scenario	Assessed scenario	MOE soil consumption rate	USDA food consumption+MOE soil consumption rate	IEUBK bioavailability adjustment	IEUBK soil-to-dust concentration ratio	USDA food consumption	Two-phase bioaccessibility	Using Health Canada TRV	USDA food consumption+two-phase bioaccessibility+Health Canada TRV
% Change	—	6% increase	3% decrease	9% decrease	10% decrease	10% decrease	11% decrease	49% decrease	60% decrease
Oral TRV (µg/kg bw/day)	1.85 (MOE)	1.85 (MOE)	1.85 (MOE)	1.85 (MOE)	1.85 (MOE)	1.85 (MOE)	1.85 (MOE)	3.7 (HC)	3.7 (HC)
Soil/dust consumption rate (mg/day)	80	100	100	80	80	80	80	80	80
Food consumption database	HC	HC	USDA	HC	HC	USDA	HC	HC	USDA
Soil-to-dust relationship	Regression equation	Regression equation	Regression equation	Regression equation	IEUBK, soil-to-dust concentration ratio (0.7)	Regression equation	Regression equation	Regression equation	Regression equation
Bioaccessibility	Soil=66% Dust=83% (one-phase)	Soil=66% Dust=83% (one-phase)	Soil=66% Dust=83% (one-phase)	Soil=66% Dust=83% (one-phase)	Soil=66% Dust=100%	Soil=66% Dust=83% (one-phase)	Soil=66% Dust=83% (one-phase)	Soil=38% Dust=43% (two-phase)	Soil=66% Dust=83% (one-phase)	Soil=38% Dust=43% (two-phase)
Bioavailability	100%	100%	100%	50% (IEUBK)	100%	100%	100%	100%	100%
Receptor group	Female Preschool Child								

Human Health Risk Assessment Results

Table 8.15 **Relative Impact to Predicted Risk from Changing Key HHRA Assumptions for Nickel in Copper Cliff**

Scenario	Assessed scenario	MOE soil consumption rate	Two-phase bioaccessibility	USDA Food Consumption+MOE Soil Consumption Rate	USDA food consumption	OEHHA oral TRV	OEHHA oral TRV+lifetime	MOE soil consumption rate+lifetime	Lifetime exposure
% Change	—	1% increase	1% increase	6% decrease	7% decrease	83% increase	30% decrease	62% decrease	62% decrease
Soil/dust consumption rate (mg/day)	80	100	80	100	80	80	80	100	80
Food consumption database	HC	HC	HC	USDA	USDA	HC	HC	HC	HC
Oral TRV (µg/kg bw/day)	20	20	20	20	20	11	11	20	20
Bioaccessibility	Soil=42% Dust=30% (one-phase)	Soil=42% Dust=30% (one-phase)	Soil=38% Dust=43% (two-phase)	Soil=42% Dust=30% (one-phase)	Soil=42% Dust=30% (one-phase)	Soil=44% Dust=31% (one-phase)	Soil=42% Dust=30% (one-phase)	Soil=42% Dust=30% (one-phase)	Soil=42% Dust=30% (one-phase)
Receptor group	Female preschool child	Female preschool child	Female preschool child	Female preschool child	Female preschool child	Female preschool child	Lifetime receptor	Lifetime receptor	Lifetime receptor

8.6 References

Andreassi M, Gioacchino MD, Sabbioni E, et al. (1998) Serum and urine nickel in nickel-sensitized women: effects of oral challenge with the metal. Contact Derm 38:5–8

Angle CR, Marcus A, Cheng IH, McIntire MS (1984) Omaha childhood blood lead and environmental lead: A linear total exposure model. Environ Res 35:160–170

Arista Laboratories Europe (Arista) (2003) U.K. smoke constituents study. Part 11: Determination of metal yields in cigarette smoke by ICP-MS & CVAAS. Report: www.the-tma.org.uk/benchmark/benchmarkresources/part11.pdf. Accessed 30 October 2006

ATSDR (2000) Toxicological profile for arsenic. U.S. Department of Health and Human Services, Public Health Service, Agency for Toxic Substances and Disease Registry, Atlanta, GA. www.atsdr.cdc.gov/toxprofiles/tp2.html. Accessed August 2005

ATSDR (2003a) Toxicological profile for nickel. Draft for public comment (update). U.S. Department of Health and Human Services, Public Health Service, Agency for Toxic Substances and Disease Registry, Atlanta, GA

ATSDR (2003b) Toxicological profile for selenium. U.S. Department of Health and Human Services, Public Health Service, Agency for Toxic Substances and Disease Registry, Atlanta, GA

ATSDR (2004a) Guidance manual for the assessment of joint toxic action of chemical mixtures. U.S. Department of Health and Human Services, Agency for Toxic Substances and Disease Registry, Atlanta, GA. www.atsdr.cdc.gov/interactionprofiles/ipga.html. Accessed August 2005

ATSDR (2004b) Toxicological profile for cobalt. U.S. Department of Health and Human Services, Public Health Service, Agency for Toxic Substances and Disease Registry, Atlanta, GA. www.atsdr.cdc.gov/toxprofiles/tp33.html. Accessed August 2005.

ATSDR (2004c) Interaction profile for: arsenic, cadmium, chromium, and lead. U.S. Department of Health and Human Services, Public Health Service, Agency for Toxic Substances and Disease Registry, Atlanta, GA. www.atsdr.cdc.gov/interactionprofiles/ip04.html. Accessed August 2005

ATSDR (2004d) Toxicological profile for copper – draft for public comment. U.S. Department of Health and Human Services, Public Health Service, Agency for Toxic Substances and Disease Registry, Atlanta, GA. www.atsdr.cdc.gov/toxprofiles/tp132.html. Accessed August 2005

ATSDR (2007) Toxicological profile for lead. U.S. Department of Health and Human Services, Public Health Service. Agency for Toxic Substances and Disease Registry, Atlanta, GA

Bernhard D, Rossmaan A, Wick G (2005) Critical review: Metals in cigarette smoke. IUBMB Life 57(12):805–809

Bruhn CM, Pangborn RM (1979) Reported incidence of pica among migrant families. J Am Diet Assoc 58:417–420

Calabrese EJ, Barnes R, Stanek E et al. (1989) How much soil do young children ingest: an epidemiologic study. Regul Toxicol Pharmacol 10:123–137

Calabrese EJ, Stanek EJ Gilbert CE, Barnes RM (1990) Preliminary adult soil ingestion estimates: results of a pilot study. Regul Toxicol Pharmacol 12:88–95

Calabrese EJ, Stanek EJ, Gilbert CE (1991) Evidence of soil-pica behavior and quantification of soil ingested. Hum Exp Toxicol 10:245–249

Chao HP, Peck SC, Wan YS (1994) Managing uncertainty: The tropospheric ozone challenge. Risk Anal 14:465–475

Clausing P, Brunekreef B, van Wijnen JH (1987) A method for estimating soil ingestion by children. Int Arch Occup Environ Health 59:73–82

Daston G, Faustman E, Ginsberg G, et al. (2004) A framework for assessing risks to children from exposure to environmental agents. Environ Health Perspect 112(2):238–256

Decou ML, Williams R, Ellis E (2001) Lead screening report, Eastside Community, Port Colborne, April – June 2001. Regional Niagara Public Health Department, August 2001

Dorland (2000) Dorland's illustrated medical dictionary, 29th edn. W.B. Saunders Company, Philadelphia, PA

Dotterud LK, Falk ES (1994) Metal allergy in north Norwegian schoolchildren and its relationship with ear piercing and atopy. Contact Derm 31:308–313

Finkel AM (1990) Confronting uncertainty in risk management: A guide for decision-makers. Center for Risk Management, Resources for the Future, Washington, DC

Finkel AM (1994) Stepping out of your own shadow: A didactic example of how facing uncertainty can improve decision-making. Risk Anal 14:751–761

Gerhardsson L, Oskarsson A, Skerfving S (1994) Acid precipitation – effects on trace elements and human health. Sci Total Environ 153:237–245

Ginsberg G, Slikker W Jr, Bruckner J, Sonawane B (2004) Incorporating children's toxicokinetics into a risk framework. Environ Health Perspect 112(2):272–283

Goyer RA, Bachmann J, Clarkson TW, et al. (1985) Potential human health effects of acid rain: report of a workshop. Environ Health Perspect 60:355–368

Goyer R, Golub M, Choudhury H, Hughes M, Kenyon E, Stifelman M (2004) Issue paper on the human health effects of metals. Risk Assessment Forum, United States Environmental Protection Agency, Washington, DC, 19 August 2004. www.cfpub.epa.gov/ncea/raf/recordisplay.cfm?deid=86119. Accessed August 2005

Gradient (1995) Exposure assessment – Placer Dome, Balmertown, Ontario, Canada. Prepared for Placer Dome Canada by Gradient Corporation, Cambridge, MA

Grandjean P (1984) Human exposure to nickel. In: Sunderman FW Jr, et al. (eds) Nickel in the human environment: Proceedings of a Joint Symposium, March 1983, Lyon, France. IARC Scientific Publication No. 53, pp. 469–480. International Agency for Research on Cancer, Lyon, France.

Health Canada (2004) Contaminated sites program. Federal contaminated site risk assessment in Canada. Part I: Guidance on human health preliminary quantitative risk assessment (PQRA) September, 2004. Health Canada, Ottawa, Ontario

Health Canada (2006) Arsenic and its compounds – PSL1. www.hc-sc.gc.ca/ewh-semt/pubs/contaminants/psl1-lsp1/arsenic_comp/arsenic_comp_3_e.html. Accessed 31 October 2006

Henrion M, Fischoff B (1986) Assessing uncertainty in physical constants. Am J Phys 54:791–798

Hoffman FO, Hammonds JS (1994) Propagation of uncertainty in risk assessments: The need to distinguish between uncertainty due to lack of knowledge and uncertainty due to variability. Risk Anal 14:707–712

Hora SC (1996) Aleatory and epistemic uncertainty in probability elicitation with an example from hazardous waste management. Reliab Eng Syst Saf 54:217–223

Hostynek JJ (2002) Nickel-induced sensitivity: etiology, immune reactions, prevention and therapy. Review article. Arch Dermatol Res 294:249–267

Hwang YH, Bornschein RL, Grote J, Menrath W, Roda S (1997) Environmental arsenic exposure of children around a former copper smelter site. Environ Res 72(1):72–81

Iyaniwura TT (2004) Individual and subpopulation variations in response to toxic chemicals: factors of susceptibility. Risk World, 1 July 2004. www.riskworld.com/Nreports/2004/Iyaniwura.htm. Accessed 20 August 2007

Jin A, Teschke K, Copes R (1997) The relationship of lead in soil to lead in blood and implications for standard settings. Sci Total Environ 208:23–40

Johnson DL, Bretsch J (2002) Soil lead and children's blood levels in Syracuse, NY, USA. Environ Geochem Health 24:375–385

Keczkes K, Basheer AM, Wyatt EH (1982) The persistence of allergic contact sensitivity: a 10 year follow-up in 100 patients. Br J Dermatol 107:461–465

Kligman AM (1966) The identification of contact allergens by human assay. The maximization test: a procedure for screening and rating contact sensitizers. J Invest Dermatol 47:393–409

Krishnan K, Brodeur J (1994) Toxic interactions among environmental pollutants: Corroborating laboratory observations with human experience. Environ Health Perspect 102(Suppl. 9):11–17

Landrigan PJ, Kimmel CA, Correa A, Eskenazi B (2004) Children's health and the environment: public health issues and challenges for risk assessment. Environ Health Perspect 112(2):257–265

Lanphear BP, Succop P, Roda S, Henningsen G (2003) The effect of soil abatement on blood lead levels in children living near a former smelting and mining operation. Public Health Rep 118:83–91

Lewin MD, Sarasua S, Jones PA (1999) A multivariate linear regression model for predicting children's blood lead levels based on soil lead levels: A study at four superfund sites. Environ Res Sect A 81:52–61

Louekari K, Mroueh UM, Maidell-Munster L, Valkonen S, Tuomi T, Savolainen K (2004) Reducing the risks of children living near the site of a former lead smeltery. Sci Total Environ 319:65–75

Menné T (1994) Quantitative aspects of nickel dermatitis. Sensitization and eliciting threshold concentrations. Sci Total Environ 148:275–281

Menné T, Christophersen J, Green A (1989) Epidemiology of nickel dermatitis. In: Maibach HI, Menné T (eds.) Nickel and the skin: Immunology and toxicology. CRC Press Inc, Boca Raton, FL, pp. 109–115

Mielke HW, Gonzales CR, Powell E, Jartun M, Mielke PW (2007) Nonlinear association between soil and blood lead of children in metropolitan New Orleans, Louisiana, 2000–2005. Sci Total Environ 388:43–53

MOE (1991) Assessment of human health risk of reported soil levels of metals and radionuclides in Port Hope. Ontario Ministry of the Environment, 117 pp.

MOE (1996) The guideline for use at contaminated sites in Ontario. Ontario Ministry of the Environment, Toronto, Ontario

MOE (1999) Deloro environmental health risk study: Overall technical summary. Ontario Ministry of the Environment, Toronto, Ontario

MOE (2001a) Survey of arsenic exposure for residents of Wawa. Goss Gilroy Inc., Ottawa, Canada

MOE (2001b) Soil investigation and human health risk assessment for the Rodney Street Community, Port Colborne, Ontario. Ontario Ministry of the Environment, Toronto, Ontario, October 2001

MOE (2001c) Fact sheet: Lead. March 2001. Ontario Ministry of the Environment, Toronto, Ontario

MOE (2004a) Ontario Ministry of the Environment Record of Site Condition Regulation, Ontario Regulation 153/04 (O. Reg. 153/0) Soil, ground water and sediment standards for use under Part XV.1 of the Environmental Protection Act, March 9, 2004. Ontario Ministry of the Environment, Toronto, Ontario

MOE (2004b) Information draft on the development of Ontario air standards for nickel and its compounds. Ontario Ministry of the Environment, Standards Development Branch. June 2004. www.ene.gov.on.ca/envision/env_reg/er/documents/2004/air%20standards/information%20drafts/PA04E0029-i.pdf. Accessed 17 April 2007

MOE (2005) Procedures for the use of risk assessment under Part XV.1 of the Environmental Protection Act. Ontario Ministry of the Environment, Standards Development Branch, October 2005. Ontario Ministry of the Environment, Toronto, Ontario

MOE (2007) Development of Ontario air standards for lead and its compounds. EBR Registry Number PA06E0006. August 31, 2007. Ontario Ministry of the Environment, Toronto, Ontario

Moore DRJ, Elliott BJ (1996) Should uncertainty be quantified in human and ecological risk assessments used for decision-making? Human Ecol Risk Assess 2:11–24

Mozzanica N, Rizzolo L, Veneroni G, Diotti R, Hepeisen S, Finzi AF (1990) HLA-A, B, C and DR antigens in nickel contact sensitivity. Br J Dermatol 122(3):309–313

Muldoon SB, Cauley JA, Kuller LH (1996) Effects of blood lead levels on cognitive function of older women. Neuroepidemiology 15(2):62–72

NAS (1993) Pesticides in the diets of infants and children. National Academy of Science. National Academy Press, Washington, DC.

Newman MC, Diamond GL, Menzie C, Moya J, Nriagu J (2004) Issue paper on metal exposure assessment. Risk assessment forum, United States Environmental Protection Agency, Washington, DC. August 19, 2004. www.cfpub.epa.gov/ncea/raf/recordisplay.cfm?deid=86119. Accessed August 2005

Nickel Institute (1997) Safe use of nickel in the workplace. www.nickelinstitute.org/index.cfm/ci_id/13028/la_id/4.htm. Accessed 31 October 2006

OEHHA (2003) Nickel and nickel compounds. Office of Environmental Human Health Hazard Assessment. California Environmental Protection Agency. www.oehha.org/ecotox.html. Accessed 20 August 2007

OJEU (2005) Directive 2004/107/EC of the European Parliament and of the Council of 15 December 2004 relating to arsenic, cadmium, mercury, nickel and polycyclic aromatic hydrocarbons in ambient air. Official Journal of the European Union – 23/3. 26 January 2005

Paté-Cornell E (2002) Risk and uncertainty analysis in government safety decisions. Risk Anal 22:633–646

Payton M, Riggs KM, Spiro A (1998) Relations of bone and blood lead to cognitive function: the VA normative aging study. Neurotoxicol Teratol 20(1):19–27

Reckhow KH (1994) Water quality simulation modeling and uncertainty analysis for risk assessment and decision making. Ecol Model 72:1–20

Ren HM, Wang JD, Zhang XL (2006) Assessment of soil lead exposure in children in Shenyang, China. Environ Pollut 144:327–335

Rowe WD (1994) Understanding uncertainty. Risk Anal 14:743–750

SARA Group (2008) Sudbury Soils Study, Volume II: Human Health Risk Assessment, Prepared by SARA Group, March, 2008. www.Sudburysoilsstudy.com Accessed 17 April 2009

Sedman R, Mahmood RS (1994) Soil ingestion by children and adults reconsidered using the results of recent tracer studies. Air Waste 44:141–144

Seilkop SK (2004) Estimation of respiratory cancer risks associated with exposure to small airborne concentrations of nickel-containing substances. Presentation to the Ontario Ministry of the Environment and Inco, 10 February 2004

Seilkop SK, Oller AR (2003) Respiratory cancer risks associated with low-level nickel exposure: an integrated assessment based on animal, epidemiological and mechanistic data. Regul Toxicol Pharmacol 37:173–190

Smith RL (1992) Elements of ecology, 3rd edn. Harper Collins Publishers Inc., pp. 423–424, 427–431

Sokal RR, Rohlf FJ (1981) Biometry. W.H. Freeman and Company, New York

Stanek EJ, Calabrese EJ (1995) Daily estimates of soil ingestion in children. Environ Health Perspect 103(3):276–285

Stanek EJ, Calabrese EJ, Zorn M (2001) Biasing factors for simple soil ingestion estimates in mass balance studies of soil ingestion. Hum Ecol Risk Assess 7(2):329–355

Steele MJ, Beck BD, Murphy BL, Strauss HS (1990) Assessing the contribution from lead in mining wastes to blood lead. Regul Toxicol Pharmacol 11:158–190

Stern AH (1994) Derivation of a target level of lead in soil at residential sites corresponding to a de minimis contribution to blood lead concentration. Risk Anal 14(6):1049–1056

U.S. EPA (1986) Guidelines for the health risk assessment of chemical mixtures. Appendix A. U.S. Environmental Protection Agency. September 24, 1986. EPA/630/R-98/002. U.S. EPA, Washington, DC

U.S. EPA (1989) Risk assessment guidance for superfund. Volume I: Human health evaluation manual (Part A). U.S. Environmental Protection Agency. December 1989. EPA/540/1-89/002. U.S. EPA, Washington, DC

U.S. EPA (1990) Supplementary guidance for conducting health risk assessment of chemical mixtures. U.S. Environmental Protection Agency. August 2000. EPA/630/R-00/002. U.S. EPA, Washington, DC

U.S. EPA (1999) Risk assessment guidance for superfund: Volume 3 – Part A, Process for conducting probabilistic risk assessment. United States Environmental Protection Agency. Office of Solid Waste and Emergency Response, Washington, DC

U.S. EPA (2000) Supplementary guidance for conducting health risk assessment of chemical mixtures. Risk Assessment Forum, United States Environmental Protection Agency. August, 2000. EPA/630/R-00/002. www.epa.gov/ncea/raf/pdfs/chem_mix/chem_mix_08_2001.pdf. Accessed August 2005

U.S. EPA (2001a) Risk assessment guidance for superfund. Volume 3, Part A – Process for conducting probabilistic risk assessment. Office of Emergency and Remedial Response, United States Environmental Protection Agency. December, 2001. EPA 540-R-02-002. www.epa.gov/superfund/RAGS3A/index.htm. Accessed 15 September 2006

U.S. EPA (2001b) Lead: Identification of dangerous levels of lead. Federal Register 40 CFR Part 745, Volume 66, Number 4. Friday January 5, 2001. U.S. Environmental Protection Agency, Washington, DC

U.S. EPA (2002a) Child-specific exposure factors handbook – Interim report. EPA-600-P-00-002B. National Center for Environmental Assessment – Washington Office, Office of Research and Development, United States Environmental Protection Agency, Washington, DC

U.S. EPA (2002b) Determination of the appropriate FQPA safety factor(s) for use in the tolerance-setting process. Office of Pesticide Programs, Office of Prevention, Pesticides, and Toxic Substances, Washington, D.C. www.epa.gov/oppfead1/trac/science/determ.pdf. Accessed 15 October 2006

U.S. EPA (2004) Framework for inorganic metals risk assessment. U.S. Environmental Protection Agency. November 2004. EPA/630/P-04/068B. www.cfpub2.epa.gov/ncea/raf/recordisplay.cfm?deid=88903. Accessed August 2005

U.S. EPA (2005) Supplemental guidance for assessing susceptibility from early-life exposure to carcinogens. Prepared by a Technical Panel of the Risk Assessment Forum, U.S. EPA, March 2005. EPA/630/R-03-003F. United States Environmental Protection Agency, Washington, DC

Van de Wiele TR, Oomen AG, Wragg J, et al. (2007) Comparison of five in vitro digestion models to in vivo experimental results: Lead bioaccessibility in the human gastrointestinal tract. J Environ Sci Health Environ Sci Eng 42:1203–1211

Von Lindern I, Spalinger S, Petroysan V, von Braun M (2003) Assessing remedial effectiveness through the blood lead:soil/dust lead relationship at the Bunker Hill Superfund Site in the Silver Valley of Idaho. Sci Total Environ 303:139–170

Walker S, Griffin S (1998) Site-specific data confirm arsenic exposure predicted by the U.S. Environmental Protection Agency. Environ Health Perspect 106(3):133–139

Warren-Hicks W, Moore DRJ (1998) Uncertainty analysis in ecological risk assessment. SETAC Press, Pensacola, FL

WHO (1991) Nickel. Environmental health criteria, No. 108. WHO, Geneva

Wilkinson DS, Wilkinson JD (1989) Nickel allergy and hand eczema. In: Maibach HI, Menné T (eds) Nickel and the skin: Immunology and toxicology. CRC Press, Boca Raton, FL, p. 133

Wolfsperger M, Hauser G, Gossler W, Schlagenhaufen C (1994) Heavy metals in human hair samples from Austria and Italy: influence of sex and smoking habits. Sci Total Environ 156(3):235–242

8.7 Appendix Chapter 8: Abbreviations

ADI, Allowable Daily Intake

ATSDR, Agency for Toxic Substances and Disease Registry

BBL, blood lead level

COC, chemical of concern

COI, community of interest

CRL, cancer risk level

CTE, central tendency estimate

EDI, estimated daily intake

EPC, exposure point concentration

ERA, ecological risk assessment

ETDI, estimated total daily intake

EU, European Union and also exposure unit

GSA, Greater Sudbury area

HHRA, human health risk assessment

HQ, hazard quotient

IEUBK, integrated exposure uptake biokinetic

ILCR, incremental lifetime cancer risk

IUR, inhalation unit risk

MDL, method detection limit

MOE, Ministry of the Environment (Ontario)

NOAEL, no observed adverse effect level

OJEU, Official Journal of the European Union
OTR, Ontario typical range
PAC, Public Advisory Committee
PM, particulate matter
PRG, preliminary remediation goal
QA/QC, quality assurance/quality control
RAF, relative application factor
REL Reference Exposure Level
RfD, reference dose
RME, reasonable maximum exposure
SARA, Sudbury area risk assessment
SF, slope factor
SIR, soil ingestion rate
SRM, standard reference material
SRML, soil risk management level
TC, technical committee
TOR, typical Ontario resident
TRV, toxicity reference value
TSCA, Toxic Substances Control Act
UCL, upper confidence level
UCLM, upper confidence level on the mean
UF, uncertainty factor
USDA, United States Department of Agriculture
USEPA, United States Environmental Protection Agency
WHO, World Health Organization

9.0 Ecological Risk Assessment Problem Formulation

DOI: 10.5645/b.1.9

Authors

Christopher Wren, PhD
MIRARCO, Laurentian University
935 Ramsey Lake Road, Sudbury, ON Canada P3E 2C6
Email: cwren@mirarco.org

Ruth N. Hull, MSc
Intrinsik Environmental Sciences Inc.
6605 Hurontario Street, Suite 500, Mississauga, ON Canada L5T 0A3
Email: rhull@intrinsikscience.com

Table of Contents

9.1 Problem Formulation Stage of the ERA 257
9.2 Background and Scope for the Sudbury ERA 257
9.3 ERA Goals and Objectives 259
9.4 Compilation and Review of Available Ecological Information 259
9.5 Definition of the Study Area for the Terrestrial ERA 260
9.6 Identification of COC for the Terrestrial ERA 260
9.7 Identification of Valued Ecosystem Components (VEC) 261
    9.7.1 VECs for Objective #1: Evaluate the Extent to Which COC are Preventing the Recovery of Regionally Representative, Self-sustaining Terrestrial Plant Communities 262
    9.7.2 VECs for Objective #2: Evaluate Risks to Terrestrial Wildlife Populations and Communities Due to COC and Objective #3: Evaluate Risks to Individuals of Threatened or Endangered Terrestrial Species Due to COC 263
    9.7.3 Species Not Recommended as VECs 267
9.8 Assessment Endpoints and Measures 268
9.9 Conceptual Model 270
9.10 Analysis Plan and Identification of Data Gaps 272
9.11 Acknowledgements 273
9.12 References 273
9.13 Appendix Chapter 9: Abbreviations 275

Tables

9.1 Summary of Rationale for Recommendation of Terrestrial VECs 264
9.2 Valued Ecosystem Components Assessed in Various Regions of the Study Area 266
9.3 Summary of VEC Assessment Endpoints and Measures 269

Figures

9.1 Conceptual Linkages of Past Impacts, COC and Wildlife Habitat with VECs 258
9.2 Study Area Subdivisions Used in the Wildlife Exposure Model 260
9.3 Selection Process for Terrestrial Valued Ecosystem Components 262
9.4 Conceptual Model for the Sudbury Terrestrial ERA 271

9.1 Problem Formulation Stage of the ERA

Ecological risk assessment (ERA) is a process that evaluates the likelihood that adverse ecological effects may occur as a result of exposure to a stressor (US EPA 1998). An ERA collects, organizes and analyzes information to estimate the likelihood of undesired effects on nonhuman organisms, generally at the individual organism, population, or community level (Suter II et al. 2000). An ERA is conducted at a contaminated site to provide information needed to make decisions concerning risk management, including remediation (CCME 1996; Suter II et al. 2000). There are four main components of an ERA:

- Problem Formulation – presents background information and the ERA approach.
- Exposure Assessment – estimates the amount of chemical to which a receptor comes into contact.
- Effects Assessment – describes the toxicity of the chemicals.
- Risk Characterization – combines the exposure and effects information into a risk prediction.

To address the many scientific and technical aspects involved in the broad scope of the Sudbury ERA, a large team consisting of consultants and academics was assembled. The number of scientists involved grew as the objectives were clarified and the analysis plans developed. This allowed the study team to draw upon a broad range of expertise, including risk assessors, local researchers, faculty members and ecologists having extensive experience with the Sudbury environment.

The first major step of the ERA was to complete the problem formulation. The problem formulation establishes the goals, scope and focus of the assessment. It is a systematic planning step that identifies the major factors to be considered and is linked to the regulatory and policy context of the assessment (US EPA 1998). The components of the problem formulation can vary somewhat between different jurisdictions and guidance documents, but generally will include the following:

- identification of the scope of the ERA
- identification of the risk assessment objectives
- review of available background information
- definition and description of the study area
- selection of Chemicals of Concern (COC)
- selection of Valued Ecosystem Components (VECs) or ecological receptors
- identification of the assessment endpoints and measures
- description of the conceptual site model, and
- identification of data gaps or missing data.

Once completed, the problem formulation became the blueprint for the detailed risk assessments that followed. Additional data and information were collected to fill data gaps identified in the problem formulation before the ERA proceeded to the next step.

9.2 Background and Scope for the Sudbury ERA

The smelting facilities and historical emissions were described in some detail in Chapter 2. The first anthropogenic impact to the region was selective harvesting of the red and white pine, which began about 1870 and continued into the 1920s. Mining and smelting activities began in the mid-1880s, and widespread clear cutting began in earnest at this time to provide fuel to roast the ore containing Ni and Cu. The original roasting yards were replaced by more modern smelting facilities in the late 1920s and early 1930s. However, the relatively low smokestacks, and lack of pollution control technology associated with these early smelters resulted in large emissions of particulate matter, SO_2 and metals that were deposited in a relatively localized area. These high volume emissions and localized deposition continued into the 1970s when more effective emission controls were implemented, higher smoke stacks were constructed and the smelter at Coniston was closed.

Environmental degradation was apparent in the early 1900s, and by the early 1940s, government representatives were already meeting with local industry to discuss the smelter emissions and damage to the forests and local agriculture. The most dramatic impacts to the vegetation were centered on the three historical smelting centers at Copper Cliff, Falconbridge and Coniston. The most severely impacted area was largely devoid of trees and soil pH was commonly <4.0. This was referred to as the 'Barrens', which covered approximately 20,000 ha by 1970. A further 80,000 ha, located slightly more distant to the smelters, were considered 'Semi-Barrens' and supported a savannah-like cover of grasses with clusters of stunted white birch and red maple.

Beginning in the mid-1970s, several stakeholders in the Sudbury region began restoration activities aimed at 're-greening' the Sudbury landscape. This included soil liming to raise soil pH, seeding with grass/legume mixtures, fertilizing and planting trees. The Sudbury re-greening activities are described later in this book in Chapter 12. There is little doubt that a combination of smelter emissions and other activities (primarily logging and forest fires) historically caused widespread loss of the original forest cover in the area. Subsequent events resulted in a large area with either bare exposed bedrock that provided little growth medium for plants, or impacted soils that prevented natural recovery of a self-sustaining diverse ecosystem. Figure 9.1 provides a conceptual model illustrating linkages between some of the known ecological stressors, and impacts on the plant community and wildlife inhabiting the area.

The Sudbury ERA recognized three fundamental aspects of the terrestrial ecosystem in the study area: (1) past activities have severely impacted the vegetation, (2) restoration and re-greening activities have been, and continue to be, very successful over broad areas, and, (3) not all areas have received re-greening treatments and, therefore, in many areas vegetation recovery has not occurred (SARA 2009).

From a regulatory perspective, the Sudbury risk assessments were based on concerns that the concentrations of several metals and metalloids in soil exceeded the Ontario Ministry of the Environment (MOE) soil quality criteria. While it was recognized that terrestrial and aquatic environments are connected through various processes and interactions, the assessment of aquatic ecological risks is not directly linked to soil quality criteria. Although there have been documented impacts to aquatic ecosystems in the Sudbury area due to smelter emissions, it was not expected that remediation or restoration of the aquatic systems would be proposed at the time of the study. Therefore, a detailed assessment of the aquatic environment was considered to be outside the scope of the Sudbury ERA.

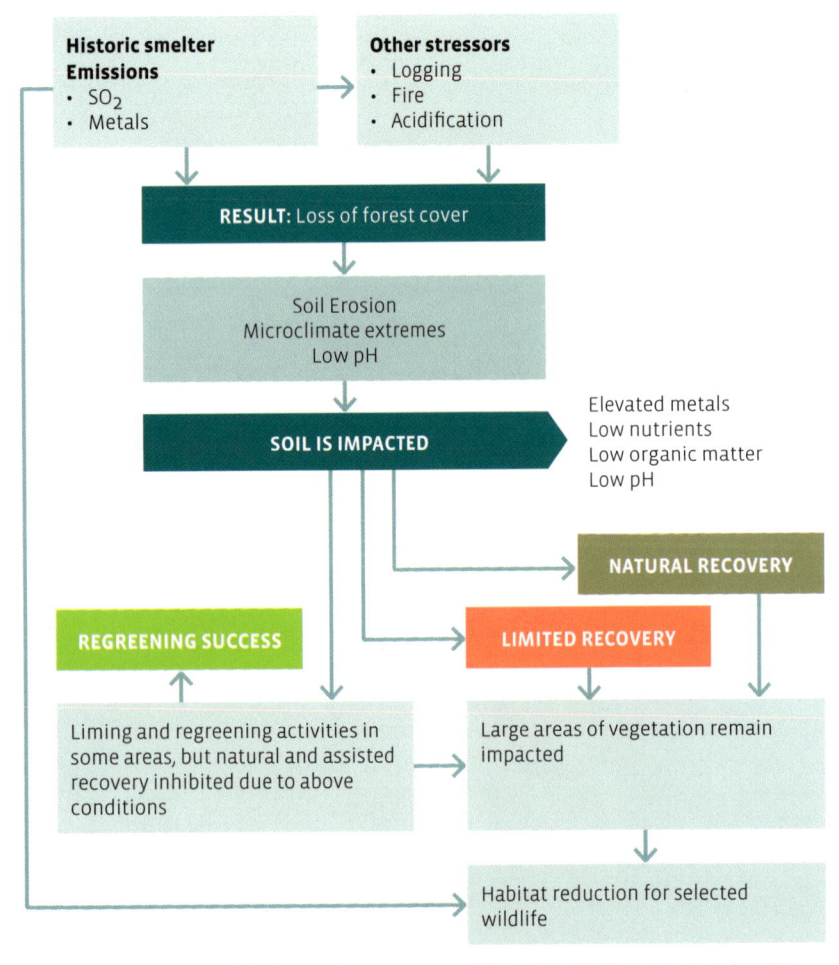

Figure 9.1 Conceptual Linkages of Past Impacts, COC and Wildlife Habitat with VECs

9.3 ERA Goals and Objectives

One of the first tasks for this study was to achieve clear agreement on, and understanding of, the goals and objectives for the ERA. These were based on the initial set of objectives articulated by the Technical Committee (TC) as tasks to be completed as part of the ERA. The goals and objectives had to recognize that the Sudbury ERA was being conducted against a backdrop of significantly reduced smelter emissions and considerable re-greening activities. The final goals for the Sudbury ERA became:

To characterize the current and future risks to terrestrial and aquatic ecosystem components due to Chemicals of Concern (COC) from the particulate emissions of Sudbury smelters; and to provide information that will support activities related to the recovery of regionally representative, self-sustaining ecosystems in areas of Sudbury affected by the COC.

To achieve the ERA goals, four specific ERA objectives were identified:

- Objective 1: Evaluate the extent to which the COC are preventing the recovery of regionally representative, self-sustaining terrestrial plant communities.
- Objective 2: Evaluate risks to terrestrial wildlife populations and communities due to the COC.
- Objective 3: Evaluate risks to individuals of threatened or endangered terrestrial species due to the COC.
- Objective 4: Conduct a comprehensive Problem Formulation for the aquatic and wetland environments in the Sudbury area to facilitate more detailed risk assessment in the aquatic/wetland ecosystems.

The problem formulation for the terrestrial ERA was common to Objectives 1, 2 and 3 and is the subject of the remainder of this chapter. The methods used to address Objective 1 were very different from the approach used to address Objectives 2 and 3, and consequently are discussed in separate chapters. Chapter 10 describes the study approach and results of Objective 1, while Chapter 11 addresses Objectives 2 and 3. Although an aquatic problem formulation was developed for the study technical reports, it is not included in this publication because the focus herein is on the terrestrial environment.

9.4 Compilation and Review of Available Ecological Information

The ecological data review involved the collection of information from relevant published scientific documents, web-based sources, and industry and government publications. A literature search was performed to determine the current state of knowledge relating to the ecological effects of metals in the greater Sudbury area. Results were reviewed to ascertain which studies best fit the scope of the ERA.

Initial literature search tasks included identifying applicable studies from existing collections of local ecological research documents. These collections were located throughout Ontario, and specifically Sudbury, including in the offices of the MOE (Ministry of the Environment), Vale, Xstrata Nickel, City of Greater Sudbury, Sudbury Public Library, and local educational institutions such as the collections at the J.N. Desmarais Library and the Centre for Environmental Monitoring (CEM) at Laurentian University. In addition, a literature search was conducted using available ecological and toxicological databases. A separate search was conducted to identify toxicological literature related to the COC and effects on terrestrial biota. Those papers were used to develop toxicological profiles which were in turn used to develop Toxicity Reference Values (TRVs) for selected VECs (See Chapter 11).

The review demonstrated that there were limited data available to parameterize the SARA (Sudbury Area Risk Assessment) wildlife exposure model with Sudbury-specific wildlife information. In addition, there were limited data available to assist in the interpretation of the results of the risk characterization (i.e. data on wildlife population trends, reproductive success, and/or evidence of local presence/absence of wildlife species). Thus, additional information was sought from other sources to incorporate local knowledge on wildlife populations. The sources of this additional information were: biological research community at Laurentian University; data from government agencies; published and anecdotal knowledge from wildlife societies; published and anecdotal knowledge from hunting societies; and web-based searches. The relevant Sudbury-specific data from all sources listed above were incorporated into the ecological exposure model, effects assessment and the risk characterization, as described in Chapter 11.

9.5 Definition of the Study Area for the Terrestrial ERA

The initial study area for the Sudbury Soils Study was defined as the area from which soil samples were collected during the 2001 Sudbury Regional Soils Project (CEM 2004). The extensive 2001 soil survey was undertaken to address data gaps in the concentrations of metals and metalloids identified by the MOE in their review of the historical soil and vegetation quality data for the Sudbury area (MOE 2001). The CEM at Laurentian University designed the regional soil sampling program to collect data from rural and remote areas with undisturbed soils. Details on the sampling methodology and results were provided in Chapters 3 and 4.

The area sampled by the CEM covered a broad geographical area. The cells of the stratified sampling plan ranged in size, with the smallest cells being closest to the smelters and in zones historically known to reflect smelter impact. The final nested sampling grid covered an area approximately 200 km × 200 km in size (40,000 km²) as shown in Chapter 3, Figure 3.1. This area was considered the study area for the ERA. To provide a more refined and meaningful interpretation of the wildlife exposure model results, the study area was subdivided into three Zones and four Communities of Interest (COI) as identified in Figure 9.2. The boundaries of the zones were defined on the basis of metal concentrations in soil and on the basis of terrain. Zone 1 is generally 'upwind' of the smelters with low metal levels in soil. Zone 2 was delineated to include the area between the three smelters containing the highest metal levels in soil. Zone 3 is south and southeast (generally downwind) of the smelters. The four Communities of Interest (i.e. Coniston, Copper Cliff, Falconbridge and Sudbury–central) were defined in the human health risk assessment (HHRA) (see Chapter 5). These Communities of Interest (COI) were urban environments and were appropriate for the assessment of some VECs which live in urban areas.

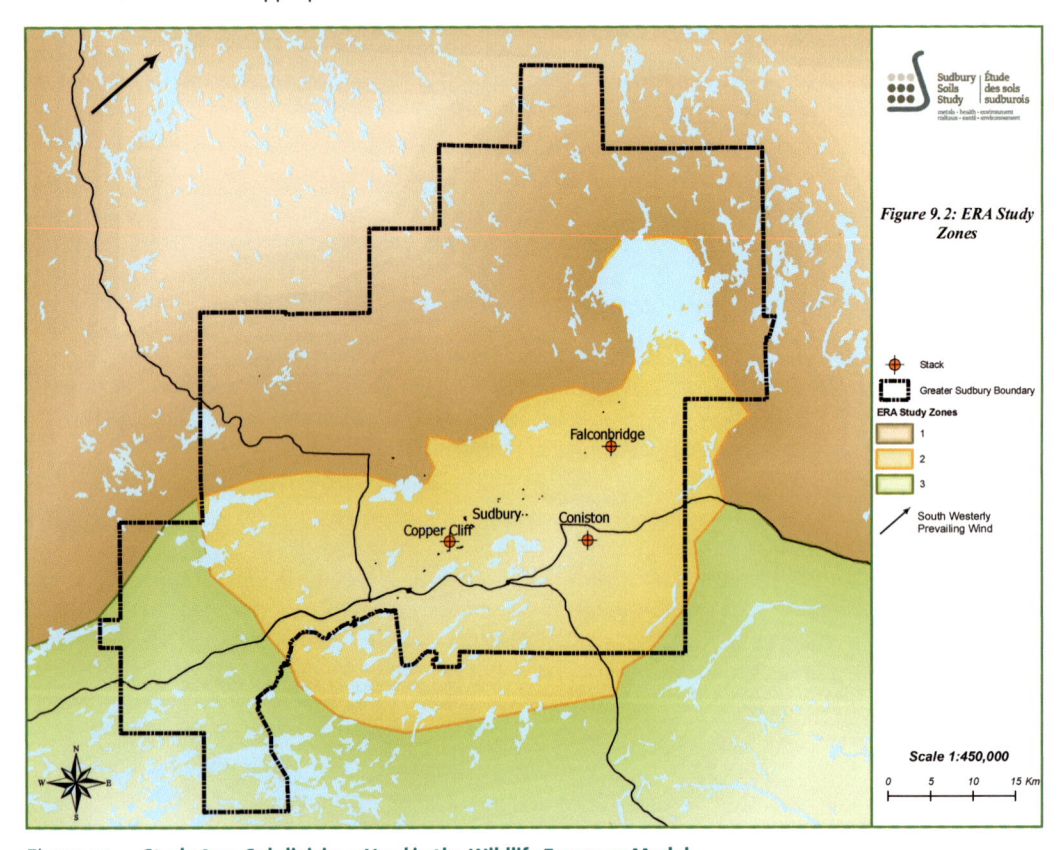

Figure 9.2 **Study Area Subdivisions Used in the Wildlife Exposure Model**

9.6 Identification of COC for the Terrestrial ERA

The primary source of COC to the terrestrial environment in this assessment was aerial deposition of particulate-associated metals and metalloids from smelter emissions. In addition, fugitive dust from tailings areas, old roasting beds and ongoing operations have contributed to metal levels in soils. In this step, the chemicals present in the study area that pose the greatest potential for exposure and risk to the terrestrial ecosystem were identified. It is common practice in ERAs to limit the number of chemicals evaluated to those chemicals that represent the greatest concern in the area under consideration. This is done because it is impractical in terms of time and cost to conduct a risk assessment for every chemical that has been found to occur in a particular area. In addition, the concentrations of many chemicals associated with a particular site may be similar to chemical concentrations found naturally in the area rather than the result of current or former activities on a site.

A complete description of the COC selection process for the Sudbury Soils Study including the ERA was provided in Chapter 3. An overview of the COC selection process is provided in this section. The selection of COC for the risk assessment was based on metal concentrations in Sudbury soils measured during the 2001 soil survey. Approximately 8400 soil samples were collected from the study area and analyzed for 20 inorganic parameters:

Aluminum (Al)	**Antimony** (Sb)	**Arsenic** (As)	**Barium** (Ba)	**Beryllium** (Be)
Calcium (Ca)	**Cadmium** (Cd)	**Cobalt** (Co)	**Copper** (Cu)	**Chromium** (Cr)
Iron (Fe)	**Magnesium** (Mg)	**Manganese** (Mn)	**Molybdenum** (Mo)	**Nickel** (Ni)
Lead (Pb)	**Selenium** (Se)	**Strontium** (Sr)	**Vanadium** (V)	**Zinc** (Zn)

The TC (Technical Committee) required that three criteria be met for the selection of COC:

- Parameter must be above the soil quality criteria published in the Ministry's *Guideline for Use at Contaminated Sites in Ontario* (the 'Guideline'; MOEE 1997).
- Parameter must be present across the study area.
- Parameter must scientifically show origin from the companies' operations.

The first three criteria for COC screening were met by As, Co, Cu, Ni, Pb and Se. The soil quality criteria in the *Guideline for Use at Contaminated Sites in Ontario* (MOEE 1997) were only applied to soils with a pH range of 5.0 to 9.0, the pH range for which the criteria were developed. According to Ontario risk assessment guidance, for soils with a pH outside this range, background soil concentrations were used as alternative screening values (MOEE 1997). Soil pH in northern Ontario is known to be naturally low (pH < 5.0) with soil pH further reduced in the study area due to the smelter emissions. There were 229 soil samples from the urban soil survey for which pH results existed. Of these, 224 (98%) had a pH >5.0. This higher pH range was attributed to homeowners amending residential soils with lime, organic matter and fertilizers. Therefore, low soil pH relative to the MOE generic criteria was considered to be an issue primarily in samples from rural, unamended sites, and not urban residential properties. Of 365 surface (0 to 5 cm) samples analyzed from the regional soils survey, soil pH was <5.0 for 347 (95%) of the samples. Therefore, the chemical database was re-screened using background concentrations as the screening criteria, for those samples with soil pH <5.0. The results of this secondary screening exercise were presented in Chapter 3, Table 3.4. Cadmium exceeded background concentrations in 29 soil samples with pH <5.0, and the 95th percentile concentration (1.7 mg/kg) was greater than the background (Table F) value (1.0 mg/kg). Cadmium can also accumulate through the food web including uptake in invertebrates (Hsu et al. 2006), plants and birds (Larison 2000) and mammals (Parker and Hamr 2001). Based on the available information, Cd was included as a COC for the ERA.

9.7 Identification of Valued Ecosystem Components (VEC)

A VEC is a species, population or community that is important to people, has economic and/or social value, is ecologically significant and can be evaluated in a risk assessment. The process of identification and selection of VECs began early in the Sudbury study with the general selection process illustrated in Figure 9.3. Background information was collected on species of plants, mammals, birds and reptiles that were reported to occur in the study area. Information was collected from existing literature, and discussions with key agencies (e.g. Ontario Ministry of Natural Resources (MNR)), researchers at Laurentian University and special interest groups (e.g. Sudbury Ornithological Society). A series of public workshops was held to obtain input from members of the public, as well as special interest groups, to identify species of special interest or concern. This step was valuable in identifying candidate VECs. The VECs for Objective #1 are described in Section 9.7.1 and those for Objectives #2 and #3 are described in Section 9.7.2.

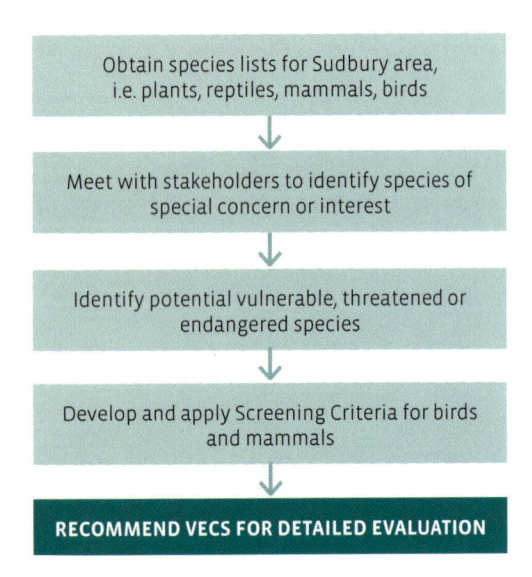

```
┌─────────────────────────────────┐
│ Obtain species lists for Sudbury area, │
│ i.e. plants, reptiles, mammals, birds  │
└─────────────────────────────────┘
              ↓
┌─────────────────────────────────┐
│ Meet with stakeholders to identify │
│ species of special concern or interest │
└─────────────────────────────────┘
              ↓
┌─────────────────────────────────┐
│ Identify potential vulnerable,  │
│ threatened or endangered species │
└─────────────────────────────────┘
              ↓
┌─────────────────────────────────┐
│ Develop and apply Screening     │
│ Criteria for birds and mammals  │
└─────────────────────────────────┘
              ↓
┌─────────────────────────────────┐
│ RECOMMEND VECS FOR DETAILED EVALUATION │
└─────────────────────────────────┘
```

Figure 9.3   Selection Process for Terrestrial Valued Ecosystem Components

9.7.1 VECs for Objective #1: Evaluate the Extent to Which COC are Preventing the Recovery of Regionally Representative, Self-sustaining Terrestrial Plant Communities

Terrestrial plants are critical components of the ecosystem because of the functions they fulfill, including: decreasing soil erosion; carbon cycling; provision of human enjoyment (hiking, berry-picking, camping, etc.); and provision of habitat to wildlife (including food, cover from predators, nesting sites, etc.). These functions are not fulfilled by single species or types of plants in an area, but by the community as a whole (i.e. trees may provide nesting sites, grasses may provide cover from predators, and understorey plants may provide seeds and berries that are eaten by wildlife). Species lists of terrestrial plants in the study area were assembled as part of the VEC selection process, but it was decided to select the forest vegetation community as a VEC rather than a single species for this ERA objective.

The advantage of selecting the vegetation community versus a single species as the VEC was highlighted in a study of impacts of one of the Sudbury smelters on local vegetation (Anand and Desrochers 2004). The authors compared a single species metric (in this case, amount of cover provided by either paper birch (*Betula papyrifera*), wavy hair grass (*Deschampsia flexuosa*) or blueberry (*Vaccinium angustifolium*)) to a community metric (in this case, Shannon diversity index, which uses the number of species and species dominance in a measure of diversity). The two approaches were evaluated at sites located at various distances from the Coniston smelter in Sudbury. The study showed that only the community metric (diversity) showed a consistent relationship (increase) with distance from the smelter (as a measure of exposure). The single species cover metric showed no relationship with distance from the smelter. The authors concluded that, although the study of single species provides some information, it may be misleading to use these data for an assessment of a larger process, such as ecosystem 'recovery'. Therefore, because no one plant species is indicative of a community, and the concern in Sudbury was for 'recovery' or protection of more than a single species, the Sudbury ERA evaluated terrestrial plants as a community (or several communities) and, therefore, it was not necessary to identify individual plant species as VECs. The blueberry was one exception and is discussed in more detail below.

Four general plant 'communities' were described for the Sudbury area (Amiro and Courtin 1981). The first is 'barren' land, completely devoid of trees, that includes the areas immediately surrounding the smelters. In the 1970s, the barren area was estimated to cover 20,000 ha. The second area is comprised of two communities, the 'transition birch' and 'transition maple' (this area is also referred to as semi-barren or the transition zone). During the 1970s, the extent of the semi-barrens was estimated to be 80,000 ha. Beyond this transition zone, the Sudbury landscape returns to what is considered the natural plant community of the region, which includes red oak (*Quercus rubra*), jack pine (*Pinus banksiana*), birch, maple (*Acer* sp.), sugar maple (*A. saccharum*), largetooth aspen (*Populus grandidentata*) and poplars (*Populus* sp.) (Courtin 1995). In 1993 (Sinclair 1996), the following plant communities were found: birch/pine community, trembling Aspen community, big-toothed aspen community, red oak community, white birch community, and an association of sites termed the

Treeless community (barrens). These general plant communities served as the starting point for addressing risks to and recovery of vegetation. Plant communities in the Sudbury area were assessed in detail for the terrestrial ERA as described in Chapter 10 of this volume.

Blueberries are widely distributed throughout the study area and have special economic and social value in Sudbury, which makes them an important VEC. Blueberries prefer well-drained, acidic soils (pH 4.0 to 5.5) and full exposure to sunlight (NB DAFA 2003), which are two conditions present in the semi-barren areas. Therefore, it is possible that some remedial activities (liming and fertilizing of soil) to promote shrub and tree growth could adversely affect blueberries in some areas. Blueberries were not specifically evaluated in the ERA but were included as a VEC for the ERA in recognition of their social-economic importance and to help ensure these considerations are not overlooked during future risk management.

Terrestrial plants within private property in residential areas, including grass, flowers, ornamental trees and vegetables, were not included as part of the ERA. Residents routinely amend their soil, as shown by the less acidic soil pH measurements for residential areas and gardens in Sudbury. In addition, there have been few complaints from residents with regard to difficulties with growing ornamental plants including grass, trees and flowers, with the exception of some past complaints directly related to SO_2 fumigation. Therefore, due to the lack of widespread concern about the growth or yield of decorative plants on private residential property, these groups of plants were not recommended as VECs for the ERA. However, metal uptake into vegetables was evaluated as part of the human health risk assessment (Chapter 6).

Soil communities contain diverse populations of invertebrates which include decomposers, predators, microbial symbionts, pathogens, and parasites. This group of organisms was not considered to have any social importance by the average citizen but they perform very important ecological functions. These organisms facilitate soil formation, organic matter breakdown, nutrient cycling, and thus contribute to overall soil fertility. They also provide food to wildlife. The quality of the soil, its fertility, and structure are essential for the maintenance of the biodiversity and dynamics of terrestrial ecosystems (Wentsel et al. 2003). It has been suggested previously that plants and soil-dwelling organisms may be more at risk to atmospheric emissions from smelters than wildlife (Welbourn 1996). Therefore, soil invertebrates were identified as VECs under this objective. The toxicity of Sudbury soil to earthworms was evaluated directly through laboratory testing. In addition, qualitative observations on the relative abundance of earthworms in the study area were made during the field investigations. The results of those studies are provided in Chapter 10. In summary, the final VECs recommended for Objective #1 were the forest ecosystem, blueberries and soil invertebrates.

9.7.2 VECs for Objective #2: Evaluate Risks to Terrestrial Wildlife Populations and Communities Due to COC and Objective #3: Evaluate Risks to Individuals of Threatened or Endangered Terrestrial Species Due to COC

Lists of avian and mammalian species present in the Sudbury area, as well as species identified as being of special interest to Sudbury residents were compiled. Consideration was also given to species that may have been present in the Sudbury area in the past, but which are no longer present due to their sensitivity to current conditions. The Ontario MNR and several Laurentian University biology professors and researchers were contacted to determine whether any such wildlife species could be identified. Threatened and endangered species and other species of concern were identified by reviewing the Federal Committee on the Status of Endangered Wildlife in Canada (COSEWIC) and Natural Heritage Information Centre (NHIC) databases and by contacting the local MNR (MNR 2003a,b, pers. comm.) office in Sudbury. Only vulnerable, threatened or endangered (VTE) species present in the study area were identified as VECs, because there could have been numerous reasons, unrelated to the smelters, for species to have become extinct or locally extirpated.

No species were reported as threatened or endangered within the City of Greater Sudbury. One species, the Wood Turtle, was listed as vulnerable for the general area. The VTE species outside, but within a 100 km radius of, the City of Greater Sudbury included: Bald Eagle (endangered), Peregrine Falcon (endangered), Red-shouldered Hawk (vulnerable) and Eastern Massasauga Rattlesnake (threatened). Selection criteria were developed and applied to reduce the candidate lists of mammalian and avian species to a manageable number, while ensuring adequate representation from all relevant trophic levels and feeding guilds. The approach to developing the VEC selection criteria was a Sudbury-specific approach modified from the approach outlined by Suter II (1993), Becker et al. (1998) and U.S. EPA (1998). Criteria used to select avian and mammalian VECs for this study included:

- VTE species (sensitive species)
- resident or reproduces in the study area (thereby exposed to COC during a sensitive life stage)
- ecological significance (a VEC which is 'ecologically significant' is one that helps sustain the natural structure, function and biodiversity of an ecosystem or its components [U.S. EPA 1998]; non-native or pest species were considered to have low ecological significance)
- identified by a stakeholder as being important
- has socio-economic importance (e.g. moose) and therefore a direct connection to the human health risk assessment (e.g. is hunted and consumed by Sudbury residents)
- had potential for high exposure to the COC (this reflects a close association with soil for terrestrial species [e.g. consumes soil invertebrates rather than flying insects], small home range, high food intake relative to body weight)
- information existed on populations in the area
- toxicity data for the COC were available for closely-related species (e.g. laboratory rat and mouse data were considered appropriate for wild small mammals such as shrews, voles, mice, rats, squirrels, moles), and
- represented a major feeding guild (and trophic level).

These ranking criteria were evaluated to determine which species would be selected as the representative VEC for a particular feeding guild. The end result of the ranking was a list of species that are commonly evaluated in an ERA and that were representative of the species and foodwebs in Sudbury. In addition, two species were added at the request of the TC (White-tailed Deer, Red Fox). A summary of the rationale for the recommendation of each of the terrestrial avian and mammalian VECs is provided in Table 9.1. The final list of wildlife VECs for Objectives 2 and 3 of the Sudbury ERA included:

- Mammals: Northern Short-tailed Shrew, Meadow Vole, Moose, White-tailed Deer, Red Fox and Beaver
- Birds: American Robin, Ruffed Grouse, and Peregrine Falcon.

Table 9.1 **Summary of Rationale for Recommendation of Terrestrial VECs**

VEC	Feeding guild and/or trophic level	Rationale for recommendation
Terrestrial plant communities	Primary producers	• plants provide various functions, including: decreasing soil erosion, carbon cycling; provision of human enjoyment (hiking, camping); provision of habitat to wildlife • the plant community is more relevant than selecting a particular species • have been studied by Laurentian University and others • toxicity data are available for individual species • site-specific data collected in this study were used to supplement data in the literature
Blueberries	Primary producers	• plentiful in the Sudbury area, have economic and social value • identified by Stakeholders as requiring special attention • have been studied by Laurentian University • habitat may be affected by restoration activities
Soil invertebrate communities	Primary consumers, decomposers	• invertebrates provide various important ecosystem functions, including: soil aeration and breakdown of organic matter; provision of food to wildlife • have been studied by Laurentian University • toxicity data in the literature were available • site-specific data collected in this study were used to supplement data in the literature

Ecological Risk Assessment Problem Formulation

VEC	Feeding guild and/or trophic level	Rationale for recommendation
Northern short-tailed shrew (*Blarina brevicauda*)	Invertebrate-eating small mammal populations; secondary consumer	• close contact with soil, small home range, and high food intake rate relative to body weight result in high exposure potential to COC • reported as plentiful in the Sudbury area (Dobbyn and Wilson 1994) • small mammals with small home range identified by Stakeholders as requiring special attention • have been studied by Laurentian University • toxicity data are abundant for small mammals (primarily rats and mice) • shrews breed in the area and serve as food to wildlife • shrews can represent other small mammals that consume soil and litter invertebrates (e.g. worms, insects, larvae) as well as omnivores such as mice
Meadow vole (*Microtus pennsylvanicus*)	Herbivorous small mammal populations; primary consumer	• close contact with soil, small home range, and high food intake rate relative to body weight result in high exposure potential to COC • reported as common in the Sudbury area (Dobbyn and Wilson 1994) • small mammals with small range identified by Stakeholders as requiring special attention • have been studied by Laurentian University • toxicity data are abundant for small mammals (primarily rats and mice) which are closely related to this species • voles breed in the area and serve ecological functions (as food to wildlife) • voles can represent other small mammals which consume plants (grass, seeds, roots, fungi, etc.) as well as omnivores such as mice
Moose (*Alces alces*) and white-tailed deer (*Odocoileus virginianus*)	Herbivorous large mammal populations; primary consumers	• elk, deer and moose are all ruminants, common in the area, and identified by Stakeholders as important. Moose and deer were recommended as VECs because both are hunted (social and economic value, link to human health) • moose forage in wetland areas on plants including aquatic species, whereas deer forage in farmland and suburban areas on various terrestrial plants • there are few toxicity data for wild ruminants, and therefore, the ERA would have to rely on toxicity data from studies with cattle • however, due to Stakeholder concern and the link to human health, these species were recommended as VECs • deer were added as a VEC specifically at the request of the TC • elk were not added as VECs because they would be exposed to similar levels of COC as deer • a study by Parker and Hamr (2001) concluded that the elk herd was not severely impacted or threatened by metal contamination • the elk population was re-introduced and occurs only in one portion of the study area
Red fox (*Vulpes vulpes*)	Omnivorous mammal populations; primary consumer through top predator	• fox are omnivorous, feeding on whatever is available, including fruit and other vegetation in summer, birds and mammals in winter, invertebrates such as grasshoppers and beetles, and even crayfish • fox were identified by Stakeholders as requiring special attention • sightings are recorded for the Sudbury area, although no population studies have been conducted • the ERA would rely on toxicity data from studies with dogs or rodents • fox were specifically requested by the TC for inclusion in the ERA, although they have an opportunistic feeding strategy; the request was that they be considered as a representative predator
American beaver (*Castor canadensis*)	Herbivorous mammal populations with link to aquatic environment; primary consumer	• live and breed in lakes throughout the study area including urban lakes near smelters, which may result in high exposure to COC in the aquatic environment • reported as plentiful in the Sudbury area (Dobbyn and Wilson 1994) • were identified by Stakeholders as requiring special attention • have been studied by Laurentian University • there are no toxicity data for this species; data for small rodents (primarily rats and mice) were used

VEC	Feeding guild and/or trophic level	Rationale for recommendation
American robin (*Turdus migratorius*)	Invertebrate-eating bird populations; secondary consumer	• is common and breeds in Sudbury • identified by Stakeholders as requiring special attention • feeds on worms and other soil and litter invertebrates which are in close contact with soils and, therefore, has potential for high exposure to COC in soil and diet • although toxicity data are not available for this species, robins are routinely evaluated in ERAs by using data from other bird species • robins may be representative of other invertebrate-eating or omnivorous birds, such as American woodcock
Ruffed grouse (*Bonasa umbellus*)	Herbivorous bird populations; primary consumer	• non-migratory, therefore occur year-round and breed within the study area • identified by Stakeholders as requiring special attention • hunted by local residents for food • have been studied by Laurentian University • ground-dwelling bird that feeds on seeds, buds, berries, insects and, therefore, has high potential for exposure to COC in soil and diet • toxicity data were available for closely-related species such as chickens and quail • ruffed grouse may be representative of other herbivorous birds
Peregrine falcon (*Falco peregrinus*)	Endangered species; carnivorous bird populations; top predator	• re-introduced to the Sudbury area • breed in the Sudbury area, and there is some local information on these birds • identified by Stakeholders as requiring special attention • this species is endangered, therefore, has special status • falcons can represent other carnivorous birds such as owls, hawks and harriers

Many of the species identified above have relatively small home ranges and could be exposed to different concentrations of COC within different study area zones, including areas within or close to the urban centers. However, wide-ranging species (moose, deer) would forage minimally within the urban areas (Coniston, Falconbridge, Copper Cliff and Sudbury Core). Therefore, it was considered appropriate to assess individual VECs for each study zone to assess different exposure scenarios, as summarized in Table 9.2.

Table 9.2 **Valued Ecosystem Components Assessed in Various Regions of the Study Area**

VEC	Falconbridge	Copper Cliff	Coniston	Sudbury Core	Zone 1	Zone 2	Zone 3
Northern short-tailed shrew	✓	✓	✓	✓	✓	✓	✓
Meadow vole	✓	✓	✓	✓	✓	✓	✓
Moose					✓	✓	✓
White-tailed deer					✓	✓	✓
Red fox	✓	✓	✓	✓	✓	✓	✓
Beaver				✓	✓	✓	✓
American robin	✓	✓	✓	✓	✓	✓	✓
Ruffed grouse		✓	✓	✓	✓	✓	✓
Peregrine falcon	✓	✓	✓	✓	✓	✓	✓

9.7.3 Species Not Recommended as VECs

Various stakeholder groups or members of the public nominated a range of species or animal groups to be considered as VECs, which were ultimately not chosen by the risk assessment team. There were two main reasons for this. First, some species recommended by a stakeholder, such as black bear, were identified because they were plentiful and actually considered a nuisance. Second, the VECs selected for evaluation in the ERA could be considered surrogates for many species because they have a similar diet. For example, stakeholders identified the bird Blue Jay as a candidate VEC. It was not included as a VEC by the study team since other birds were selected to represent herbivorous and insectivorous birds. Other species or groups not chosen as VECs included soil microbes, reptiles, agricultural species and pets. For the purpose of transparency, the rationale for excluding these species groups was documented and is outlined below.

Soil Microbial Community

The soil microbial community *per se* was not evaluated as a VEC. However, the functions it provides (i.e. litter decomposition) was considered as an important measure that can influence plant community success. Various soil parameters were included as measures for the plant community VEC (e.g. soil nutrient levels, litter decomposition rates) rather than assessing the soil microbial community as an independent VEC. Litter decomposition by the soil microbial community was assessed by undertaking field trials using litter bags (see Chapter 10).

Consideration of the function of the soil microbial community rather than the diversity of the community is supported by work by the U.S. EPA (2003) and a study in Sudbury (Anand et al. 2003). The Anand et al. (2003) study evaluated soil (excluding litter) microorganism diversity and dynamics and found few correlations between the microbial community and soil metal levels along a transect south from Coniston. Although Pb concentrations were significantly correlated to microbial diversity, water-soluble manganese to microbial assemblage, and water-soluble zinc to microbial diversity, the authors concluded that the impact of metals on soil microbes was 'not remarkable'. In addition, although changes in soil microbial community and function can be measured (Maxwell 1995), the importance of these changes to the plant community is not clear, at least in a quantitative way. Soil fungi play important roles in soil function, particularly in decomposition and nutrient dynamics. Mycorrhizal fungi play a particularly important role as they are symbiotic with higher plants, considerably enhancing nutrient uptake into plant tissue as well as playing other roles, depending on type. Metals have been shown to impact both the number of fungal fruiting bodies and the structure of the fungal community (Gunn 1995). However, simple enumeration of the extent of mycorrhizal infection is not necessarily a valid measure of impact. In summary, soil mycorrhizal fungi were not included as a VEC for the risk assessment.

Reptiles

Reptiles were not selected as VECs for a number of reasons. First, the lack of metal toxicity data for reptiles prevented their direct evaluation via exposure and risk modeling. Toxicity tests could not be conducted because there are no standardized tests for reptiles native to Ontario (Meyers-Schone 2000). In many cases, an ERA will focus on a reptilian species only when there is a threatened or endangered reptile at a site (Meyers-Schone 2000). The only vulnerable, threatened or endangered reptile found within the City of Greater Sudbury was the Wood Turtle (vulnerable), which is found in the Vermillion River system. Sudbury is on the edge of the Wood Turtle's Canadian range and one of the main threats to its survival is habitat degradation (Litzgus 1996). Although one of the most terrestrial members of this turtle family, it frequents streams, creeks and rivers, preferring clear streams with moderate currents (Litzgus 1996). Because of these factors, and the fact that the Vermillion River, where this species is found, is primarily upwind of emissions from the smelters, the Wood Turtle was not selected as a VEC.

The Eastern Massasauga Rattlesnake (threatened) has been sighted within a 100 km radius of Sudbury, southwest of the City of Greater Sudbury, but it is not common. This snake is carnivorous, preying primarily on small rodents and birds and, therefore, is expected to receive a relatively low exposure to metals in soils. Therefore, the Eastern Massasauga Rattlesnake was not selected as a VEC.

Agricultural Species

Agricultural species, including a variety of livestock and crops, are present within the Sudbury area. The main farming area is located in the northwestern portion of the City of Greater Sudbury which is predominantly upwind of the smelters. The local office of the Ontario Ministry of Agriculture and Food (OMAF), and the local offices of the MOE, Vale and Xstrata Nickel were contacted for information pertaining to agricultural species and possible concerns related to smelter emissions. They informed the SARA Group that no one has come forward with an official complaint with regard to impacts to agricultural crops or livestock from smelter emissions within the past decade, and previous concerns were related to SO_2. Therefore, due to the low exposure potential, lack of evidence of an impact and relationship with human health rather than ecological integrity, agricultural species were not selected as VECs for the ERA.

Pets

Several residents were concerned for the health of their pets, such as cats and dogs, due to smelter emissions. Similar to livestock, pets were expected to receive lower exposure to emissions than wildlife species, because they typically live independent of the natural environment in man-made structures and consume food that is not obtained from the local ecosystem. Pets are not routinely assessed by ERA or HHRA and no standard methods exist to assess risks to pets. The ERA did assess risks to some wildlife that are similar to some pets (fox are related to dogs). Therefore, pets were not selected as separate VECs for the ERA.

9.8 Assessment Endpoints and Measures

An assessment endpoint is defined as an explicit expression of the environmental values to be protected (Suter II 1989; U.S. EPA 1998). The endpoint must be an important property of the system that can be measured or estimated, and is not a policy goal (Suter II et al. 2000). Two components are needed to define assessment endpoint, first an *entity*, and second an *attribute* or *characteristic* (Sergeant 2002). The entity generally is a species (i.e. American robin) or community (i.e. terrestrial plants). The attribute or characteristic of that entity is important to protect and which is potentially at risk (i.e. growth, reproduction, diversity). Assessment endpoints are neutral (U.S. EPA 1998) and, therefore, they should not contain words such as 'protect', 'maintain', or 'restore' or indicate a direction for change, such as 'loss' or 'increase'. The assessment endpoint entity helps to define the scope and character of the ERA and its attribute determines what to measure (Sergeant 2002).

The assessment endpoint chosen for the plant communities was presence of a self-sustaining forest ecosystem. For blueberries, the assessment endpoints were survival, growth and yield of blueberries. For soil invertebrate communities, the assessment endpoints were survival and reproduction of soil and litter biota, including earthworms. The assessment endpoints for threatened and endangered wildlife were slightly different from those for other wildlife. The assessment endpoints for threatened and endangered wildlife (peregrine falcon) were survival, growth and reproduction of *individual* peregrine falcons in the City of Greater Sudbury and surrounding area. The assessment endpoints for the remaining terrestrial wildlife VECs were survival, growth and reproduction to allow *population persistence* in the City of Greater Sudbury and surrounding area. That is, the goal for threatened and endangered species was protection of individuals, while the goal for other wildlife was protection of populations. Wildlife could be affected directly (as a result of direct toxicity of the COC on survival, growth or reproduction), or indirectly (as a result of decreased habitat suitability, or COC effects on food resources). Both direct and indirect toxicity to wildlife are addressed in Chapter 11.

Assessment endpoints must be measurable directly or by using surrogate measures (U.S. EPA 1998). For example, presence of loons may be assessed directly because loon surveys have been undertaken for approximately 30 years in the Sudbury area. However, few population surveys have been completed/published for small mammals such as shrews and voles in Sudbury (Robitaille and Linley 2006). Therefore, if the assessment endpoints are not directly measurable, then other 'measures' may be used to evaluate the risk to the assessment endpoints. There are three categories of measures (U.S. EPA 1998):

1. Measure of Exposure: a measure of chemical presence and movement in the environment and its contact with the VEC. Examples: concentrations of COC in soil, and concentrations of COC in wildlife dietary items.
2. Measure of Effect: a measure that describes a change in a characteristic of a VEC in response to a chemical to which it is exposed. Examples: number of shrews in an area (density); and laboratory mammalian toxicity test data from the literature.
3. Measure of Ecosystem and VEC Characteristics: measures that influence the behavior and location of VECs, the distribution of a chemical, and life-history characteristics of the VECs that may affect exposure or response to the chemical. Examples: home range; and habitat requirements and preferences for the VEC.

Different measures were used to assess each VEC, depending upon the availability of data and the level of uncertainty in these data. Several measures were used to evaluate the lines of evidence for the plant and invertebrate communities, and these are described in Chapter 10. Measures for wildlife VECs are described in Chapter 11. A summary of the VECs, assessment endpoints and measures of exposure and effect is provided in Table 9.3.

Table 9.3 **Summary of VEC Assessment Endpoints and Measures**

VEC	Assessment endpoint	Measures of exposure and effect
Terrestrial plant communities	Presence of a self-sustaining forest ecosystem	• Concentration of COC in soil • Soil characterization parameters (i.e. organic matter, fertility) • Site-specific laboratory toxicity test results (northern wheatgrass, red clover, goldenrod and white spruce) • Site-specific plant community survey data (various measures of site biodiversity, ecological integrity, long-term site productivity, and soil and water conservation) • Site-specific measurement of decomposition and nutrient cycling
Blueberries	Survival, growth and yield	• Concentration of COC in soil
Soil invertebrate communities	Survival and reproduction	• Concentration of COC in soil • Site-specific laboratory toxicity test results (earthworm, rates of litter decomposition)
Avian and mammalian wildlife	Survival, growth and reproduction to allow population persistence	• Concentration of COC in soil • Concentration of COC in diet • Wildlife toxicity data from the literature • Population information from the Sudbury area
Peregrine falcon (threatened and endangered wildlife)	Survival, growth and reproduction of individual organisms	• Concentration of COC in soil • Concentration of COC in diet • Wildlife toxicity data from the literature • Information from the Sudbury area on individual peregrine falcons

9.9 Conceptual Model

A conceptual model is a written description and a visual representation of the relationships between VECs and the COC to which they may be exposed. A major focus of the conceptual model is the development of a series of working hypotheses for determining exposure pathways (U.S. EPA 1998; CCME 1996). Conceptual models can serve three purposes: 1) clarification of assumptions concerning the situation being assessed, 2) as a communication tool for conveying those assumptions, and 3) providing a basis for organization and completion of the risk assessment (Suter II 1999). Conceptual models are powerful learning and communication tools when initiating an ERA, and they are easily modified as the ERA progresses and data gaps are filled. Conceptual models include identification and discussion of contaminated media, chemical migration pathways, exposure pathways and VECs. Equally important, conceptual models may exclude some exposure pathways, receptors or other aspects because they are considered unimportant, unlikely, or outside the scope of the assessment (Suter II 1993). The conceptual model diagram for plants, terrestrial invertebrates and wildlife for the Sudbury ERA is provided in Figure 9.4.

Wildlife VECs may be exposed to chemicals via several potential exposure pathways, such as ingestion, inhalation and dermal contact. Chemicals may be ingested by consumption of food and water, and by incidental ingestion of soil or sediment. Chemicals may be inhaled if they are volatile, or if they are components of fine particulate matter. Dermal exposure occurs when chemicals are absorbed through the skin from soil, water or sediment. However, dermal exposure was assumed to be negligible for birds and mammals. Feathers on birds and fur on mammals reduce dermal exposure by limiting the contact of skin with chemicals in environmental media (Sample et al. 1997). In addition, metals do not cross the dermis, and are unlikely to be absorbed through skin (Watters et al. 1980). Inhalation exposure is also assumed to be negligible for birds and mammals, because the COC (metals and metalloids) generally have low volatility, and resuspension of fine soil particles is minimized when there is adequate vegetative cover. Relevant inhalation toxicity data also are lacking for wildlife, which makes it difficult to assess ecological risks from this exposure route, especially for birds, but also for mammals where the toxicity endpoint often is not related to the ecological assessment endpoints of concern (lung irritation as opposed to reproduction). Therefore, only exposure via ingestion was evaluated in the ERA. This was consistent with Environment Canada (1994) guidance which acknowledged ingestion as the major pathway of concern for wildlife at contaminated sites.

Ecological Risk Assessment Problem Formulation

Figure 9.4 **Conceptual Model for the Sudbury Terrestrial ERA**

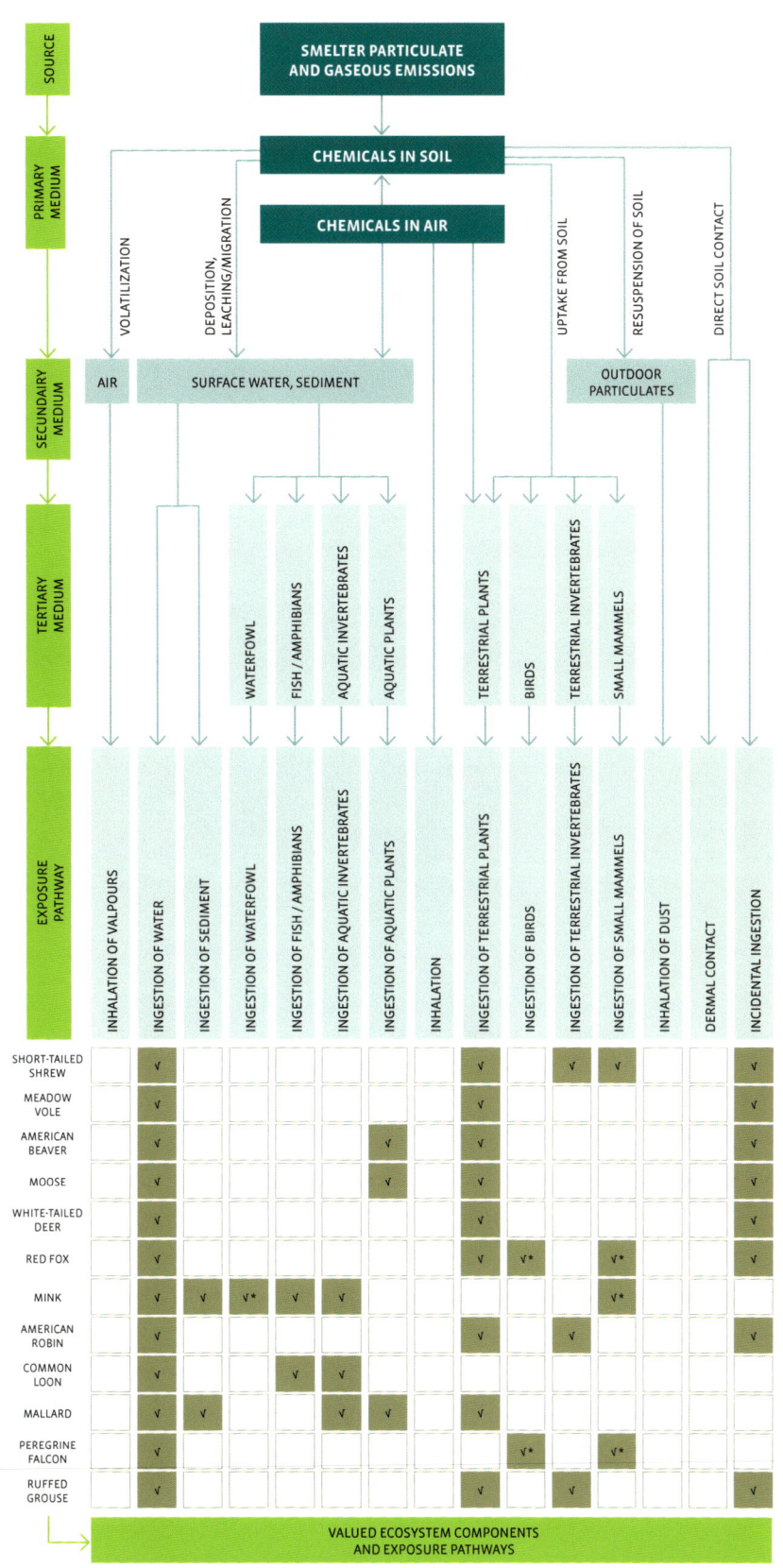

* Although red fox, mink and peregrine falcons consume birds (waterfowl, in the case of mink) in addition to small mammals, the model conservatively assumed that this part of the diet consisted of small mammals only, due to the availability of models to predict body burden in small mammals. Check marks indicate pathway was assessed in the era; blank cells indicate pathway was not assessed. Mink, loon and mallard are assessed as part of the aquatic problem formulation.

9.10 Analysis Plan and Identification of Data Gaps

It was recognized early in the ERA that different study approaches would be required to address ERA Objective #1 compared with ERA Objectives #2 and #3. This section provides a brief introduction to the analysis plans to fill data gaps for the ERA Objectives, building on the information presented on the assessment and measurement endpoints as summarized above in Table 9.3, development of the conceptual model (Figure 9.4), and the review of available ecological information (Section 9.4).

Objective #1: Evaluate the Extent to Which COC are Preventing the Recovery of Regionally Representative, Self-sustaining Terrestrial Plant Communities

The purpose of this objective was to determine what role the COC from the smelters had in inhibiting recovery of vegetation in the study area, and to identify whether other causal factors were also involved. This information can then be used to help guide future reclamation activities. The study team determined that a combined field and laboratory program was necessary to collect the data required to identify those environmental factors or variables that may be inhibiting ecosystem recovery. The study approach included collection of data from four distinct 'lines of evidence' (LOE). These data would then be evaluated using a weight-of-evidence (WOE) approach. The LOE were:

- physical and chemical soil characterization
- toxicity testing with terrestrial species in the laboratory
- a detailed ecological plant community assessment, and
- an assessment of decomposition using *in situ* litter bags.

The approach used to evaluate Objective #1 was a combination of a retrospective risk assessment and a predictive risk assessment. The ERA was retrospective in that vegetation impacts had already visibly occurred within the study area. The ERA not only documented these impacts, but attempted to determine the causal factors contributing to the impacts. The ERA was also predictive, in that there had been significant reductions in the source of the chemical stressors (atmospheric emissions) over the past three decades, so other questions addressed by the ERA were whether the existing metal concentrations (the COC) in Sudbury soils were preventing vegetation growth and recovery, and the extent to which soil amendments to increase pH would be beneficial to plant communities.

Limited information existed for each of the LOE identified above from previous scientific studies. The scope of these other studies was either restricted in their geographic coverage or metals included in the studies, or both. The 2001 soil survey provided an excellent database of metal concentrations in Sudbury soils over a broad study area (Chapters 3 and 4) but the parameters measured did not include nutrients, organic carbon and a wide range of other parameters that are important indicators of the potential of soil to support plant growth. Therefore, critical data gaps existed for all four LOE that had to be addressed to evaluate this Objective. A substantial amount of effort and resources were directed at filling these data gaps, and those studies are described in detail in Chapter 10.

Objective #2: Evaluate Risks to Terrestrial Wildlife Populations and Communities Due to COC; and Objective #3: Evaluate Risks to Individuals of Threatened or Endangered Terrestrial Species Due to COC

The approach used to characterize risk to terrestrial wildlife populations, and individuals of threatened or endangered species was essentially the same, so these two objectives were considered together. This evaluation followed a more traditional ERA approach of completing a detailed exposure assessment and hazard assessment for each of the VEC and COC combinations within the study area. Based on the results of the wildlife modeling, and the review of the available information on wildlife and habitat in the Sudbury area, it was determined that wildlife field surveys were not required as part of the risk assessment. Risks were characterized using a probabilistic modeling approach. One data gap that was identified included the measurement of COC levels in dietary items for terrestrial wildlife. In particular, data were lacking for plants and terrestrial invertebrates that are consumed by a wide range of small mammals and birds. To fill this data gap, samples were collected from a number of locations representing a range of soil metal concentrations. The study methodology and results of the sample collections are described in Chapter 11. Those data were used in modeling exposure of different VECs to the COC.

9.11 Acknowledgements

The authors thank the many people who contributed their knowledge of the ecosystems in and around Sudbury, allowing us to complete a thorough problem formulation. This includes many researchers at Laurentian University (Dr Glenn Parker, Dr Peter Beckett, Dr Graeme Spiers, Dr John Gunn, Mr Bill Keller, Mr George Morgan, Dr J.F. Robitaille, Mr Chris Bloome, Dr David Lesbarrères, Dr Jacqueline Litzgus), the Ontario Ministry of Natural Resources, Environment Canada, Sudbury & District Health Unit, City of Greater Sudbury, Sudbury Naturalists, Science North, Ducks Unlimited, Ruffed Grouse Society of Canada, the Technical Advisory Committee, and the people of the City of Greater Sudbury who participated in our Have Your Say workshops and otherwise spoke to us about the wildlife of the area.

9.12 References

Amiro BD, Courtin GM (1981) Patterns of vegetation in the vicinity of an industrially disturbed ecosystem, Sudbury, Ontario (Canada). Can J Bot 59:1623–1639

Anand M, Desrochers RE (2004) Quantification of restoration success using complex systems concepts and models. Restor Ecol 12(1):117–123

Anand M, Okonski KM, Levin A, McCreath D (2003) Characterising biocomplexity and soil microbial dynamics along a smelter-damaged landscape gradient. Sci Total Environ 311:247–259

Becker J, Brandt C, Dauble D (1998) Species selection for an ecological risk assessment of the Columbia River at the Hanford site, Washington, USA. Environ Toxicol Chem 17(11): 2354–2357

CCME (1996) A framework for ecological risk assessment: General guidance. ISBN 0-662-2426-3. Canadian Council for Ministers of the Environment, Winnipeg, Manitoba, 32 pp

CEM (2004) Metal levels in the soils of the Sudbury smelter footprint. Draft report to Inco and Falconbridge. Submitted by Centre for Environmental Monitoring, Laurentian University, Sudbury, Ontario, 1 March 2004

Courtin GM (1995) Birch coppice woodlands near the Sudbury smelters: Dynamics of a forest monoculture. In: Gunn JM (ed) Restoration and recovery of an industrial region, pp. 233–246. Springer-Verlag, New York

Dobbyn JE, Wilson N (1994) Atlas of the mammals of Ontario. ISBN 1-896059-02-3. Federation of Ontario Naturalists, Toronto, Ontario

Environment Canada (1994) A framework for ecological risk assessment at contaminated sites in Canada: Review and recommendations. Scientific Series No. 199. ISBN 0-662-22156-7. Environment Canada, Ecosystem Conservation Directorate, Evaluation and Interpretation Branch, Ottawa, ON

Gunn JM (ed) (1995) Restoration and recovery of an industrial region: Progress in restoring the smelter-damaged landscape near Sudbury, Canada. Springer-Verlag, New York, 358 pp

Hsu MJ, Selvaraj K, Agoramoorthy G (2006) Taiwan's industrial heavy metal pollution threatens terrestrial biota. Environ Pollut 143(2):327–334

Larison JR (2000) Cadmium toxicity among wildlife in the Colorado Rocky Mountains. Nature 406:181–183

Litzgus JD (1996) Status report on the wood turtle, Clemmys insculpta, in Canada. Committee on the Status of Endangered Wildlife in Canada. 59 pp. www.speciesatrisk.gc.ca/search/speciesDetails_e.cfm?SpeciesID=286. Accessed 10 September 2006

Maxwell CD (1995) Acidification and metal contamination: Implications for the soil biota of Sudbury. In: Gunn JM (ed) Restoration and recovery of an industrial region, Chapter 17. Progress in restoring the smelter-damaged landscape near Sudbury, Canada. Springer-Verlag, New York, NY

Meyers-Schone L (2000) Ecological risk assessment of reptiles. In: Sparling DW, Linder G, Bishop CA (eds) Ecotoxicology of amphibians and reptiles, Chapter 14b. SETAC Technical Publications Series, Society of Environmental Toxicology and Chemistry. SETAC Press, Pensacola, FL

MNR (2003a) Personal communication between Jan Linquist and Brenda Harrow, a fish and wildlife biologist at the Sudbury District Ministry of Natural Resources office, 16 June 2003. VTE stats within City of Greater Sudbury

MNR (2003b) Personal communication between Jan Linquist and Randy Staples at the Sudbury District Ministry of Natural Resources office, 31 July 2003. VTE stats within 100 km radius of Sudbury based on NRVIS database

MOE (2001) Metals in soil and vegetation in the Sudbury area (Survey 2000 and additional historic data). Report Number SDB-045-3511-2001. Ecological Standards and Toxicology Section, Standards Development Branch, Ontario Ministry of the Environment, Toronto, Ontario

MOEE (1997) Guideline for use at contaminated sites in Ontario. ISBN-0-7778-4052-9. Ontario Ministry of Environment and Energy, Toronto, Ontario

NB DAFA (2003) Growth and development of the wild blueberry. New Brunswick Department of Agriculture, Fisheries and Aquaculture, Fredericton, NB. www.gnb.ca/0171/10/0171100026-e.asp. Accessed 10 September 2007

Parker GH, Hamr J (2001) Metal levels in body tissues, forage and fecal pellets of elk (Cervus elaphus) living near the ore smelters at Sudbury, Ontario. Environ Pollut 113:347–355

Robitaille JF, Linley RD (2006) Structure of forests used by small mammals in the industrially damaged landscape of Sudbury, Ontario, Canada. For Ecol Manage 225:160–167

Sample BE, Aplin MS, Efroymson RA, Suter II GW, Welsh CJE (1997) Methods and tools for estimation of the exposure of terrestrial wildlife to contaminants. ORNL/TM-13391. Oak Ridge National Laboratory, Oak Ridge, TN

SARA Group (2009) Sudbury soils study. Volume III – Ecological risk assessment. Final Report. Prepared by the SARA Group, Guelph, Ontario, March 2009 Accessed 16 July 2010

Sergeant A (2002) Ecological risk assessment: history and fundamentals. In: Paustenbach DJ (ed) Human and ecological risk assessment: theory and practice, pp. 369–442. John Wiley and Sons Inc., New York

Sinclair A (1996) Floristics, structure and dynamics of plant communities on acid, metal-contaminated soils in the Sudbury Area. MSc thesis, Laurentian University

Suter II GW (1989) Ecological endpoints. In: Warren-Hick W, Parkhurst BR, Baker JSS (eds) Ecological assessment of hazardous waste sites: a field and laboratory reference document. EPA 600/3-89/013. Corvallis Environmental Research Laboratory, Corvallis, OR

Suter II GW (1993) Ecological risk assessment. Lewis Publishers, Chelsea, MI, 538 pp

Suter II GW (1999) A framework for assessment of ecological risks from multiple activities. Hum Ecol Risk Assess 5(2):397–413

Suter II GW, Elfroymson RA, Sample BE, Jones DS (2000) Ecological risk assessment for contaminated sites. Lewis Publishers, CRC Press, Boca Raton, FL, 438 pp

U.S. EPA (1998) Guidelines for ecological risk assessment. EPA/630/R-95/002F. Risk Assessment Forum, US Environmental Protection Agency, Washington, DC

U.S. EPA (2003) Assessment of whether to develop ecological soil screening levels for microbes and microbial processes. Guidance for developing ecological soil screening levels (Eco-SSLs). OSWER Directive 92857-55, November 2003. U.S. Environmental Protection Agency, Washington, DC

Watters RL, Edgington DN, Hakonson TE, et al. (1980) Synthesis of the research literature. In Hanson WC (ed) Transuranic elements in the environment. Technical Information Center/U.S. Department of Energy

Welbourn P (1996) A review of the direct and indirect effects on wildlife from copper/zinc smelters and refinery releases into aquatic or terrestrial ecosystems. Final Draft, 14 May 1996. Prepared by Welbourn Consulting for National Wildlife Research Centre, Environment Canada

Wentsel RS, Beyer WN, Edwards CA, Kapustka LA, Kuperman RG (2003) Effects of contaminants on soil ecosystem structure and function. In: Lanno RP (ed) Contaminated soils: From soil-chemical interactions to ecosystem management. Society of Environmental Toxicology and Chemistry (SETAC). SETAC Press, Pensacola, FL, 444 pp

9.13 Appendix Chapter 9: Abbreviations

CCME, Canadian Council of Ministers of the Environment
CEM, Centre for Environmental Monitoring
COC, chemical of concern
COI, community of interest
COSEWIC, Committee on the Status of Endangered Wildlife in Canada
ERA, ecological risk assessment
HHRA, human health risk assessment
LOE, line of evidence
MNR, Ministry of Natural Resources
MOE, Ministry of the Environment (Ontario)
MOEE, Ministry of Environment and Energy
NHIC, Natural Heritage Information Centre
OMAF, Ontario Ministry of Agriculture and Food
SARA, Sudbury Area Risk Assessment (Group)
SO$_2$, sulfur dioxide
TC, technical committee
TRV, toxicity reference value
USEPA, United States Environmental Protection Agency
VEC, valued ecosystem component
VTE, vulnerable, threatened and endangered
WOE, weight-of-evidence

DOI: 10.5645/b.1.10

10.0 ERA Objective #1: Are the Chemicals of Concern Preventing the Recovery of Terrestrial Plant Communities?

Authors

Christopher Wren, Ph.D.
MIRARCO,
935 Ramsey Lake Road, Sudbury, ON Canada P3E 2C6
Email: cwren@mirarco.org

Mary-Kate Gilbertson, M.Sc.
AECOM,
512 Woolwich Street, Guelph, ON Canada
Email: mary-kate.gilbertson@aecom.com

Graeme Spiers, Ph.D.
MIRARCO, Laurentian University,
935 Ramsey Lake Road, Sudbury, ON Canada P3E 2C6
Email: gspiers@mirarco.org

Peter Beckett, Ph.D.
Laurentian University,
935 Ramsey Lake Road, Sudbury, ON Canada P3E 2C6
Email: pbeckett@laurentian.ca

Devon Stanbury, M.Sc.
Escape Designs,
64 Galt St, Guelph, ON Canada
Email: devonstanbury@escapedesigns.com

Maureen Kershaw, M.Sc.
Faculty of Natural Resources Management,
Lakehead University,
955 Oliver Road, Thunder Bay, ON Canada P7B5E1
Email: maureen.kershaw@lakehead.ca

Table of Contents

10.1 Introduction	280
10.2 Overview of Study Approach	281
10.2.1 General Overview	281
10.2.2 Lines of Evidence	281
10.3 Site Selection Approach	285
10.4 Soil Sampling and Analysis	286

    10.4.1 Soil Collection 286
    10.4.2 Analytical Methods 287
    10.4.3 Chemical Analysis 287
    10.4.4 Quality Assurance and Quality Control 289
    10.4.5 Physical and Chemical Results 290
10.5 Soil Characterization LOE 294
    10.5.1 The Ranking Approach 294
10.6 Plant Community Line of Evidence (LOE) 296
    10.6.1 Defining a Self-Sustaining Forest Ecosystem 296
    10.6.2 Approach to Plant Community Assessment 297
    10.6.3 Plant Community Results 299
    10.6.4 Site Ranking Based on Plant Community LOE 301
10.7 Soil Toxicity Line of Evidence 303
    10.7.1 Selection of Test Species 303
    10.7.2 Toxicity Test Methods 306
    10.7.3 Toxicity Testing Results 306
    10.7.4 Site Ranking Based on Toxicity Testing Results 310
10.8 Litter Decomposition Line of Evidence (LOE) 313
    10.8.1 Rationale 313
    10.8.2 Methods 314
    10.8.3 Results and Site Ranking 315
10.9 Final Site Ranking and Integration of LOE 316
    10.9.1 Final Site Ranking Approach 317
10.10 Interactions Between Lines of Evidence 318
    10.10.1 Statistical Approach 318
    10.10.2 Creating Independent Soil Variables 318
    10.10.3 Statistical Analysis 320
10.11 Site Ranking Trends 322
10.12 The Effect of Historical Liming on Site Characteristics 323
    10.12.1 Background 323
    10.12.2 Comparison of COC Concentrations 324
    10.12.3 Soil Characterization 324
    10.12.4 Plant Community 326
    10.12.5 Soil Toxicity Comparison 328
    10.12.6 Decomposition 329
    10.12.7 Summary of Comparison of Limed to Untreated Site 329
    10.12.8 The Role of pH in Objective #1 329
10.13 Applying Results to the Broader Study Area 330
10.14 Influence of Distance from Smelter on Site Ranking 330
10.15 Characteristics of Impacted and Reference Sites 332
    10.15.1 Site Chemistry 332
    10.15.2 Plant Community 333
10.16 Uncertainties Related to Objective #1 334
10.17 Conclusions 337
10.18 Acknowledgements 337
10.19 References 338
10.20 Appendix Chapter 10: Abbreviations 341

Tables
10.1 Summary of Data Collected for Each Line of Evidence (LOE) 282
10.2 Parameters Measured in Soil and Analytical Method 288
10.3 Total COC Concentrations (mg/kg) and pH of Soil at the ERA Sites 290
10.4 Summary of Total COC Concentrations (mg/kg) at Test and Reference Sites 291
10.5 Water Leach COC Concentrations (mg/kg) and pH in Site Soils 292
10.6 Summary of Soil pH, Organic Matter and Cations in Site Soils 293
10.7 Summary of Nutrients in Site Soils 293
10.8 Range of Soil Chemistry Parameters for Site Ranking 295
10.9 Summary of Test Site Ranking Based on Soil Chemistry 296

10.10 Number of Plant Species at Each Test Site ... 299
10.11 Site Ranking for the Plant Community Assessment LOE ... 303
10.12 Test Species and Endpoints Used for the Toxicity Testing LOE ... 305
10.13 Site Ranking for the Toxicity Testing LOE in Natural Soil ... 313
10.14 Summary of the Final Site Rankings after the Integration of All LOE ... 317
10.15 Groupings of Soil Chemistry Parameters for Statistical Analyses ... 320
10.16 Toxicity Endpoints in Multiple Linear Regression Analysis ... 321
10.17 Copper Cliff Sites in Order of Total and Water Leachable Metals and Distance from Smelter ... 323
10.18 Metal Levels (mg/kg) in Soil from Limed and Non-limed Sites ... 324
10.19 Physical and Chemical Parameters at limed and untreated Sites ... 325
10.20 Plant Community Indicators at Limed and Untreated Sites ... 327
10.21 Range of Site Distances (km) from Smelters Relative to Impact ... 331
10.22 Range of Total COC Concentrations (mg/kg) at Different Sites ... 332
10.23 Characteristics of Plant Communities Relative to Impact Ranking ... 333

Figures

10.1 Overview of Field Studies for Each Line of Evidence ... 281
10.2 Overall Approach Used to Evaluate Objective #1 ... 285
10.3 Location of ERA Field Sites ... 286
10.4 Total Cu and Ni Concentrations in Soil Versus Distance from Nearest Smelter ... 291
10.5 Plant Species Richness versus Distance from the Nearest Smelter ... 300
10.6 Framework for Ranking Plant Communities ... 302
10.7 Number of Juvenile Earthworms Produced in Natural and pH-amended Soils ... 307
10.8 White Spruce Seedlings in Copper Cliff (CC) Soils (Photo from SRC 2005) ... 308
10.9 White Spruce Grown in Reference and Soils along the Coniston Transect
      (Photo by M. Moody, SRC 2005) ... 308
10.10 Goldenrod Seedling Growth in Falconbridge Soils (Figure from SRC 2005) ... 308
10.11 Mean Biomass (mg) of Red Clover Grown in Natural and pH-amended Soil ... 309
10.12 Mean Biomass (mg) of Northern Wheatgrass in Natural and pH-amended Soil ... 309
10.13 Wheatgrass Grown in Natural and pH-amended Soils from a Reference and a Test Site
      (Courtesy of Soil Toxicology Laboratory, Biological Assessment and Standardization Section,
      Environment Canada, Ottawa) ... 310
10.14 Site Ranking Approach for Soil Toxicity Testing LOE ... 312
10.15 Birch Leaves in Nylon Mesh Litter Bags on Forest Surface ... 314
10.16 Final Biomass Loss (% weight) of Birch Leaf Litter at the Reference and Exposure Sites ... 316
10.17 Distribution of ERA Field Sites and Overall Ranking ... 318
10.18 Photographs of the Limed Site (Left) and Non-treated Site (Right) ... 326
10.19 Biomass (mg) of Plants Grown in Soil from Limed Site and Untreated Sites ... 328
10.20 Distance from Nearest Smelter for Sites at Each Impact Level ... 331
10.21 Total Cu and Ni Concentrations in Soils Relative to Site Ranking ... 333
10.22 Conceptual Linkages of Historical Emissions and Other Factors Leading to Current Soil Conditions ... 335

10.1 Introduction

This chapter addresses Objective #1 of the ERA, which was to evaluate the extent to which the chemicals of concern (COC) were preventing the recovery of regionally representative, self-sustaining, terrestrial plant communities in the Sudbury region. This was a key objective of the Sudbury ERA, as it dovetails with the re-greening and reclamation activities that have been ongoing in the Sudbury area for almost four decades. The re-greening initiatives and programs are described in Chapter 12. While considerable progress has been made on the 're-greening' of the Sudbury landscape, significant portions of the region have not recovered, or have not recovered to what is considered to be their full ecological potential. In addition, biodiversity in the reclaimed areas remains low compared with a natural forest ecosystem. Therefore, it was the goal of this objective of the ERA to determine whether the COC in soils are inhibiting recovery of the vegetation, and to identify if other causal factors are involved. This information could then be used to help guide future reclamation activities.

Risk assessment often uses toxicity values derived from the general literature to predict the potential toxic effect of metals to soil organisms including plants. However, there were three primary environmental variables/conditions in the Sudbury area that rendered the use of literature values insufficient to address this objective:

1. Metal mixtures were present in Sudbury soils.
2. Sudbury soils have a low soil pH (which affects the toxicity of metals).
3. The effects of conditions 1 and 2 to plant species and communities relevant to Sudbury have not been previously fully examined.

Ecological communities are an aggregation of populations consisting of all plant, animal and microbial populations that occur in the same time and place and that interact physically, chemically and/or behaviorally. Although the community itself is what the risk managers ultimately aim to protect, it is not possible to study all components and, therefore, functional groups must be selected. Doyle et al. (2003) suggested that plants and soil dwelling organisms might be more at risk to atmospheric emissions from smelters than wildlife.

It can be a challenge to determine whether an ecosystem is impaired, and to what extent. The complexity of direct and indirect interactions between physical, biological and chemical components with varying temporal and spatial scales requires the use of multiple assessment approaches, with consequent need to integrate the diverse data collected (Burton et al. 2002b). To answer Objective #1 and to address the Sudbury specific conditions noted above, it was determined that a combined field and laboratory program was necessary to collect the data required to identify those environmental factors or variables that may be inhibiting ecosystem recovery. A weight-of-evidence (WOE) approach was utilized in which a variety of data were collected to produce four distinct 'lines of evidence' (LOE). The WOE approaches reported in the literature vary broadly and there was no standard guideline to describe how a WOE process should be conducted (Hull and Swanson 2006). Weighting or integrating multiple LOE into a conclusion does not remove uncertainty. Rather, it should provide a sound, transparent process for reducing uncertainty by integrating the best available scientific information available at the time (Burton et al. 2002a). The effectiveness and accuracy of any WOE approach is heavily dependent on five factors (Burton et al. 2002a):

- The quality of the data
- The quality of the study design
- The expertise of the principal investigators
- The severity of the impairment (greater is easier to detect)
- A matching of objectives and data.

A great deal of effort was expended for this Objective to determine: a) if there were ecosystem impacts within the study area; b) what the relative magnitude of the impacts were; and c) what the causative factors were. In this regard, it was more similar to a retrospective risk assessment, compared to a more predictive risk assessment approach used for Objectives #2 and #3 (Chapter 11). Previous activities and stresses in the study area resulted in widespread loss of the forest vegetation. Impacts to the vegetation began in the 1880s starting with logging for marketable timber and fuel for the early roast beds for Ni and Cu production. The early smelting operations released large quantities of low-level SO_2 and metals which further impacted and prevented recovery of the natural vegetation (Freedman and Hutchinson 1980a). The lack of vegetation subsequently caused soil erosion resulting in a landscape consisting of barren and semi-barren areas as described earlier in Chapter 2. The development of industrial areas into barren open landscapes due to deposition of airborne pollutants was described and documented for many sites around the globe by Kozlov and Zvereva (2007). In Sudbury, it was not until atmospheric emissions began to be significantly curtailed in the 1970s that conditions improved sufficiently to enable the beginning of restoration activities (see Chapter 12).

10.2 Overview of Study Approach

10.2.1 General Overview

To address Objective #1, detailed data for four LOE were collected. Each of these LOE was evaluated independently to determine impact at each site. Next, the interactions between the LOE were statistically evaluated. Finally, the LOE were integrated using a WOE approach to determine whether the concentrations of COC in the soil were impeding recovery of a self-sustaining forest system.

Each of these factors was considered in the approach used to address Objective #1. Four types of data, or LOE, were collected:

- physical and chemical soil characteristics
- toxicity testing with terrestrial species in the laboratory
- a plant community assessment, and
- an assessment of decomposition using *in situ* litter bags.

Detailed data and samples for each LOE were gathered from 22 study sites (18 test sites, three reference sites and one historically limed and treated site) during an intensive field and laboratory program in 2004 and 2005. Total Cu and Ni concentrations and soil pH were the primary criteria used to guide site selection. The sequence of steps involved in site selection, site characterization and data collection for each LOE is illustrated in Figure 10.1.

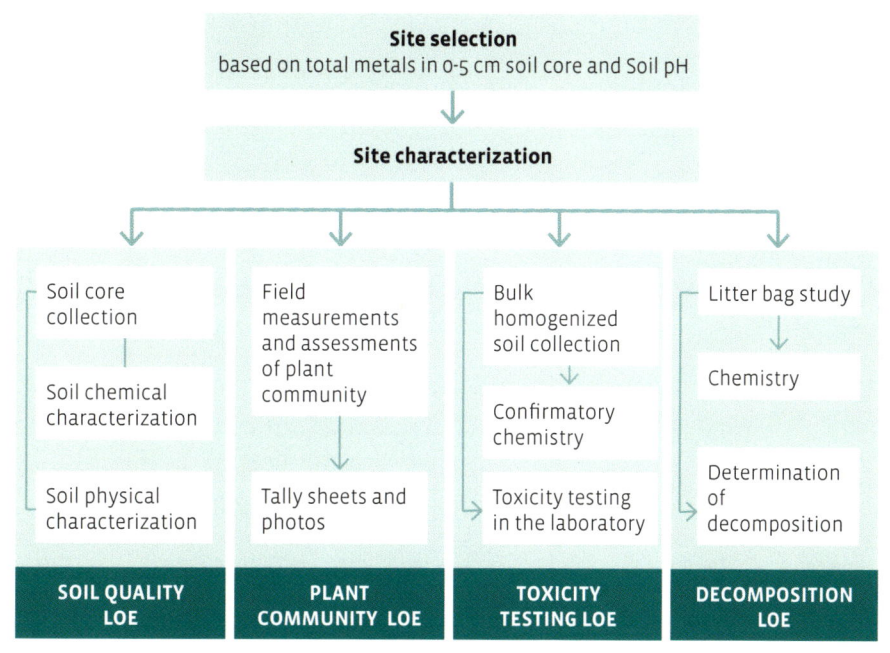

Figure 10.1 **Overview of Field Studies for Each Line of Evidence**

10.2.2 Lines of Evidence

For each of the LOE, a ranking approach was developed to assign a level of impact to each of the measured variables, and then to each site. The ranking approach was largely based upon comparing data from the test sites to the reference sites. A final site ranking was given to each site by combining the ranks of the individual LOE. Each LOE was weighted differently depending upon a variety of factors, such as ecological significance and the uncertainty related to the LOE. All of the LOE were evaluated separately and then integrated to determine whether there was concurrence between the various LOE. Concurrence between the LOE was examined to determine whether the COC were likely causing any observed impairment. Using a WOE approach of this kind gave greater weight to endpoint agreement. Lack of concurrence did not necessarily mean one LOE is inaccurate; rather, it may simply reflect the complexity of the system. Table 10.1 summarizes the list of variables and data collected for each LOE. Overall, several hundred measurements were taken, counted or analyzed at each field site and used to quantify over 70 parameters. The process of data evaluation and integration followed a three-step procedure:

- Step 1: the information from each LOE at every site was evaluated, and then the four LOE were integrated to determine a final impact rank for each site
- Step 2: the interactions between the LOE were statistically examined; specifically, whether soil physical and chemical parameters, including metals, were related to the other three LOE
- Step 3: results of each LOE were evaluated qualitatively to identify the environmental factors most likely related to the observed impacts at the test sites

Table 10.1 **Summary of Data Collected for Each Line of Evidence (LOE)**

Physical and chemical soil characterization			Plant community assessment	
Category		Parameter	Tallies	Parameter
Soil development		Pedon classification	Broad plant inventory	Trees
		Particle size		Tall shrubs
		Bulk density		Low shrubs
Chemical	Metals	Total metals (HNO$_3$ extraction)		Herbs
		Water leach (plant available fraction)		Graminoids
	pH and conductivity	pH in water		Pteridophytes
		pH (CaCl$_2$)		Bryophytes
		Electrical conductivity		Lichens
	Organic matter	Carbon	Percent cover	% Vegetation cover
		Total nitrogen		% Ground cover
		Total sulfur	Tree and tall shrub assessment	Species present
	Soil exchange complex chemistry	CEC		% Cover
		Potassium		Average height
		Sodium		Growth form
		Calcium		# Snags
		Magnesium		Diameter
		Manganese		% Mortality and dieback
	Fertility	Nitrogen	Coarse woody material	Species
		Phosphorous		Length
		Potassium		Diameter
		Nitrate/Nitrite		Degree of decomposition
		Sulfur	Ecosite classification	Terrain
		Avail. iron		% Slope
		Avail. manganese		Soil depth
		Ammonium		Soil texture
		Avail. magnesium		% Bedrock
				Dominant understory
				Dominant overstory
				% Canopy cover

Soil toxicity testing			Decomposition
Category	Species	Parameter	Parameter
Tree	White Spruce	Root length	Rate of Decomposition
		Root mass	
		Shoot length	
		Shoot mass	
Monocot	Northern Wheatgrass	Root length	
		Root mass	
		Shoot length	
		Shoot mass	
Dicot	Red Clover and Goldenrod	Root length	
		Root mass	
		Shoot length	
		Shoot mass	
Invertebrate*	Earthworm	No. of juveniles	
		Mass of juveniles	

* Invertebrate results were not used in the final site ranking.

Step 1: Evaluation of Individual LOE

- Each LOE was evaluated and ranked at each site relative to reference site conditions, or to other criteria developed for a particular measured variable. This evaluation was completed independent of the metal concentrations of the soil or knowledge of the site soil conditions. The following procedure was applied to each of the LOE at each site:
- The LOE information at the reference sites was evaluated to establish whether the reference site conditions were indicative of a 'typical' northern Ontario site, and to determine whether the reference sites were similar
- The LOE information from each test site was compared to the reference sites using a ranking system based on parameters and criteria appropriate to each LOE to determine an impact ranking for each criterion

An overall impact rank was established for each LOE by integrating the ranks of all criteria.

Each rank was assigned a color code (green, yellow, red) to help evaluate and illustrate trends in the data. At the completion of Step 1, each site had a rank for each separate LOE (four ranks in total). A final overall ranking was then assigned to each site by integrating the individual LOE rankings. The evaluation system consisted of three possible ranks:

Rank	Description
Green	Low to not impacted
Yellow	Moderately impacted
Red	Severely impacted

Step 2: Evaluation of Interactions between the LOE

Independent of the conclusions found at the end of Step 1, the data from each LOE were compared to each other using statistical techniques. The aim of this evaluation was to determine whether the various LOE were related to each other. Two statistical approaches were used:

- Multiple linear regression analyses were used to determine whether there was a relationship between soil chemical parameters and soil toxicity
- A canonical correspondence analysis (CCA) was used to determine whether there was a relationship between the soil chemistry parameters, the plant community at the site and decomposition endpoints

Step 3: Determine Whether Metals in Soil Are the Most Likely Cause of Impairment

To determine whether metals were the most likely cause of observed impairment, all of the sites were identified by color according to their final rank (from Step 1). The sites were then grouped by transect and organized according to the concentration of total metals, water available metals and distance from the smelter. Other factors identified from the evaluation of the soil physical and chemical characterization LOE were also tabulated and compared. The three-step procedure is illustrated in Figure 10.2.

STEP 1

Evaluation of each LOE

| LOE Data Collection and analysis | → | Evaluation of reference sites — Establish the reference sites as representative of 'typical' north eastern Ontario forested community. | → | Evaluation of test sites — Compare the test sites to the reference sites. |

Final rank site based on integration of four LOEs ← **Rank** LOE for each site

- FINAL RANK: low to not impacted ← GREEN: low to not impacted
- FINAL RANK: moderately impacted ← YELLOW: moderately impacted
- FINAL RANK: severely impacted ← RED: severely impacted

STEP 2
Evaluation of interactions between LOEs
Use statistical analysis to determine whether the lines of evidence are related to each other.

STEP 3
Determine if metals are causing impact
Determine whether metals in the soil or other factors are the most likely cause of impact.

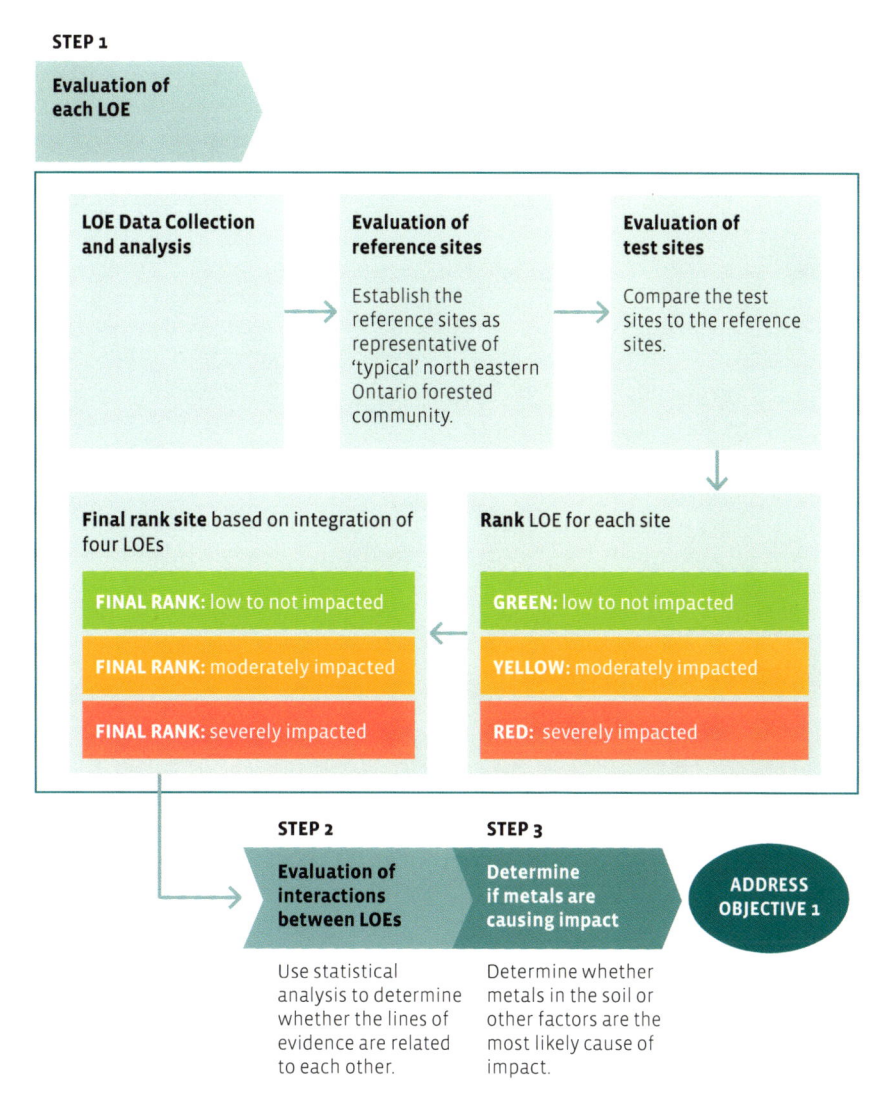

ADDRESS OBJECTIVE 1

Figure 10.2   **Overall Approach Used to Evaluate Objective #1**

10.3 Site Selection Approach

A total of 22 study sites (18 test sites, one historically limed site and three reference sites) were selected within the study area. The term 'test' site was used to refer to an exposure study site with elevated soil metal levels or other conditions not considered indicative of background conditions. The sites were selected on the basis of: total Cu and Ni concentrations and soil pH between 4.0 and 5.0 in surface soil. Due to the re-greening initiatives, large areas of Sudbury have been treated with lime to increase soil pH. These areas were avoided during site selection. However, for comparison purposes, one historically limed site adjacent to an unlimed area was also included. The three reference sites had similar soil texture and pH to the test sites but contained background concentrations of metals. It should be noted that plant community composition was not considered until the sites were already selected and, therefore, did not factor into the initial site selection. The sites (Figure 10.3) were located on transects radiating from the smelters:

- Copper Cliff (CC) transect: seven test sites
- Falconbridge (FB) transect: five test sites
- Coniston (CON) transect: six test sites and one historically limed and re-greened site
- Reference (REF): three reference sites.

Once site identification was allocated, it was not transferred to another site even if that site was excluded because it failed to meet the selection criteria. As a result, the numeric continuity of the site numbering was sometimes interrupted (for example, there is no REF-01 as this site was rejected because the soil textural analysis and pH indicated it was not a useful reference site). The reference sites were between 28.6 and 41.3 km from the nearest smelter.

The study sites generally corresponded to the location of the three smelters and the prevailing wind patterns (south west to north east). Conversely, the Copper Cliff transect was established against the general prevailing wind pattern as the City of Greater Sudbury was located in the intended direction. The term 'transect' was used loosely to describe the orientation of the study sites as the sites were not oriented along a straight line or 'transect' as logistical constraints such as site access or lack of soil strongly influenced actual site selection. The primary goal was to obtain a series of sites associated with each smelter that provided sufficient soil to study and provided a gradient of metal concentrations related to distances from each smelter.

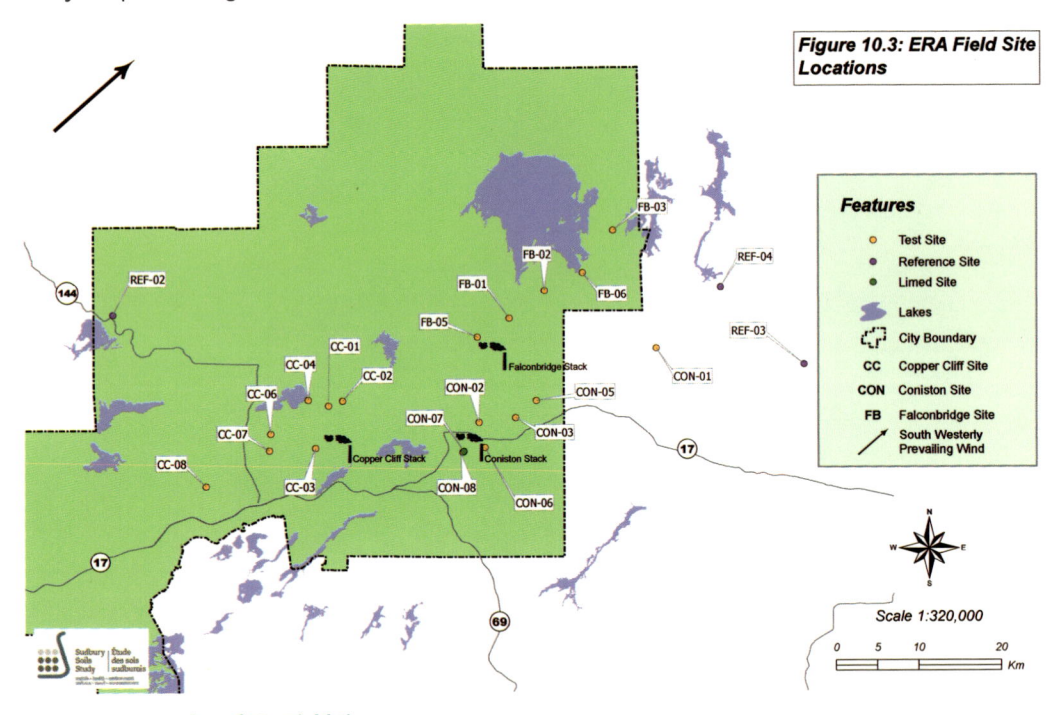

Figure 10.3 **Location of ERA Field Sites**

10.4 Soil Sampling and Analysis

The primary focus of this ERA Objective was to assess the role of the COC (As, Cd, Cu, Co, Ni, Pb and Se) in soil in limiting plant community development. However, soil is a complex matrix of organic and inorganic constituents. The physical attributes of soil, such as particle size, texture, and proportion of constituents (sand, silt and clay) all interact to define the quality of soil as a growth medium. Detailed sample collection and handling procedures were developed before any field work and documented in Standard Operating Procedures (SOPs) to ensure consistency of approach and that all methods were properly recorded. A detailed description of the analytical methods was provided in appendices to the ERA final report (www.sudburysoilsstudy.com). Soil was collected at each site for physical and chemical characterization and for use in toxicity testing. Soil core samples were collected for the majority of the physical and chemical analyses. Large quantities of soil were required for toxicity testing and these were referred to as 'bulk or homogenized soil'.

10.4.1 Soil Collection

Composite soil samples consisting of >50 cores at each site were collected from three depths: 0–5, 5–10 and 10–20 cm (or to corer refusal). The method followed guidance provided by Environment Canada (2005a) and was similar to that used in the Regional Soil Survey described previously in Chapter 3. The 0–5 cm soil core samples were submitted to the laboratory for analysis of total metals and pH. A large quantity of soil was collected and mixed or homogenized for use in toxicity testing. The bulk soil consisted of a well-mixed sample of the 0–5 cm site soil collected from a variety of shallow test pits within 50 m of the center of the site. The 0–5 cm layer was removed, sieved and homogenized. Before the buckets were sealed, one representative sample was collected by combining an equal amount of soil from each of the buckets to obtain a mass of 400 g. This sample was submitted for total metal and pH analysis to compare the characteristics of the homogenized soil to those of the core samples also collected at the site.

10.4.2 Analytical Methods

10.4.2.1 Physical Analysis

Pedon Classification

The pedon layers at each site were visually examined in test pits which provided a description of the size of the 'O' horizon, which consists of the litter (L), fermentation (F) and humic (H) portions, and the 'A' horizon, which was characterized by having a large amount of organic material. The size of each of these horizons gave an indication of how well plants might grow in the soil and how mobile chemicals and nutrients might be.

Particle Size

Particle size distribution in the 0–5 cm core samples was determined after the methods of Sheldrike and Wang (1993). The particle size analysis provided a breakdown of the proportion of sand, silt and clay as defined by the textural classification of the soil. Soil texture affects a range of physical and chemical properties of the site soil and has important implications for soil fertility and the binding of metals.

Bulk Density

Bulk density (g/cm^3) of the soil was measured to determine soil structure, total pore space and the degree of packing of the soil particles. Moisture content of the soil was also determined from this sample.

10.4.3 Chemical Analysis

Parameters Measured

A summary of the chemical parameters measured at each site, the analytical facility and the method used is provided in Table 10.2. The methods and a brief rationale for the measurement are provided in the following sections. Before analysis, the soil was thoroughly mixed, sieved and homogenized. The samples were then dried, ground and split into six portions for different analysis.

Metals

Both total metal and water leach metal levels were measured in the 0–5 cm soil core samples. For the total metal analysis, the samples were digested using concentrated nitric acid and microwave heating and then analyzed by ICP-MS. The nitric acid extraction (total metals) was included to allow direct comparison to MOE soil guidelines and other soil survey data as this is a standard approach for determining metal concentrations in soil (Nolan et al. 2003). Because total metal concentration provides only partial information with regard to potential bioavailability, analysis was also conducted on water extracts from the same 0–5 cm core samples. The analysis of the water extracted metal concentrations provided a measure of the 'labile' metal pool within the soil sample. For this method, the soil was mixed with water at a ratio of 1 part soil to 5 parts water. The mixture was shaken vigorously and the supernatant analyzed for metal concentration by ICP-MS. The supernatant metal concentration was then converted from mg/l to mg/kg. A water extraction isolates the water-soluble and labile metals (or free ions), which are typically viewed as the readily bioavailable fraction (Courchesne et al. 2006). The rationale for choosing a water extraction compared to a weak acid or salt extraction was based on the supporting literature that a water extract would best represent the fraction of metal that is immediately available to the plant or soil organism (Sanders 1982; Reddy et al. 1995; Sauve et al. 1997; Els Smit and Van Gestel 1998; Kuinto et al. 1999).

pH and Conductivity

Soil pH was determined by two separate methods in all of the 0–5 cm core samples and the bulk soil: pH of a water/soil slurry, and calcium chloride buffer pH. The pH measured in water is considered to be the most accurate measure of the pH found in the field (Hendershot et al. 1993). The use of $CaCl_2$ pH is common in agricultural situations because the measurement is less dependent on recent fertilizer or liming events. The $CaCl_2$ method generally produces a soil pH about 0.5 units lower than the water slurry method.

Soil Organic Matter

Organic materials exert a profound influence on every facet of the nature of soil (Troeh and Thompson 2005). The chemical composition of humus can be considered from the point of view of its elemental constituents, which are primarily carbon, hydrogen, oxygen and nitrogen and to a lesser extent sulfur. To gain an estimate of the soil organic matter, total carbon, organic carbon, inorganic carbon, total sulfur and total nitrogen were measured in the 0–5 cm core samples.

Soil Exchange Complex Chemistry

Soil possesses electrostatic charges that counter (exchangeable) ions to form the exchange complex. The cation-exchange capacity (CEC) is a measure of the amount of ions that can be absorbed, in an exchangeable fashion, on the negatively charged sites in the soil (Bache 1976). The higher the CEC of the soil, the more ability the soil has to hold onto plant available soil nutrients and thus not lose them through leaching (Schroth and Sinclair 2003). The ability of a soil to adsorb cations has very important implications for soil fertility (Schroth and Sinclair 2003) and for the ability of soil to sorb metals due to ion exchange ability (Lanno 2003). The CEC in soil samples was measured by the ammonium acetate method and the exchangeable cations were determined using the method identified in Table 10.2.

Fertility Analysis

All 13 of the essential macronutrients (N, P, K, Ca, Mg, S) and micronutrients (Fe, Cu, Mn, Zn, B, Cl and Mo) required for plants to complete their life cycle were analyzed in the 0–5 cm core samples. In addition, fertility analysis (P, K, Mn and Mg) was performed on the bulk soil used for toxicity testing at the start and finish of each test.

Table 10.2 **Parameters Measured in Soil and Analytical Method**

Category	Parameter	Units	Soil sample analyzed	Method
Metals	Total metals (HNO_3)	mg/kg	0–5 cm and homogenized	Microwave digest by: Method 3051 ICP-MS by: SW846, Method 6020
	Water leach (plant available metals)	mg/kg	0–5 cm	Water leach extraction 1:5 sample/water ratio Analysis by ICP-MS by: SW846, Method 6020
pH and conductivity	pH in water	pH units	0–5 cm and homogenized soil	Modified APHA-4500
	Soil pH in 0.01 M $CaCl_2$	pH units	0–5 cm and homogenized soil	Carter 16.3 M.R. Carter (1993)
	Electrical conductivity	µS/cm	0–5 cm	Modified APHA-2510
Organic matter	Total, organic and inorganic carbon	% dry	0–5 cm	ASTM E1915-01
	Total nitrogen			
	Total sulfur			

Category	Parameter	Units	Soil sample analyzed	Method
Soil exchange complex chemistry	Cation exchange capacity	cmol(+)/kg	0–5 cm	Carter 19.4 M.R. Carter (1993)
	Potassium	cmol(+)/kg	0–5 cm	Ion chromatography modified SW846-9056
	Sodium			
	Calcium			
	Magnesium			
	Manganese			
Fertility analysis	Nitrogen: nitrate, nitrite and ammonium as N	mg/kg	0–5 cm	Nitrate/nitrite: ion chromatography modified SW846-9056 Ammonium as N: Flow analysis modified APHA-4500.
	Calcium, phosphorous, and sulfur	mg/kg	0–5 cm	Water leach extraction 1:5 sample/water ratio Analysis by ICP-MS by: SW846, Method 6020
	Available iron and manganese	mg/kg	0–5 cm	DPTA-extractable Carter 11 M.R. Carter, Ed. 1993.
Fertility analysis in homogenized soil	Phosphorous	%	Homogenized soil start and finish of toxicity test	Sodium bicarbonate extractable P
	Potassium	mg/kg		Ammonium acetate extractable
	Water leach magnesium	mg/kg		Ammonium acetate extractable
	Water leach manganese	mg/kg		Sodium bicarbonate extractable Mn

10.4.4 Quality Assurance and Quality Control

The quality assurance and quality control (QA/QC) procedures were established before sample collection and analysis. All laboratories used in this study were accredited by the Canadian Association of Environmental Analytical Laboratories (CAEAL) and by the Standards Council of Canada (SCC). The methodology used by the laboratory for the cleaning, preparation and analysis of the sample was established before sample delivery. QC samples included field duplicate soil samples and analysis of certified reference material (CRM). Since soil can be heterogeneous, a difference of 30% or less between duplicate samples was considered acceptable. If the percent difference was greater than 30%, the cause of the difference between the samples was investigated.

The analysis of the duplicate samples provided a measure of the variability of the soil metal content and the physical and chemical characteristics of the soil to ensure that each site was adequately represented by the sampling method. Fourteen duplicate core samples were submitted from three randomly selected field sites which represented over 10% of the sites. The analysis showed that the level of variability between samples within sites was considered acceptable. The CRM used for metal analysis was purchased from the National Institute of Standards and Technology and consisted of San Joaquin Soil (CRM 2709). CRM samples were submitted to the laboratory for analysis on four occasions to check the accuracy of the results given by the laboratory. The measured concentrations of 14 elements in each submission were compared to the certified values for the reference material. More than 70% of the submissions differed from the certified value by less than 30%. Similarly, the percent recovery for all but the same four elements was greater than 75%.

10.4.5 Physical and Chemical Results

Bulk density of soils at the reference sites displayed a relatively narrow range from 0.72 to 0.88 g/cm³. Soil bulk density at the test sites ranged from 0.43 to 1.45 g/cm³. Higher bulk densities were likely an indication of the exposed B horizons (mineral soils) at the test sites which tended to display more erosion. Lower bulk density reflects a more complete soil profile, signifying that the surface soil included both organic and mineral soil material.

The results of the ICP-MS scan of total (HNO_3 extractable) COC levels in the 0–5 cm core samples for each site are provided in Table 10.3 and summarized in Table 10.4. Soil pH ($CaCl_2$ method) in the test sites ranged from 3.22 to 4.05. The pH at site CON-07 was higher (6.45) because it was limed in about 1978. Soil pH at the reference sites was relatively low and ranged from 3.59 to 4.14 (Table 10.3).

Table 10.3 **Total COC Concentrations (mg/kg) and pH of Soil at the ERA Sites**

Site	pH ($CaCl_2$)	As	Cd	Co	Cu	Pb	Ni	Se
CON-01	3.44	9.5	0.28	5.51	76	28	77	0.85
CON-02	3.76	12.7	0.17	9.01	195	15.0	138	1.0
CON-03	3.61	28	0.24	11.5	191	35	112	0.92
CON-05	3.59	11.4	0.44	11.0	118	15.1	92.9	0.7
CON-06	4.03	2.1	0.12	9.4	48.7	4.6	70.2	0.3
CON-07[a]	6.45	7.2	0.15	10.2	240	11.0	255	1.1
CON-08	3.96	5.2	0.15	10.9	107	9.1	132	0.89
FB-01	3.21	117	0.99	23.3	655	162	422	5.6
FB-02	4.05	45	1.17	48.4	320	83	325	3.4
FB-03	3.64	10.9	0.28	4.84	87	28	78	1.1
FB-05	3.86	41	0.26	10.3	215	33	140	1.2
FB-06	3.48	26	0.61	11.7	200	61	179	1.7
CC-01	3.81	46	1.26	26.7	960	70	700	6.2
CC-02	3.95	44	0.67	35.8	611	53	511	4.7
CC-03	3.81	72	0.61	41.5	1000	99.5	1100	10.5
CC-04	3.81	29	0.93	21.8	441	49	386	2.7
CC-06	3.85	15.5	0.43	9.9	144	17.2	103	1.5
CC-07	3.61	26	0.52	14.0	303	38	200	2.4
CC-08	3.62	9.6	0.27	7.81	97	29	77.5	1.4
REF-02	3.59	4.6	0.28	4.87	42	33	46	1.0
REF-03	4.14	2.66	0.23	11.5	18.7	14	40	0.48
REF-04	3.6	5.85	0.17	5.35	39.3	18.6	38.9	0.75

[a] CON-07 is the historically limed site, consequently the pH is higher than in the other test sites.

The concentration of Cu, Ni and Se tended to be higher along the Copper Cliff transect (Table 10.4). Maximum soil concentrations were 1100 mg/kg for Ni, and 1000 mg/kg for Cu (at CC-03). In comparison, total Ni concentrations at the reference sites ranged from 39 to 46 mg/kg, while total Cu levels ranged from 19 to 42 mg/kg. Most COC levels were lower along the Coniston transect. This was not surprising since the Coniston smelter was closed in 1972. In addition, the Coniston sites tended to be more eroded and have less soil. Along the Coniston transect, total Ni levels ranged from 70 to 255 mg/kg, while total Cu ranged from 49 to 240 mg/kg. The maximum Ni and Cu levels for this transect were found at CON-07, which is the historically limed and re-greened site and is located close to the smelter. This limed site (CON-07) was immediately adjacent to a non-limed site (CON-08) and was used to compare vegetation and toxicity relative to soil pH and metal bioavailability (see Section 10.14).

Table 10.4 **Summary of Total COC Concentrations (mg/kg) at Test and Reference Sites**

Metal	Copper Cliff Range	Copper Cliff Mean	Falconbridge Range	Falconbridge Mean	Coniston Range	Coniston Mean	Reference Range	Reference Mean
As	26–72	34.5	11–117	48	2.1–28.0	10.8	2.6–5.8	4.4
Cd	0.27–1.26	0.67	0.26–1.1	0.66	0.12–0.44	0.22	0.17–0.28	0.23
Co	7.8–41.5	22	4.8–48.4	19.7	5.5–11.5	9.6	4.8–11.5	7.24
Cu	97–1000	508	87–655	211	76–240	139	18.7–42	33.3
Pb	29–99	51	28–162	73.4	4.6–35	16.8	18.6–33.0	21.8
Ni	77–1100	440	78–422	229	77–255	125	39–46	41.6
Se	1.4–10.5	4.2	1.2–5.6	2.6	0.3–0.92	0.82	0.48–1.0	0.74

Nickel concentrations along the Falconbridge transect ranged from 78 to 422 mg/kg (Table 10.4), while total Cu levels ranged from 87 to 655 mg/kg. The maximum levels of Ni, Cu, As, and Se along this transect were found at site FB-01 (Table 10.3). The lowest total metal levels were present at FB-03, the test site furthest from the smelter. Arsenic concentrations tended to be higher in soil along the Falconbridge transect compared with the other areas due to higher As in the ore processed at this smelter as discussed in Chapter 2. Soil metal concentrations were elevated within about 15 km from the nearest smelter and then leveled off. The pattern of metal concentration relative to distance from the nearest smelter is illustrated for Ni and Cu in Figure 10.4.

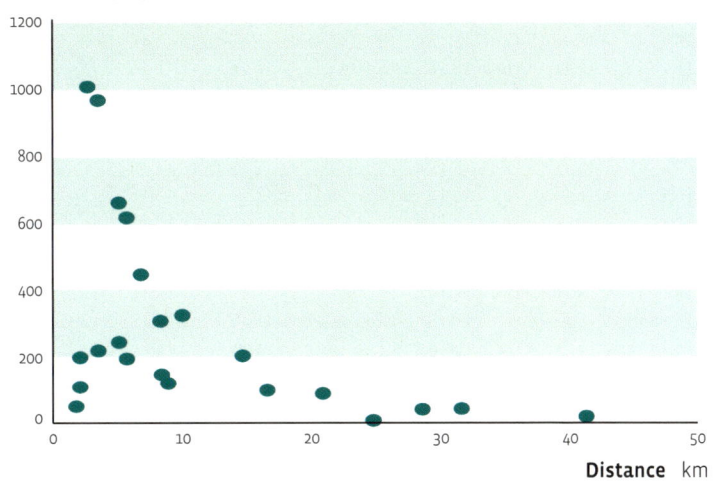

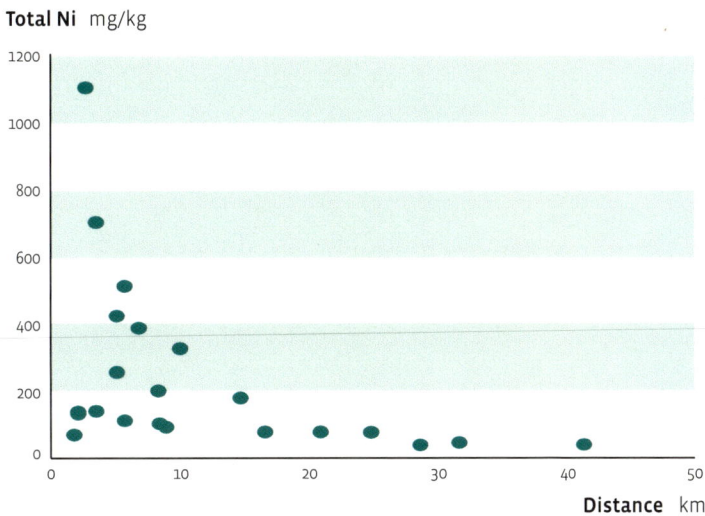

Figure 10.4 **Total Cu and Ni Concentrations in Soil Versus Distance from Nearest Smelter**

10.4.5.1 Water Leach Metals

Plant bio-available metal levels were estimated using a water leach extraction. The COC concentrations from the water leach extraction and pH for the soil cores from each of the sites are provided in Table 10.5. The concentration of metal levels determined by water leach was up to 500 times lower than the total metal concentrations using HNO_3 extraction. The water leach results were converted from μg/L to mg/kg to allow for comparison with other extractions. The conversions were made using the results from the leachate (μg/L) multiplied by the standard volume (L) and the result is divided by the soil weight corrected for moisture (g). Soil moisture content was 26.8% and the soil mass used was 20 g.

The maximum soil Ni (water leach) concentration was 2.55 mg/kg at CON-08, while the highest Cu level was 1.74 mg/kg at the historically limed site, CON-07 (Table 10.5). For the Falconbridge transect, the highest Ni and Cu concentrations were 1.99 mg/kg and 1.47 mg/kg, respectively (FB-01, FB-02). The highest Ni and Cu concentrations were observed along the Copper Cliff transect (Table 10.5), where they were 2.71 mg/kg for Ni, and 1.73 mg/kg for Cu (CC-02). In comparison, Ni concentrations at the reference sites ranged from 0.04 to 0.15 mg/kg, while Cu levels ranged from 0.01 to 0.08 mg/kg.

Table 10.5 Water Leach COC Concentrations (mg/kg) and pH in Site Soils

Site	pH (CaCl$_2$)	Ar	Cd	Co	Cu	Pb	Ni	Se
CON-01	3.44	0.02	0.01	0.01	0.46	0.03	0.29	0.01
CON-02	3.76	<0.01	0.00	0.02	0.27	<0.01	0.40	<0.01
CON-03	3.61	0.03	0.00	0.01	0.44	<0.01	0.23	<0.01
CON-05	3.59	0.01	0.00	0.01	0.09	<0.01	0.38	<0.01
CON-06	4.03	<0.01	0.00	0.03	0.03	<0.01	0.98	<0.01
CON-07[a]	6.45	0.09	0.00	0.08	1.74	0.07	1.76	<0.01
CON-08	3.96	0.04	0.00	0.08	0.38	<0.01	2.55	<0.01
FB-01	3.21	0.04	0.02	0.08	1.07	0.01	1.99	<0.01
FB-02	4.05	0.25	0.01	0.06	1.47	0.08	1.16	0.04
FB-03	3.64	0.01	0.00	0.02	0.08	<0.01	0.58	<0.01
FB-05	3.86	0.01	0.00	0.02	0.12	<0.01	0.31	0.01
FB-06	3.48	0.14	0.00	0.02	0.61	0.08	0.26	<0.01
CC-01	3.81	0.03	0.01	0.04	0.52	<0.01	1.01	0.01
CC-02	3.95	<0.01	0.02	0.07	1.73	<0.01	2.71	<0.01
CC-03	3.81	<0.01	0.01	0.05	1.00	<0.01	1.56	0.02
CC-04	3.81	0.02	0.01	0.03	0.46	<0.01	1.67	0.01
CC-06	3.85	<0.01	0.00	0.02	0.17	<0.01	0.70	<0.01
CC-07	3.61	0.01	0.00	0.02	0.20	<0.01	0.45	<0.01
CC-08	3.62	<0.01	0.00	0.01	0.12	<0.01	0.24	<0.01
REF-02	3.59	<0.01	0.00	0.01	0.03	<0.01	0.12	<0.01
REF-03	4.14	<0.01	0.00	0.00	0.01	<0.01	0.04	<0.01
REF-04	3.60	0.01	0.00	0.01	0.08	0.01	0.15	<0.01

[a] CON-07 is the historically limed site.

10.4.5.2 Other Soil Variables

The soil pH, conductivity, organic matter content, cation and CEC values measured at the test and reference sites are summarized in Table 10.6. The test and reference sites have a pH (water slurry) which is within the 4–5 range. Soil pH as indicated by the $CaCl_2$ method is slightly lower. The higher pH at the limed site (CON-07) reflects the liming activity that took place at this site. Several parameters were higher at the reference sites relative to the test sites including total N, total and organic carbon, Ca and Mg.

Table 10.6 **Summary of Soil pH, Organic Matter and Cations in Site Soils**

Parameter	Range of test sites			Coniston limed site (CON-07)	Reference sites
	Copper Cliff	Falconbridge	Coniston		
pH (water/slurry)	4.19–4.81	4.1–4.77	4.34–4.60	7.19	4.04–4.88
pH ($CaCl_2$)	3.61–3.95	3.21–4.05	3.44–4.03	6.45	3.59–4.14
Conductivity (µS/cm)	16–58.4	16.9–112.4	15.5–65.7	44.1	24.6–41.1
Total N	0.13–0.3	0.1–0.35	0.03–0.27	0.09	0.23–0.34
Total S	<DL–0.13	<DL	<DL	<DL	<DL
Total C	2.29–5.59	1.76–7.73	0.46–3.75	1.83	4.24–7
Organic C	2.29–5.59	1.76–7.73	0.45–3.75	1.71	4.18–6.93
Inorganic C	0–0.21	0–0.05	0–0.12	0.12	0–0.07
CEC	16.8–52.1	11–124	11–45.4	14.6	27.4–29.1
K	0.12–0.28	0.12–0.57	0.11–0.38	0.16	0.15–0.24
Na	0.02–0.06	0.01–0.1	0.02–0.12	0.054	0.03–0.04
Ca	0.15–2.1	0.24–3.83	0.11–1.8	9.4	0.38–2.8
Mg	0.05–0.3	0.05–0.81	0.05–1	1.3	0.18–0.72
Mn	<DL–0.2	<DL–0.32	<DL–0.21	<DL	<DL

<DL indicates concentration was less than the method detection limit.

10.4.5.3 Fertility

The nutrient concentrations at the test and reference sites are summarized in Table 10.7.

Table 10.7 **Summary of Nutrients in Site Soils**

Nutrient	Measurement	Unit	Range by transect			Limed site (CON-07)	Reference sites
			Copper Cliff	Falconbridge	Coniston		
Nitrogen	Total N	% dry	0.13–0.3	0.1–0.35	0.04–0.27	0.09	0.23–0.34
	Nitrate	mg/kg	<DL–22.5	<DL–35.5	<DL–3.28	<DL	<DL–23.2
	Nitrite	mg/kg	<DL	<DL	<DL	<DL	<DL
	Ammonia	mg/L	0.01–11.5	0.2–36	0.01–3.2	0.10	0.45–3.5
Phosphorous	Total P	mg/kg	550–2500	240–700	180–850	180	208–501
	Water leach P	µg/L	<DL	<DL–3.01	0.003–2.12	310	<DL–0.17

Nutrient	Measurement	Unit	Copper Cliff	Falconbridge	Coniston	Limed site (CON-07)	Reference sites
Sulfur	Total S	% dry	<DL–0.13	<DL	<DL	<DL	<DL–0.03
	Water leach S	µg/L	<DL–0.02	0.02–0.06	0.01–0.06	4100	<DL–0.01
Iron	Total Fe	mg/kg	861–1864	676–1876	222–2442	9300	919–1256
	Water leach Fe	µg/L	<DL–1.23	0.32–22.5	<DL–225	33,000	0.5–3.1
Manganese	Total Mn	mg/kg	19–205	4.8–266	12.3–109	130	38–103
	Water leach Mn	µg/L	0.29–3.5	0.38–4.2	0.07–1.5	190	0.31–0.72

<DL indicates concentration was less than the method detection limit.

10.5 Soil Characterization LOE

The primary focus of the study was directed at the concentrations of COC in soil. However, soil is a complex matrix of organic and inorganic constituents and these other factors can also influence plant growth. The site ranking associated with this LOE focused on a variety of physical and chemical soil characteristics measured to assess the soil as a growth medium. The ranking approach compared the test site soils to soil quality parameter ranges that were established using literature review values, reference site values and best professional judgment. A final ranking for each test site was based on soil characteristics, which did not include the COC. At each site, the following categories of soil quality parameters were assessed: organic matter, soil exchange complex chemistry, and fertility.

The amount of organic matter in soil is a function of the addition of fresh material and the decomposition rate. Despite the fact that plants do not require organic matter as such for growth and development, it is still considered one of the most important components of soil fertility because of its influence on a wide range of soil properties and processes (Gregorich et al. 1994; Schroth and Sinclair 2003). The quality of soil organic matter at each site was ranked relative to baseline conditions documented in the reference sites. Key soil organic matter components evaluated include total carbon and nitrogen levels. Soil exchange chemistry parameters include the CEC, the concentrations of Ca, Mg, the Ca/Mg ratio and base saturation. Of these parameters, CEC was considered the strongest determinant of soil quality.

For soil fertility, 13 nutrient elements are considered essential plant nutrients. Each of these nutrients has a critical function and is required in varying amounts in plant tissue. Macronutrients (N, P, K, Ca, Mg and S) are required in the largest amount in plants while micronutrients (Fe, Cu, Mn, Zn, Bo, Mo, Cl) are required in relatively smaller amounts. The overall fertility of the sites was evaluated by examining the concentrations of the macronutrients: N, P, K and Mg; and of the micronutrients: Fe and Mn. The macronutrients were considered stronger determinants of soil quality than the micronutrients.

10.5.1 The Ranking Approach

Test sites were ranked on soil characterization results following three steps.

1. The range of soil quality parameter values at undisturbed reference sites were defined based on data collected at the reference sites and information from the literature for un-impacted forested regions of north-eastern Ontario
2. Literature and reference values were then used to define numerical parameter ranges representing moderate and severe impact for each soil quality parameter
3. Soil quality values at each test site were then compared to the parameter ranges to assign an impact rank for each parameter (low, moderate, severe). Then, best professional judgment was used to give each site an overall ranking based on the total soil characteristics. Based on the above evaluation of soil quality parameters, each test site was placed into one of three categories:

Rank		Description
Green	Low to no impact	The majority of the soil quality parameters at these sites fell within the 'green' parameter range. The soil was considered a good medium for plant growth
Yellow	Moderately impacted	The majority of the soil quality parameters at these sites fell within the 'yellow' parameter range. Some soil quality parameters appeared to be affecting the performance of the soil as a medium for plant growth
Red	Severely impacted	The majority of the soil quality parameters at these sites fell within the 'red' parameter range. Some soil quality parameters appeared to be seriously impacting soil potential and were likely limiting plant growth

The numerical ranges established for each soil quality parameter are presented in Table 10.8. These ranges were based on regionally representative values from the reference sites, literature values and best professional judgment. Sites that contained soil conditions within the undisturbed or reference conditions were considered to be in the 'low to not impacted' or 'green' category.

Table 10.8 **Range of Soil Chemistry Parameters for Site Ranking**

Rank	Low quality soil (red)	Moderate quality soil (yellow)	High quality soil (green)
Organic matter (g/100 g)			
Total C	<3	3–3.9	>3.9
Total N	<0.1	0.11–0.21	>0.22
Soil exchange complex chemistry (cmol(+)/kg)			
Cation exchange capacity	<19	20–24	>25
Calcium	<0.24	0.25–0.39	>0.4
Magnesium	<0.1	0.1–0.15	>0.15
Ca/Mg ratio	<1.4	1.5–2.9 or >6	3–5.9
Base saturation (%)	<1.9	2–4.9	>5
Fertility (mg/kg)			
N as ammonium	<0.19	0.2–0.39	>0.4
Extractable P	<5	5–7.9	>8
Extractable K	<44	45–64	>65
Extractable Fe	<499	500–749 or >1800	750–1800
Extractable Mn	<10	10–24 or >200	25–200
Fe/Mn	<5	5–14 or >50	15–50
Extractable Mg	<15	15–25	>25

The three reference sites were ranked 'green', indicating that the soil characteristics present at the reference sites were typical of forested areas of north-eastern Ontario and were representative of conditions conducive to plant growth. The majority of the test sites were ranked as moderately impacted (yellow), indicating that at least some of the soil quality parameters measured at these sites were limiting plant growth (Table 10.9). Some sites on both the Copper Cliff and Coniston transects were ranked severely impacted, indicating that the growing conditions at these sites were unlikely to support healthy plant growth. Two sites on the Falconbridge transect were ranked low to not impacted, indicating that the soil conditions at these sites were comparable to the conditions at the reference sites.

Table 10.9 **Summary of Test Site Ranking Based on Soil Chemistry**

Site	Rank	Distance from associated smelter
CC-01	Yellow	Copper Cliff 5.3 km
CC-02	Red	Copper Cliff 5.7 km
CC-03	Red	Copper Cliff 2.7 km
CC-04	Yellow	Copper Cliff 6.8 km
CC-06	Yellow	Copper Cliff 8.4 km
CC-07	Yellow	Copper Cliff 8.3 km
CC-08	Yellow	Copper Cliff 16.6 km
CON-01	Yellow	Coniston 24.8 km
CON-02	Red	Coniston 2.1 km
CON-03	Yellow	Coniston 5.7 km
CON-05	Yellow	Coniston 8.9 km
CON-06	Yellow	Coniston 1.8 km
CON-08	Red	Coniston 2.1 km
FB-01	Yellow	Falconbridge 5.1 km
FB-02	Green	Falconbridge 10 km
FB-03	Yellow	Falconbridge 20.9 km
FB-05	Yellow	Falconbridge 3.5 km
FB-06	Green	Falconbridge 14.7 km

10.6 Plant Community Line of Evidence (LOE)

10.6.1 Defining a Self-Sustaining Forest Ecosystem

To assess relative impact on vegetation at the test sites, it was first necessary to establish a reference for comparison. A definition of a self-sustaining forest ecosystem was developed for the purpose of this study to create a reference base from which to assess impact. A self-sustaining forest ecosystem is an assemblage of plants, with a treed over-story, that occurs with a degree of predictability for any given time since disturbance on any given topographic position, soil type and aspect within a climatic zone. In self-sustaining forest communities, ecosystem processes and functions such as energy flow, production, nutrient cycling, reproduction, regeneration and decomposition are not impaired. Topography, soil structure, texture and nutrients are important determinants of species composition and forest structure.

The structural and functional components of a natural regional forest ecosystem are predictable. They include the complexity of the tree, shrub and ground layers that provide habitat for mammals, birds and invertebrates. Topography, soil structure, texture, nutrients and moisture are important determinants of species composition and forest structure. Studies have assessed the rate of community development in forested regions after effects such as fire, logging and erosion and have reported that in a system where soils are not impairing regeneration, the forest canopy can be closed (i.e. tree crowns touch) 30 years after the event has ceased (Chambers 1995). The approach used to rank the plant community data considered in part whether the site was regenerating along this predictable pattern. The metal levels, low pH and erosion at some of the test sites represent ongoing perturbations, which could interrupt a predictable pattern of vegetation recovery. At these sites, once a plant community becomes established, it may be dominated by acid and metal tolerant species.

10.6.2 Approach to Plant Community Assessment

The terrestrial plant community assessment surveys were conducted in August and early September of 2004, with some additional work done in the summer of 2005. At each site, a 100 m × 100 m square plot was established and the site characteristics were recorded. The size of plot was similar to those established by Natural Resources Canada to monitor the effects of acid precipitation and, more recently, climate change. The plots were large enough to include some of the natural variability that occurs in the Sudbury Region, where plant distribution is often clumped and interspersed with forest openings on rocky outcrops. At the same time, the plot was placed where site conditions and plant cover were relatively uniform in terms of dominant tree cover, mode of deposition, range of soil depth, terrain and stoniness.

The 100 m × 100 m plot size optimized the sampling effort and permitted the site description to refer to relatively uniform site conditions in forested ecosystems. The plot was also established with a perimeter buffer strip of at least 10 m to avoid edge effects from roads, agricultural areas or other potentially influencing factors. Sub-sampling units were established to record species presence and cover in the understory, ground cover and downed woody debris.

Each site was classified by ecosite. The same format from the Ontario provincial classification system was adopted to name the ecosite types and plant communities: dominant over-story species, dominant understory species and a terrain modifier. Linkages were also made to earlier plant community classifications for the Sudbury area (Amiro and Courtin 1981; James and Courtin 1985; Sinclair 1996). There were five major components to the field work performed for the plant community assessments: 1) Broad plant inventory, 2) Percent cover assessment, 3) Detailed tree and tall shrub assessment, 4) Coarse woody material assessment, and 5) Ecosite classification. Each of these is described in the following text.

10.6.2.1 Broad Plant Inventory

The aim of the broad plant inventory was to produce a detailed list of plant species growing at each site. The parameters measured included species from the following groups: trees; tall shrubs; low shrubs; herbs; graminoids; ferns and fern allies; and bryophytes and lichens. Any unknown species were collected and keyed out in the lab using Gleason and Cronquist (1991), Soper and Heimburger (1990), Dore and McNeill (1980) for grasses; Crum (1976) for bryophytes; and Brodo and Sharnoff (2001) for lichens, with reference checks with herbarium specimens. The presence of all species was recorded within a 10 m width along four 50 m transects radiating north, east, south and west from the central staked area. The design increased the probability that the survey covered the range of local site conditions that occurred in the plot, with a total surveyed area of approximately 2000 m^2.

10.6.2.2 Percent Cover Assessment

The aim of the percent cover assessment was to collect quantitative information about the relative abundance of a variety of plant species and site conditions. To assess the percent cover, a 25 m transect was established from a northeast to southwest direction over the staked area. Again, the transect design was used rather than a square format to increase the probability of describing the range of plant community conditions that occurred at the site. This design was a standard approach on terrain that consists of a mosaic of variable soil depths over bedrock or variable conditions. Square or circular plots are more common in areas of uniform conditions. Along this transect, 25 quadrats each 1×1 m square were established; the 1 m square quadrat size is the standard for assessing the bryophyte/lichen, herb and low shrub layers (Brower et al. 1997). In each quadrat, the percentage cover of each plant species present (adult trees and tall shrubs were generally not included in this assessment) was estimated. In addition, the percentage ground cover of non-vegetative substrate, including bedrock, gravel/cobbles, bare soil, leaf litter, surface crusts and woody debris (<7.5 cm in diameter pieces) was estimated to the nearest 5%.

10.6.2.3 Detailed Tree and Tall Shrub Assessment

At both ends of the 25 m transect used to determine the percent cover assessment, a 10 m × 10 m plot was established to record the trees and shrubs existing at the site. This is the standard plot size for assessing tall shrubs and trees (Brower et al. 1997). Within each plot, the following parameters were assessed:

- all species were identified
- the strata (canopy or understory) were identified for each species
- the number of specimens from each species and strata was counted
- the percentage cover was estimated for each species in both strata
- the average height was estimated for each species, and
- the average diameter at 30 cm and diameter at breast height (DBH) were recorded for each species.

In addition, a separate, thorough assessment of the mortality and percentage dieback for trees and tall shrubs was conducted in the summer of 2005. For this assessment, a 10 m × 10 m plot was established 25 m northeast of the stake, and another plot 25 m southwest of the stake. The plot locations were adjusted slightly to ensure that they fell within the dominant ecosite type for the site. The following parameters were collected in each plot:

- the total percent tree cover in the plot
- identification of all species in the plot
- the growth form (i.e. single-stem vs. multi-stemmed) for each species
- the height of each tree
- estimation of the % crown mortality
- the number of dead stems (per clump if growth form was coppiced)
- the number of live stems (per clump if growth form was coppiced), and
- any snags present at the site were noted, with the species identified.

10.6.2.4 Assessment of Coarse Woody Material

The coarse woody material at the site was assessed along a transect established from the southwest to the northeast corner of the initial 10 m × 10 m tree and shrub assessment areas. Along each transect, each stump or piece of downed woody debris greater than 7.5 cm in diameter was recorded in terms of species (if possible), length, diameter and degree of decomposition. The degree of decomposition was established using decay classes developed in Sollins (1982). This measure is important because it indicates suitable seedbeds for many native species and, returns organic matter into the ecosystem providing a niche for microflora and fauna, which are indicators of healthy ecosystems.

10.6.2.5 Ecosite Classification

Each site was classified according to the dominant plant community present. The following information was collected at each site:

- Terrain (level, undulating, rolling, hilly), and an estimate of the average % slope
- Soil depth (very shallow <10 cm, shallow 10–30 cm, moderately deep 30–60 cm, deep >60 cm)
- Soil type according to textural analysis (loam, silt loam, loamy coarse sand, etc.)
- Percentage cover of exposed bedrock
- Dominant understory, including dominant species of tall shrub, low shrub (note also presence/absence of blueberry species), herbaceous, pteridophyte, graminoid, bryophyte (note also presence/absence of *Polytrichum* species), and lichen (note also presence/absence of *Cladina* species)
- Dominant overstory, with species composition
- Percentage canopy cover category (<30%, 30–50%, 50–70%, >70%), along with a visual estimate of actual percentage.

Using the above information, an ecosite description was provided for the site, including a subdominant community where necessary. The communities were named based on the dominant overstory, dominant understory, and a description of the terrain. The parameters measured are those used by the Ontario Ministry of

Natural Resources Ecosystem Classification System (Chambers et al. 1997). The ecosite classification provides a common language ecosystem classification description for each site that is recognizable by ecologists familiar with the Ecological Land Classification (ELC) system in Ontario. The ecosite classification provides a succinct description of the ecological community present at each site.

10.6.3 Plant Community Results

A large quantity of data was collected during the plant community survey. This information was analyzed in a variety of ways to gain an overall ranking for each site. An overview of some of the salient results is presented below.

10.6.3.1 Broad Plant Inventory

In total, 297 plant species (31 ferns and fern allies, 26 grasses and sedges, 109 herbaceous species, 29 lichen, 28 low shrubs, 26 mosses, 30 tall shrubs and 18 tree species) were identified. The number of species at the test sites ranged from 21 to 82, and the number of species at the reference sites ranged from 57 to 89. A summary of the number of species within each group at each site is provided in Table 10.10.

Table 10.10 **Number of Plant Species at Each Test Site**

Site	Ferns and fern allies	Grasses and sedges	Herbaceous	Lichens	Mosses	Low shrubs	Tall shrubs	Trees	Total species
CC-01	1	3	9	14	7	6	4	5	49
CC-02	0	4	0	11	3	4	1	7	30
CC-03	0	4	0	7	4	2	0	4	21
CC-04	2	5	11	6	7	5	9	8	53
CC-06	6	4	17	13	9	7	12	4	72
CC-07	6	5	10	12	7	7	7	7	61
CC-08	6	9	18	20	8	6	6	9	82
CON-01	3	2	11	10	6	4	5	6	47
CON-02	2	5	6	13	7	5	4	4	46
CON-03	4	6	9	12	8	6	9	4	58
CON-05	3	4	6	15	7	5	10	5	55
CON-06	3	12	13	17	7	5	4	7	68
CON-07[a]	2	8	29	20	7	1	7	6	80
CON-08	2	4	7	15	2	3	3	5	41
FB-01	2	1	1	2	2	4	2	6	20
FB-02	5	2	10	11	8	6	6	7	55
FB-03	5	1	8	17	11	9	6	6	63
FB-05	2	3	5	9	3	2	3	7	34
FB-06	1	3	5	18	10	5	2	8	52
REF-02	5	3	13	8	7	6	8	7	57
REF-03	4	6	26	1	10	7	13	6	73
REF-04	4	3	13	22	15	11	12	9	89

[a] This site was historically limed and replanted.

Figure 10.5 illustrates the trend of plant species richness versus distance from the nearest smelter. The number of plant species present at the sites closer to the smelters was generally lower on the Copper Cliff and Falconbridge transects than for sites farther away. This trend was not observed with the Coniston transect. The lack of relationship at the Coniston sites may be a result of the low total metal levels in the soils and/or other factors such as soil erosion. Some other observations relative to distance from the smelters included:

- There was no relationship for the number of lichens, bryophytes, herbs or low shrub species and distance to the smelter
- There was a higher percentage of bare rock and soil near the smelters
- There were lower numbers of lichens and moss species near the smelters which indicate that conditions were not suitable to support these species
- There was a lower percentage of leaf litter (a component of soil fertility) close to the smelters

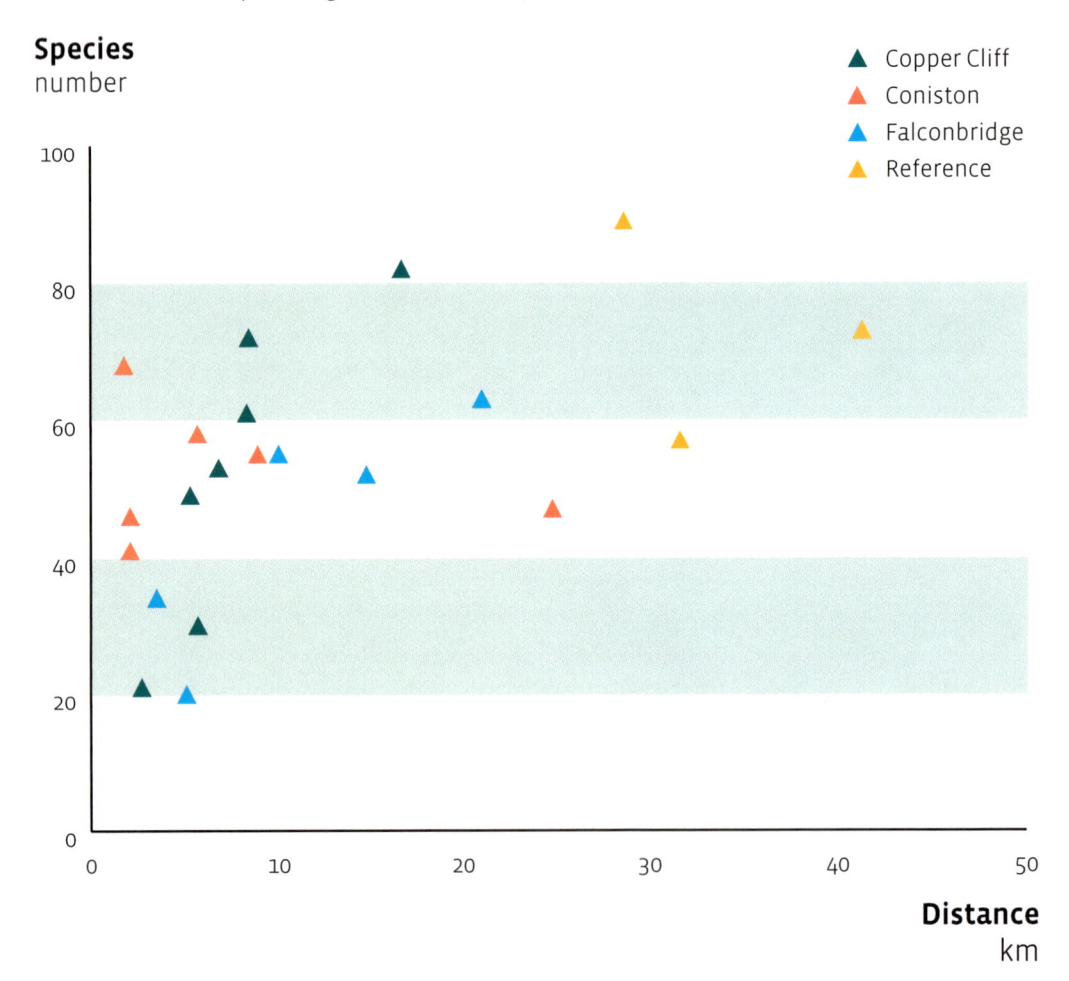

Figure 10.5 **Plant Species Richness versus Distance from the Nearest Smelter**

Along the Copper Cliff transect, there was a noticeable difference in species from site to site. There were no ferns, fern allies or herbaceous plants observed at CC-02 or CC-03. The number of tree species observed along this transect ranged from 4 to 9. Lichens and mosses had the greatest species richness at all sites apart from CC-04 where herbaceous plants and shrubs had the greatest species richness. Plant species that were considered acid-metal tolerant were also categorized at each site. Sites along the Copper Cliff transect had an average of 13 acid-metal tolerant indicator species (ranging from 10 to 18). The Coniston transect had a greater representation of plant types at all sites. Sites on the Coniston transect had an average of 13 acid-metal tolerant indicator species (ranging from eight indicators at CON-01 to 17 at CON-06). The distribution of plant species was different at the Falconbridge sites compared to the other two transects. The number of tree species (6–8) was similar to the other two transects. Sites on the Falconbridge transect had an average of 10 acid-metal tolerant indicator species (ranging from five indicators at FB-02 to 14 at FB-03).

The reference sites were generally similar to the test sites in their distribution of plant species, with herbaceous plants, lichens and mosses and shrubs having the greatest species richness. The number of tree species was similar to the other three transects with 6–9 tree species recorded at the reference sites. The average number of acid-metal tolerant indicator species at the reference sites was 7 (ranging from 4 to 10). The greatest species richness was observed at REF-04 with almost 90 different species recorded, while FB-01 had the least with only 20 species present. The range of total number of species for the transects were: Copper Cliff, 21–82 species; Coniston, 41–80 species; and Falconbridge, 20–62 species. At the three reference sites, the total number of species ranged from 58 to 89.

10.6.4 Site Ranking Based on Plant Community LOE

The plant community assessment provided a compelling line of evidence for the ERA since the living vegetation at a site reflected the total integration of all site conditions (i.e. climate, soil quality, physical and chemical characteristics). The objective of this LOE was to provide detailed documentation of the relative condition or diversity of the plant community at each test site by comparing them to the reference sites. For the purposes of the plant community assessment, all information with regard to the soil metal levels and other soil chemistry was deliberately not provided to the team members responsible for interpretation of the ecological data to reduce the potential for bias in the plant community assessment. For this LOE, the plant community results were evaluated in terms of four ecologically significant criteria:

Site Biodiversity:

This criterion provided a snapshot of the number and distribution of species in the community. As such, it was a biological index of the current overall status of the community. It did not provide any evaluation of the 'quality' of the community in terms of the integrity of the species present or the presence/absence of all components in the community, nor did it provide a strong indication of the sustainability of the community over time. However, it could be considered as a rough estimate of the resilience of the community to stresses.

Ecological Integrity:

This criterion provided an index of the number of different species present in the community and their growth habits. For example, ecological integrity took into consideration the presence or absence of invasive species, the presence or absence of acid-metal tolerant indicator species, and the presence or absence of a skewed species distribution, thereby providing an indication of potentially unusual site conditions. This criterion measured the completeness of the community and the integration among the parts.

Long-term Productivity:

This criterion took into consideration how well species were growing on a site and considered whether this growth was sustainable into the future. For example, the indication of stressors, such as insects and disease, is considered. In addition, long-term productivity provided insight into the actual growing capacity of the site as reflected in a surrogate for biomass production (height of trees, density of trees and shrubs). It also provided an indication of the availability of organic debris in the community, which provides a slow release of organic-based nutrients.

Soil and Water Conservation:

This criterion reflected the integrity of the growing medium at the site. In addition, assessment of this criterion provided an indication of the degree to which water is held on the site to support growth. The properties of the soil define regeneration success potential by promoting or limiting plant growth.

The selection of these criteria was based on the biological criteria defined by the Canadian Council of Forest Ministers (CCFM 2000) to measure Canada's progress in the sustainable management of its forests as well as the criteria used to assess biodiversity within Canada's biodiversity strategy, a framework adopted by national and provincial parks and protected areas (Environment Canada 1995). A number of indicators were assigned to each of the four criteria (Figure 10.6). Links between criteria were defined and, in some cases, indicators addressed multiple values under different criteria. The indicators were selected because they were considered to be good descriptors of the relative status of the community with respect to the criteria, they were easily measured, and they could be used for future monitoring of changes to communities.

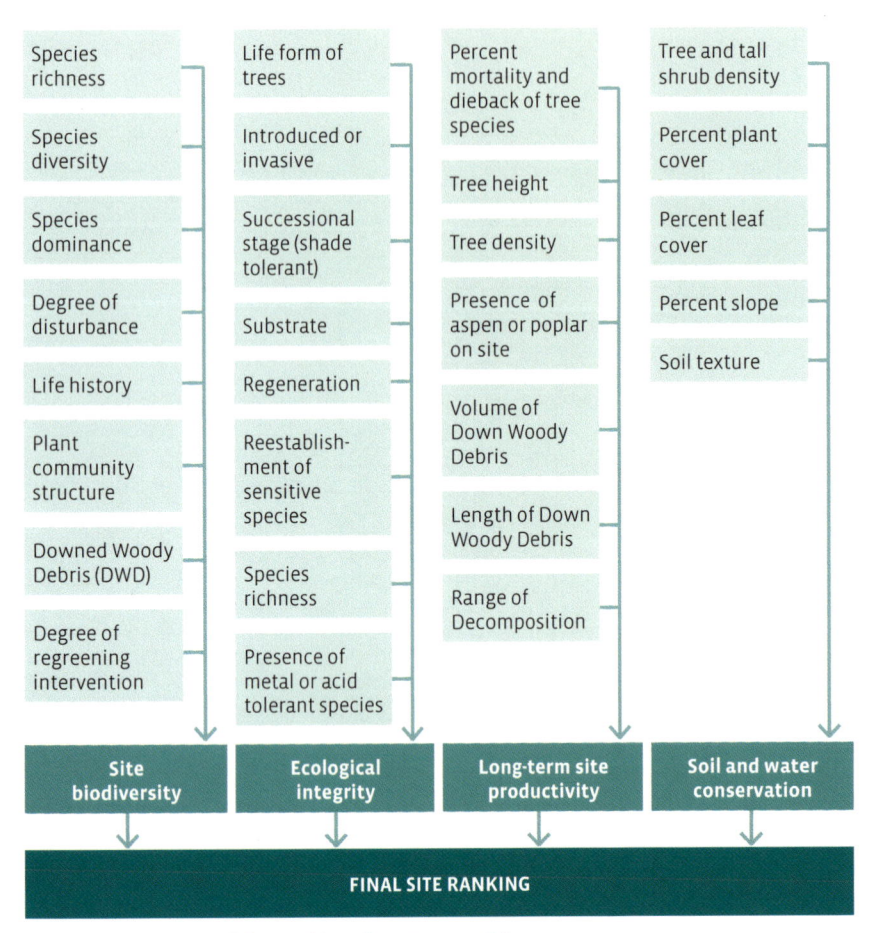

Figure 10.6 Framework for Ranking Plant Communities

Site biodiversity indicators:
- Species richness
- Species diversity
- Species dominance
- Degree of disturbance
- Life history
- Plant community structure
- Downed Woody Debris (DWD)
- Degree of regreening intervention

Ecological integrity indicators:
- Life form of trees
- Introduced or invasive
- Successional stage (shade tolerant)
- Substrate
- Regeneration
- Reestablishment of sensitive species
- Species richness
- Presence of metal or acid tolerant species

Long-term site productivity indicators:
- Percent mortality and dieback of tree species
- Tree height
- Tree density
- Presence of aspen or poplar on site
- Volume of Down Woody Debris
- Length of Down Woody Debris
- Range of Decomposition

Soil and water conservation indicators:
- Tree and tall shrub density
- Percent plant cover
- Percent leaf cover
- Percent slope
- Soil texture

→ **FINAL SITE RANKING**

The application of ranking followed a stepwise process:

- Indicators were selected within each of the four ecological criteria (site biodiversity; ecological integrity; long-term site productivity; and soil and water conservation)
- The indicator results were calculated for each site and assigned a rank based on comparison to the reference sites
- Using best professional judgment, a rank was given to the criterion, based on the associated indicator ranking results. Ecology experts considered the ranks given to the indicators, as well as their impression of the condition of the site to assign criteria ranks

Once all criteria were ranked, each test site was evaluated using best professional judgment to determine whether it was impacted or not. For the purposes of evaluating the results of the plant community assessment, each site was ranked as follows:

Rank		Description
Green	Low to not impacted	The site was representative of a complete forest ecosystem in that eco-district of Ontario
Yellow	Moderately impacted	The site was not representative of a complete forest ecosystem. It was an ecosystem in transition and was showing signs of recovery or decline
Red	Severely impacted	The site was not at all representative of a complete forest ecosystem, and showed very few signs of recovery

Each reference site was evaluated in terms of the ecological criteria and indicators to determine whether it represented the range of plant communities typifying the Ontario Eco-districts. The Sudbury region covers three eco-districts. Evaluation of the reference sites showed them to be satisfactory comparison sites for the range of plant communities typifying a transitional vegetation zone where boreal jack pine, balsam fir and spruce forest elements mix with Great Lakes St. Lawrence Forest hardwoods and mixed-forest elements. The reference sites did emphasize that not every indicator had to be rated high for a plant community to be natural and self-sustaining. The evaluation revealed that the test sites contained a diversity of plant communities. The majority of the test sites were ranked severely impacted, indicating that the plant community was not representative of a complete forest ecosystem and showed very few signs of recovery (Table 10.11). Some sites were ranked moderately impacted, indicating that the site was not representative of a complete ecosystem but was an ecosystem in transition, showing signs of either recovery or decline. There was only one test site (FB-02) where the plant community was considered similar (low impact) to the reference sites.

Table 10.11 **Site Ranking for the Plant Community Assessment LOE**

Site	Rank
CC-01	Red
CC-02	Red
CC-03	Red
CC-04	Red
CC-06	Yellow
CC-07	Red
CC-08	Red
CON-01	Yellow
CON-02	Red
CON-03	Yellow
CON-05	Red
CON-06	Red
CON-08	Red
FB-01	Red
FB-02	Green
FB-03	Yellow
FB-05	Yellow
FB-06	Yellow

10.7 Soil Toxicity Line of Evidence

10.7.1 Selection of Test Species

Ecological communities are an aggregation of populations consisting of all plants, animals and microbes that occur in the same time and place and that interact physically, chemically and/or behaviorally. Although the community itself is what the risk managers ultimately aim to protect, it was not possible to study all components of the ecosystem. The rationale for conducting toxicity testing was that the soils in Sudbury contain more than one metal and the soil pH is below the range (5–9) addressed in the generic MOE soil criteria (MOE 1997). By conducting toxicity testing in soil collected from the sites, it was possible, under standardized laboratory conditions, to determine whether the soil was toxic to plant and invertebrate test species irrespective of other environmental conditions (microclimate, moisture, depth to bedrock, etc.). This testing in 'natural' soil from the site (soil collected from the site but not amended in any way) was used to rank the soil toxicity LOE. No attempt was made to calculate a specific toxicity value (i.e. LC_{50}, EC_{20}, etc.) to correspond with a given metal concentration since the soils represented a mix of metals, low pH and combination of other site-specific factors that strongly influenced soil toxicity. Rather, the results of toxicity tests from the 'exposure' sites were compared to results from the reference site tests, to compare the relative toxicity between sites.

All of the selected reference and test sites had soil pH levels between 4 and 5. A pH of less than 5 is known to be limiting for growth in some plants and survival in some invertebrates (Winterhalder 1995; Troeh and Thompson 2005). Concurrent to the testing in natural soil, tests were also completed using pH-amended soil (soil collected from the site and the pH raised to a standardized level). The aim of the amendments was to determine whether soil pH was a limiting factor to plant growth or invertebrate survival and reproduction. However, it should be noted that, when soil pH is raised, the availability of metals decreases, therefore, it was not possible to completely separate the effect of pH on plant growth from the effect of metals. The testing in the pH-amended soils was not used in the overall site ranking but rather to investigate uncertainties surrounding soil pH. As mentioned beforehand, site CON-07 that was historically limed and re-planted with seeds and trees was also studied. The limed site was adjacent to an untreated site (CON-08) and all of the toxicity tests were conducted on the CON-07 soil so that the role of historic liming could be evaluated. The results of those toxicity tests are presented in Section 10.13.

More than one test species was used for toxicity testing. This is sometimes referred to as a battery approach. Lanno (2003) and Environment Canada (2005b) outlined some of the considerations involved with the selection of suitable test organisms for use as a terrestrial test battery for evaluation of soil toxicity at contaminated sites. A battery approach decreases the uncertainty related to the toxicity testing which is important in risk assessment. Ideally, the species used for the toxicity testing should be native to the Sudbury region and have soil toxicity test methods already developed for them. Unfortunately, the vast majority of standardized Canadian toxicity tests were for species of agricultural importance rather than native species. For this study, it was deemed important that the test species be representative of ecologically relevant native species or groups because it has not been demonstrated that crop species serve as surrogates for non-crop species found in nature. The test species used for the Sudbury toxicity testing were selected by considering the following factors:

- the sensitivity of the species to metals and pH
- the availability of the seeds or culture animals
- the species should be representative of the range of species found in Sudbury ecosystems
- the species should be easy to maintain and culture under laboratory conditions, and
- ideally, established toxicity test protocols should exist.

First, it was necessary to determine whether the species could be cultured in the laboratory and if they were sensitive to metals and low soil pH. Therefore, a phased approach to soil toxicity testing was used as follows:

- Step 1: Preliminary toxicity tests were conducted on potential test species in two site soils (one high, one low metal concentration) to determine which species would be suitable for further testing.
- Step 2: Detailed toxicity testing with select species in soils from all test and reference sites
- Step 3: Use of the Step 2 results to rank test sites on the basis of soil toxicity.

These steps are described in the following section.

10.7.1.1 Step 1: Preliminary Test Species Screening

The objective of Step A was to establish that the proposed toxicity tests could be viably performed in Sudbury soil and to determine whether any observed effects could be attributed to soil metals or pH level. A site with high metal concentrations (as determined by Cu and Ni), and a site with low metal concentrations with comparable pH levels were selected for the Step 1 testing. The high metal site selected was CC-03 (total Cu 948 mg/kg and total Ni 1100 mg/kg) and the low metal site was a reference site (REF-02: Cu 41 mg/kg and Ni 32 mg/kg). A total of six plant species and two invertebrates, a soft-bodied soil-dwelling species and an arthropod, were evaluated during the preliminary testing. The species tested were as follows:

- Trembling aspen (*Populus tremuloides*)
- Black spruce (*Picea mariana*)
- White spruce (*Picea glauca*)
- Northern wheatgrass (*Elymus lanceolatus*)
- Goldenrod (*Solidago canadensis*)
- Red clover (*Trifolium pratense*)
- Springtail (*Folsomi candida*)
- Earthworm (*Eisenia andrei*)

Bulk homogenized soil samples were used in all toxicity tests. An internal control soil (artificial soil) was used with each test run so that the viability of the organisms was assessed. In most cases, the organisms grew well in the artificial soil, indicating that the seeds, invertebrates and test conditions were acceptable. During the preliminary toxicity testing, it was observed that springtail survived well in the test site soil and was not sensitive to either elevated metal levels or low pH. Therefore, it was dropped from further use in detailed soil evaluation. All three tree species grew relatively well but aspen displayed poor growth in the reference soils and results were highly variable so it was dropped from further consideration. Both black and white spruce grew well in the soils and appeared sensitive to metals and/or low soil pH. White spruce was ultimately selected over black spruce for more detailed soil evaluation as it has been used in the Sudbury re-greening and re-planting programs. Based on the preliminary toxicity testing results, it was decided to undertake more detailed testing on all ERA site soils using one invertebrate (earthworm), one tree species (white spruce), a herbaceous native plant species (goldenrod), a dicotyledonous plant species (red clover) and a monocotyledonous species (northern wheatgrass).

10.7.1.2 Step 2: Detailed Soil Toxicity Testing

Following the results of Step 1, the recommended test battery included the five test species, measurement endpoints and experimental conditions outlined in Table 10.12. Toxicity tests were conducted for these species at all sites in natural site soil. In addition, tests were conducted concurrently in pH-amended soils with northern wheatgrass, red clover and earthworms. Before tests commenced, soils were hydrated using deionized water to approximately 70% of their water holding capacity (WHC) to achieve a uniformly moist, crumbly texture. Measurements of soil WHC, moisture, pH and conductivity were made at the beginning and conclusion of all tests.

Table 10.12 **Test Species and Endpoints Used for the Toxicity Testing LOE**

Ecological component	Test species	Endpoint	Test conditions
Trees	White spruce	Root length	Natural site soil
		Root mass	
		Shoot length	
		Shoot mass	
Herbaceous (monocot)	Northern wheatgrass	Root length	Natural site soil and pH-amended site soil
		Root mass	
		Shoot length	
		Shoot mass	
Herbaceous (dicot)	Red clover	Root length	Natural site soil and pH-amended site soil
		Root mass	
		Shoot length	
		Shoot mass	
Herbaceous (dicot)	Goldenrod	Root length	Natural site soil
		Root mass	
		Shoot length	
		Shoot mass	
Invertebrate	Earthworm	Survival	pH-amended site soil and natural site soil
		# juveniles	
		Mass of juveniles	
		Growth	

10.7.2 Toxicity Test Methods

10.7.2.1 Earthworm Test Methods

Earthworms are a useful test species as they have short life cycles so it is possible to conduct reproductive trials in the laboratory. They also represent an important ecological niche and are recommended for use in evaluation of contaminated soils. The existing test protocol of Environment Canada (2005a) was used for this study. Ten test containers (replicates) were prepared for each site with a volume of approximately 350 g soil. Containers with earthworms were maintained at 20±2°C with a photoperiod of 16 h light and 8 h dark. A sample of 20 adult worms were weighed at the beginning of the test and randomly distributed among the replicate containers. Two sexually mature adults were placed in each replicate container. The adults were removed on day 35 and the survival and weight of remaining adults were determined. The tests were terminated on day 63 when the number of surviving juveniles and their weight were measured. Earthworm tests were also conducted simultaneously in soils amended with $CaCO_3$ (calcium carbonate) to raise the pH to about 5.2 to simulate land treatment.

10.7.2.2 Plant Test Methods

Tests were carried out in temperature and light controlled environmental chambers at a constant temperature of 24±3°C following the protocol of Environment Canada (2005b). A light/dark photoperiod of 16:8 h was used. Soil moisture was supplemented as necessary with deionized water. Five replicate treatments were prepared per site soil. The duration of the tests varied between species: red clover, 14 days; northern wheatgrass, 21 days; and 8 weeks for white spruce and goldenrod. The criterion for seedling emergence was a minimum shoot height of 3 mm above the surface of the soil. At the conclusion of each test, each plant container was carefully inverted and representative samples of soil collected for moisture, pH and conductivity or other chemical analysis. Seedlings were gently washed to remove soil particles and shoot and root lengths (mm) recorded. Once shoot and root lengths were determined, they were separated with a scalpel. The shoots and roots were dried then pooled for dry weight (biomass) determination. Toxicity tests in pH-amended soil were also conducted using northern wheatgrass and red clover using the same protocols as those performed in natural site soil.

10.7.2.3 Role of Soil pH

Soil pH was not a chemical of concern in this risk assessment, but is known to have a major influence on metal availability and the suitability of soils for plant growth. Within the Sudbury region, the range of soil pH was below that which is typically considered suitable for plant growth (pH < 5.0). The pH of soil indicates the activity of hydrogen ions held on the clay and organic matter particles. The pH scale is logarithmic based on the 'powers of ten' so pH 6.0 is 10 times more acidic than pH 7.0, while pH 5.0 is 100 times more acidic than pH 7.0. Acidity influences a whole range of soil characteristics such as the nature of the variable charge, nutrient availability, microbial activity and the release of metals (Ashman and Puri 2002).

The degree of soil acidity has a direct influence on plant growth. Many Boreal forest ecosystems have very acidic soils where the pH values may be 4 or less. Due to industrial emissions in the region, certain Sudbury area soils were acidified from an expected normal pH range of 4.5 to 5.5 to levels of pH 3.2 to 4.4. The role of pH in soil toxicity was evaluated in two ways: a) toxicity tests were conducted in natural and pH-amended soils, and b) the characteristics of a historically limed field site (CON-07) were compared to an adjacent non-limed site (CON-08).

10.7.3 Toxicity Testing Results

The toxicity tests generated a significant amount of data from the combination of five test species, 20 endpoints, 22 sites plus natural and pH-amended soils. Some data from the toxicity tests are presented here, but more importantly, the results were used to formulate an overall toxicity testing LOE ranking for each site.

Adult earthworms survived in all of the natural reference and test site soils. However, there was no reproduction or survival of progeny in any of the natural (not pH-adjusted) test soils, and only minor reproduction in the natural reference sites. Therefore, the natural site soils were consistently toxic to earthworm reproduction. When the site soils were amended with $CaCO_3$ to raise the pH to about 5.2, reproduction and survival of young earthworms increased significantly. Thus, earthworm production was very sensitive to pH, and less sensitive to metals at the range of metal concentrations in this study. Figure 10.7 illustrates the mean number of juvenile worms produced from the soils along each transect (CC = Copper Cliff; CON = Coniston; FB = Falconbridge; REF = Reference sites) in natural and pH-amended soils. Within a transect, earthworm reproduction tended to be lower in site soils with higher metal concentration but the relationship was not consistent.

Live juveniles
number

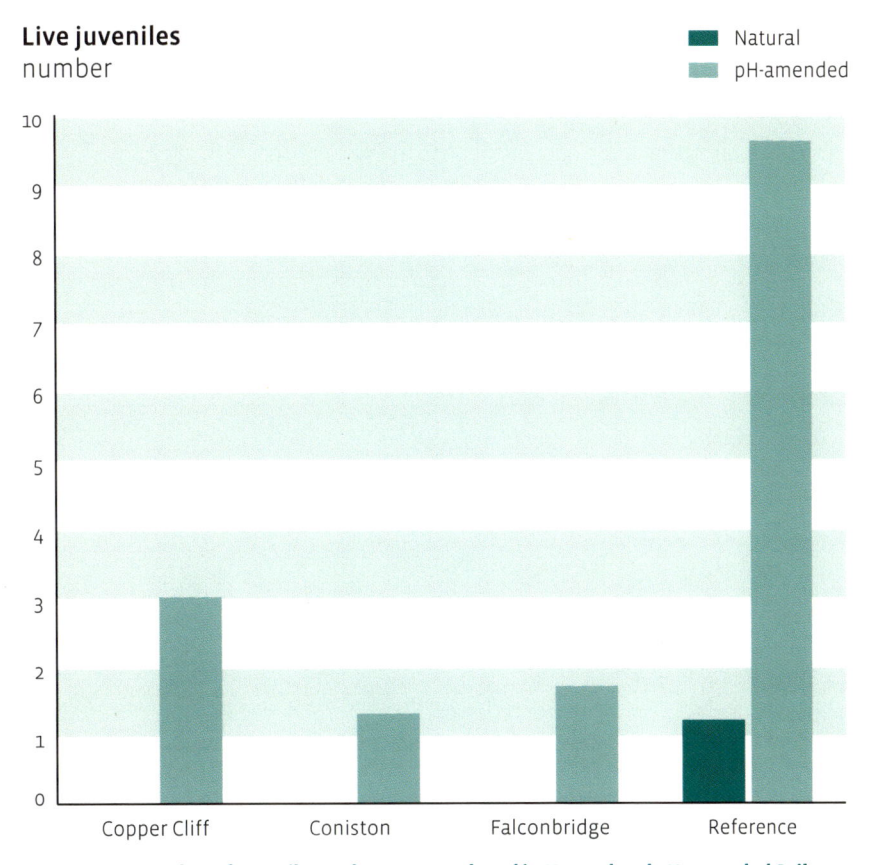

■ Natural
■ pH-amended

Figure 10.7 **Number of Juvenile Earthworms Produced in Natural and pH-amended Soils**

Ultimately, the earthworm reproductive test results were not used in site ranking or the toxicity test LOE. The reason for not using these test results was because there was no production of progeny in any natural soils from any of the three exposure transects. There was production of progeny in pH-amended soils but there was no difference between sites that could be attributed to metal concentrations. Therefore, the earthworm tests demonstrated that the worms were very sensitive to soil pH, but the results did not yield information that could be used to differentiate, or rank, the test sites. Spurgeon et al. (2006) demonstrated that earthworm reproduction was very sensitive to metals, soil pH and their interactions.

Overall, growth and development of all plant species was reduced in the natural test soils relative to soils from the reference sites. Different endpoints responded differently between species and between transects so it was not possible to choose a single endpoint over another as an indicator of stress. This was not considered a limitation as Kapustka (1997) pointed out that phytotoxicity can be expressed in one or more plant endpoints. Spurgeon et al. (2006) also reported for earthworms that different endpoints showed different strengths of relationships with soil metal fractions suggesting different dependencies of individual parameters on soil chemistry. In general, total biomass (combination of shoot and root biomass) tended to be the most sensitive measure of plant growth in different site soils. Figure 10.8 shows the different endpoints (i.e. shoot growth, shoot weight) of white spruce grown in soil collected from the Copper Cliff sites. The figure shows there was considerable variability within replicates within a site and between sites along a transect. The photographs of white spruce grown in soils from the Coniston transect (Figure 10.9) clearly illustrate the differences in growth rate observed between different site soils even within a transect. Figure 10.10 illustrates two different endpoints for goldenrod grown in soils collected from the Falconbridge sites. Similar to the white spruce, there was considerable variability between test sites, and even within replicates from a single site. There was no published test protocol or guidance document for goldenrod but it is a native species that is widely distributed. It was considered to be a useful test species for this study.

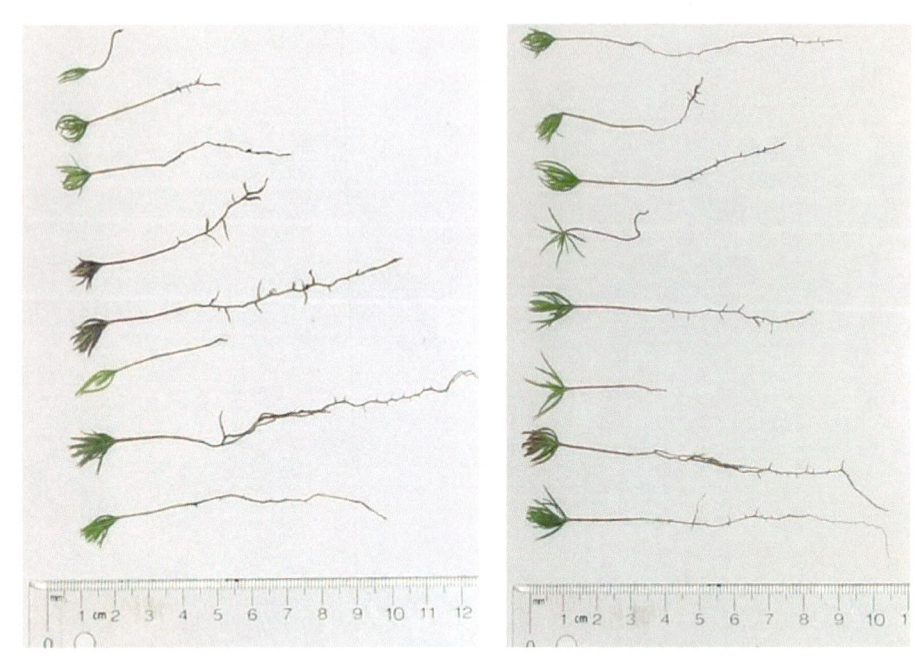

Figure 10.8   **White Spruce Seedlings in Copper Cliff (CC) Soils (Photo from SRC 2005)**

Figure 10.9   **White Spruce Grown in Reference and Soils along the Coniston Transect (Photo by M. Moody, SRC 2005)**

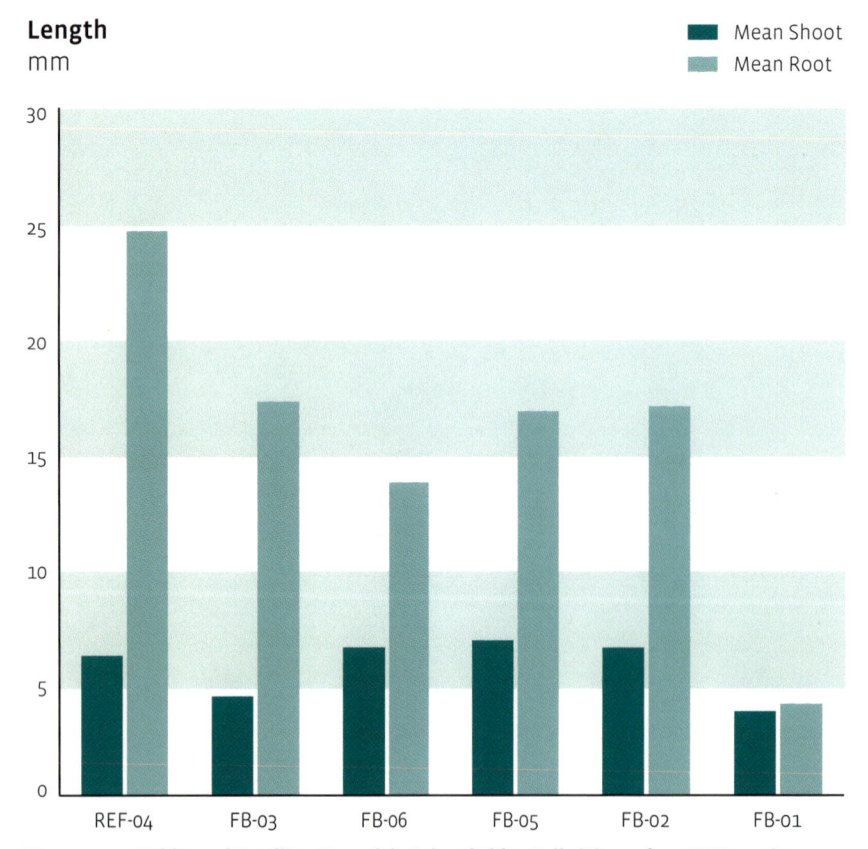

Figure 10.10 **Goldenrod Seedling Growth in Falconbridge Soils (Figure from SRC 2005)**

Total biomass of red clover and northern wheatgrass grown in natural and pH-amended soils is shown in Figures 10.11 and 10.12, respectively. Surprisingly, the growth of red clover was similar in natural soils from the test sites to growth in natural soils from the reference sites (Figure 10.11). Raising soil pH significantly increased growth of red clover in soils from Copper Cliff and the reference sites. Total biomass of northern wheatgrass was higher in natural soils from the reference site compared with the test soils (Figure 10.12). Raising soil pH significantly improved growth of wheatgrass in the test soils such that, overall, it was similar to that in pH-amended soils from the reference sites.

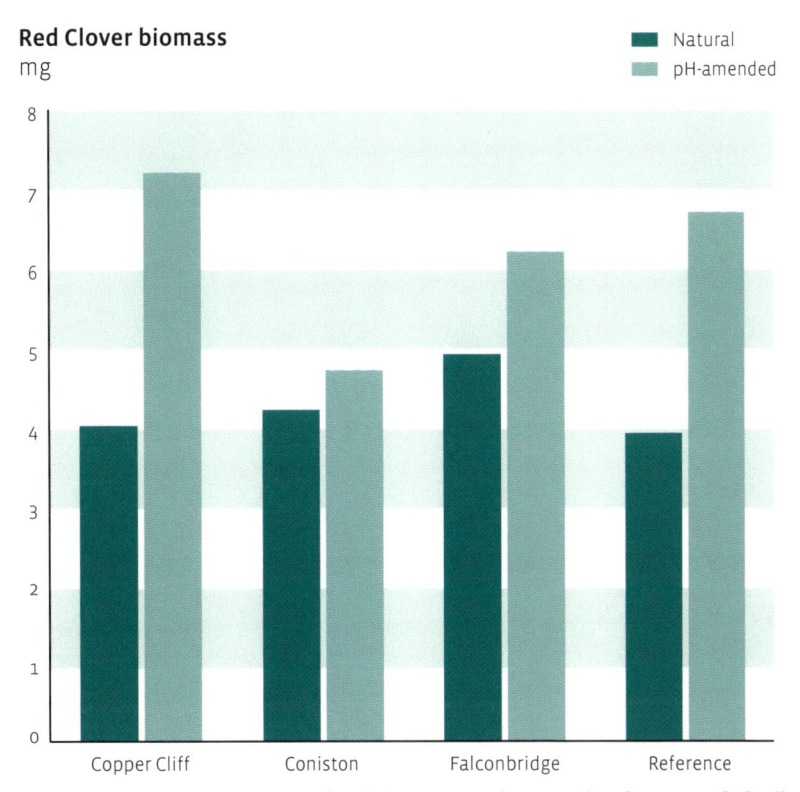

Figure 10.11 **Mean Biomass (mg) of Red Clover Grown in Natural and pH-amended Soil**

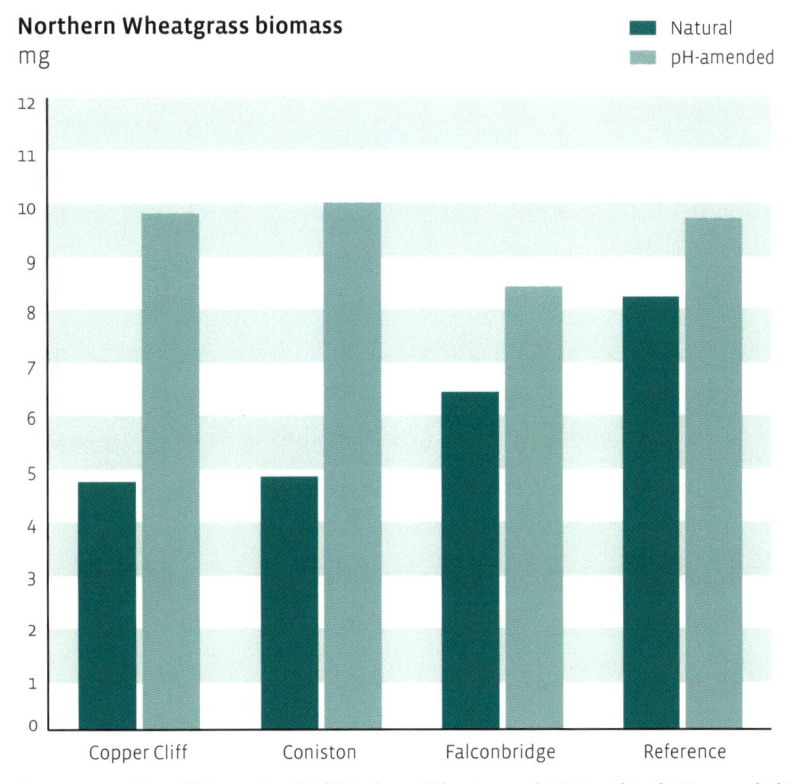

Figure 10.12 **Mean Biomass (mg) of Northern Wheatgrass in Natural and pH-amended Soil**

The effect of site soil and low pH on development of northern wheatgrass is shown in the photographs provided in Figure 10.13. The benefit of raising soil pH in the reference soil can be seen with more rigorous root growth in the higher pH (5.18) soil compared to root growth in natural pH (3.56) soil. Growth of roots and shoots was visibly inhibited in soil with elevated metals and low pH (CC-03) relative to reference soil. Total Ni and Cu levels in CC-03 soil were 1100 and 1000 mg/kg, respectively. The benefit of raising soil pH in CC-03 soil was apparent (photo far right) although metal toxicity was not totally alleviated.

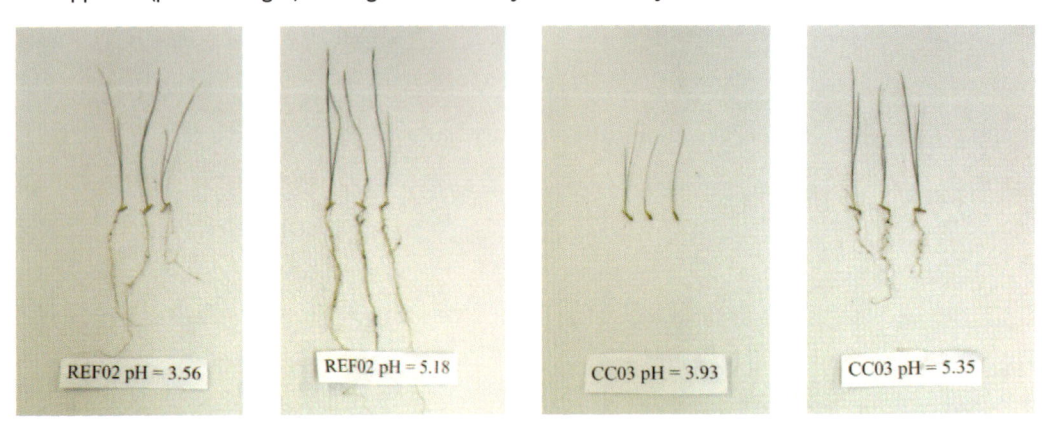

Figure 10.13 **Wheatgrass Grown in Natural and pH-amended Soils from a Reference and a Test Site (Courtesy of Soil Toxicology Laboratory, Biological Assessment and Standardization Section, Environment Canada, Ottawa)**

As mentioned previously, no attempt was made in this study to calculate a numerical expression of soil toxicity (i.e. LC_{50}, EC_{50}) based on the concentration of single metals. The reason was twofold. First, a mixture of metals was present so that calculation of a single metal toxicity value would be potentially very misleading. Second, soil type differed between each site such that soil variables that are known to influence metal toxicity (i.e. pH, organic matter, CEC) were also different. In themselves, these would have affected soil toxicity irrespective of the metal concentrations. Various studies have emphasized the importance of soil properties on metal toxicity to plants and soil invertebrates. Bradham et al. (2006) studied Pb toxicity to earthworms in 21 different types of soil. At a concentration of 2000 mg/kg, earthworm mortality ranged from 0 to 100% depending on the soil used in the test. This represents an enormous range of response. The authors concluded that pH was the single most important soil property affecting Pb toxicity. Rooney et al. (2006) evaluated the toxicity of Cu to barley and tomatoes in 18 different soils. In that study, the EC_{50} (Cu concentration causing 50% growth inhibition) varied by 15-fold (36 to 536 mg/kg) for barley and 39-fold (22 to 851 mg/kg) for tomatoes, in different soil types. These authors determined that soil Ca and CEC levels were the best predictors of Cu toxicity to plants.

Feisthauer et al. (2006) conducted a battery of toxicity tests on three samples of Sudbury soils. Chronic exposure to the soils caused adverse effects to plant growth and invertebrate survival where the pattern of toxicity followed a metal concentration gradient. The range of metal concentrations in that study (i.e. Ni 45–435 mg/kg; Cu 38–640 mg/kg) were within the range of metal levels in soils used for toxicity testing in the current study. The authors also acknowledged that soil toxicity was likely due to a combination of factors including metals, low soil pH, organic matter, CEC, Al and Mg levels. Kapustka et al. (2006) calculated a mean EC_{20} for Co and Ni based on plant growth endpoints using three different plant species in two soil types. The mean EC_{20} for Co was 30.6 mg/kg, and was 27.9 mg/kg for Ni. These toxicity levels were lower than the Ontario provincial soil quality guidelines for Co (40 mg/kg) and Ni (150 mg/kg). The Ni EC_{20} was also lower than the mean Ni concentration (36.1 mg/kg) observed in parent (i.e. background) soil from the Sudbury area. The authors acknowledged that seed emergence and mortality were less sensitive than growth parameters and were not used to calculate the EC_{20}.

10.7.4 Site Ranking Based on Toxicity Testing Results

The objective of the toxicity testing LOE was to determine whether the performance of test species in soils collected from the test sites was inhibited relative to the reference sites. Each site was given a ranking for the toxicity testing LOE. The performance of the test species was assessed independently of the metal concentration of the site soil. An overview of the approach and a summary of the resulting ranks are presented in the following sections.

10.7.4.1 Reference Site Evaluation

Most of the standard toxicity test species and protocols were developed for agricultural soils, not for soils from north-eastern Ontario boreal forests. The Sudbury soils differ markedly from agricultural soils in that they displayed low soil pH and low mineral and nutrient content. Therefore, various procedures were undertaken to verify the performance of the toxicity tests in the reference soils. These evaluations included an evaluation of the performance of the organisms in artificial soil to provide baseline measurements; an evaluation of the sensitivity of the organisms to pH in artificial soil; and a comparison of the performance of the organisms in the reference soils to the artificial soil with a comparable pH. These evaluations established that the soil from the reference sites provided an adequate baseline for comparison to the test site soils.

The results also demonstrated that, in the absence of pH as an influencing factor, performance of the test species in the reference soils was sometimes limited by other variables. This finding did not affect the usefulness of the reference soils as a comparison to the test soils but did demonstrate that the chosen test species did not always perform well in Sudbury soils. In the absence of more regionally appropriate test species developed for forested regions, the battery of species selected for this study was considered to be the most appropriate option available at the time. Although there was variation between the performances of the test species in the soil from the three reference sites, no one reference site stood out as particularly poor; some reference sites were excellent for one species but not for another. For comparative purposes, a mean of the values for each endpoint for the three reference sites was determined and was referred to as REFmean. The REFmean value provided a baseline for comparison with the test sites and was considered indicative of the average performance of the test species in soil from forested regions of the Sudbury area.

10.7.4.2 Test Site Evaluation

Unlike the other LOE, two separate approaches were used to independently rank the test sites based on soil toxicity results:

Approach 1:

The toxicity data from each of the test sites were statistically compared to the results from each of the three reference soils (REF-02, REF-03 and REF-04) using an analysis of variance (ANOVA) to determine whether there was a significant ($P < 0.05$) difference among treatment means. Using the results of an ANOVA analysis, the toxicity test results from each site were ranked by endpoint, then by species to eventually produce a site rank for Approach 1.

Approach 2:

The toxicity results from each of the test sites were compared to the mean of the toxicity endpoint from the three reference soils, referred to as REFmean. The comparison of the test sites to REFmean was ranked by endpoint, then by species to eventually produce a site rank for Approach 2. The REFmean approach was used to eliminate some of the variability observed for some endpoints and some test species between reference sites. The two approaches were combined to give the overall ranking for the site. This process is illustrated in Figure 10.14.

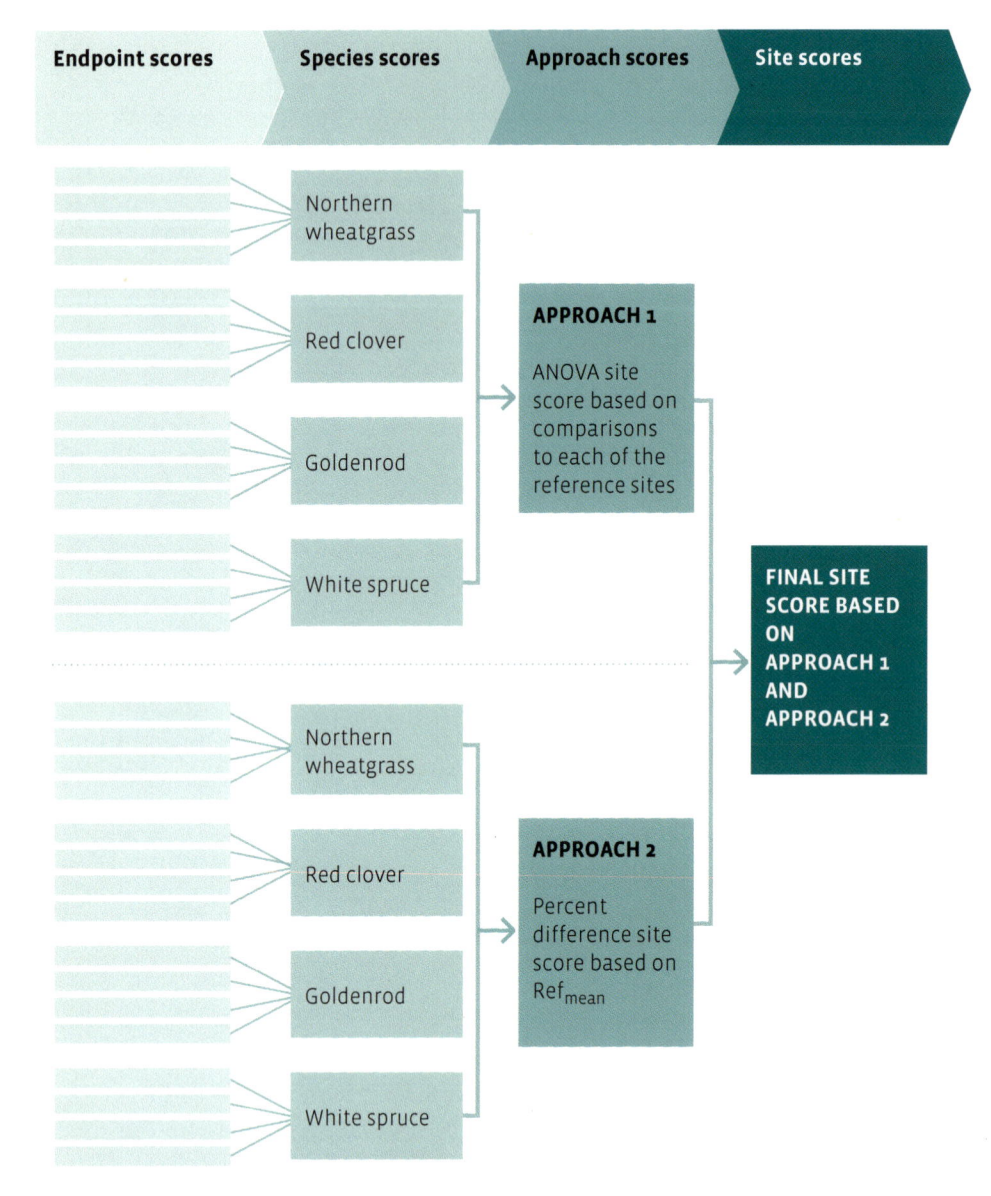

Figure 10.14 **Site Ranking Approach for Soil Toxicity Testing LOE**

The performance of the test organisms in the test site soils was compared to the reference site soils and the test sites were classified into one of three ranks: severely impacted (red), moderately impacted (yellow), or low to not impacted (green) as described in the following box.

Rank		Description
Green	Low to not impacted in comparison to the reference sites	The majority of the test species at these sites performed the same as or better than the test species at the reference sites
Yellow	Moderately impacted in comparison to the reference sites	The majority of the test species at these sites performed at a level that was slightly lower than that observed at the reference sites. Some component of the soil appeared not to promote the measured endpoints (growth or reproduction) of the test species
Red	Severely impacted in comparison to the reference sites	The majority of the test species at these sites performed at a level that was much lower than that observed at the reference sites. Some component of the soil appeared to seriously impact the measured endpoints (growth or reproduction) of the test species

10.7.4.3 Overall Site Ranking Results

The two approaches were weighted equally in the overall ranking for each test site. If the approaches provided identical rankings, then no further evaluation was required. If the two methods were not in agreement, the site was given a split ranking (such as red/yellow or yellow/green) to illustrate the separate rankings (Table 10.13). Five of the seven test sites along the Copper Cliff transect were ranked 'severely impacted' by approach #1, and six of seven were ranked 'severely impacted' by the second approach. The sites on the Coniston transect were ranked either 'severely impacted', or between 'severely impacted' and 'moderately impacted'. The exception to this was site CON-01, which was ranked 'moderately impacted' by both ranking methods. Among the Falconbridge sites, FB-01 was ranked 'severely impacted', FB-02 and FB-03 were given a split rank between 'severely impacted' and 'moderately impacted' and the remaining two sites were given 'low to moderately impacted' ranks.

Table 10.13 **Site Ranking for the Toxicity Testing LOE in Natural Soil**

Site	Rank Approach 1	Rank Approach 2
CC-01	Red	Red
CC-02	Red	Red
CC-03	Red	Red
CC-04	Red	Red
CC-06	Yellow	Red
CC-07	Red	Red
CC-08	Yellow	Yellow
CON-01	Yellow	Yellow
CON-02	Red	Red
CON-03	Yellow	Red
CON-05	Yellow	Red
CON-06	Red	Red
CON-08	Red	Red
FB-01	Red	Red
FB-02	Yellow	Red
FB-03	Yellow	Red
FB-05	Green	Yellow
FB-06	Green	Green

10.8 Litter Decomposition Line of Evidence (LOE)

10.8.1 Rationale

The process of litter decomposition is critical for maintaining soil fertility and productivity. The microbial community is an important component of soil biota which are involved with the breakdown of organic material. Microbial communities perform a range of critical soil processes and play essential roles in maintaining overall soil ecosystem functions (Zak et al. 1994). Through the decomposition of litter, nutrients return to the soil where they again become available to the plant ecosystem. Also, through the production of humus, the soil biota contribute to soil texture, water holding capacity and mineral complexation (Maxwell 1995). A litter decay rate that is too slow or too fast can have negative effects, such as nutrient losses or poor growth conditions (Andersson 2005).

A number of studies have examined the effect of elevated metal levels in soils relative to microbial activity and rates of litter decomposition (Freedman and Hutchinson 1980b; Kelly and Tate 1998; McEnroe and Helmisaari 2001; Kautz et al. 2001; Johnson and Hale 2004; Anderson et al. 2009). Elevated concentrations of some metals, such as Cu and Ni, in soil have been linked to reduced rates of litter decomposition and an increase in the litter layer on forest floors (Andersson 2005). Breymeyer et al. (1997) suggested that metal pollution of litter had a greater negative effect on decomposition than corresponding pollution of soil, implying that palatability of the litter substrate was the key limiting factor rather than direct metal toxicity to microbes in the soil. However, experiments conducted by Kautz et al. (2001) measured decreased microbial biomass and activity in soil samples collected adjacent to a magnesite plant in the Slovak Republic. The microbial activity was related to distance from the source and to metal levels in the soil. Regardless of the mechanism of effect, if litter decomposition is inhibited in forest soils, the biogeochemical cycling of nutrients can be significantly impacted. Since it has been shown that metals in soils can have a deleterious effect on soil microbial activity, the rate of leaf litter decomposition was recognized as a vital function in the forest system and, as such, was included as an LOE in the evaluation of Objective #1.

10.8.2 Methods

To assess the rate of litter decomposition, an *in situ* litter bag study was conducted. The objective of this study was to measure the mass loss of leaf litter in *in situ* litter bags containing white birch (*Betula papyrifera*) leaves over a 13-month period at the test and reference sites. The study approach was based on the work of Johnson and Hale (2004) and the European Guidance Document: *Effects of Plant Protection Products on Functional Endpoints in Soil (EPFES)* (Römbke et al. 2003). The EPFES guidance document approach is for bags buried in agricultural settings. These basic methods were adjusted using the findings from Johnson and Hale (2004) to design a study appropriate to a forested region in north-eastern Ontario. Birch leaves were used instead of straw, the bags were left on the forest surface instead of being buried and the duration of the study was a full year.

The litter material used was white birch foliage collected from one site in the fall of 2004 from five individual trees within a 10 m radius. Approximately 10 g (fresh weight) of leaf litter was placed separately into nylon mesh bags (Figure 10.15). The bags were placed at the sites in October 2004 with the last bags retrieved in November 2005. At each site, 25 litter bags were placed on the forest floor in a 5×5 block design, with additional bags at the reference sites. Figure 10.15 shows the following steps: a) preparation of white birch leaves; b) weighing birch leaves; c) birch leaves in nylon mesh litter bags; and, d) litter bags on forest surface.

Five bags from each site were collected at 7, 8, 9, 10, and 13 months from the initial time of placement in October 2004. Litter bags were placed at a total of 20 sites. Two of the test sites were not used in the decomposition study due to limited access. Decomposition was estimated as the loss of dry mass over time. The mass loss at each site was calculated as the difference between the dry weight of the leaves at the beginning and end of the study. Upon retrieval of the litter bags, debris and plant material were removed from the exterior of the litter bags and the contents were washed to remove surface deposition, before drying. This washing procedure was kept as gentle as possible to minimize the amount of leaching or abrasion.

Figure 10.15 **Birch Leaves in Nylon Mesh Litter Bags on Forest Surface**

The loss of dry weight was used to calculate a decomposition rate constant (k=slope of regression line of mass loss over time) for each site. The k values from test sites were then compared to k values from the reference sites. A predetermined 'reference criterion' was developed before conducting the study based on the work of Johnson and Hale (2004) and the EPFES guideline. It was decided that the % mass loss at each reference site should be at least 50% over the course of the study. The decomposition results (k value) at each test site were compared to the mean of the reference sites (REFmean) by ANOVA. If the k value at the test site was not significantly different ($P \geq 0.05$), the test site was considered to be low or not impacted (green rank), if the k value was statistically different (where $0.01 \leq P \geq 0.05$), the test site was considered moderately impacted (yellow rank), and if $P \geq 0.01$, the test site was considered severely impacted. The test site ranks provide an indication of the ability of the microbial communities at each site to decompose organic matter. The three rank categories were as follows:

Rank		Description
Green	Low to not impacted in comparison to the reference site mean	There was no difference between the rate of decomposition or the amount of mass loss at the test site when compared to the mean of the reference sites
Yellow	Moderately impacted in comparison to the reference site mean	The rate of decomposition or mass loss was impacted in comparison to the mean of the reference sites
Red	Severely impacted in comparison to the reference site mean	The rate of decomposition or mass loss was significantly impacted in comparison to the mean of the reference sites

10.8.3 Results and Site Ranking

The amount of decomposition at the three reference sites was considered to represent the natural decomposition variation within forested areas in north-eastern Ontario. With the exception of two exposure sites, CC-04 and FB-05 (low impact), decomposition at the test sites was ranked either moderately or severely impacted compared to the rate of decomposition at the reference sites. The overall mean mass loss at the test sites was lower than at the reference sites (Figure 10.16). The biomass loss at the reference sites (mean, 60%; range 51-73%) was higher than the test sites at Coniston (mean 43%; range 35-50%), Copper Cliff (mean 48%; range 43-58%) and Falconbridge (mean 46.6%; range 38-52%). These data indicate that the rate of litter decomposition was reduced by soil conditions found at the test sites.

In earlier studies, Freedman and Hutchinson (1980b) measured litter biomass and decomposition rates using leaf litter bags at contaminated sites near the Copper Cliff smelter. In that study, leaf litter decomposition was reduced at sites closer to the smelter compared with distant background sites, but the rate of decomposition after 850 days (65%) was much higher than observed in the current study at either the exposure or reference sites. Freedman and Hutchinson (1980b) also reported higher litter standing crop closer to the smelter, and reduced CO_2 production which was thought to be related to decreased microbial activity. More recently, Johnson and Hale (2004) studied the decomposition of birch leaf litter at metal-contaminated sites (soil Ni: 670 mg/kg, Cu: 1070 mg/kg, and Pb: 141 mg/kg) as well as uncontaminated sites near Sudbury. The maximum mass loss of leaf litter at the reference sites at 12 months was about 50% and no further increase in mass loss occurred between months 12 and 18. In contrast, the biomass loss at the contaminated site was only about 20%.

McEnroe and Helmisaari (2001) reported reduced decomposition of needles and fine roots of Scots pine (*Pinus sylvestris*) at sites in proximity to a smelter complex in SW Finland compared with more distant reference sites. Anderson et al. (2009) reported that the effects of metal contamination on microbial biomass and activity were pronounced in soils near the historical Anaconda smelter, Montana, USA. Reduced soil microbial activity in litter was reported near a smelter in Sweden (Nordgren et al. 1985) as well as downwind of the smelter at Rouyn-Noranda, Quebec (Dumontet et al. 1992). Depressed soil microbial activity was also reported in a forest floor near a Cu-Ni smelter in Finland (Fritze et al. 1989). Soil properties can also affect the toxicity of metals to soil microbial activity as previously noted with plants. Oorts et al. (2006) studied the toxicity of Cu and Ni to microbial processes in 16 different soil types. The authors observed that Cu toxicity thresholds increased (i.e. toxicity decreased) with increasing soil organic content, while Ni toxicity was lowered by increasing CEC and clay content. Soil pH was found to have an insignificant effect on total Cu or Ni toxicity due to counteracting effects on metal sorption and uptake by the microorganisms.

The results of this and other studies indicate that direct or indirect measures of the soil microbial community may be valuable indicators of ecosystem stress. Evaluation of the soil microbial community or microbial activity such as litter decomposition rates are also a useful component of ecological risk assessment. This important ecosystem component should be taken into consideration when planning and implementing risk management measures.

Final mean litter biomass loss
percentage

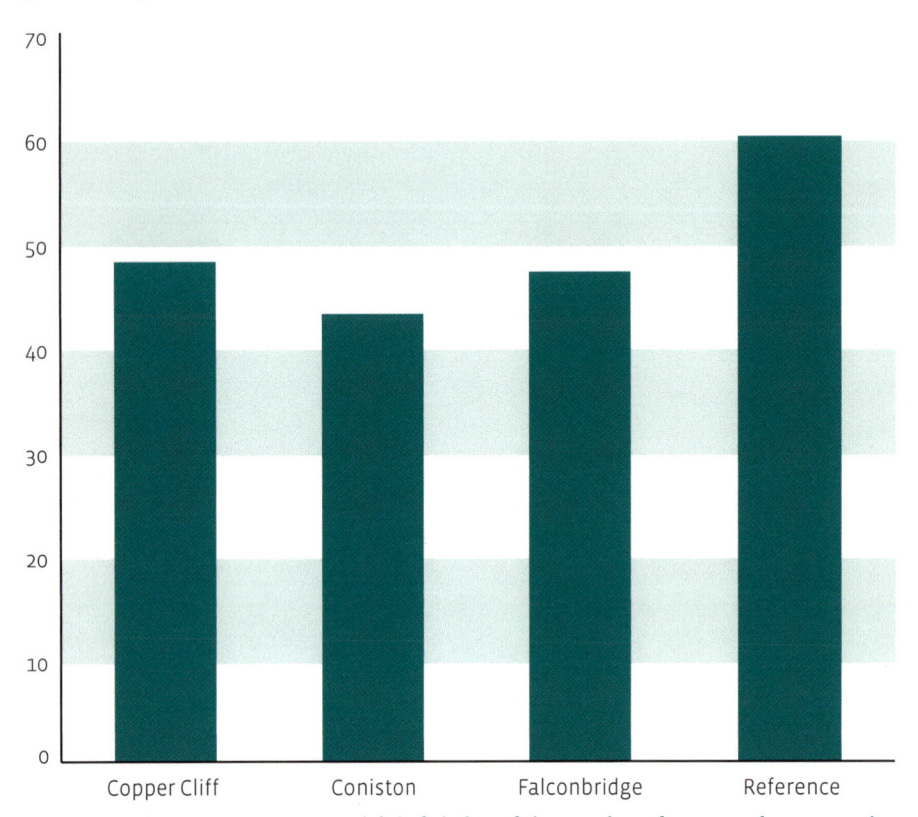

Figure 10.16 **Final Biomass Loss (% weight) of Birch Leaf Litter at the Reference and Exposure Sites**

10.9 Final Site Ranking and Integration of LOE

Four different lines of evidence were used to assess Objective #1 of the Sudbury ERA. The use of multiple LOE in the assessment of ecosystem impairment minimizes the occurrence of false-positive and false-negative conclusions (Rutgers and den Besten 2005), however, it requires that the LOE be integrated, generally with some form of WOE (weight-of-evidence) approach (Burton et al. 2002b). Chapman et al. (2002) described a WOE analysis as a determination of possible ecological impacts based on multiple LOE, incorporating judgments concerning the quality, extent and congruence of the data. The WOE framework should be logical, transparent, understandable by lay personnel and should appropriately distinguish between hazard and risk. Some WOE frameworks were available in the literature, although these were more common for sediments (e.g. sediment quality triad) and aquatic systems than for terrestrial systems (Rutgers and den Besten 2005). A series of 10 papers on the WOE approach were reviewed including: Burton et al. (2002a,b), Batley et al. (2002), Chapman et al. (2002), Reynoldson et al. (2002a,b) and Smith et al. (2002). Chapman et al. (2002) described the five general categories of WOE frameworks: indices, statistical summarization, scoring systems, logic systems, and best professional judgment.

This assessment used a combination of the above methods which provided a comprehensive, logical and transparent approach to evaluate a range of ecosystem variables and functions. The preceding sections described how each test site was ranked for each LOE. This section describes how the four LOE were integrated to provide a final ranking for each site. The process of distilling down a large volume of diverse data into relatively simple categories of 'red', 'yellow' or 'green' (severely, moderately, or low to not impacted) was analogous to using a biological index. Biological indices have been used widely for environmental monitoring and environmental assessments. The final ranking approach used in this study summarized a large volume of data into results that facilitate decision-making by risk managers.

10.9.1 Final Site Ranking Approach

The overall site ranking incorporated the four LOE and used the following considerations:

- The plant community assessment was considered the most significant LOE, because it reflected the actual current ecological condition of the sites
- The toxicity testing LOE was given the second most significant weighting
- More weight was given to the results of the toxicity testing and plant community than to the other LOE due the large amount of data generated and number of variables examined. If these LOE were both ranked severely impacted, then the site was considered severely impacted
- Soil characterization was considered throughout the ranking of each of the three other LOE but not given as much weight as the plant community or toxicity results
- Although very ecologically significant, decomposition was weighted less than the other LOE because the litter bag study only measured one variable, and was based on a modified test protocol.

Using the approach outlined above, the site ranking for each LOE as well as the final rank are summarized in Table 10.14. The spatial distribution of the test sites and their ranking is illustrated in Figure 10.17.

Table 10.14 Summary of the Final Site Rankings after the Integration of All LOE

Site	Plant community assessment	Toxicity testing Approach 1	Toxicity testing Approach 2	Soil characterization	Decomposition assessment	Final rank
CC-01	Red	Red	Red	Yellow	Red	**Red**
CC-02	Red	Red	Red	Red	Red	**Red**
CC-03	Red	Red	Red	Red	N/A	**Red**
CC-04	Red	Red	Red	Yellow	Green	**Red**
CC-06	Yellow	Yellow	Red	Yellow	Red	**Yellow**
CC-07	Red	Red	Red	Yellow	Red	**Red**
CC-08	Red	Yellow	Yellow	Yellow	Yellow	**Yellow**
CON-01	Yellow	Yellow	Yellow	Yellow	Red	**Yellow**
CON-02	Red	Red	Red	Red	Red	**Red**
CON-03	Yellow	Yellow	Red	Yellow	Red	**Yellow**
CON-05	Red	Yellow	Red	Yellow	N/A	**Red**
CON-06	Red	Red	Red	Yellow	Red	**Red**
CON-08	Red	Red	Red	Red	Red	**Red**
FB-01	Red	Red	Red	Yellow	Red	**Red**
FB-02	Green	Yellow	Red	Green	Red	**Yellow**
FB-03	Yellow	Yellow	Red	Yellow	Yellow	**Yellow**
FB-05	Yellow	Green	Yellow	Yellow	Green	**Yellow**
FB-06	Yellow	Green	Green	Green	Red	**Yellow**

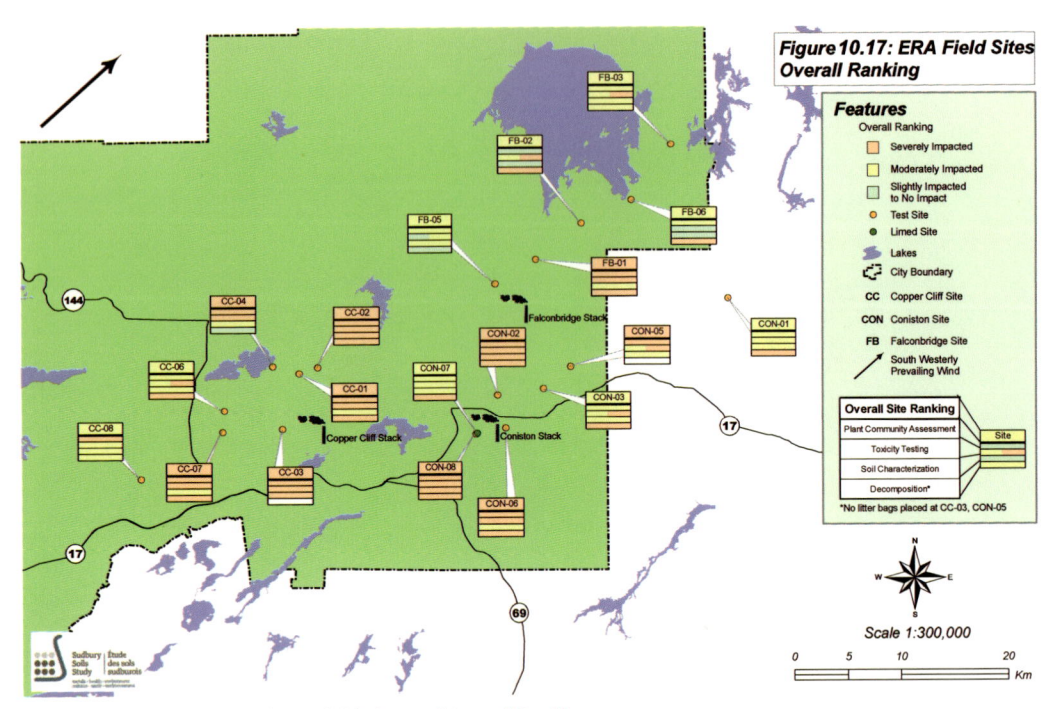

Figure 10.17 **Distribution of ERA Field Sites and Overall Ranking**

10.10 Interactions Between Lines of Evidence

10.10.1 Statistical Approach

Two statistical approaches were applied to examine interactions between the LOE as follows:

- A multiple linear regression analysis was used to determine whether there was a relationship between soil chemical parameters and soil toxicity
- A canonical correspondence analysis (CCA) was used to determine whether there was a relationship between the soil chemistry parameters and the plant community parameters.

Before undertaking the statistical analyses, the measurement variables were reduced to a manageable number as described in the next section.

10.10.2 Creating Independent Soil Variables

The data set for this objective included over 60 variables for only 22 sites, which introduced a considerable amount of auto-correlation between variables. This auto-correlation prevented a simple multiple linear regression approach from being utilized. To minimize the auto-correlation in the large data set, the information contained in these variables was grouped into factors. The data were first standardized by converting the raw data to Z-scores (Standard Ecological Variables) to normalize the central location and average variability of the data set. This Z-score transformation did not affect the skewness and kurtosis observed in the original data. The pooling of the variables into factors was based on a separation of variable type, variable source (i.e. anthropic or geogenic), textural or fertility relationships. The soil chemistry results (presented in Section 10.4) were pooled into different groupings, and each group was considered an independent variable. To achieve this, related soil parameters were grouped together as described below. The chemical parameters included in each group are summarized in Table 10.15.

Technogenic Load Factor (TLF)

The total concentrations of the COC (As, Cd, Co, Cu, Pb, Ni, Se) in soil cores by digestion with concentrated nitric acid, followed by ICP-MS analysis of resultant solutions were used to develop this factor. All raw data were standardized and then pooled into this one factor.

Geogenic Load Factor (GLF)

The total concentration of the metals in the soil matrix not assumed to be influenced by addition as aerosols from regional smelting activities obtained by acid digestion with concentrated nitric acid and ICPMS analysis of resultant solutions was used to develop this factor. The metals in this suite included Al, Ba, Be, B, Ca, Cr, Fe, La, Mg, Mn, P, V and Zn. All raw data were standardized and then pooled into this one factor.

Technogenic Bioavailability Factor (TBF)

The concentration of the chemicals of concern (As, Cd, Co, Cu, Pb, Ni and Se) obtained by a simple water extraction and analysis of the resultant solutions by ICP-MS analysis were used to develop this factor. Bioavailable S, a component of the acidic rainout and a probable soluble salt product in many of the smelter fallout materials was included in a second factor, TBF+S, to allow the influence of available S, probably as sulfate ion, to be examined. All raw data were standardized and then pooled into this one factor.

Geogenic Bioavailability Factor (GBF)

The concentration of the metals obtained by a simple water extraction of the soil matrix, with analysis of the resultant solutions by ICP-MS, was used to develop this factor. The abundances of these elements were not assumed to be influenced by addition as aerosols from regional smelting activities. The metals included in this suite are Al, Ba, B, Ca, Fe, Mg, Mn, S, and Zn. Although S may be both of geogenic and smelter origin, the element was included in the initial factor calculation because, to date, this element was excluded from all of the discussions in the risk assessment process. However, a second (GBF-S) factor was included in the results to allow the influence to be examined. All raw data were standardized and then pooled into this one factor.

pH Factor (pHF)

The determination of pH in both water and calcium chloride extractions is a common measurement in the environmental sciences, but the actual importance and value of the measurement are commonly overvalued. The potentially toxic effect of high levels of metals is due to their speciation in the solution phase, with actual species distribution being commonly dependent on the pH of the soil solution. Thus, the pH factor is computed in a manner identical to that for all other variable groups. However, because pH is a measure of hydrogen ion activity in solution, additional factors, namely TBF+pH and FF+pH, were also computed to examine the effect of hydrogen ion activity on toxicological response. All raw data were standardized and then pooled into the pH factor.

Cation Exchange Factor (CECF)

Both cation exchange capacity (CEC) and exchangeable cations (K, Na, Ca, Mg) were used to generate this factor. All raw data were standardized and then pooled into this one factor.

Organic Matter Factor (OMF)

The key parameters available for analysis were the total amounts of C and N. These parameters were standardized and pooled into this one factor.

Parent Material Depositional Factor (PMDF)

The soil parent materials included in this study were of till, glaciofluvial and glaciolacustrine origin. The textural and bulk density measurements reflect the mode of deposition of these soil-forming materials; all textural measurements were included with the density estimate in the calculation of this factor. All raw data were standardized and then pooled into this one factor.

Fertility Factor (FF)

A series of routine agronomic soil fertility analyses completed on all samples from this study were used to develop a soil fertility factor. These data included available nitrate, extractable P, K and Mg. Levels of diethylenetriaminepentaacetic acid (DPTA) extractable Fe and Mn were also included in the calculation of the fertility related factors. The data obtained by water extraction for ammonia (as N) from the bioavailability assessment were also included. The estimates of DPTA extractable iron and manganese were not included in the calculation of FF. A factor including the DPTA data (FF+DPTA) and a DPTA factor (DPTA) were also calculated to allow examination of the various extractants individually. All raw data were standardized and then pooled into these factors as appropriate.

Table 10.15 **Groupings of Soil Chemistry Parameters for Statistical Analyses**

Grouping	Soil chemistry parameter(s)	Description
Technogenic Load Factor (TLF)	As, Cd, Co, Cu, Pb, Ni, Se	The concentrations of the chemicals of concern (COC) obtained by total metal analysis
Geogenic Load Factor (GLF)	Al, Ba, Be, B, Ca, Cr, Fe, La, Mg, Mn, P, V, Zn	The total concentration data for the metals (total metal analysis) in the soil matrix, not assumed to be influenced by addition as aerosols from regional smelting activities
Technogenic Bioavailability Factor (TBF)	As, Cd, Co, Cu, Pb, Ni, Se	The concentration of COC obtained by a simple water extract with analysis of the resultant solutions by ICP-MS analysis. TBF was also combined with sulfur (TBF+S), a measure of the bioavailable sulfur, a component of the acidic rainout and a probable soluble salt product in many of the smelter fallout materials. The comparison between TBF and TBF+S allows the influence of available sulfur, probably as sulfate ion, to be examined. TBF was also combined with the pH factor (TBF+pHF)
Geogenic Bioavailability Factor (GBF)	Al, Ba, B, Ca, Fe, Mg, Mn, S, Zn	The concentration of accurately quantified metals obtained by a simple water extract of the soil matrix. The abundance of these elements was not assumed to be influenced by addition as aerosols from regional smelting activities. A second factor without sulfur (GBF-S) was included in the results
pH Factor (pHF)	pH	The pH factor (both water and calcium chloride extracts) was computed in a manner identical to that for all other variable groups. However, because pH is but a measure of hydrogen ion concentration in solution, additional factors, namely TBF+pH and FF+pH, were also computed to enable the effect of hydrogen ion concentration on toxicological response to be examined in exactly the manner as the water-soluble bioavailable ionic species
Cation Exchange Factor (CECF)	Cation exchange capacity	The data for both cation exchange capacity (CEC) and exchangeable cations (K, Na, Ca, Mg) were standardized, pooled and averaged to calculate this factor
Organic Matter Factor (OMF)	C, N	This factor consisted of total amounts of carbon and nitrogen in organic matter
Parent Material Depositional Factor (PMDF)	Bulk density, soil (sand, silt, clay)	The soil parent materials included in this study are of till, glaciofluvial and glaciolacustrine origin. Measures such as soil texture and bulk density reflect the mode of deposition of these soil-forming materials
Fertility Factor (FF)	N, nitrate, P, K, Mg	A series of routine agronomic soil fertility analyses were used to develop a soil fertility factor. These data included water extractable ammonia (as N), available nitrate, extractable phosphorus, potassium, and magnesium. This factor was also combined with the pH factor called FF+pHF
Extractable Metals (DPTA)	Fe, Mn	This factor includes DPTA extractable Fe and Mn. This factor was also combined with the fertility factor (called DPTA+FF)

10.10.3 Statistical Analysis

Analysis 1: Relationship between Physical and Chemical Parameters and Toxicity Endpoints

A multiple linear regression (MLR) was used to examine relationships between soil chemical parameters and soil toxicity. The toxicological endpoints of interest were terrestrial plant growth (i.e. root and shoot length) and reproductive success of earthworms (i.e. number of juveniles) as listed in Table 10.16. Preliminary examination of the results showed that inclusion of the root and shoot weight did not influence the results of the MLR analysis. The combination of endpoints and soil chemistry groupings provided 11 different combinations that were evaluated (Table 10.16).

The parameters were standardized using Z scores and then made into variables by pooling all values and calculating an average. The mean of the standardized values for each of the defined groups was used in the multiple linear regression analysis. This method of grouping increased the degrees of freedom and decreased the co-linearity between many of the measured soil chemistry parameters. Models were run in stepwise fashion to identify significant variables and to eliminate additional variables that did not contribute to the R^2 of F value up to a maximum of eight variables.

Table 10.16 Toxicity Endpoints in Multiple Linear Regression Analysis

Species	Endpoint	Type of Soil	Type of Normalized Data
Goldenrod	Root length	Natural	Standardized Ecological Variables (Z-scores)
Goldenrod	Shoot length	Natural	Standardized Ecological Variables (Z-scores)
Red clover	Root length	Natural	Standardized Ecological Variables (Z-scores)
Red clover	Shoot length	Natural	Standardized Ecological Variables (Z-scores)
Northern wheatgrass	Shoot length	Natural	Standardized Ecological Variables (Z-scores)
White spruce	Shoot length	Natural	Standardized Ecological Variables (Z-scores)
Northern wheatgrass	Root length	pH-amended	Standardized Ecological Variables (Z-scores)
Northern wheatgrass	Shoot length	pH-amended	Standardized Ecological Variables (Z-scores)
Red clover	Root length	pH-amended	Standardized Ecological Variables (Z-scores)
Red clover	Shoot length	pH-amended	Standardized Ecological Variables (Z-scores)
E. andrei	Number of juveniles	pH-amended	Standardized Ecological Variables (Z-scores)

Over 160 combinations of different equations or models were generated to examine the relationships between the endpoints measured in the toxicity tests and the physical and chemical soil parameters. After reviewing these models, the following observations were provided:

- The technogenic load factor (TLF), which includes the COC, was inversely related to the root and shoot length of goldenrod, northern wheatgrass and red clover in natural site soil
- The fertility of the site soil was a positive factor in the growth of goldenrod and red clover
- White spruce behaved differently from the other plants. In most situations, statistically significant regression models could not be developed but where they could, organic matter (OM) and the soil matrices other than the COC (OM and geogenic load factor (GLF)) were the only significant variables in the models; OM was positively related while the GLF was inversely related to the plant endpoints
- Parent material depositional factor (PMDF) was consistently present in the models for goldenrod, northern wheatgrass and red clover indicating a positive relationship
- The shoot length of white spruce had an inverse relationship to pH (pH was present in nine out of 11 models). Lower pH was associated with increased shoot length. No explanation was apparent for this relationship
- Depending on the plant type, either fertility (positive relationship) or pH and fertility combined (inverse relationship) contributed to the shoot length for goldenrod, red clover and white spruce
- For the number of earthworm juveniles in pH-amended soil, only four of 11 combinations of parameters could be developed into models with significant interactions. Where the models were established, the geogenic metals (GLF) had an inverse relationship, while pH had the most influence (positive) on earthworm production.

In almost all scenarios, the COC and soil properties (TLF and PMDF) were inversely related to the growth of plants in the toxicity testing LOE. Many of the models included either soil fertility (FF) or pH (pHF) as important factors. Although results for white spruce were quite different compared to the other plant species, the results indicate that the concentration of metals in the site soil along with pH, soil texture and fertility were the primary factors related to the toxicity endpoints measured.

Analysis 2: Evaluation of the Relationship between Plant Community and Physical and Chemical Parameters

Analysis 1 (MLR) identified the main physical and chemical factors related to the toxicity endpoints measured during the toxicity testing. The aim of Analysis 2 was to determine which factors were related to the plant community assemblage at the study sites. Canonical correspondence analyses (CCA) (ter Braak 1986) was used to determine whether the parameters identified in Analysis 1 were also related to the plant communities at the sites. Canonical correspondence analyses were performed using percent cover of plant types (trees, tall shrubs, low shrubs, forbs and cryptograms) and TLF, PMDF and FF variables. PMDF did not conform to the normality assumption required for CCA and so a non-parametric method known as non-metric multidimensional scaling (MDS) was performed in parallel with the CCA. MDS is free of assumptions of normality but is unconstrained by environmental factors. Concordance between the CCA and MDS would suggest that CCA results were valid and that violations of the assumption of normality of the CCA method were not grave enough to warrant abandoning the results. Species and sites scores were scaled for plotting. The significance of results was determined using 1000 Monte Carlo permutations.

The hypothesis that plant assemblage composition was related to TLF and FF was supported by CCA. The first two canonical axes explained 37.8% of the total variance in the data. TLF was significantly correlated ($P<0.01$) with the first two canonical axes ($R^2=0.432$) as was FF ($P=0.050$, $R^2=0.287$). PMDF was not a significant explanatory factor for the plant assemblages. Similarly, MDS ordination uncovered nearly identical relationships between species, sites and the factors TLF and FF. The general distribution of species, sites and their relationships to factors differed little between the CCA and MDS. TLF was significantly correlated ($P=0.019$) with the first two canonical axes ($R^2=0.381$) as was FF ($P=0.034$, $R^2=0.333$). The metals of concern (TLF) were significantly related to the structure of plant communities at the study sites. Sites with the highest TLF scores (high COC levels) had some of the lowest TLF scores. Trees and low shrubs had higher relative percent cover when TLF was high, while forb cover was negatively related to TLF. Soil nutrients, or fertility (FF), were also a significant determinant of plant assemblage structure.

Summary of Statistical Analysis

Analysis 1 showed that the COC and soil properties were inversely related to the growth of plants from the toxicity testing. Other factors identified as important were soil fertility (positive relationship) and pH (inverse relationship). Alternatively, Ca was positively related to those endpoints. Other factors that were identified as important were As, Pb and Cu (all inversely related). The results for the tree species tested (white spruce) were quite different compared to the other plant species and no models could be derived to describe the root and shoot lengths of white spruce.

Analysis 2 found that COC (factor TLF) was significantly related to the structure of plant communities at the study sites. Trees and low shrubs had higher relative percent cover when TLF was high while forb cover was negatively related to TLF. This result mirrors the finding in Analysis 1, providing agreement between the laboratory studies and the field survey. As with the toxicity testing, soil nutrients were also a significant determinant of plant assemblage structure. The soil properties (PMDF) were not a significant explanatory factor for the plant assemblages.

The combination of these analyses show that, at the 22 sites established to evaluate Objective #1 of the ERA, the level of COC in the soil was related to the toxicity to plants as measured in toxicity tests and to the structure of the plant communities surveyed at the sites. Fertility was also a factor of importance.

10.11 Site Ranking Trends

The purpose of this final step was to determine whether the concentration of metals at the test sites were the most likely cause of ecosystem impairment and to determine whether there are other factors that may be contributing to the overall toxicity (or lack of toxicity) of the site soil. For each transect, the sites were placed in order according to their total metal concentration (from highest to lowest concentration), water leach metals (highest to lowest concentration) and distance from the smelter (closest to furthest away). This approach provided a visual representation of the data to qualitatively examine the role of total metals, water leachable metals and distance from the smelter with site ranking. Results from sites along the Copper Cliff transect were used as an example in Table 10.17. The concentrations of water leachable Pb and Se were below detection limits in most cases so they were not included in the table.

Along the Copper Cliff transect, the sites that had the highest total and water-extracted metal concentrations were always the sites that were ranked as severely impacted. The one site along the Copper Cliff transect that was ranked moderately impacted (CC-08) was the site with the lowest metal concentrations and was furthest away from the smelter. Two of the sites were eroded (CC-03 and CC-01) with poor soil development, but both of these sites also had high metal levels.

Table 10.17 **Copper Cliff Sites in Order of Total and Water Leachable Metals and Distance from Smelter**

Total metals (descending from high to low concentration)							Water leachable metals (descending from high to low concentration)					Distance from smelter (closest to farthest)
Ni	Cu	As	Cd	Co	Pb	Se	Ni	Cu	As	Cd	Co	
CC-03	CC-03	CC-03	CC-01	CC-03	CC-03	CC-03	CC-02	CC-02	CC-01	CC-02	CC-03	CC-03
CC-01	CC-01	CC-01	CC-04	CC-02	CC-01	CC-01	CC-04	CC-03	CC-04	CC-01	CC-02	CC-01
CC-02	CC-02	CC-02	CC-02	CC-01	CC-02	CC-02	CC-03	CC-01	CC-07	CC-04	CC-01	CC-02
CC-04	CC-04	CC-04	CC-03	CC-04	CC-04	CC-04	CC-01	CC-04	CC-02	CC-03	CC-04	CC-04
CC-07	CC-07	CC-07	CC-07	CC-07	CC-07	CC-07	CC-06	CC-07	CC-03	CC-06	CC-07	CC-07
CC-06	CC-06	CC-06	CC-06	CC-06	CC-08	CC-06	CC-07	CC-06	CC-06	CC-07	CC-06	CC-06
CC-08	CC-08	CC-08	CC-08	CC-08	CC-06	CC-08	CC-08	CC-08	CC-08	CC-08	CC-08	CC-08

Data from the other two transects were organized similar to that for the Copper Cliff transect. The detailed results are not shown but the general observations are discussed herein. Along the Coniston transect, the sites with the highest metal levels were generally the ones that were most impacted. Where this was not the case, the site tended to be highly eroded. Many of the sites along this transect were severely to moderately eroded. Historically, it was likely that the metal levels and SO_2 levels were high at these sites, resulting in a loss of vegetation, which led to a loss of soil due to erosion. When the soil was lost at these sites, the metals were also lost. The relict soil layers that remain are nutrient deficient and lack organic matter.

Four of the five sites along the Falconbridge transect were ranked as moderately impacted and there appeared to be an association between the metals in the soil and the level of impact. The two sites closest to the Falconbridge smelter in a downwind direction had the highest metal levels. One of these sites (FB-01), was ranked as severely impacted, while the other, FB-02, was considered moderately impacted. Although metals were elevated at both of these sites, the pH and Ca levels were higher at FB-02, which likely reduced the bioavailability of metals at this site.

When the results of the statistical analyses and site rankings were evaluated together, it appeared that the COC in the Sudbury region were related to level of site impact, and impeding the recovery of a self-sustaining plant community. Other factors identified as important were soil fertility, pH of the soil, concentration of Ca and the levels of organic matter in the soil.

10.12 The Effect of Historical Liming on Site Characteristics

10.12.1 Background

Similar information was collected at a historically limed site (CON-07) and an adjacent site not subjected to any treatment (CON-08). Collection of these data allowed a detailed comparison and evaluation of the effect of the historic pH-amendment (liming) and replanting, assuming other site conditions were similar given their close proximity to each other. The impact of anthropogenic activities to vegetation in the Sudbury region has been examined in detail over the past three decades (Winterhalder 1983, 1996; Courtin 1994) and described earlier in Chapter 2. The term 're-greening' was used to describe the reclamation activities designed to re-establish vegetation cover on industrially damaged land in the Sudbury region. Studies in the 1970s showed that liming of the soil raised pH sufficiently to reduce soil toxicity and facilitated growth and survival of grasses on many test sites throughout the city. A review of the past and present re-greening efforts in the Sudbury region is presented in Chapter 12.

Dolomitic limestone (Ca and Mg carbonate) was used to reduce soil acidity on chosen lands. The direct effect of liming was to change soil pH but this indirectly changed several soil properties that vary with pH (Troeh and Thompson 2005). From 1978 to 1984, CON-07 was treated through liming, fertilizing, and seeding. Dolomitic limestone was applied to the site at a concentration of 10 tonnes/ha. Fertilizer (Type 5-20-20) containing 5% N, 8.7%

elemental P, and 16.6% elemental K was applied to the site at a concentration of 40 kg/ha. The high amount of P in the fertilizer was necessary for promoting grass germination and growth. The site was then seeded (40 kg seeds/ha) with a seed mixture comprised of 80% grasses and 20% legumes. In 1986, 1995, and 1997, the second stage of the Land Reclamation Program was carried out with the planting of small trees. The major tree species planted throughout the site during these years included jack pine (*Pinus banksiana*), red pine (*Pinus resinoasa*), and white pine (*Pinus strobus*). A small number of red oak (*Quercus rubra*) was also planted in 1986.

10.12.2 Comparison of COC Concentrations

Total metal (nitric acid extractable) and water leach concentrations of COC in the 0–5 cm core samples from the limed (CON-07) and non-treated site (CON-08) are shown in Table 10.18. Total and water extractable Cu and Ni were elevated at both CON-07 and CON-08 compared to the concentrations measured at the three reference sites. The total COC concentrations of As, Cd, Co, Pb and Se at CON-07 and CON-08 were quite low and well below MOE soil quality guidelines. The most striking difference between the two sites was in the concentrations of Cu and Ni, which were 77% and 64%, respectively, higher at CON-07. This may be reflective of the highly eroded state of CON-08.

The water leach metal concentrations are one possible measure of 'plant bioavailable' metal concentrations. In general, water leach metal levels followed a similar pattern to the total metal levels, apart from Ni, where the water leach concentration was 37% higher at CON-08 compared to CON-07.

Table 10.18 **Metal Levels (mg/kg) in Soil from Limed and Non-limed Sites**

Metal	CON-07 total metals (mg/kg)	CON-08 total metals (mg/kg)	CON-07 water leach metals (mg/kg)	CON-08 water leach metals (mg/kg)
As	7.2	5.2	0.09	0.04
Cd	0.15	0.15	<DL	<DL
Co	10.2	10.9	0.08	0.08
Cu	240	107	1.73	0.38
Pb	11	9.1	0.07	<DL
Ni	255	132	1.76	2.55
Se	1.1	0.89	<DL	<DL

<DL indicates concentration was less than the method detection limit.

10.12.3 Soil Characterization

Comparison of the chemical properties at the two adjacent sites revealed numerous differences that were likely affecting the recovery of the forested community. The historic liming and re-greening activities at CON-07 resulted in the pH of the soil being significantly (almost 50%) higher compared to the untreated site (Table 10.19). In terms of particle size, the soils from CON-07 and CON-08 were quite similar; both were silt loams with bulk density values in the same range (CON-07, 1.33 g/cm^3; CON-08, 1.32 g/cm^3). There were definite differences in soil development between the two sites. At CON-07, the soil was imperfectly drained and there was evidence of moderate erosion. However, the relict subsurface mineral soil horizons were well developed and a thin Ah horizon, the product of cultivation and organism activity, was present at the site. At CON-08 (not treated site), the imperfectly drained soil was highly eroded and it was evident that the litter fermentation humus (LFH) horizon was inadequate for seedling germination and growth. The surface mineral horizons (Ae) were also completely eroded from the site with the structural integrity of the remaining mineral horizons at this site being very poor.

Table 10.19 Physical and Chemical Parameters at limed and untreated Sites

	CON-07 (treated)	CON-08 (not treated)
Soil properties		
Particle size	Silt loam	Silt loam
Bulk density (g/cm³)	1.33	1.32
Soil pH (water/slurry in cores)	7.19	4.5
Organic matter (%)		
Total N	0.1	0.03
Organic C	1.7	0.82
Soil exchange complex chemistry		
Cation exchange (cmol(+)/kg)	15	11
Calcium (cmol(+)/kg)	9	0.82
Magnesium (cmol(+)/kg)	1	1
Ca/Mg ratio	7	0.8
Fertility parameters (mg/kg)		
N as ammonium	0.68	0.01
Extractable P	7	42
Extractable K	68	64
Extractable Fe	321	547
Extractable Mn	38	28
Extractable Mg	227	14

Both of the sites were lacking organic matter, as measured by the total N (CON-07, 0.1% and CON-08, 0.03%) and organic carbon levels (CON-07, 1.7% and CON-08, 0.82%) when compared to the ranges measured at the reference sites. Ashman and Puri (2002) considered soil C and N levels less than 2% and 0.1%, respectively as being low. The CEC of the two sites was similar (CON-07, 15 cmol(+)/kg and CON-08, 11 cmol(+)/kg) although these sites were much lower than the reference site CEC values (Reference CEC range 27–29 cmol(+)/kg). The Ca levels at CON-07 (9 cmol(+)/kg) were 167% higher than at CON-08 (0.82 cmol(+)/kg) and much higher than at any of the reference sites (range 0.38–2.8 cmol(+)/kg), reflecting the past liming activities that took place. Exchangeable Ca in soil has an important relationship to soil pH and the availability of several macro- and micronutrients. Calcium is a structural component of plant cell walls and is, therefore, vital in plant growth (Troeh and Thompson 2005). Plants that are grown in Ca-deficient soil will often be stunted which was observed in the plant community at CON-08.

Soils rarely contain enough nitrogen for maximum plant growth. Generally, younger plants need more nitrogen than older plants. Organic matter, which contains nitrogen, must be at least partially decomposed before the nitrogen is available. Microbial action gradually decomposes the organic matter producing the ammonium ion, a form of nitrogen that is readily assimilated by plants. The available N as ammonium at CON-07 (0.68 mg/kg) was comparable to the level found at some of the reference sites (0.45–3.49 mg/kg) but nearly 200% higher than the level found at CON-08 (0.01 mg/kg). Magnesium is contained in dolomitic limestone, the liming application applied to CON-07. It was not surprising, therefore, to find that the extractable Mg levels were almost 16 times higher at the limed site than the unlimed site (Table 10.20). Mg is vital to the process of photosynthesis as it is a component of chlorophyll (Troeh and Thompson 2005). The Fe levels at both the limed and unlimed sites were relatively low (CON-07, 321 mg/kg; CON-08, 547 mg/kg) when compared to the reference sites (918–1256 mg/kg). However, the levels at the unlimed site (CON-08) were about 50% higher than at CON-07.

In summary, soil chemistry (irrespective of metal levels) was different between the two sites. The effect of past liming activities was still apparent and contributed to better growing conditions at CON-07 compared to the unlimed site (CON-08).

10.12.4 Plant Community

At each site, observational data were collected as part of five major ecological components: a broad plant inventory, percent cover assessment, detailed tree and tall shrub assessment, assessment of coarse woody debris and an ecosite classification. The data were interpreted and placed into one of four ecological criteria: site biodiversity, ecological integrity, long-term site productivity and soil and water conservation. The comparison of the plant community at CON-07 and the adjacent unlimed site, CON-08 (Figure 10.18) is discussed in the following section. The comparison of the plant communities at CON-07 and CON-08 revealed very different structures (Table 10.20). Although the sites are adjacent to each other (the center stakes being less than 500 m apart), the two communities displayed completely different characteristics. CON-07 represented a site that had a plant community in transition, while CON-08 showed clear indications of being impacted (Figure 10.18).

Figure 10.18 **Photographs of the Limed Site (Left) and Non-treated Site (Right)**

The plant community at CON-07 was dominated by scattered, stunted poplar trees and a continuous cover of herbaceous species. The species diversity at the site was comparable to, or higher than, reference site values for many indices. The site biodiversity, ecological integrity and long-term productivity were all ranked as moderately impacted but the soil and water conservation of the site was considered good. Conifer cover was virtually absent in the community despite the planting of white and jack pine over 20 years earlier. This lack of species represents a gap in the ecological integrity of the site as the lack of conifers would provide no protective winter covering to ameliorate the site from high winds and cold temperatures. In addition, there were a very large number of non-native species, accounting for over half of the ground cover at this site, which is indicative of the history of liming and seeding on this site. A high number of metal- and acid-tolerant species were also recorded at this site. There was good evidence of successful regeneration of white birch, balsam poplar and trembling aspen in the understory.

There was only limited evidence of dieback at the site. Not surprisingly, maximum tree height at this site was low and downed woody debris (DWD) was absent at this site. This represents another gap in the ecosystem, whereby long-term site productivity is potentially in jeopardy due to the lack of an adequate supply of organic material to the soil. However, trembling aspen and balsam poplar were abundant at the site, suggesting that this was a relatively productive site. The high tree density values also support this interpretation. The silt loam-textured soils consisted of highly erosion- and frost-prone silts and clays. However, the level terrain, continuous plant cover and high density of trees greater than 1 m in height minimized the risk of rapid soil water/soil solution flow-through in the silt loam soils. During soil collection, the field workers discovered earthworms at this limed site. No attempt was made to identify or quantify abundance but it was clear that a thriving and successful earthworm population existed.

Table 10.20 **Plant Community Indicators at Limed and Untreated Sites**

	Limed (CON-07)	Untreated (CON-08)
Soil biodiversity		
Species richness (total species)	80	41
Species diversity (H' value – all species)	1.6	1.2
Species dominance (% cover)	22	26
Degree of disturbance (proximity to barren and semi-barren areas)	High	High
Life history (% perennial species)	83	100
Plant community structure (# of strata)	5	5
Downed woody debris (count)	0	0
Degree of re-greening intervention (proximity to disturbed and re-greened areas)	High	High
Ecological integrity		
Life form of trees (% cover – canopy/% cover – understory)	0/2.25	0/2.25
Introduced or invasive (% non-native and potentially invasive species)	56	17
Successional stage – shade tolerant (# indicator species)	6	1
Substrate (% total mineral substrate)	6.92	20.48
Regeneration (% seedlings, saplings/% tree species in understory)	5/33	2/2
Reestablishment of sensitive species (# good/# intermediate/# poor indicators)	6/10/4	2/7/5
Species richness (total species)	80	41
Presence of metal or acid tolerant species (# of indicators)	12	14
Long-term site productivity		
Permanent mortality and dieback of tree species (% overall dieback)	12	No trees present
Tree height (m)	6	No trees present
Tree/tall shrub density (# per hectare)	6700	1500
Presence of aspen or poplar on site (height-canopy/height-understory/# trees in canopy/# trees in understory)	6/1/39/34	No trees present
Volume of downed woody debris (average volume)	0	0
Length of downed woody debris (average length)	0	0
Range of decomposition (# classes represented)	0	0
Soil and water conservation		
Tree and tall shrub density (# per hectare)	5650	0
Percent plant cover (surface soil retention index)	100	81
Percent leaf litter cover (%)	76	4
Percent slope (%)	<5	10
Soil texture	Silt loam	Silt loam

In contrast, CON-08 displayed many characteristics of a stressed plant community. The site biodiversity, ecological integrity, site productivity and soil and water conservation were all ranked severely impacted and at risk. The analysis showed low species richness and species diversity. Downed woody debris, trees and tall shrubs were completely absent, representing a major gap in the biodiversity of the site. This gap would leave the site vulnerable to the loss of surface soils due to erosion and to losses in soil quality due to the very limited seasonal additions of organic matter. The plant community structure was characterized by few species within each layer and the complete absence of trees. Conifer cover was negligible, and there was a scarcity of shade-tolerant indicator species and cryptogam, indicators of improved site conditions. In addition, there were high numbers of metal- and acid-tolerant species and very little evidence of regeneration. Of major concern was that over 20% of the surface was exposed, unproductive mineral substrate. As a result, the risk of soil loss through surface erosion remained high and subsoil water retention would be limited due to the absence of a shrub and tree cover. In contrast to CON-07, only 500 m away, no earthworms were found at this site.

10.12.5 Soil Toxicity Comparison

Soil pH at the untreated CON-08 site was 4.45, while the soil pH at CON-07 was 7.21. As a result, all of the toxicity test plant species performed much better at CON-07 than at CON-08. For instance, the performance of goldenrod was fourfold higher at CON-07. A summary of the toxicity test results in the historically limed soil from CON-07 and natural site soil from CON-08 is shown in Figure 10.19.

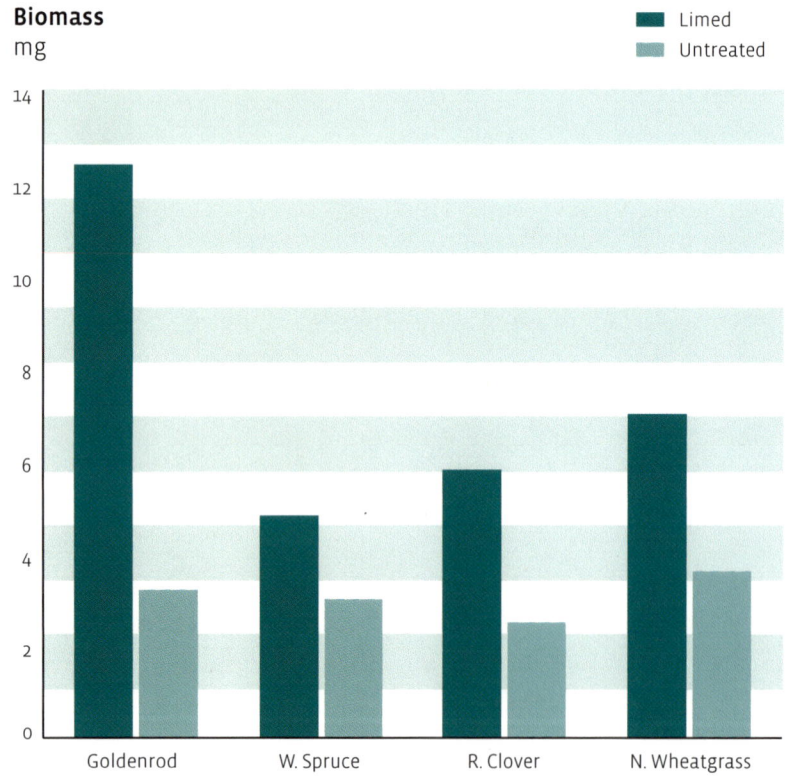

Figure 10.19 **Biomass (mg) of Plants Grown in Soil from Limed Site and Untreated Sites**

To further investigate the influence of pH on plant growth, site soil from CON-08 was amended with calcium carbonate to raise the pH from 4.4 to 5.2 ± 0.2. The toxicity of the pH-amended soil was examined using red clover, northern wheatgrass and earthworms. Raising pH of the CON-08 soil increased growth of both species, in particular, northern wheatgrass. The final total biomass of red clover grown in the natural and pH-amended soil was 2.5 and 3.1 mg, respectively. The final total biomass of northern wheatgrass in natural and pH-amended soil was 3.6 and 8.6 mg, respectively.

In earthworm tests, all adults survived in the natural soils from both CON-07 and CON-08. Worms reproduced successfully in CON-07 soils, with an average of 1.8 juveniles per adult, but no reproduction took place in CON-08 soil. When the soil from CON-08 was amended to increase pH, the survival rate for adult earthworms was high but the earthworms were unable to successfully reproduce. Thirty percent of the earthworms in the

CON-08 pH-amended soil were missing their reproductive organs midway through the 63-day reproduction test. Earthworm survival rate was an insensitive measure of overall health. Earthworms are resilient and will live in all types of conditions. However, they will only reproduce in good conditions. The earthworms from the CON-07 site test were healthy, and were able to successfully reproduce.

In summary, it is evident that all test species in natural soil performed better at CON-07 than at CON-08. When the pH at CON-08 was raised, the plant growth for northern wheatgrass improved beyond the performance at CON-07. Alternatively, for red clover, plant growth did improve but not to the same extent. Earthworms were sensitive to CON-08 soil and did not reproduce in the natural or pH-amended soil from this site. The toxicity testing results show that pH is not the only factor limiting plant growth in this area. They also suggest that liming or pH amendment affects plant species differently, which is an important consideration for future re-greening and planting strategies.

10.12.6 Decomposition

Decomposition is a vital function in a forest ecosystem. The process of litter decomposition is critical for maintaining site fertility and productivity by returning nutrients to the soil where they become available to plants. To measure decomposition, a year-long study was initiated with the objective to measure the mass loss of leaf litter, in *in-situ* litter bags containing white birch (*Betula papyrifera*). Decomposition, as measured by mass loss, was evaluated by comparing the decomposition rate at each site to the calculated mean rate of the three reference sites (REFmean). This comparison provided a measure of the ability of the site microorganisms to decompose organic matter. At the limed site (CON-07), 53% of the mass of the leaves was lost over the course of the 13-month study whereas, at the unlimed site, only 35% of the mass of the leaves decomposed. The rate of decomposition (k) per year was calculated as 0.55 g/g/year (dry weight) at CON-07 and 0.26 g/g/year (dry weight) at CON-08. These data indicate that litter decomposition was inhibited at the untreated site compared with the treated site (CON-07). When compared to REFmean, the rate of decomposition at both CON-07 and CON-08 was significantly lower, indicating a severely impacted microorganism system.

10.12.7 Summary of Comparison of Limed to Untreated Site

The plant communities at CON-07 and CON-08, although in proximity to each other, were remarkably different. The limed site (CON-07) showed evidence of being a site in transition, while CON-08 was ranked as severely impacted. The past liming and re-greening activities have helped to establish a diverse plant community, with the introduction of essential minerals (Ca, Mg), providing a viable seed source, and increasing the soil pH thereby decreasing metal availability. Although CON-07 is not as productive as the established reference sites, the data collected from the four LOE indicate that it is on its way to re-establishing itself, compared to CON-08, and that the re-greening activities employed within the Sudbury region were working.

On the other hand, without the addition of lime, seed source or strategic tree planting, CON-08 retained its barren appearance and its status as a severely impacted site. Soil erosion, lack of organic matter and poor community structure all showed that the site remains impacted. These results indicate that a variety of factors are contributing to the lack of recovery at CON-08 including: low soil fertility, low pH, lack of a growth medium and the increased bioavailability of metals in the soil.

10.12.8 The Role of pH in Objective #1

Although soil pH was not a COC in this study, it was recognized that soil pH played a significant role in affecting metal bioavailability and metal toxicity as well as direct soil toxicity. A considerable amount of effort was devoted to examining the role of soil pH and its interaction with metal toxicity in this study. Although the studies conducted to examine these interactions were not intended to be exhaustive or definitive, they did provide valuable information. Some of the salient findings were:

- The pH in Sudbury soils was low enough (either naturally or with additional depositional effects) to inhibit the growth of some of the toxicity test plant species
- The low pH of Sudbury soil totally inhibited earthworm reproduction
- Raising soil pH to 5.2 improved measured plant toxicity endpoints of the two plant species tested in the majority of instances. The higher pH also resulted in the onset of earthworm reproduction in the site soil from many sites
- Other soil variables (e.g. fertility, organic content, soil development) also significantly influenced the growth potential of the test site soils.

10.13 Applying Results to the Broader Study Area

The results of the final site ranking derived from the 22 field sites were extrapolated to the larger study area using remote sensing techniques. This approach assumed that that there was an association between ground cover characteristics and the final site impact rankings. The satellite images used in this analysis covered 9238 km². Areas were identified that had similar spectral signatures (i.e. similar ground cover characteristics) as the impacted (red and yellow) and reference (green) sites identified from the Objective #1 studies. Screening of these areas was achieved based upon the spectral image of the natural features as well as filtering techniques to remove questionable pixels and, therefore, reduce the uncertainty associated with the analysis. Approximately 85% of the area (7956 km²) could not be classified according to the ranking system. The unclassified areas represented regions that were discarded in the analysis for various reasons and consist of a variety of land uses including lakes, wetlands, industrial areas and urban centers as well as regions that were screened out to increase certainty in the analysis. Within the classified areas (1281 km²), 19% of this area was identified as red or severely impacted, 31% as yellow or moderately impacted, and 49% as green or corresponding to the reference sites. Using this approach suggested that within the areas that could be classified, up to 50% of the land was moderately or severely impacted.

On the ground, each pixel represents an area of 30 m by 30 m. One of the major limitations of using the remote sensing approach is that the level of resolution required for discrimination of subtle differences between sites could not be achieved. However, use of remote sensing techniques to identify potential areas of impact was considered a reasonable approach that provided a starting point for planners and scientists. The maps and images produced from this approach could be used as a qualitative guide to identify areas to focus remediation and monitoring efforts. Ground truthing of the site classification is required at each location before any future work is undertaken. Some of these locations can be used as benchmarks to evaluate the rate of vegetation change through ongoing monitoring and observation. Mapping support can be improved over time through additional ground observations, and with the use of higher resolution imagery to refine the areas of impact.

10.14 Influence of Distance from Smelter on Site Ranking

The field study was not designed to accurately measure the spatial extent of impacts from the smelters, but some generalizations were possible based on the results. The test and reference sites were selected primarily based on soil Cu and Ni levels. Suitable reference sites were all greater than 25 km from the nearest smelter. The relationship between site ranking and distance from the smelter is illustrated in Figure 10.20.

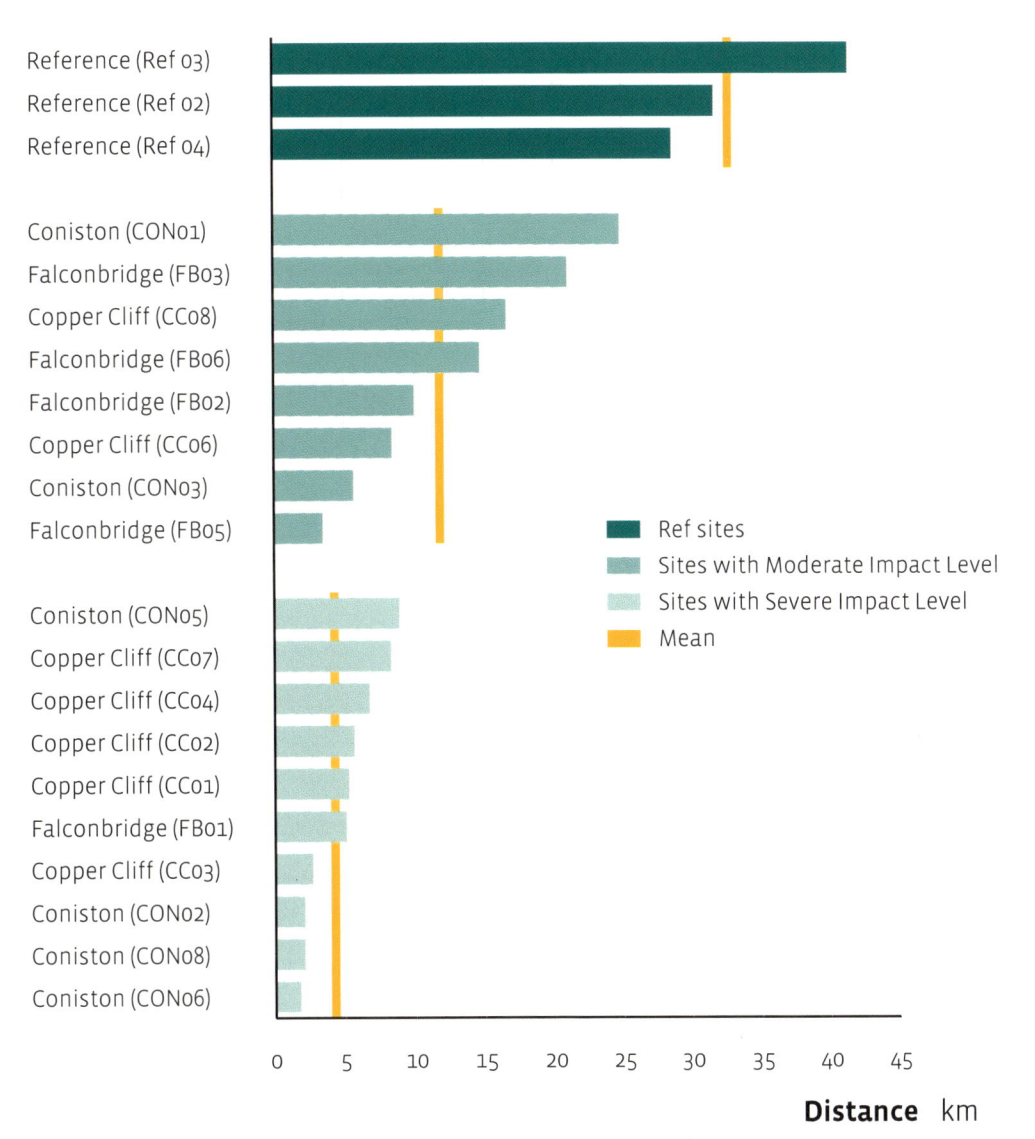

Figure 10.20 **Distance from Nearest Smelter for Sites at Each Impact Level**

None of the test sites were given a final 'green' or low impact ranking, although at some test sites individual LOE did have low impact rankings. The 'moderately' impacted sites ranged from 3.5 to 25 km from the nearest smelter (Table 10.21) although most (7/8) were over 5 km from the smelter, with five out of eight sites greater than 8 km away. The 'severely' impacted sites ranged from 1.8 to 8.9 km from a smelter, with 80% being within 5 km of a smelter. These distances could be used for general guidance to help focus risk management activities, recognizing that there will be exceptions and that site-specific conditions must be taken into account. All three reference sites were located well outside of the 'peanut'-shaped area referred to as the semi-barrens (Figure 2.8). Some of the test sites were also located outside the semi-barren area, although this boundary was not a factor in site selection. These results clearly demonstrate that ecological effects extended beyond the area previously identified as the semi-barrens.

Table 10.21 **Range of Site Distances (km) from Smelters Relative to Impact**

	Reference	Moderate impact	Severe impact
Copper Cliff	31.6	8.4–16.6	2.7–8.3
Coniston	41.3	5.7–24.8	1.8–8.9
Falconbridge	28.6	3.5–20.9	5.1

There was also an association between distance from the smelter and some plant community variables (such as species richness, reestablishment of sensitive lichen species and percentage leaf litter). The number of plant species tended to increase with increasing distance away from the smelter as previously shown in Figure 10.5.

10.15 Characteristics of Impacted and Reference Sites

The degree to which an ecosystem was considered impaired can be determined by comparing its key structural and functional components to those of a healthy system. Based on the study results, some measurements of the plant community and soil quality parameters were characterized to provide guidance for future assessments to determine level of site impact. Plant community and soil chemistry were two lines of evidence that were directly measured in the field, and through laboratory soil analysis. The other lines of evidence used in the ERA (soil toxicity testing, litter bag decomposition) involved lengthy and costly tests, and could not be readily incorporated into a field site assessment procedure.

10.15.1 Site Chemistry

The ranges of values for soil chemical parameters associated with the different site classifications are presented in Table 10.22. These characteristics may be used as a tool for the preliminary identification of impact levels. Soil samples must be collected in the field and returned to the laboratory to obtain these types of data. However, the information is fundamental to classifying the relative impact or risk associated with a site, and can be useful to help identify appropriate soil treatment strategies for future risk management. The ranges of COC concentrations in soil associated with site ranking are presented in Table 10.22. Although there are some patterns, there is considerable overlap and variability in soil metal concentrations between site rankings.

Table 10.22 **Range of Total COC Concentrations (mg/kg) at Different Sites**

COC	Reference site Mean	Reference site Range	Moderately impacted Mean	Moderately impacted Range	Severely impacted Mean	Severely impacted Range
As	4.37	2.7–5.9	23.2	9.5–45	36.5	2.1–117
Cd	0.23	0.17–0.28	0.4	0.24–1.2	0.59	0.12–1.3
Co	7.24	4.9–11.5	14	4.8–48	20.3	9.01–41
Cu	33.3	18.7–42.0	166	76–320	444	48.7–1000
Pb	21.9	14.0–33.0	39.3	17.2–83	51.5	4.6–162
Ni	41.6	38.9–46.0	136	77–325	376	70.2–1110
Se	0.74	0.48–1.0	1.5	0.85–3.4	3.5	0.3–10.5

Figure 10.21 clearly shows that, while the highest metal levels were always associated with severe impacts to the plant community, severe impacts were also present at eroded sites with very low metal levels. Thus, factors other than only soil metal concentrations must be considered when describing the extent of impact at a site.

Total Cu and Ni in soils
mg/kg

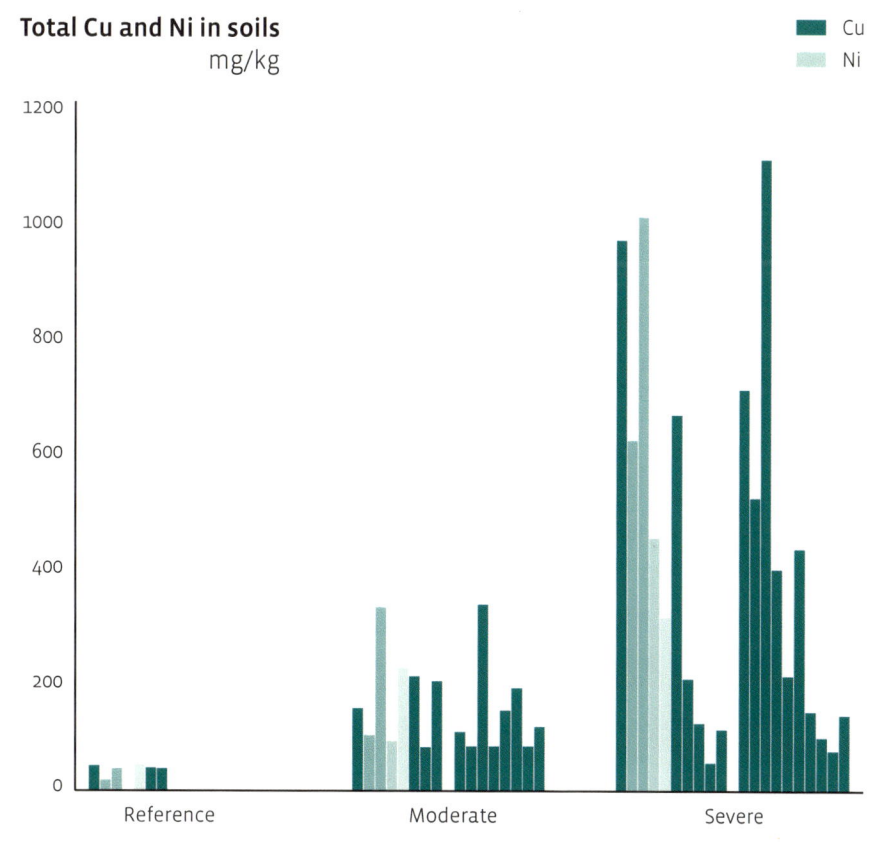

Figure 10.21 **Total Cu and Ni Concentrations in Soils Relative to Site Ranking**

10.15.2 Plant Community

Some of the important characteristics of the plant community that could be quantified for different levels of impact were extracted from the plant LOE. These characteristics are summarized in Table 10.23 and could be used in the field as a guide to help identify the severity of site impact. For example, a self-sustaining system tended to be composed of 50 or more plant species. Sites with less than 50 species may be considered impacted. Impacted sites tend to have trees shorter than 10 m, and have up to 60% bare rock or soil which is often eroded. Non-impacted sites can have high numbers of shade-tolerant, perennial, and sensitive species, whereas impacted sites have few. Impacted sites tend to be dominated by a single species, which may be non-native and invasive. More examples of plant community characteristics that may be used in the field as a tool for preliminary identification of impact levels are provided in Table 10.23.

Table 10.23 **Characteristics of Plant Communities Relative to Impact Ranking**

Characteristic	Low to not impacted	Moderately impacted	Severely impacted
Life history (perennial analysis)	Sites tend to have approximately 50 or more perennial species	Sites tend to have fewer than 50 perennial species	
Species dominance	Sites tend to have less than 20% cover by a single species	Sites tend to have more than 20% cover by a single species	
Conifer cover	Sites tend to have near-complete canopy cover	Sites tend to have 0 to 50% canopy cover, with 0 to 10% understory cover	Sites tend to have 0 to 5% combined canopy and understory cover

Characteristic	Low to not impacted	Moderately impacted	Severely impacted
Introduced and invasive species	Sites have negligible cover of non-native and potentially invasive species	Sites tend to have 0 to 50% cover of non-native and potentially invasive species combined	
Shade tolerance	Sites tend to have 10 to 15 shade-tolerant species	Sites tend to have 5 to 10 shade-tolerant species	Sites tend to have 0 to 5 shade-tolerant species
Percent cover of mineral substrate	Sites tend to have no bare rock or soil	Sites tend to have 0 to 10% of bare rock or soil	Sites have 0 to 60% of bare rock or soil
Reestablishment of sensitive species	Sites tend to have 5 to 10 good conditions indicator species	Sites tend to have 0 to 5 good conditions indicator species	
Acid- and metal-tolerant indicators	Sites tend to have 5 to 10 acid- and metal-tolerant indicator species	Sites tend to have 10 to 15 acid- and metal-tolerant indicator species	
Maximum tree height	Sites tend to have a maximum tree height of 10 to 14 m	Sites tend to have a maximum tree height of less than 10 m	
Total vegetation cover	Sites tend to have a surface soil retention index of 100	Sites tend to have a surface soil retention index of 95 to 100	Sites tend to have a surface soil retention index of 40 to 95
Leaf litter cover	Sites tend to have 85 to 90% leaf litter cover	Sites tend to have 65 to 85% leaf litter cover	Sites tend to have 0 to 65% leaf litter cover

10.16 Uncertainties Related to Objective #1

An important consideration of any risk assessment is to identify uncertainties associated with the methodology, available information and results. These areas of uncertainty were subjectively evaluated as part of this study and discussed with the purpose of providing confidence in the final results and conclusion. Risk managers need to be aware of the uncertainty surrounding the study conclusions so they can make recommendations and decisions accordingly. With respect to the amount of information collected for this study and the availability of supporting documentation, the study scientists felt there were no other tests or information that could be collected that would have changed the final site ranking designations. Each LOE from the Objective #1 study contained uncertainties which are briefly discussed below along with some other possible confounding factors that could contribute to uncertainty.

Plant Community LOE

There was a high level of confidence in the study approach and metrics used for this LOE. There was some variability in field measurements, but this was minimized by using the same field biologists at all sites, and the quality control procedures implemented by the expert ecologists.

Toxicity Testing LOE

There was a moderate level of certainty for this LOE. Multiple test species were evaluated with standardized test protocols to address species variability and different sensitivity. However, these protocols were not designed for the Sudbury-specific test species, which is a source of uncertainty. Uncertainty was also associated with quantifying the interactions between pH and metal bioavailability, and with the inability of the testing procedures to separate these two factors. The invertebrate tests were also a source of uncertainty because one test did not show any adverse responses (springtails), while the other did not produce consistent results (earthworms).

Litter Decomposition LOE

There was a moderate level of uncertainty for this LOE as only one variable (% biomass loss over time) was measured, using methods modified from agricultural soils for use in a boreal forest ecosystem. Sampling variability was minimized by having numerous replicate samples at 20 different sites.

Soil Chemistry LOE

There was a high level of confidence with this LOE as the methodology provided relatively precise data and numerous samples were collected, including many QA/QC measures. There was some inherent variability because natural soils tend to be very heterogeneous. There was some uncertainty in the ranges developed for soil chemistry values for reference soils since there was no existing recognized classification of 'normal' or 'reference' conditions for soils in north-eastern Ontario. The categories of high, medium and low quality soils were based upon a mixture of literature values, reference site conditions and the professional opinion of soil scientists familiar with the Sudbury region.

Selection of Reference Sites

The Sudbury region is in a transitional zone between the Great Lakes–Saint Lawrence Forest and the Boreal Forest ecological regions. It is also where four climatic zones intersect. Therefore, the reference sites did not completely represent all of the test sites, nor were all of the climatic zones entirely represented.

Sulfur Dioxide

Sulfur and SO_2, like pH, were not considered COC in this study but were known to have a significant effect on the landscape. In addition to the thousands of tonnes of metal particulates emitted over more than a century of smelting in the Sudbury area, more than 100 million tonnes of SO_2 were released into the atmosphere. The poor air and soil quality resulting from historic SO_2 emissions most certainly affected the region's forests. The possible direct and indirect effects of SO_2 fumigation are conceptually shown Figure 10.22. Due to technological advances in processing sulfur ores (Chapter 2), current SO_2 emissions are less than 10% of those 30 years ago. The current emission levels were not considered to be directly affecting area vegetation. However, the plant community was affected by historical SO_2 emissions which caused vegetation kills in the past, and contributed to a sequence of events that continue to impact the plant community. Historic SO_2 emissions were, therefore, a potential confounding factor for the purposes of the objective of evaluating the impact of COC in preventing the recovery of regionally representative self-sustaining terrestrial plant communities.

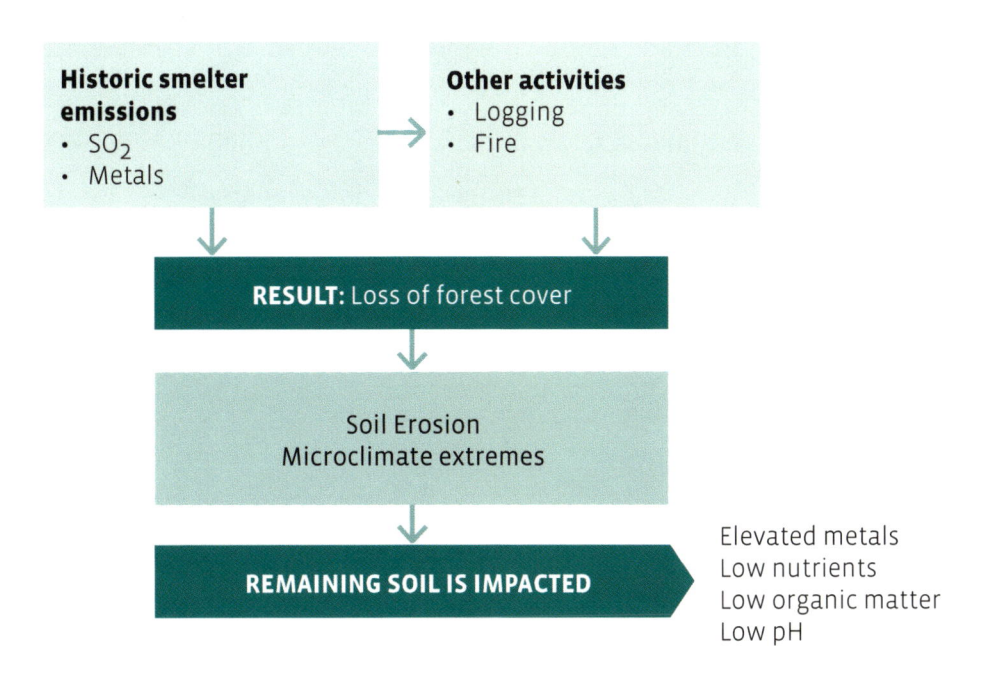

Figure 10.22 **Conceptual Linkages of Historical Emissions and Other Factors Leading to Current Soil Conditions**

Other Confounding Factors

Deforestation (from logging, forest fires) was another historic factor responsible in part for the status of the present forest community in Sudbury. While it could not be definitively quantified, deforestation must also be considered when determining the causal factors of impact. The loss of vegetation resulted in extensive soil loss which has impeded vegetation recovery in the Sudbury area. This mass scale erosion resulted in the loss of organic matter and the crucial topsoil layers that provide the medium for plant growth, leaving areas with either exposed rock or relic soil layers that are deficient in nutrients and organic matter.

Metal toxicity is strongly influenced by pH. In particular, low pH increases the bioavailability (and hence toxicity) of cationic metals. While some areas of low pH soils were directly associated with SO_2 deposition, the lowest soil pH levels were not in regions where SO_2 deposition was traditionally the highest. Therefore, low pH is a natural characteristic of the region, and must be taken into consideration when determining soil toxicity and causal relationships.

Bioavailability/Bioaccessibility

Plant growth may be more related to the concentrations of bioavailable metals rather than total metal concentrations. However, this study focused on total metal levels (which are the basis of regulatory soil quality guidelines). At present, there is no standardized method of determining metal bioavailability that is widely accepted by the scientific community. Some partial extraction techniques such as mild water leach were used in this study for a preliminary assessment of 'plant bioavailable' metal levels, but the resultant data were no better correlated to plant community metrics or soil toxicity results than total metal levels. The exact bioavailable metal concentrations and relationship with plant community structure and toxicity results remained an area of uncertainty in this study but did not affect the results or conclusions.

Split Rankings for Toxicity Testing LOE

There was considerable variability in the ability of the test species to perform in the soil from the various reference sites. Therefore, a mean of the values for each endpoint for the three reference sites was established and was referred to as REFmean. Consequently, two approaches were used to evaluate the toxicity test data for the test sites. The use of two statistical approaches to evaluate the data increased the robustness of the toxicity testing LOE and increased the confidence of the final site ranking based on soil toxicity and was not considered an area of high uncertainty.

Aluminum Toxicity

Aluminum is a major soil constituent, but it is not a plant nutrient. At low pH, it can inhibit plant growth. Under acidic conditions (pH < 5.5), Al dissolves and becomes readily available for plant uptake, creating a potentially toxic environment. However, under neutral or basic conditions (pH ≥ 7), it precipitates and is no longer a threat to plant health. Given the low pH of soils in the Sudbury area, it is possible that Al toxicity was contributing to the observed ecosystem impairment at the test sites. Hutchinson and Whitby (1977) and Cox and Hutchinson (1981) suggested that Al may have been a major selective factor in plant establishment in the Sudbury area. Winterhalder (1996) also re-iterated that Al ions were likely present at phytotoxic concentrations in Sudbury soils. More recently, Feisthauer et al. (2006) acknowledged that Al, along with other confounding soil factors, likely influenced metal availability and, hence toxicity, of Sudbury soils to invertebrates and plants. Al was not selected as a COC for the Sudbury ERA because it did not meet the selection criteria. Therefore, Al was only considered as a possible confounding factor and a source of uncertainty in the assessment. The available information suggested that Al toxicity could be an issue in Sudbury soils due to low pH. This was a confounding factor in the assessment and an area of uncertainty.

Color Ranking Approach

The color ranking approach used in this study was essentially a visual aid used to interpret a large quantity of combined data. Warren-Hicks and Moore (1998) state that it is often desirable to combine sets of existing physical, chemical or biological data to develop a more comprehensive characterization of exposure or toxicity, and discuss some of the uncertainties associated with this process in ERA. The color ranking system did not provide quantitative or numeric values to determine overall ranking, which caused some uncertainty with regard to precision of the final site rank. The authors could have used a numeric system for site ranking but felt that numeric designation implied more precision than was considered appropriate. The WOE stepwise process used provided decision makers with all of the pertinent information necessary for risk management. It was never assumed that risk managers would treat all 'red' sites the same. Rather, the delineation of a severely impacted ranking (red) enables risk managers to prioritize sites, categorize remediation strategies, and identify areas that could be addressed. The areas predicted to be at risk need to be verified through ground-truthing and supplemental data collection. Therefore, there was very low probability of making the wrong risk management decision due to uncertainty with the ranking system.

10.17 Conclusions

A considerable amount of effort was devoted to the study design for this Objective and selection of test and reference sites. The sampling locations were reflective of the distribution and concentration of COC in the Greater Sudbury study area. Using data collected from the LOE, eight of the 18 test sites were ranked as moderately impacted with the other 10 test sites ranked as severely impacted relative to the reference sites. The use of multiple lines of evidence in a weight of evidence approach was taken to reduce the uncertainty associated with a single LOE. The spatial extent of vegetation impacts extended beyond the boundary of areas previously considered to be directly impacted by smelter emissions.

The role of soil pH as a confounding factor in this study was examined by conducting soil toxicity tests in natural (low pH) and pH-amended soils, as well as collecting and comparing detailed field measurements from a historically limed site and an adjacent site that was not treated. These studies showed that low soil pH impacted plant growth directly, and it interacted with soil metals to contribute to soil toxicity. The liming and re-greening activities have helped to establish a diverse plant community with the introduction of essential minerals (Cs, Mg), a viable seed source and planted trees. Because soil pH influences metal speciation and bioavailability, it was not possible to totally separate the relative role of pH from metal toxicity.

The results of the studies carried out to evaluate Objective #1 showed that the concentrations of COC have impacted the plant communities, and are continuing to impede the recovery of a self-sustaining forest community in the Sudbury region. However, other environmental variables and soil conditions were also contributing to inhibit ecosystem recovery and all these factors are intertwined.

This study clearly identified that impacts to the landscape are still pronounced and extensive, despite almost 40 years of re-greening activities. This helped act as a catalyst to stimulate a new phase of risk management activities and development of a Biodiversity Action Plan (see Chapter 12) for the Sudbury area. The data will be used to help design future land treatment strategies and to guide areas for future planning and treatment. The detailed ecological data gathered at the test sites can be used as a starting point for monitoring natural recovery and succession in years to come. In addition, the conditions documented at the reference sites can be used as a benchmark to monitor natural succession or ecological changes occurring as a result of other stressors such as climate change.

10.18 Acknowledgements

Field crew for soil collection and ecological work was conducted by Kara Hearne, Robert Price and Caitlin Meanwell among others. Janice Lindquist provided valuable field and laboratory logistics. Toxicity testing was competently carried out by Dr Gladys Stephenson, Natalie Feisthauer, Stantec Consulting, Guelph, Ontario, Mr Rick Scroggins, Leanne Vandervillet and Juliska Princz, Environment Canada Biological Methods Division, Ottawa, and Mary Moody, Saskatchewan Research Council (SRC). The litter bag study was implemented by Dave Marshell. Samples were analyzed by Testmark Laboratories, Garson, Ontario. Particle size distribution measurements were completed at the Soil and Nutrient Laboratory, University of Guelph. Steve Gautreau undertook many hours of statistical analysis on the data and never lost his sense of humor. The insightful comments of Paul Welsh were appreciated. The authors would also like to recognize the unfailing support and wisdom of June Gilbertson for MKG.

10.19 References

Amiro BD, Courtin GM (1981) Patterns of vegetation in the vicinity of an industrially disturbed ecosystem, Sudbury, Ontario. Can J Bot 59:1623–1639

Anderson JA, Hooper MJ, Zak JC et al. (2009) Characterization of the structural and functional diversity of indigenous soil microbial communities in smelter-impacted and non-impacted soils. Environ Toxicol Chem 28:534–541

Andersson C (2005) Litter decomposition in the forest ecosystem – influence of trace elements, nutrients and climate. ESS Bull 3(1):4–17

Ashman MR, Puri G (2002) Essential soil science: A clear and concise introduction to soil science. Blackwell Science, London, 198 pp

Bache BW (1976) The measurement of cation exchange capacity of soils. J Sci Food Agric 27:273–280

Batley GE, Burton GA, Chapman PM et al. (2002) Uncertainties in sediment quality weight-of-evidence (WOE) assessments. Hum Ecol Risk Assess 8(7):1517–1547

Bradham KD, Dayton EA, Basta NT et al. (2006) Effects of soil properties on lead bioavailability and toxicity to earthworms. Environ Toxicol Chem 25:769–775

Breymeyer A, Kegorski M, Reed D (1997) Decomposition of pine litter organic matter and chemical properties of upper soil layers: transect studies. Environ Pollut 98:361–367

Brodo I, Sharnoff SD (2001) Lichens of North America. Yale University Press, New Haven, CT, USA, 795 pp

Brower JE, Zar JH, von Ende CN (1997) Field and laboratory methods for general ecology, 4th edn. WCB McGraw Hill, Boston, MA, USA, 273 pp

Burton GA Jr, Batley GE, Chapman PM et al. (2002a) A weight-of-evidence framework for assessing sediment (or other) contamination: improving certainty in the decision-making process. Hum Ecol Risk Assess 8(7):1675–1696

Burton GA Jr, Chapman PM, Smith EP (2002b) Weight-of-evidence approaches for assessing ecosystem impairment. Hum Ecol Risk Assess 8(7):1657–1673

Carter MR (ed) (1993) Soil sampling and methods of analysis. Lewis Publishers, Boca Raton, FL, USA, 823 pp

CCFM (2000) Criteria and indicators of sustainable forest management in Canada. National Status 2000. Canadian Council of Forest Ministers, Ottawa, ON, 122 pp

Chambers BA (1995) Successional trends by site type in Northeastern Ontario. Ed, T.G. McCarthy. NEST Technical Report TR-009. Ontario Ministry of Natural Resources, Timmins, 64 pp

Chambers B, Naylot B, Nieppola J, et al. (1997) Field guide to forest ecosystems of central Ontario. SCSS Field Guide FG-01. Queen's Printer, Bracebridge, ON, 200 pp

Chapman PM, McDonald BG, Lawrence GS (2002) Weight-of-evidence issues and frameworks for sediment quality (and other) assessments. Hum Ecol Risk Assess 8(7):1489–1515

Courchesne F, Kruyts N, Legrand P (2006) Labile zinc concentration and free copper ion activity in the rhizosphere of forest soils. Environ Toxicol Chem 25(3):635–642

Courtin GM (1994) The last 150 years: a history of environmental degradation in Sudbury. Sci Total Environ 148: 99–102

Cox RM, Hutchinson TC (1981) Environmental factors influencing the rate of spread of the grass *Deschampia cespitosa* invading areas around the Sudbury nickel-copper smelters. Water Air Soil Pollut 16:83–106

Crum H (1976) Mosses of the Great Lakes forest. University Herbarium, University of Michigan, Ann Arbor, MI, 404 pp

Dore WG, McNeill J (1980) Grasses of Ontario. Monograph 26. Agriculture Canada. Canadian Government Pub Centre, Canada, 566 pp

Doyle PJ, Gutzman DW, Bird GA et al. (2003) An ecological risk assessment of air emissions of trace metals from copper and zinc production facilities. Hum Ecol Risk Assess 9(2):607–636

Dumontet S, Dinel H, Levesque PEN (1992) The distribution of pollutant heavy metals and their effects on soil respiration and acid phosphatase activity in mineral soils of Rouyn-Noranda region, Quebec. Sci Total Environ 121:231–245

Els Smit C, Van Gestel CAM (1998) Effects of soil type, prepercolation, and ageing on bioaccumulation and toxicity of zinc for the spring tail *Folsomia candida*. Environ Toxicol Chem 17(6):1132–1141

Environment Canada (1995) Canada's biodiversity strategy. Environment Canada, Ottawa, ON

Environment Canada (2005a) Biological test method: Tests for toxicity of contaminated soil to earthworms (*Eisenia andrei, Eisenia fetida* or *Lumbricus terrestris*). EPS 1/RM/43. Method Development and Applications Section, Environmental Technology Centre, Environment Canada

Environment Canada (2005b) Biological test method: Test for measuring emergence and growth of terrestrial plants exposed to contaminants in soil. EPS 1/RM/45. Method Development and Applications Section, Environmental Technology Centre, Environment Canada

Feisthauer NC, Stephenson GL, Princz JI et al. (2006) Effects of metal-contaminated forest soils from the Canadian shield to terrestrial organisms. Environ Toxicol Chem 25:823–835

Freedman B, Hutchinson TC (1980a) Long term effects of smelter pollution at Sudbury, Ontario, on forest community composition. Can J Bot 58:2123–2140.

Freedman B, Hutchinson TC (1980b) Effects of smelter pollutants on forest leaf litter decomposition near a nickel-copper smelter at Sudbury, Ontario. Can J Bot 58:1722–1736

Fritze H, Niini S, Mikkola K et al. (1989) Soil microbial effects of a Cu-Ni smelter in southwestern Finland. Biol Fertil Soils 8:87–94

Gleason HA, Cronquist A (1991) Manual of vascular plants of northeastern United States and adjacent Canada. New York Botanical Garden, New York, USA, 910 pp

Gregorich EG, Carter MR, Angers DA et al. (1994) Towards a minimum data set to assess soil organic matter quality in agricultural soils. Can J Soil Sci 74:367–385

Hendershot WH, Lalande H, Duquette M (1993) Soil reaction and exchangeable acidity. In: Carter MR (ed) Soil sampling and methods of analysis, pp. 141–145. Lewis Publishers, Boca Raton, FL

Hull RN, Swanson SM (2006) Sequential analysis of lines of evidence – An advanced weight-of-evidence approach for ecological risk assessment. Integr Environ Assess Manag 2(4):302–311

Hutchinson TC, Whitby LM (1977) The effects of acid rainfall and heavy metal particulates on a boreal forest ecosystem near the Sudbury smelting region of Canada. Water Air Soil Pollut 7:421–438

James GI, Courtin GM (1985) Stand structure and growth form of the birch transition community in an industrially damaged ecosystem, Sudbury, Ontario. Can J For Res 15:809–817

Johnson D, Hale B (2004) White birch (*Betula papyrifera Marshall*) foliar litter decomposition in relation to trace metal atmospheric inputs at metal-contaminated and uncontaminated sites near Sudbury, Ontario and Rouyn-Noranda, Quebec, Canada. Environ Pollut 127:65–72

Kapustka LA (1997) Selection of phytotoxicity tests for use in ecological risk assessments. In: Wang W, Goruch J, Hughes JS (eds) Plants for environmental studies, pp. 515–548. Lewis Publishers, Boca Raton, FL, USA

Kapustka LA, Eskew D, Yocum JM (2006) Plant toxicity testing to derive ecological soil screening levels for cobalt and nickel. Environ Toxicol Chem 25:865-874

Kautz G, Zimmer M, Zach P et al. (2001) Suppression of soil microorganisms by emissions of a magnesite plant in Slovak Republic. Water Air Soil Pollut 125:121–132

Kelly J, Tate RL (1998) Effects of heavy metal contamination and remediation on soil microbial communities in the vicinity of a zinc smelter. J Environ Qual 27:609–617

Kozlov MV, Zvereva EL (2007) Industrial barrens: extreme habitats created by non-ferrous metallurgy. Rev Environ Sci Biotechnol 6:231–259

Kuinto T, Saeki K, Oyaizu H et al. (1999) Influence of copper forms on toxicity to microorganisms in soils. Ecotoxicol Environ Safety 44:174–181

Lanno RP (2003) Contaminated soils: From soil-chemical interactions to ecosystem management. SETAC, Pensacola, FL, USA, 445 pp

Maxwell CD (1995) Acidification and metal contamination: implications for the soil biota of Sudbury. In: Gunn JM (ed) Restoration and recovery of an industrial region, pp. 219–231. Springer Verlag, New York

McCall J, Gunn J, Struik H (1995) Photo interpretive study of recovery of damaged lands near the metal smelters of Sudbury, Canada. Water Air Soil Pollut 85:847–852

McEnroe N, Helmisaari H (2001) Decomposition of coniferous forest litter along a heavy metal pollution gradient, southwest Finland. Environ Pollut 113:11–18

MOE (1997) Guidelines for use at contaminated sites in Ontario. ISBN-0-7778-4052-9. Ontario Ministry of Environment, Toronto, Ontario, Canada

Nolan AL Lombi E, McLaughlin MJ (2003) Metal bioaccumulation and toxicity in soils – Why bother with speciation? Aust J Chem 56:77–91

Nordgren A, Baath E, Soderstrom B (1985) Soil microfungi in an area polluted by heavy metals. Can J Bot 63:448–455

Oorts K, Ghesquiere U, Swinnen K et al. (2006) Soil properties affecting the toxicity of $CuCl_2$ and $NiCl_2$ for soil microbial processes in freshly spiked soils. Environ Toxicol Chem 25:836–844

Reddy KJ, Wang L, Gloss SP (1995) Solubility and mobility of copper, zinc and lead in acidic environments. Plant Soil 53:53–58

Reynoldson TB, Smith EP, Bailer AJ (2002a) A comparison of three weight-of-evidence approaches for integrating sediment contamination data within and across lines of evidence. Hum Ecol Risk Assess 8(7):1613–1624

Reynoldson TB, Thompson SP, Milani D (2002b) Integrating multiple toxicological endpoints in a decision-making framework for contaminated sediments. Hum Ecol Risk Assess 8(7):1569–1584

Römbke J, Heimbach F, Hoy S et al. (eds) (2003) Effects of plant protection products on functional endpoints in soil. EPFES, Lisbon, 24–26 April 2002. Society of Environmental Toxicology and Chemistry (SETAC), Pensacola, FL, USA, 109 pp

Rooney CP, Zhao FJ, McGrath SP (2006) Soil factors controlling the expression of copper toxicity to plant in a wide range of European soils. Environ Toxicol Chem 25:725–732

Rutgers M, den Besten P (2005) Approach to legislation in a global context: The Netherlands perspective – soils and sediments. In: Thompson KC, Wadhia K, Loibner AP (eds) Environmental toxicity testing. Blackwell Publishing and CRC Press

Sanders JR (1982) The effect of pH upon the copper and cupric ion concentrations in soil solutions. J Soil Sci 33:679–689

Sauve S, McBride MB, Norvell WA et al. (1997) Copper solubility and speciation of in situ contaminated soils: Effects of copper level, pH and organic matter. Water Air Soil Pollut 100:133–149

Schroth G, Sinclair FL (eds) (2003) Trees, crops and soil fertility concepts and research methods. CABI, Wallingford, UK

Sheldrike BH, Wang C (1993) Particle size distribution. In: Carter MR (ed) Soil sampling and methods of analysis, pp. 499–512. Canadian Society of Soil Science. Lewis Publishing, Boca Raton, FL, USA

Sinclair A (1996) Floristics, structure and dynamics of plant communities on acid, metal-contaminated soils. MSc thesis, Laurentian University, Sudbury, ON

Smith EP, Lipkovich I, Ye K (2002) Weight-of-evidence (WOE): quantitative estimation of probability of impairment for individual and multiple lines of evidence. Hum Ecol Risk Assess 8(7):1585–1596

Sollins P (1982) Input and decay of coarse woody debris in coniferous stands in western Oregon and Washington. Can J For Res 12:18–28Soper JH, Heimburger ML (1990) Shrubs of Ontario. Royal Ontario Museum, Toronto, ON, 495 pp

Spurgeon DJ, Lofts S, Hankard PK et al. (2006) Effect of pH on metal speciation and resulting metal uptake and toxicity for earthworms. Environ Toxicol Chem 25:788–796

SRC (2005) Testing of twenty two soils for the Sudbury area risk assessment using boreal forest plants. Report to Cantox Environmental Inc. SRC Publication No. 11892-3C05. Saskatchewan Research Council, Environment and Mineral Division, Saskatoon, SK, Canada

ter Braak CJF (1986) Canonical correspondence analysis: a new eigenvector technique for multivariate direct gradient analysis. Ecology 67:1167–1179

Troeh FR, Thompson LM (2005) Soils and soil fertility, 6th edn. Blackwell Publishing Professional, Iowa, USA, 489 pp

Warren-Hicks WJ, Moore DRJ (eds) (1998) Uncertainty analysis in ecological risk assessment. SETAC Special Publication. SETAC, Pensacola, FL, USA, 227 pp

Winterhalder K (1983) The use of manual surface seeding, liming and fertilization in the reclamation of acid metal-contaminated land in the Sudbury, Ontario mining and smelting region of Canada. Environ Technol Lett 4:209–216

Winterhalder K (1995) Natural recovery of vascular plant communities on the industrial barrens of the Sudbury area. In: Gunn J (ed) Restoration and recovery of an industrial region: the smelter-damaged landscape near Sudbury, Canada, pp. 93–102. Springer-Verlag, New York

Winterhalder K (1996) Environmental degradation and rehabilitation of the landscape around Sudbury, a major mining and smelting area. Environ Rev 4:185–224

Zak JC, Willig MR, Moorhead DL et al. (1994) Functional diversity of microbial communities: A quantitative approach. Soil Biol Biochem 26:1101–1108

10.20 Appendix Chapter 10: Abbreviations

ANOVA, analysis of variance
CAEAL, Canadian Association of Environmental Analytical Laboratories
CC, Copper Cliff
CCA, canonical correspondence analysis
CCFM, Canadian Council of Forest Ministers
CEC, cation exchange capacity
CECF, cation exchange factor
CEM, Centre for Environmental Monitoring
COC, chemical of concern
CON, Coniston
CRM, certified reference material
DBH, diameter at breast height
DL, detection limit
DPTA, diethylenetriaminepentaacetic acid
DWD, downed woody debris
EC$_{20}$, effective concentration that results in 20% effect in test organism
ELC, ecological land classification
EPFES, effects of plant protection products on functional endpoints in soil
ERA, ecological risk assessment
FB, Falconbridge
FF, fertility factor
GBF, geogenic bioavailability factor
GLF, geogenic load factor
ICP-MS, inductively coupled plasma mass spectroscopy
LC$_{50}$, lethal concentration to kill 50% of test organisms
LFH, litter fermentation humus layer
LN, log normal
LOAEL, lowest observed adverse effect level
LOE, line of evidence
MDL, minimum detection limit
MDS, multidimensional scaling

MLR, multiple linear regression
MOE, Ministry of the Environment (Ontario)
MOH, Medical Officer of Health
OM, organic matter
OMF, organic matter factor
pHF, pH factor
PMDF, parent material depositional factor
QA/QC, quality assurance/quality control
REF, reference
SARA, Sudbury Area Risk Assessment
SOP, standard operating procedure
TBF, technogenic bioavailability factor
TLF, technogenic load factor
WHC, water holding capacity
WOE, weight of evidence

DOI: 10.5645/b.1.11

11.0 Ecological Risk Assessment: Evaluating Risks to Terrestrial Wildlife

Authors

Ruth N. Hull, M.Sc.
Intrinsik Environmental Sciences Inc.,
6605 Hurontario Street, Suite 500, Mississauga, ON Canada L5T 0A3
Email: rhull@intrinsikscience.com

Karl Bresee, B.Sc., PBD, P.Biol.
Intrinsik Environmental Sciences Inc.,
Suite 1060, 736 8th Avenue SW, Calgary, Alberta, T2P 1H4
Email: kbresee@intrinsikscience.com

Table of Contents

11.1 Introduction	345
11.2 Exposure Assessment	345
11.2.1 Wildlife Exposure Modeling Methodology	345
11.2.2 Concentrations of COC in Water, Soil and Dietary Items	351
11.2.3 Concentrations of COC in Wildlife Food Items	356
11.3 Effects Assessment	360
11.3.1 Toxicity Reference Values	360
11.3.2 General Information on Status of Mammals and Birds in the Sudbury Area	363
11.4 Risk Characterization	365
11.4.1 Zone 1 Results	366
11.4.2 Copper Cliff Results	368
11.4.3 Risk Characterization Summary	369
11.5 Uncertainty Analysis	371
11.5.1 Surface Water Concentrations	372
11.5.2 Soil Concentrations	372
11.5.3 Medium-to-Biota Uptake Models	372
11.5.4 Dietary Composition and Use of Surrogates	372
11.5.5 Wildlife Receptor Variables	373
11.5.6 Soil Bioaccessibility	373
11.5.7 Model Uncertainty	373
11.5.8 Toxicity Reference Values (TRVs)	373
11.6 Summary and Conclusions	374
11.7 Acknowledgements	374
11.8 References	374
11.9 Appendix Chapter 11: Abbreviations	379

Tables

11.1 Input Variables for Northern Short-tailed Shrew Exposure Analyses 347
11.2 Summary of Dietary Item Assumptions Used in the Exposure Model 349
11.3 Comparison of Sudbury Bioaccessibility (%) Assumptions with Two Other Studies 350
11.4 Statistical Summary of COC Concentrations in Surface Water (µg/L) 351
11.5 Bioconcentration Factors Used for Estimating COC Concentrations in Aquatic Plants 352
11.6 Soil COC Concentrations (95% UCLM, mg/kg) for ERA Zones 353
11.7 Valued Ecosystem Component Home Ranges 354
11.8 Measured COC Concentrations (mg/kg dw) in Soil and Biota from the Study Area 356
11.9 Predictive Models Used to Estimate COC Concentrations in Insects 358
11.10 Predictive Models Used to Estimate COC Concentrations in Roots 359
11.11 Predictive Models Used to Estimate COC Concentrations in Shoots 359
11.12 Zone 1 to 3 Predictive Models Used to Estimate COC Concentrations in Earthworms 359
11.13 Urban Predictive Models Used to Estimate COC Concentrations in Earthworms 360
11.14 Toxicity Reference Values (TRVs) Used in the Sudbury ERA 361
11.15 Probability of ER > 1.0 for Zone 1 366
11.16 Probability of ER > 1.0 for Copper Cliff 368
11.17 Summary of ERs Where There Was a >10% Probability of an ER>1.0 370
11.18 Definition of Variability Qualifiers 371
11.19 Definition of Uncertainty Qualifiers 372

Figures

11.1 Framework Used to Model Exposure of Wildlife to Chemicals of Concern 345
11.2 Study Area Subdivisions Used in the Wildlife Exposure Model 352
11.3 Distribution of Ni Concentrations in Surface Soil (0 to 5 cm) in Zone 2 355
11.4 Measured Ni Concentrations in Soil vs. the Mean Concentration Defined by Confidence Limits 355
11.5 Correlation of Ni in Grasshoppers with Deschampsia Shoots 357
11.6 Distribution of Se Concentrations in Soil (0 to 5 cm) in Zone 1 367
11.7 Cumulative Probability of Exposure and ERs for Individual Meadow Voles Exposed to Se in Zone 1 367
11.8 Cumulative Probability of Exposure and ERs for American Robins Exposed to Se in Copper Cliff 369

11.1 Introduction

There were two ERA objectives related to terrestrial wildlife in the Sudbury study: Objective #2: Evaluate risk to terrestrial wildlife populations and communities due to Chemicals of Concern (COC), and Objective #3: Evaluate risks to individuals of threatened or endangered terrestrial species due to the COC. These objectives were similar, but under common guidance, explicit consideration must be given to species with special status, i.e. those that are threatened or endangered. The approach to estimating risk to threatened and endangered species is similar to other species, except the assessment is aimed at the individual organism level of organization (similar to human health risk assessment (HHRA)) as opposed to the population level of organization. The problem formulation for these objectives was completed in Chapter 9. This chapter presents the remaining components of the wildlife ERA: exposure assessment, effects assessment and risk characterization.

11.2 Exposure Assessment

The purpose of this section is to describe the general approach and methods used to estimate exposure of terrestrial wildlife to the COC. This section begins with a description of the exposure model that was used to estimate total daily intake of each COC by selected wildlife species. Input variables for the exposure model were established using life history information for the valued ecosystem components (VECs) and concentrations of COC in prey, soil and water. Model input parameters (e.g. COC concentration, VEC characteristics) were defined as probabilistic distributions to characterize uncertainty and variability in exposures where information was adequate. Monte Carlo analysis was then performed to propagate input variable uncertainties through the exposure model. Figure 11.1 depicts the general framework for the exposure assessment.

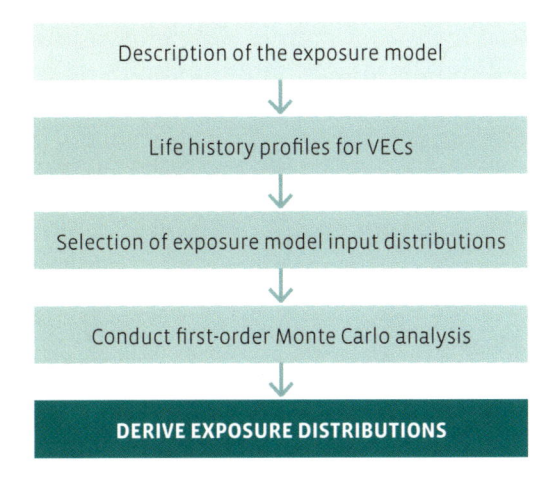

Figure 11.1 Framework Used to Model Exposure of Wildlife to Chemicals of Concern

11.2.1 Wildlife Exposure Modeling Methodology

There are two general approaches to exposure modeling: deterministic and probabilistic. Deterministic approaches are methods in which all biological, chemical, physical, and environmental parameter values are assumed to be constant, and one number is selected for each parameter. Probabilistic approaches are methods in which important biological, chemical, physical, and environmental parameters are assumed to vary or are uncertain, and therefore, are specified using distributions of probable values. The use of probabilistic methods in risk analysis is growing rapidly and U.S. EPA has produced guidance on how to conduct such analyses for Superfund and other contaminated site programs (U.S. EPA 2001b). The benefit of using probabilistic methods in exposure assessment is that it gives the risk assessor the ability to fully characterize exposure, rather than providing a best estimate or a conservatively biased estimate of exposure. By including the entire distribution for exposure, all exposures are considered and all of the data and information collected to characterize a situation are included. Input distributions for the exposure analyses were generally assigned as follows:

- lognormal distributions for variables that were right skewed with a lower bound of zero and no upper bound
- beta distributions for variables bounded by zero and one
- normal distributions for variables that were symmetric and not bounded by one (e.g. body weight (BW)), and
- point estimates for minor variables or variables with low coefficients of variation.

Monte Carlo analysis was used to quantify uncertainty. This is a technique whereby parameter values were drawn at random from defined input probability distributions, combined according to a model equation, and the process repeated iteratively until a stable distribution of solutions results. It is most useful when input distributions are known reasonably well. The primary goal of a Monte Carlo analysis in this assessment was to quantitatively characterize the uncertainty and variability in estimates of exposure. A secondary goal was to identify key sources of variability and uncertainty, and to quantify the relative contribution of these sources to the overall range of wildlife exposure model results.

The wildlife exposure model contained multiple variables, some of which may be correlated. The assumption of independence may be inappropriate, because dependencies can affect the estimates of exposure. If correlations are not accounted for, the variance and the tails of the exposure distribution may be poorly estimated. The current approach included an assumption of perfect covariance (e.g. when the diet consists of two prey items, the proportion of one item in the diet is equal to one minus the proportion of the other item). The proportion of each particular dietary item was normalized to ensure that total diet summed to 1.0 in every iteration of the Monte Carlo analyses. The exposure concentrations for dietary media were positively correlated with soil, water or sediment, due to the use of trophic transfer models. Where independence of variables seemed intuitively obvious (e.g. COC concentration in the prey item and proportion of that item in the diet), independence was assumed.

Intake Model

The wildlife exposure modeling relied on the use of a total daily intake model. The primary focus of the model was on ingestion of food (prey), soil, sediment and water, which are generally the most important exposure pathways for wildlife (Moore and Caux 1997; Moore et al. 1999). Thus, the wildlife exposure assessments did not include the dermal or inhalation routes of exposure in the model calculations. The wildlife exposure model followed the general form:

$$TDI = \left[(\sum_{i=1}^{n} FIR_i \cdot C_i \cdot P_i) + (SIR \cdot C_s \cdot RAF) + (WIR \cdot C_w) \right]$$

where:

TDI = Total daily intake (mg/kg bw/day)

FIR_i = Normalized food intake rate of ith food item in the diet (kg dw/kg bw/day)

C_i = Concentration in ith food item (mg/kg dw)

C_s = Concentration in soil or sediment (mg/kg dw)

C_w = Concentration in water (mg/L)

SIR = Normalized soil or sediment intake rate (kg dw/kg bw/day)

WIR = Normalized water intake rate (L/kg bw/day)

RAF = Relative absorption factor (unitless)

P_i = Proportion of the ith food item in the diet

This general exposure model was customized for each VEC to reflect diet, foraging range, and habitat preferences. Extensive literature searches were conducted and data collected to describe life history for each VEC and to determine the appropriate model inputs. Sudbury-specific information was used when available. Each of these inputs is discussed briefly below. As an example, the parameter values and distributions used in the probabilistic exposure model for the northern short-tailed shrew (*Blarina brevicauda*) are shown in Table 11.1.

Table 11.1 Input Variables for Northern Short-tailed Shrew Exposure Analyses

Input variable	Distribution	Parameters	References	Notes
BW (kg)	Normal	Mean=0.0223 SD=0.00287	U.S. EPA 2004a	Values for female shrews collected from Housatonic River, MA area
Assimilation efficiency – earthworms	Beta	$\alpha=80, \beta=12$ Scale=1		Parameters assumed to be the same as for terrestrial invertebrates
Assimilation efficiency – terrestrial invertebrates	Beta	$\alpha=80, \beta=12$ Scale=1	Barrett and Stueck 1976; Grodzinski and Wunder 1975	
Assimilation efficiency – small mammals	Beta	$\alpha=84, \beta=20$ Scale=1.05	Litvaitis and Mautz 1976; Vogtsberger and Barrett 1973; Grodzinski and Wunder 1975	
Assimilation efficiency – vegetation	Beta	$\alpha=60, \beta=19$ Scale=1.11	Grodzinski and Wunder 1975; Drozdz 1968	Assumed fruits, seeds and nuts based on observations by Hamilton (1941)
Gross energy (kcal/kg dw) – earthworms	Normal	Mean=4600 SD=360	Cummins and Wuycheck 1971; Thayer et al. 1973; Golley 1961	
Gross energy (kcal/kg dw) – terrestrial invertebrates	Normal	Mean=5600 SD=160	Cummins and Wuycheck 1971; Collopy 1975; Bell 1990	Assumed equal to measured values for grasshoppers, crickets and beetles
Gross energy (kcal/kg dw) – small mammals	Normal	Mean=5000 SD=1300	Górecki 1975; Golley 1960; Koplin et al. 1980	
Gross energy (kcal/kg dw) – vegetation	Normal	Mean=3650 SD=1300	Golley 1961; Karasov 1990; Dice 1922; Robel et al. 1979; Drozdz 1968	Assumed seeds and fruits
Proportion in diet – earthworms	Beta	$\alpha=5, \beta=5$ Scale=0.48	Whitaker and Ferraro 1963; Linzey and Linzey 1973; Hamilton 1941; Eadie 1944, 1948	Each iteration of Monte Carlo analysis corrects P_i so that combined items sum to one
Proportion in diet – terrestrial invertebrates	Beta	$\alpha=20, \beta=20$ Scale=1.24		
Proportion in diet – small mammals	Beta	$\alpha=3, \beta=3$ Scale=1.36		
Proportion in diet – vegetation	Beta	$\alpha=3, \beta=3$ Scale=0.144		
Proportion of soil in diet	Point estimate	0.03	U.S. EPA 2005	
Proportion time in area	Point estimate	1	–	Area assessed much larger than shrew foraging range

Body Weight (BW)

BW was not used in the model directly, but was a required variable in allometric models (e.g. Nagy et al. 1999) to estimate food intake or free metabolic rates (FMRs; see below). BWs for each of the representative wildlife species were obtained from the literature. Adult female BWs were used in the VEC exposure modeling because many of the toxicological endpoints in this assessment were reproductive and developmental.

Sediment/Soil Intake Rate (SIR)

U.S. EPA (1993, 1999a, 2005) and Sample and Suter (1994) estimated sediment and soil consumption rates for a large number of wildlife species based on reviews of the literature. When data were unavailable for a particular species, a soil ingestion rate equal to the soil ingestion rate of the most closely related species, as determined by faunal class, size and feeding behavior, was assumed. The SIR was expressed as a proportion of the rate of the overall FIR.

Water Intake Rate (WIR)

The WIR is based on an allometric relationship (Calder and Braun 1983). The relationship relates avian or mammalian BW to daily water flux rate. The daily water flux rate or turnover rate for birds was estimated as follows:

$$WIR\ (L/d) = a \cdot BW(kg)^b$$

where WIR is the drinking rate (L/day), and BW is the body weight (kg). The slope (a) and power (b) parameters for birds were 0.059 and 0.67, respectively. The slope (a) and power (b) parameters for mammals were 0.099 and 0.9, respectively. WIR was then normalized by dividing the average daily water flux by the BW of the species of interest (i.e. L/kg bw/day).

Food Intake Rate (FIR)

Food intake rate (FIR) is a function of many factors, including the free-living (or field) metabolic rate (FMR), the energy devoted to growth and reproduction, and composition of the diet (U.S. EPA 1993). FMR represents the total energy requirements for animals in the wild, including thermoregulation, feeding, reproduction and predator avoidance. Traditional methods for estimating FIR for use in ERA use allometric equations (e.g. Nagy 1987; Nagy et al. 1999; U.S. EPA 1993). These equations rely on general dietary composition assumptions and estimates of FMR to calculate FIR as a simple function of body weight. Data on FIR were only available for a few species, primarily due to the difficulties in measuring intake for free-ranging wildlife. Food intake rate of each prey item (i) was derived from FMR using the following equation:

$$FIR_i\ (kg/kg\ bw/day) = \frac{FMR}{AE_i \cdot GE_i}$$

where:

FMR = Normalized free metabolic rate (kcal/kg bw/day)

AE_i = Assimilation efficiency of the ith food item (unitless)

GE_i = Gross energy of the ith food item (kcal/kg dw)

As most of the VECs in this ERA did not have measured metabolic rate data available for free-ranging organisms, estimates of FMR were derived from the models developed by Nagy et al. (1999). The estimated FMRs for the meadow vole, white-tailed deer, red fox, and beaver were based on the regression analysis of the data for herbivorous mammals and the northern short-tailed shrew was based on regression analysis of data for omnivorous mammals. FMRs for birds also were derived from various models: the peregrine falcon was based on the regression equation for carnivorous birds; the American robin estimate was based on the regression equation for all passerine birds; and the ruffed grouse estimate was based on the regression equation for all birds. An experimentally measured FMR was available and used for moose (*Alces alces*). The exposure model for moose used the measured yearly average FMR derived by Renecker and Hudson (1985). The suitability of this value was examined and found to fall within the confidence intervals of the FMR derived using the allometric equation in Nagy et al. (1999) for herbivorous mammals. The general form of the model for free metabolic rate was:

$$FMR\ (kcal/d) = \frac{10^a \cdot BW(g)^b}{4.1875}$$

For the Monte Carlo analyses, FMR was estimated using a probabilistic approach by incorporating a distribution for BW, rather than a point estimate, and by incorporating the uncertainty resulting from lack of model fit (LMF) in the fitted allometric relationship (i.e. a normal distribution parameterized with mean of zero and a standard deviation calculated as the square root of the unexplained sum of squares from the fitted allometric regression model). The slope (a) and power (b) parameters and unexplained sum of squares were based on the regression analysis of the data reported in Nagy et al. (1999), assuming an underlying normal distribution for the latter.

Proportions of Dietary Items (P_i)

Extensive literature searches were conducted to locate data and information on the dietary preferences of the wildlife species assessed. The information in the literature on dietary preferences was evaluated to determine the relevance to the VECs. Some wildlife species have dietary preferences that can include a large number of different prey items. Therefore, only dietary items that comprised at least 5% of the total diet of each species were included in the exposure model. It is practically impossible to sample each dietary item that an animal consumes on a frequent basis and with sufficient coverage to fully characterize the chemical concentrations in every diet item across the entire Sudbury study area. To compensate for this lack of information, the wildlife exposure model used surrogate models or best available and scientifically defensible information to estimate the exposures that wildlife would receive. Table 11.2 provides a complete list of the dietary items and, where necessary, the dietary item in the exposure model that was used as a surrogate. Site-specific uptake models for the grass *Deschampsia* were used for all plant dietary items. Similarly, site-specific uptake models for grasshoppers were used for all terrestrial invertebrate dietary items apart from earthworms.

Table 11.2 Summary of Dietary Item Assumptions Used in the Exposure Model

Dietary item	Exposure model assumption	Estimation method
Amphibians	Fish	Measured in study area
Aquatic invertebrates	Benthic invertebrate	Predicted with literature-derived models
Aquatic plants or vegetation	Aquatic plant	Predicted with literature-derived models
Bark and tree stems	*Deschampsia* shoot	Measured in study area
Birds	Small mammal	Predicted with literature-derived models
Crayfish	Benthic invertebrate	Predicted with literature-derived models
Dicot shoots	*Deschampsia* shoot	Measured in study area
Earthworms	Earthworm	Measured from toxicity tests using soil from the study area
Fish	Fish	Measured in study area
Forbs	*Deschampsia* shoot	Measured in study area
Fruits	*Deschampsia* shoot	Measured in study area
Grasses	*Deschampsia* shoot	Measured in study area
Leaves	*Deschampsia* shoot	Measured in study area
Mammals	Small mammal	Predicted with literature-derived models
Monocot shoots	*Deschampsia* shoot	Measured in study area
Nuts	*Deschampsia* shoot	Measured in study area
Roots	*Deschampsia* root	Measured in study area
Roots and tubers	*Deschampsia* root	Measured in study area
Seeds	*Deschampsia* shoot	Measured in study area
Small mammals	Small mammal	Predicted with literature-derived models
Terrestrial inverts	Insect (grasshopper)	Measured in study area

Dietary item	Exposure model assumption	Estimation method
Terrestrial plants	*Deschampsia* shoot	Measured in study area
Twigs and buds	*Deschampsia* shoot	Measured in study area
Vegetation	*Deschampsia* shoot	Measured in study area
Waterfowl	Small mammal	Predicted with literature-derived models

Relative Absorption Factor (RAF)

COC in media consumed by wildlife are not 100% absorbed; to assume so would result in an overestimation of exposure and risk (U.S. EPA 2001a). The relative absorption factor (RAF) is the variable used to incorporate bioaccessibility information in ERA (Menzie et al. 2000). Bioaccessibility refers to the mass fraction of a substance that is converted to a soluble form and is, therefore, potentially available for uptake, under conditions of the external part of the membrane of interest (Owen 1990). RAF adjusts the absorption of a chemical from an exposure medium to that of the absorption of the chemical used in the toxicity study used to derive the TRV. Bioaccessibility was measured in soil for each COC as part of the HHRA (Chapter 7) and the RAFs that were developed from those analyses were used in the wildlife exposure modeling.

A comparison of the RAFs applied in the Sudbury ERA to the bioaccessibility assumptions in a U.S. EPA (2001a) ERA for a mining/smelting site (the Coeur d'Alene ERA), and the RAFs recommended by U.S. Department of Defense (U.S. DOD 2003) is presented in Table 11.3. In most cases, the assumption used in this ERA was similar to or more conservative than that used in the Coeur d'Alene ERA or recommended by U.S. DOD.

Table 11.3 **Comparison of Sudbury Bioaccessibility (%) Assumptions with Two Other Studies**

Chemical	Sudbury RAFs for soil ingestion only[a]	Sudbury HHRA[b] (95UCLM)	Coeur d'Alene total oral exposure bioavailability factors[c]	U.S. DOD Recommended RAFs[d]
As	37	37	60	10–50
Cd	33	NA	9	33
Co	27	27	NA	NA
Cu	65	65	40	NA
Pb	60	60	26	60
Ni	40	40	15	<10
Se	35	35	NA	NA

[a] Diet RAF assumed to be 1 because most data used to derive TRVs were based on dietary studies.
[b] Bioaccessibility of metals in soil derived from in vitro studies (see Chapter 7); bioaccessibility of Cd in soil was not determined.
[c] U.S. EPA (2001a); Bioavailability factor applied to total oral exposure (diet+soil+water); no RAF was estimated. NA indicates no bioavailability factors were available.
[d] U.S. DOD (2003); RAF for exposure from soil ingestion. NA indicates no RAF available.

In addition, several studies have determined the bioavailability of As and Pb in various media at different types of site (Ruby et al. 1999). The As RAFs for five soils around various smelters (Murray, Palmerton, Anaconda) ranged from 20% to 52%. The National Environmental Policy Institute (NEPI 2000) listed soil RAFs between 10% and 50% for As at contaminated sites. The Sudbury-specific As RAF of 37% is similar to the RAFs determined around other smelter and contaminated sites. The Pb RAF for six soils around various smelters (Jasper, Murray, Palmerton, Bartlesville) ranged from 35% to 71%. The Sudbury-specific Pb RAF of 60% is within the range of the RAFs determined at the other smelter sites.

11.2.2 Concentrations of COC in Water, Soil and Dietary Items

This section provides details with regard to the selection, analysis and characterization of media concentrations used to predict exposure in the ERA. Concentrations of COC vary spatially and temporally in all media. The VECs forage over distances ranging from tens of meters to greater than 10 km. Thus, individuals tend to integrate spatial variation in the tissue concentrations of their prey over time. In the Monte Carlo analysis, it was assumed that the spatially and temporally averaged exposure estimate did not vary between individuals foraging in the same area. Therefore, estimates of the central tendency (i.e. arithmetic means) were used in the exposure model as an expression of the spatial and temporal averaging of concentrations of COC in prey tissues, for VECs with large home ranges (U.S. EPA 1999a). Due to the uncertainty associated with estimating the true average concentration for a site, the 95% upper confidence limit of the arithmetic mean (95UCLM) should be used for this variable (U.S. EPA 1992). The arithmetic mean was defined by the UCLM based on ProUCL Version 3.0 Software (U.S. EPA 2004b).

Surface Water Concentrations

Concentrations of COC in surface water were represented by total metal concentrations measured in 30 lakes in the core area of the City of Greater Sudbury as part of the Urban Lakes Study in 2003. Non-volume-weighted, tygon tube composite samples were collected at mid-summer from a single deep basin within each lake. Water chemistry analyses were conducted by the Ontario Ministry of the Environment (Keller et al. 2004). The data are summarized in Table 11.4.

Table 11.4 Statistical Summary of COC Concentrations in Surface Water (µg/L)

Parameter	As	Cd	Co	Cu	Pb	Ni	Se
Mean	0.37	0.41	0.98	12.5	5.5	61.8	0.25
Standard error	0.04	0.021	0.15	1.2	0	4.8	0
Median	0.25	0.4	0.75	11	5.5	55.5	0.25
Standard deviation	0.22	0.11	0.82	6.5	0	26.3	0
Minimum	0.25	0.30	0.75	3.0	5.5	21	0.25
Maximum	1	0.65	4.90	35	5.5	114	0.25
Sample size	30	30	30	30	30	30	30
95UCLM	0.45	0.45	1.29	14.9	na	71.6	na

The complete data set for each COC was used to establish the distribution that most accurately characterized the levels of each COC in surface water. The distributions were based on the full range of the measured concentrations because of the limited number of samples that were applied to the entire study area. Surface water quality was required in the terrestrial ERA as drinking water for VECs. The distribution of COC in surface water was also used to estimate aquatic plant concentrations as forage for moose and beaver. Water-to-algae bioconcentration factors (BCFs) provided by U.S. EPA (1999c) were used to estimate the distributions of metal concentrations in aquatic plants. The following equation was used to estimate plant concentrations:

$$C_{ap} = C_w \times BCF_{ap}$$

where:

C_{ap} = Concentration of chemical in aquatic plant (mg/kg dw)

C_w = Concentration of dissolved metal in water (mg/L)

BCF_{ap} = Water-to-aquatic-plant bioconcentration factor [(mg chemical/kg dw)/(mg chemical /L water)]

The BCFs used to derive the distributions of aquatic plant concentrations for the exposure model are provided in Table 11.5. To use the BCFs in the ERA exposure model, all values were converted from a wet weight basis to a dry weight basis by multiplying by a factor of 2.92 (U.S. EPA 1999b). The conversion factor assumed a moisture content of 65.7% for aquatic plants and conservatively used total metal concentrations rather than the prescribed dissolved metal concentrations for water-to-algae BCFs.

Table 11.5 **Bioconcentration Factors Used for Estimating COC Concentrations in Aquatic Plants**

COC	Recommended value (ww basis)	Value (dw basis)	Comment/basis
As	293	856	Geometric mean of three empirical values
Cd	782	2283	Geometric mean of six empirical values
Co	61	178	No data available; assumed equal to nickel
Cu	541	1580	Geometric mean of five empirical values
Pb	1706	4982	Geometric mean of three empirical values
Ni	61	178	Geometric mean of four empirical values
Se	1845	5387	Geometric mean of three empirical values

Soil Concentrations

For the purpose of the ERA, the study area was subdivided into three broad zones and four Communities of Interest (COI). The latter included Sudbury Centre, Coniston, Copper Cliff and Falconbridge (Figure 11.2). The boundaries of the zones were defined on the basis of metal concentrations in soil and on terrain characteristics. The COI were defined in the HHRA (Chapter 6) and were used to estimate exposures for urban VECs only (i.e. short-tailed shrew, meadow vole, red fox, American robin, ruffed grouse, and peregrine falcon).

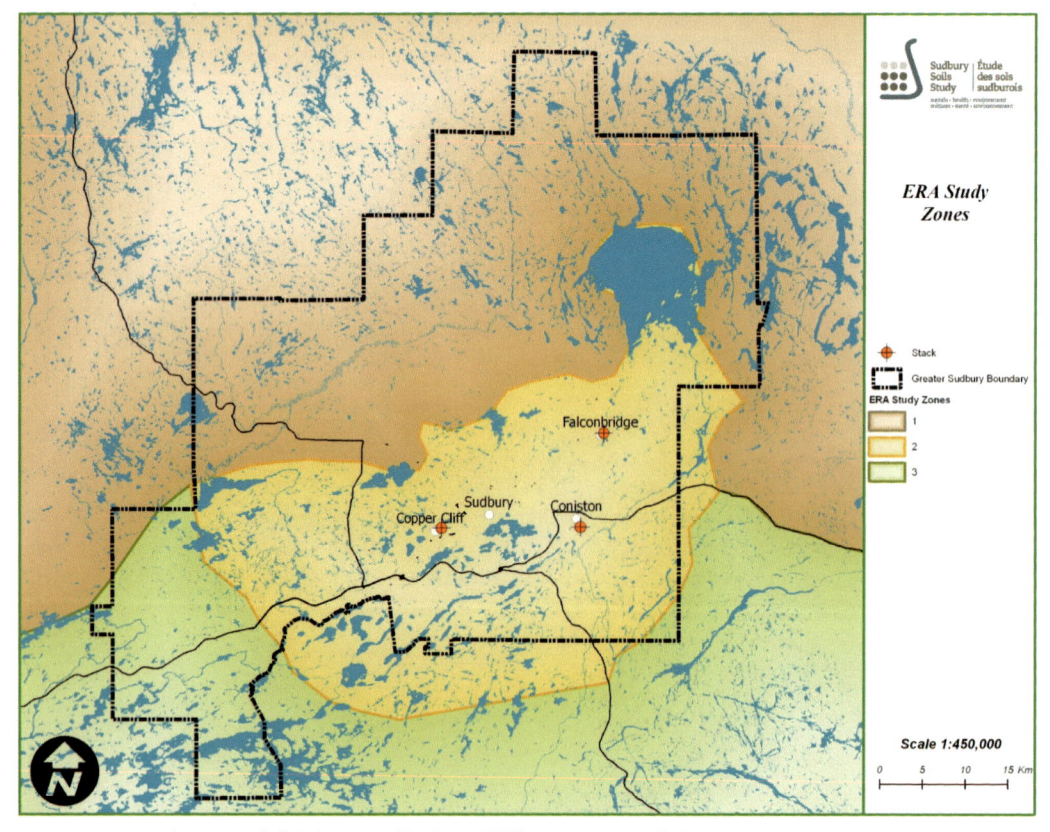

Figure 11.2 **Study Area Subdivisions Used in the Wildlife Exposure Model**

Soil samples were collected in each of the three zones at depths ranging from 0 to 5 cm, 5 to 10 cm, and 10 to 20 cm. Metal concentrations in the surface soil (0 to 5 cm) and soil-at-depth (0 to 10 cm) for each zone and COI were used to estimate exposure for VECs. Surface soil (0–5 cm) concentrations were used to predict exposure in wildlife that may incidentally ingest soil. Soil-at-depth (0–10 cm) concentration data were used to estimate exposure in wildlife that consume dietary items in direct contact with the soil (e.g. American robins that consume earthworms). That is, soil-at-depth concentrations were used to predict metal concentrations in insects, small mammals, plant roots and shoots, and worms. The 0 to 10 cm depth was chosen for the modeling of uptake into dietary items because it represents the A horizon of the soil around Sudbury. The LFH and A horizon is that part of the soil profile where most biological activity by soil organisms and plants is often found. The 95% UCLM concentrations for metals in the soils from all ERA zones are presented in Table 11.6. Cadmium exposures to VECs in urban areas were not assessed because Cd only exceeded the screening criterion in a single sample with pH<5.0. Therefore, Cd was not considered a COC for the urban areas.

It was necessary to deal with values that were below the minimum detection limit (MDL) for some metals. For all datasets (zone/COI, layer and metal combinations) that contained values below the MDL, fill-in values were calculated using a probability regression (or plot) method. This method is considered the most robust for minimizing error in non-detect-sample estimates (Gilliom and Helsel 1986; Helsel 1990, 2006; Huybrechts et al. 2002).

Table 11.6 **Soil COC Concentrations (95% UCLM, mg/kg) for ERA Zones**

Metal	Zone 1	Zone 2	Zone 3	Coniston	Copper Cliff	Falconbridge	Sudbury Centre
As (0–5 cm)	7.5	26.7	5.9	11.9	19.5	75.6	6.8
As (0–10 cm)	7.1	19.9	5.9	12.3	21.9	92	7.4
Cd (0–5 cm)	0.9	0.9	0.7	NA	NA	NA	NA
Cd (0–10 cm)	0.9	0.9	0.7	NA	NA	NA	NA
Co (0–5 cm)	7.5	19.9	10.0	19.6	35.9	50	11.5
Co (0–10 cm)	7.1	14.3	9.6	17.6	30	45	10.4
Cu (0–5 cm)	113	536	92	349	1520	806	168
Cu (0–10 cm)	77.2	343	58	328	1180	767	156
Pb (0–5 cm)	45	60	48	57	91	97	30
Pb (0–10 cm)	30.4	38	31	51.5	85	92	28.6
Ni (0–5 cm)	123	477	118	470	1081	850	182
Ni (0–10 cm)	79.4	306	77	440	934	824	172
Se (0–5 cm)	2.0	2.3	1.8	1.1	8.3	2.8	1.3
Se (0–10 cm)	1.7	2.9	1.8	1.1	6.2	2.9	1.2

NA = not applicable as Cd was not considered a COC in the urban zones.

The wildlife exposure model characterizes the distribution of chronic or long-term exposures that individuals in a 'wildlife population' might experience on a daily basis in a specific area or region. To do this, the exposure model must consider the spatial distribution of measured concentrations in relation to the home range of each VEC. If the home range of the VEC was large, then exposure to the mean concentration was considered more appropriate. In this case, large home range VECs are exposed to the mean through random foraging. However, consideration must be given to the uncertainty in the mean; therefore, the arithmetic mean was defined by the UCLM estimated using ProUCL Version 3.0 software (U.S. EPA 2004b).

To ensure that exposures of COC to wildlife VECs with small home ranges were not under-predicted, the full distribution of measured concentrations in soil was used to predict the potential for adverse exposures. This method was considered to be more ecologically relevant for small home range VECs exposed to metal concentrations in soil and dietary media. Predicted exposures assumed that individuals in the population were homogeneously distributed in the exposure unit. Table 11.7 provides a summary of the home ranges for the wildlife VECs and identified those selected for evaluation using the full distribution of measured concentrations in each exposure unit. Chronic exposures and the potential for adverse effects for the remaining VECs were evaluated with the mean and associated confidence limits. Small home range VECs were defined as those with home ranges less than 1 hectare (ha).

Table 11.7 **Valued Ecosystem Component Home Ranges**

VEC	Home range (ha)	Distribution used for estimating exposures to metals in soil
American robin	0.1 to 0.8	Full distribution
Beaver	0.7 to 17.6	Mean and confidence limits
Deer	<100	Mean and confidence limits
Falcon	1500 to 4300	Mean and confidence limits
Fox	100 to 1000	Mean and confidence limits
Grouse	2 to 12.2	Full distribution
Moose	220 to 600	Mean and confidence limits
Shrew	0.024 to 0.5	Full distribution
Vole	<0.1	Full distribution

As an example, the full distribution of measured concentrations of Ni in soil in Zone 2 is provided in Figure 11.3. Figure 11.4 illustrates the two cumulative distributions of soil Ni concentrations for Zone 2: 1) the mean and confidence limits, and 2) the full distribution of measured concentrations. The mean was used for large home range VECs, and the distribution for the full dataset was used for small home range VECs. A large home range VEC (e.g. moose) or population of large home range VECs in Zone 2 could be exposed to the distribution of measured concentrations in Figure 11.3. However, due to the large home range requirements of a moose, it is highly unlikely that any given moose would be exposed to concentrations at the high end of the range on a chronic or long-term basis. Those concentrations would only be encountered by a moose on an infrequent basis. For example, there is a 6% chance that a moose would encounter a Ni concentration in surface soil greater than 1000 mg/kg, if all areas in the exposure unit were of equal habitat quality. Chronic exposures to the moose therefore are represented by exposure to the mean Ni concentration in soil. The uncertainty surrounding the mean is represented by the distribution illustrated by the darker line in Figure 11.4.

A small home range VEC (e.g. shrew) or population of small home range VECs in Zone 2 could be exposed to the distribution of measured concentrations in Figure 11.3. Because of the small home range of the shrew (i.e. 0.024 to 0.5 ha; Table 11.6), it was likely that shrews could be exposed to concentrations at the high end of the range on a chronic basis. Therefore, use of the full distribution of measured concentrations was appropriate and ecologically relevant for VECs with small home ranges. The distribution of each COC in each study zone was calculated for the detailed exposure modeling.

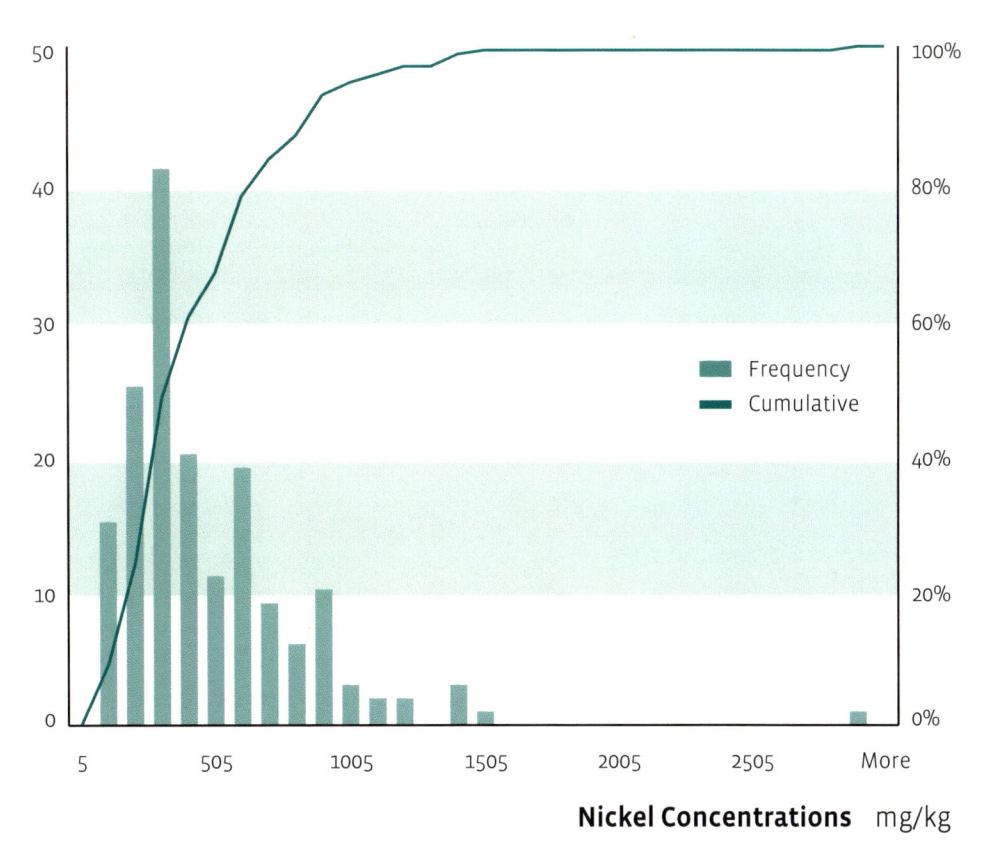

Ecological Risk Assessment: Evaluating Risks to Terrestrial Wildlife

Figure 11.3 **Distribution of Ni Concentrations in Surface Soil (0 to 5 cm) in Zone 2**

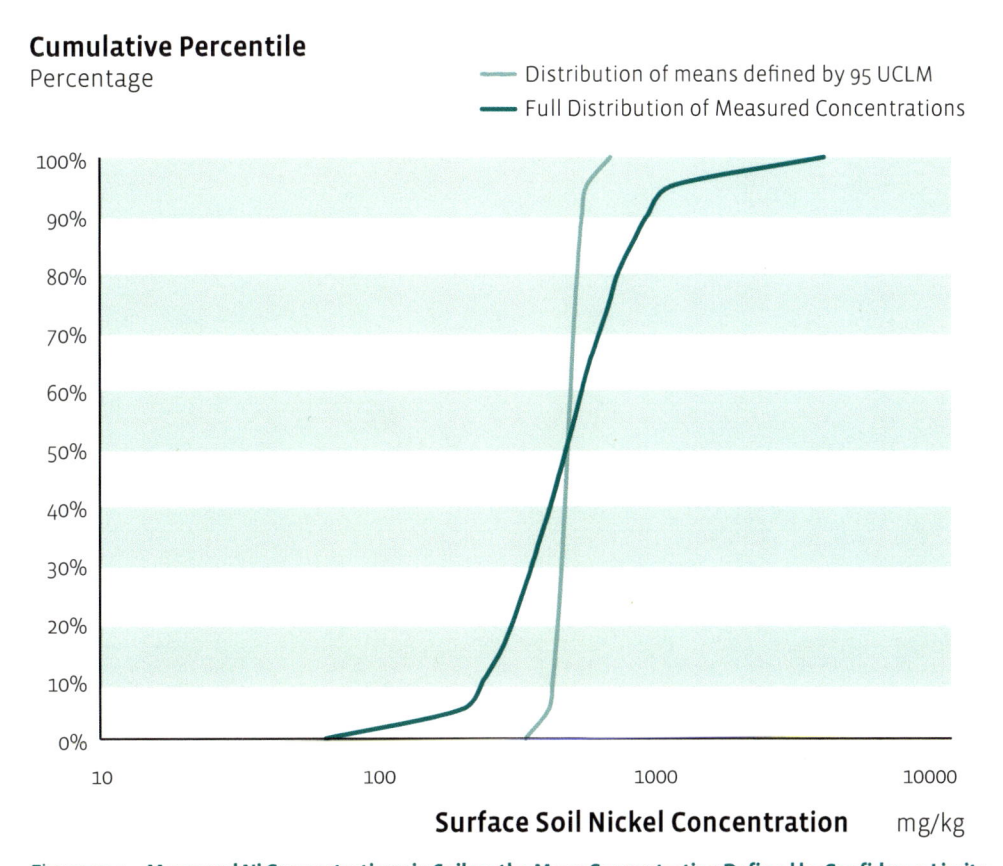

Cumulative Percentile
Percentage

Figure 11.4 **Measured Ni Concentrations in Soil vs. the Mean Concentration Defined by Confidence Limits**

11.2.3 Concentrations of COC in Wildlife Food Items

Select wildife dietary items were sampled to provide Sudbury-specific data for the wildlife exposure model. Samples of soil, a typical forage plant species and terrestrial invertebrates were collected from 17 locations in the study area. Sampling sites were generally represented by the open barren and red maple/white birch community types described by Amiro and Courtin (1981). Soil depth rarely exceeded 40 cm and this only occurred in isolated pockets. Soils were collected randomly in each sample area. The soil samples were collected at two depths (0 to 5 and 5 to 10 cm). Grasses of the genus *Deschampsia* were chosen as the target vegetation because they were abundant in the Sudbury area and both the roots and shoots are known to be components of the diets of herbivorous small mammals in the region. The original objective was to collect earthworms as the representative terrestrial invertebrate. However, following site reconnaissance, it was clear that very few earthworms were present in the Sudbury area. Grasshoppers of the *Melanoplus* genus were collected because they were abundant and a common dietary item for wildlife. All samples were cleaned, dried and submitted for metal analysis to an accredited laboratory following a detailed sampling procedure and a full analytical QA/QC program.

The concentrations of COCs in surface soils from this small study fell within the range of values reported for the more extensive 2001 soil sampling program (Table 11.8). Soil at the sampling sites represented a concentration gradient for the COCs, with a variety of soil textures. Levels of all COCs in *Deschampsia* root were higher than in shoots (Table 11.8). Metal levels tended to be lower in shoots than in soils. Concentrations of the COCs measured in soil, grasshoppers, and grass roots and shoots at the 17 sites are summarized in Table 11.8.

Table 11.8 **Measured COC Concentrations (mg/kg dw) in Soil and Biota from the Study Area**

COC	Statistic	Soil	Grasshopper	*Deschampsia* root	*Deschampsia* shoot
As	Mean	16.0	0.7	13.4	0.6
As	Min	1.0	0.2	0.1	0.007
As	Max	48.0	2.3	80.4	1.8
As	n	17	17	17	15
Cd	Mean	0.4	0.5	2.1	0.3
Cd	Min	0.2	0.1	0.3	0.04
Cd	Max	0.7	1.4	5.9	0.6
Cd	n	20	20	21	20
Co	Mean	9.5	0.8	12.5	0.4
Co	Min	3.1	0.2	1.2	0.05
Co	Max	21.0	2.6	53.3	1.4
Co	n	15	17	17	15
Cu	Mean	197.0	95.6	228.0	10.4
Cu	Min	28.0	38.7	15.4	1.9
Cu	Max	556.0	225.0	753.0	35.0
Cu	n	17	17	17	15
Pb	Mean	29.0	0.8	31.3	2.2
Pb	Min	10.0	0.02	2.8	0.2
Pb	Max	64.0	2.7	95.7	7.8
Pb	n	17	17	17	15

Ecological Risk Assessment: Evaluating Risks to Terrestrial Wildlife

COC	Statistic	Soil	Grasshopper	*Deschampsia* root	*Deschampsia* shoot
Ni	Mean	131.0	39.4	292.0	22.8
Ni	Min	27.0	6.9	30.0	7.7
Ni	Max	350.0	77.6	920.0	44.3
Ni	n	17	17	17	15
Se	Mean	1.0	1.4	2.2	0.5
Se	Min	0.1	0.2	0.1	0.08
Se	Max	3.0	2.6	7.1	1.5
Se	n	17	17	17	15

Concentrations of the COCs in grasshoppers were generally not correlated with soil or grass concentrations. The concentration of total Ni in grasshoppers was an exception as it was positively correlated (r^2=0.706; p<0.05) with the Ni content of *Deschampsia* shoots (Figure 11.5).

Grasshopper Ni
mg/kg

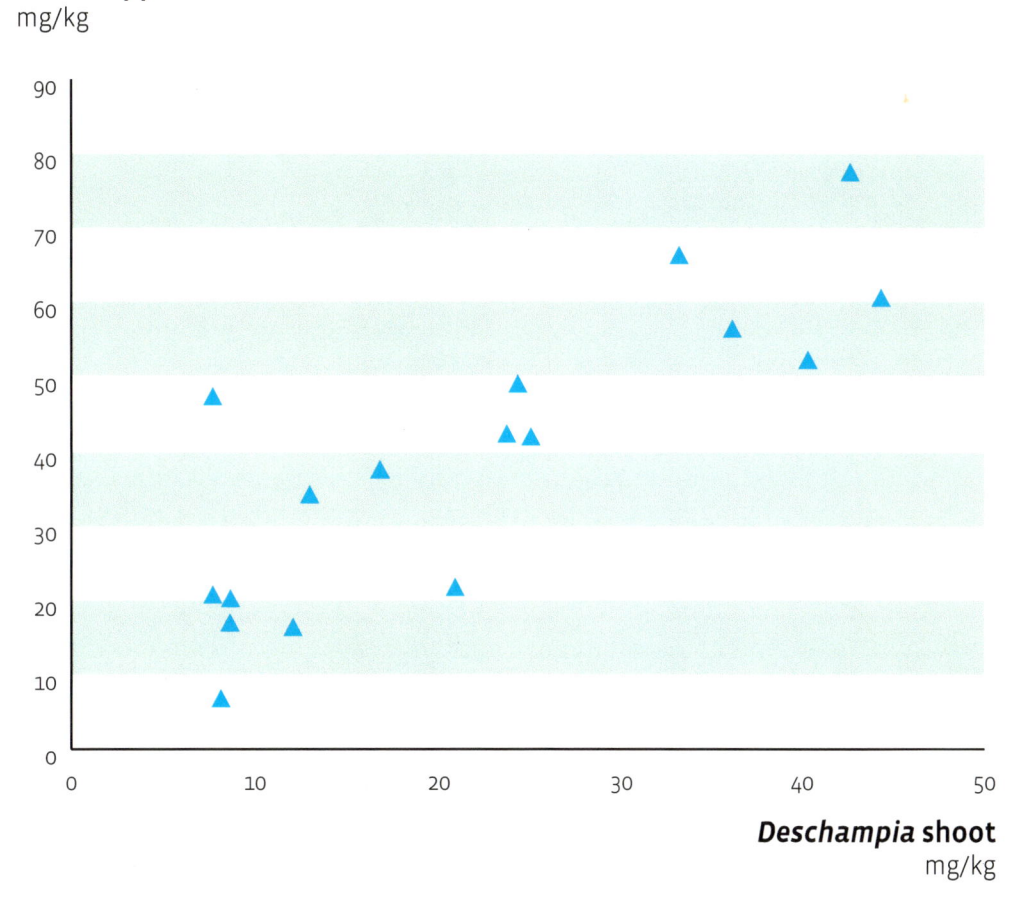

***Deschampia* shoot**
mg/kg

Figure 11.5   **Correlation of Ni in Grasshoppers with Deschampsia Shoots**

The predictive models provided in Tables 11.9 to 11.13 were used with the distributions of soil concentrations to derive the distributions of dietary concentrations for the wildlife exposure model. Empirical data on the concentration of COC in earthworms, grasshoppers and shoots and roots of *Deschampsia* were used to develop these uptake models. Two types of models were identified to estimate the potential distribution of metal concentrations in food items: 1) bioconcentration factor (BCF) models; and, 2) regression models. The following bioconcentration factor equation was used to estimate concentrations:

$$C_i = C_s \times BCF_i$$

where:

C_i = Concentration of chemical in food 'i' (mg/kg dw)

C_s = Concentration of metal in soil (µg/g or mg/kg)

BCF_i = Soil to food 'i' bioconcentration factor [(mg chemical/kg dw)/(mg chemical/kg soil)]

The following linear, and transformed power-function regression models were also used to estimate concentrations in wildlife foods:

$$C_i = a + b \times C_s + E \qquad \text{Linear Equation}$$

$$C_i = EXP\left[a + b \times \ln(C_s) + E\right] \qquad \text{Power Function Equation}$$

where:

C_i = Concentration of chemical in food 'i' (mg/kg dw)

a = Regression model intercept

b = Regression model slope

C_s = Concentration of metal in soil (µg/g)

E = Regression model root mean square error

Most uptake models were based on empirical or site-specific data and are indicated by a 'site-specific' comment in the tables. However, in certain circumstances, literature-based models were required and are referenced accordingly. Because soils in urban areas were likely amended to increase pH, earthworm models for the urban environments used the tissue data from toxicity tests with pH-adjusted soils. Earthworm models for the three zones used tissue data from the toxicity tests with unadjusted soils. Models to estimate uptake into small mammals were taken from Sample et al. (1998). The distribution of COC in soil was used to estimate concentrations of metals in wildlife foods, and to then estimate dietary exposures. The following abbreviations are common to the following tables: E = error from lack of model fit; N = normal distribution; LN = lognormal distribution.

Table 11.9 **Predictive Models Used to Estimate COC Concentrations in Insects**

COC	Model	Parameters	Comment
As	Linear regression	a = 0.33; b = 0.0238; E = N(0,0.542)	Site-specific
Cd	Bioconcentration factor	LN(1.49, 1.26)	Site-specific
Co	Power-function regression	a = −1.89; b = 0.679; E = N(0,0.562)	Site-specific
Cu	Power-function regression	a = 2.47; b = 0.40; E = N(0,0.293)	Site-specific
Pb	Bioconcentration factor	LN(0.0298, 0.0266)	Site-specific
Ni	Power-function regression	a = 0.580; b = 0.630; E = N(0,0.471)	Site-specific
Se	Linear regression	a = 1.09; b = 0.324; E = N(0,0.498)	Site-specific

Table 11.10 **Predictive Models Used to Estimate COC Concentrations in Roots**

COC	Model	Parameters	Comment
As	Linear regression	$a=-7.64; b=1.31; E=N(0,12.7)$	Site-specific
Cd	Bioconcentration factor	$LN(5.64, 5.38)$	Site-specific
Co	Bioconcentration factor	$LN(1.43, 2.1)$	Site-specific
Cu	Power-function regression	$a=3.50; b=0.00648; E=N(0,0.753)$	Site-specific
Pb	Bioconcentration factor	$LN(1.11, 0.962)$	Site-specific
Ni	Linear regression	$a=32.2; b=1.98; E=N(0,220)$	Site-specific
Se	Linear regression	$a=0.717; b=1.44; E=N(0,1.92)$	Site-specific

Table 11.11 **Predictive Models Used to Estimate COC Concentrations in Shoots**

COC	Model	Parameters	Comment
As	Linear regression	$a=0.196; b=0.0282; E=N(0,0.45)$	Site-specific
Cd	Bioconcentration factor	$LN(0.84, 0.67)$	Site-specific
Co	Bioconcentration factor	$LN(0.0507, 0.0559)$	Site-specific
Cu	Bioconcentration factor	$LN(0.0908, 0.132)$	Site-specific
Pb	Bioconcentration factor	$LN(0.105, 0.129)$	Site-specific
Ni	Linear regression	$a=11.8; b=0.0845; E=N(0,11.1)$	Site-specific
Se	Linear regression	$a=0.256; b=0.235; E=N(0,0.365)$	Site-specific

Table 11.12 **Zone 1 to 3 Predictive Models Used to Estimate COC Concentrations in Earthworms**[a]

COC	Model	Parameters	Comment
As[b]	Power-function regression	$a=-1.42; b=0.706; E=N(0,1.21)$	Sample et al. 1998
Cd[c]	Power-function regression	$a=0.452; b=0.172; E=N(0,0.11)$	Site-specific
Co[d]	Bioconcentration factor	$LN(0.139, 0.087)$	Sample et al. 1998
Cu[e]	Power-function regression	$a=-0.311; b=0.372; E=N(0,0.474)$	Site-specific
Pb[f]	Linear	$a=-1.21; b=0.181; E=N(0,1.46)$	Site-specific
Ni[g]	Power-function regression	$a=-2.37; b=0.701; E=N(0,0.661)$	Site-specific
Se[h]	Bioconcentration factor	$LN(0.62, 0.55)$	Site-specific

[a] Measured concentrations of metals in soil and earthworms used in toxicity tests.

[b] Site-specific data observed within range of Oak Ridge dataset (Sample et al. 1998) and site-specific regression model was not significant ($p=0.27$); therefore used Oak Ridge regression model instead.

[c] Site-specific model determined not significant ($p=0.12$); however regression model used instead because Oak Ridge model (Sample et al. 1998) predicts concentrations at least two times higher than concentrations observed in study area.

[d] Site-specific model not significant ($p=0.15$) and measured data found within range of literature models; used Oak Ridge model (Sample et al. 1998).

[e] Site-specific model found significant ($p=0.008$).

[f] Site-specific model found significant ($p<0.0001$).

[g] Site-specific model found significant ($p<0.001$).

[h] Site-specific model was not significant ($p=0.35$); used default site-specific bioconcentration factor model instead because Oak Ridge model (Sample et al. 1998) is observed to over-predict concentrations in the study area.

Table 11.13 **Urban Predictive Models Used to Estimate COC Concentrations in Earthworms**[a]

COC	Model	Parameters	Comment
As[b]	Linear	$a=0.961; b=0.0469; E=N(0,1.13)$	Site-specific
Co[c]	Linear	$a=0.893; b=0.0474; E=N(0,0.558)$	Site-specific
Cu[b]	Linear	$a=1.27; b=0.0308; E=N(0,2.79)$	Site-specific
Pb[b]	Linear	$a=-0.0247; b=0.0292; E=N(0,0.362)$	Site-specific
Ni[b]	Linear	$a=-0.0256; b=0.0262; E=N(0,1.78)$	Site-specific
Se[d]	Linear	$a=0.563; b=0.192; E=N(0,429)$	Site-specific

[a] Measured co-located concentrations of metals in soil and earthworms used in toxicity tests.
[b] Site-specific model found significant ($p<0.0001$).
[c] Site-specific model significant ($p=0.006$).
[d] Site-specific model found significant ($p=0.004$).

In summary, to estimate the exposure of Sudbury wildlife to the COC, the relevant dietary parameters were summarized, model input parameters were selected, and input distributions for abiotic (soil, sediment, and water concentrations) and biotic (dietary, intake rates) variables were determined. Monte Carlo analysis was then conducted to generate output distributions for each combination of COC and VEC. The output distributions were combined with the relevant toxicity data summarized in the next section to characterize risks to VECs.

11.3 Effects Assessment

The effects assessment includes two main components: the derivation of toxicity reference values (TRVs) in Section 11.3.1 and the review of local field data on bird and mammal populations (Section 11.3.2).

11.3.1 Toxicity Reference Values

Toxicity reference values (TRVs) are chemical doses equivalent to acceptable exposure levels for each VEC. If exposure levels were below a TRV, then no unacceptable risks were expected for the VEC. If exposure levels exceeded a TRV, it did not automatically imply risks to that VEC. Rather, further assessment was conducted and assumptions re-evaluated to determine whether potential effects were likely. This section describes the steps taken to derive TRVs for avian and mammalian wildlife VECs under Objectives #2 and #3 of the Sudbury ERA. The derivation process began with a comprehensive search and review of toxicological literature related to the COC and VECs of interest. The following study characteristics were used to identify studies that were most desirable for deriving TRVs:

- toxicity data derived from species with a similar gut physiology
- feeding studies (dietary exposures are more relevant than drinking water or gavage studies)
- studies with multiple exposure levels (such that a dose–response curve could be generated, or a bounded lowest observed adverse effect level (LOAEL) or no observed adverse effect level (NOAEL) could be determined)
- chronic exposure studies (defined as 1 year and 10 weeks for mammalian and avian species, respectively, or conducted during a sensitive life stage such as during reproduction or development; Sample et al. 1996), and
- studies that evaluated a reproductive endpoint, or alternatively, studies that measured survival or growth

TRVs were selected or derived according to the following order of preference:

- IC_{20} (inhibition concentration, or concentration (or dose) resulting in a 20% effect on test organisms) was selected first
- A LOAEL was selected if no IC_{20} was available, and
- A NOAEL was selected if no IC_{20} or LOAEL was available

The exception to the above ranking of effect levels was that NOAELs were selected preferentially for vulnerable, threatened or endangered (VTE) species. Although the authors recognized limitations to the use of the NOAEL (e.g. adverse effects may have been observed but were not statistically different from the control), this method followed MOE regulatory policy (MOE 2005) and was considered to be a conservative approach. The 20% effect level has been used for many years in both aquatic and terrestrial ERAs (Suter et al. 1995; BC MELP 1998). The IC_{20} values in U.S. EPA (2001a) were used to evaluate risks to wildlife exposed to metals arising from mining and smelting emissions (the Coeur d'Alene ERA). The Coeur d'Alene ERA underwent extensive peer review by the National Academy of Sciences in the United States (NAS 2005).

TRVs are chemical- and species-specific. In the past, the IC_{20}, LOAEL or NOAEL values from the test species were adjusted to the wildlife species of interest using a scaling factor from Sample et al. (1996) or Sample and Arenal (1999). However, these scaling factors were based on acute data only, and their applicability to chronic data was unknown and dose scaling is no longer recommended (Allard et al. 2010). The alternative is to apply an uncertainty factor to the toxicity data if the test species and wildlife species do not have similar gut physiology. An uncertainty factor of 3 was selected to extrapolate from mouse or rat data to fox, if more appropriate data were not available. A factor of 3 was selected because it results in a more conservative TRV than would have been obtained using the traditional method of dose-scaling by body weight (Sample and Arenal 1999), without being unreasonably conservative. After their review of available data, ep&t (1996) recommended that a factor between 3 and 10 be used to extrapolate from less sensitive to more sensitive species, because 'it is highly likely that interspecific variation in toxic response will be less than 10-fold.' There were no data to suggest that there was a large difference in sensitivity between bird species exposed to metals.

For Cd and Co, the U.S. EPA (2005) Ecological Soil Screening Level (Eco-SSL) documents were reviewed for relevant toxicity data (other COC did not have Eco-SSLs when TRVs were being selected or derived for this ERA). Uncertainty factors were not applied to the Eco-SSL values because they were generally derived from several studies using different test species, from which a geometric mean of the NOAELs or LOAELs could be calculated. Table 11.14 presents the TRV for each VEC and COC combination.

Table 11.14 **Toxicity Reference Values (TRVs) Used in the Sudbury ERA**

COC	VEC	Test species	Test species effect level (mg/kg/day)	IC_{20}, LOAEL or NOAEL	TRV (mg/kg/day)	Reference
As	Ruffed grouse	Mallard	14	IC_{20}	14	U.S. EPA 2001a derived from Stanley et al. 1994
	American robin	Mallard	14	IC_{20}	14	
	Peregrine falcon	Mallard	9	NOAEL	9	
	Northern short-tailed shrew	Mouse	10	IC_{20}	10	Byron et al. 1967
	Meadow vole	Mouse	10	IC_{20}	10	
	Moose	Cow	1.8	NOAEL	1.8	Lambou and Lin 1970
	White-tailed deer	Cow	1.8	NOAEL	1.8	
	Red fox	Dog	1	NOAEL	1	Byron et al. 1967
	Beaver	Mouse	10	IC_{20}	10	
Cd	Ruffed grouse	Chicken	3.1	LOAEL	3.1	U.S. EPA 2005 geomean of 35 studies
	American robin	Chicken	3.1	LOAEL	3.1	
	Peregrine falcon	Chicken	1.5	NOAEL	1.5	
	Northern short-tailed shrew	Rodents, sheep, pig	5.4	LOAEL	5.4	U.S. EPA 2005 geomean of 28 studies
	Meadow vole	Rodents, sheep, pig	5.4	LOAEL	5.4	
	Moose	Rodents, sheep, pig	5.4	LOAEL	5.4	
	White-tailed deer	Rodents, sheep, pig	5.4	LOAEL	5.4	
	Red fox	Rodents, sheep, pig	5.4	LOAEL	5.4	
	Beaver	Rodents, sheep, pig	5.4	LOAEL	5.4	

COC	VEC	Test species	Test species effect level (mg/kg/day)	IC$_{20}$, LOAEL or NOAEL	TRV (mg/kg/day)	Reference
Co	Ruffed grouse	Chicken	16	LOAEL	16	U.S. EPA 2005 geomean of 11 studies
	American robin	Chicken	16	LOAEL	16	
	Peregrine falcon	Chicken	7.6	NOAEL	8	U.S. EPA 2005 geomean of five studies
	Northern short-tailed shrew	Rodents, pig, cow	19	LOAEL	19	U.S. EPA 2005 geomean of 14 studies
	Meadow vole	Rodents, pig, cow	19	LOAEL	19	
	Moose	Rodents, pig, cow	19	LOAEL	19	
	White-tailed deer	Rodents, pig, cow	19	LOAEL	19	
	Red fox	Rodents, pig, cow	19	LOAEL	19	
	Beaver	Rodents, pig, cow	19	LOAEL	19	
Cu	Ruffed grouse	Chicken	60	IC$_{20}$	60	U.S. EPA 2001a derived from Mehring et al. 1960
	American robin	Chicken	60	IC$_{20}$	60	
	Peregrine falcon	Chicken	47	IC$_{10}$	47	
	Northern short-tailed shrew	Rat	80	Max. tolerable level	80	NRC 1980
	Meadow vole	Rat	80	Max. tolerable level	80	
	Moose	Cow	4	LOAEL	4	Jenkins and Hidiroglou 1989
	White-tailed deer	Cow	4	LOAEL	4	
	Red fox	Mink	19	IC$_{20}$	19	U.S. EPA 2001a derived from Aulerich et al. 1982
	Beaver	Rat	80	Max. tolerable level	80	NRC 1980
Pb	Ruffed grouse	Chicken	10	IC$_{20}$	10	U.S. EPA 2001a derived from Edens and Garlich 1983
	American robin	Chicken	10	IC$_{20}$	10	
	Peregrine falcon	Chicken	3	IC$_5$	3	
	Northern short-tailed shrew	Rat	80	LOAEL	80	Azar et al. 1973
	Meadow vole	Rat	80	LOAEL	80	
	Moose	Cow	5	LOAEL	5	Demayo et al. 1982
	White-tailed deer	Cow	5	LOAEL	5	
	Red fox	Dog	13	NOAEL	13	U.S. EPA 2001a based on Penumarthy et al. 1980
	Beaver	Rat	80	LOAEL	80	Azar et al. 1973
Ni	Ruffed grouse	Mallard	77	LOAEL	77	Cain and Pafford 1981
	American robin	Mallard	77	LOAEL	77	
	Peregrine falcon	Mallard	18	NOAEL	18	
	Northern short-tailed shrew	Rat	50	LOAEL	50	Ambrose et al. 1976
	Meadow vole	Rat	50	LOAEL	50	
	Moose	Cow	25	NOAEL	25	O'Dell et al. 1970
	White-tailed deer	Cow	25	NOAEL	25	
	Red fox	Beagle	80	LOAEL	80	Ambrose et al. 1976
	Beaver	Rat	50	LOAEL	50	Ambrose et al. 1976

COC	VEC	Test species	Test species effect level (mg/kg/day)	IC₂₀, LOAEL or NOAEL	TRV (mg/kg/day)	Reference
Se	Ruffed grouse	Chicken	0.4	LOAEL	0.4	Ort and Latshaw 1978
			0.6	LOAEL	0.6	
			0.8	LOAEL	0.8	
	American robin	Mallard, chicken, kestrel	0.4	LOAEL	0.4	Heinz et al. 1989; Ort and Latshaw 1978; Santolo et al. 1999
		Mallard, chicken, kestrel, owl	0.9	Geomean of LOAELs	0.9	16 LOAELs (range: 0.4 to 1.8 mg/kg/day)
		Screech owl, black-crowned night heron	1.8	LOAEL	1.8	Smith et al. 1988; Wiemeyer and Hoffman 1996
	Peregrine falcon	Kestrel	0.2	NOAEL	0.2	Santolo et al. 1999
			0.4	LOAEL	0.4	
	Northern short-tailed shrew	Mice	0.76	LOAEL	0.3	Schroeder and Mitchener 1971
					0.8	
		Rats	1.7	NOAEL	1.7	Lijinski et al. 1989
	Meadow vole	Mice	0.76	LOAEL	0.3	Schroeder and Mitchener 1971
					0.8	
		Rats	1.7	NOAEL	1.7	Lijinski et al. 1989
	Moose	Cows	0.1	LOAEL	0.1	NRC 1980
	White-tailed deer	Cows	0.1	LOAEL	0.1	
	Red fox	Mice	0.76	LOAEL	0.1	Schroeder and Mitchener 1971
					0.3	
		Rats	1.7	NOAEL	0.5	Lijinski et al. 1989
	Beaver	Rats	0.76	LOAEL	0.3	Schroeder and Mitchener 1971

11.3.2 General Information on Status of Mammals and Birds in the Sudbury Area

11.3.2.1 Mammals

With the exception of moose, there are no formal processes to track mammal populations in Ontario. Therefore, there was little quantitative information available on mammalian populations in the Sudbury area. To get a general sense of the reproductive success and population trends for the Sudbury mammalian populations, several sources of information, including anecdotal information, were consulted.

Northern Short-tailed Shrew

Naturalists and researchers note that short-tailed shrews are abundant in the Sudbury area (Dobbyn et al. 1994; Schoenefeld 2004 pers. comm.; Robitaille and Linley 2006). Professor Robitaille's research at Laurentian University indicates that short-tailed shrews were found in all habitat types studied; in particular, a positive relationship was found between percent herb cover and short-tailed shrew abundance (Robitaille and Linley 2006).

Meadow Vole

It is natural for meadow vole populations to experience large fluctuations in abundance (U.S. EPA 1993). Naturalists report that voles are common in the Sudbury area (Dobbyn et al. 1994; Schoenefeld 2004 pers. comm.). They are also known to be present in urban environments (Riegert et al. 2007). The preferred habitat of voles is grassed areas.

Beaver

Although there are many beavers in the Sudbury area, with lower densities to the north, no specific abundance data are available through the Ontario Ministry of Natural Resources (OMNR; Biscaia 2005 pers. comm.). However, according to OMNR officials (OMNR, Wildlife Management Unit 42), there was approximately $60,000 worth (pelt value) of nuisance beaver trapped from the Sudbury area each year (Biscaia 2005 pers. comm.). According to naturalists and the research community, beavers are common in the Sudbury area (Dobbyn et al. 1994; Schoenefeld 2004 pers. comm.; Robitaille 2005 pers. comm.).

Moose

Moose are hunted in the Sudbury area and the moose populations are followed by the OMNR. The moose population was thought to be stable in Sudbury Wildlife Management Unit 41, at approximately 2300 animals in the Sudbury district (Alkins 2005 pers. comm.). The population is productive, with numerous cows bearing twin calves; however, the population was approximately 30% below the OMNR target population for the area, of 3316 moose (Alkins 2005 pers. comm.; Deschamps 2006 pers. comm.). Recruitment into the herd is low due to the hunting pressure, which, in addition to some recent hard winters, may be the cause of the below-capacity population (Alkins 2005 pers. comm.). Due to the large home range of moose, these animals may wander in and out of the Sudbury area, and herd sizes within the WMUs do not represent a population estimate of the number of moose within the greater Sudbury area (Biscaia 2005 pers. comm.).

While-tailed Deer

Human activities such as the cutting and burning of forest blocks, seeding of agricultural crops, winter feeding of cattle, reduction of competitors, and hunting restrictions on white-tailed deer have helped to increase its northward and westward ranges in Canada (Passmore 1990). In particular, logging in the forest favors deer by opening the high canopy and encouraging new growth on the forest floor. It can be beneficial to the deer if cover is left in place to provide shelter from the deep snow. Within its range in Canada, the average density of the white-tailed deer exceeds three deer per square kilometer (Passmore 1990). The white-tailed deer can reproduce quickly; a healthy herd is capable of doubling its numbers during one favorable year (Passmore 1990). Most of the predation on fawns occurs during the first few weeks of life. White-tailed deer are common and are currently hunted in the Sudbury area.

Red Fox

The two main limiting factors to red fox populations are competition with other canids, especially coyotes, and seasonal limits on food availability (U.S. EPA 1993). However, the major causes of mortality in red foxes are typically non-natural, such as hunting and road kill (U.S. EPA 1993). The OMNR does not have specific population numbers for the red fox, but populations were thought to have peaked in the Sudbury area in 2001/2002. Since then, however, population management and rabies outbreaks have lowered numbers (Biscaia 2005 pers. comm.). Local researchers have also noted the presence of red fox in the Sudbury area (Robitaille and Linley 2006).

11.3.2.2 Birds

There are approximately 300 species of birds known to occur in Sudbury (Whitelaw 1989) of which 183 breed in the City of Greater Sudbury (Monet and Boucher 2005). Christmas bird count data for Sudbury were obtained from the National Audubon Society (2005). Because the count occurs in the winter, migratory species (such as the American robin) were not counted. A review of the available information suggested that the populations of bird species were increasing. In addition, due to diversity of habitat in the City of Greater Sudbury including forest cover, wetlands and lakes, Monet and Boucher (2005) concluded that the City's forests are large enough to support a diverse and representative bird community.

American Robin

The American robin is well adapted to habitat disturbances. The loss of forests and the growth of urban areas have both served to increase its breeding habitat (Long 2005; Sallabanks and James 1999). Robins return to the same breeding area each year (Long 2005). The Breeding Bird Atlas (1981 to 1985) (Cadman et al. 1987) noted virtually no change in robin abundance in any part of Ontario since 1967. Preliminary data indicate that the robin was breeding in all areas in the City of Greater Sudbury (Monet and Boucher 2005) and local naturalists report that robins are abundant in the Sudbury area.

Ruffed Grouse

Younger forests provide the best cover and food sources, and the ruffed grouse prospers when forests are intermittently subjected to clear-cutting or fires (Ruffed Grouse Society 2003). Thus, the re-greening and naturally regenerating areas (particularly new growth of aspens and poplars) of Sudbury may provide optimum habitat for the ruffed grouse. Ruffed grouse populations follow a predictable 10-year cycle where population numbers fluctuate yearly due to interacting factors that are poorly understood (Ruffed Grouse Society 2003). Grouse were extirpated from portions of the Sudbury area but the population has since recovered and it is currently a hunted species (Abbey 2004 pers. comm.). Data show that the ruffed grouse was breeding in the City of Greater Sudbury in 2000 to 2005 (Monet and Boucher 2005).

Peregrine Falcon

DDT and organochlorine pesticides caused the extirpation of peregrine falcons in most of North America in the 1950s and 1960s (Blomme and Laws 1995). The Breeding Bird Atlas (1981–1985) noted that breeding without human help had not been observed in Ontario since 1963 (Cadman et al. 1987). A total of 32 peregrine falcons were re-introduced to the Sudbury area during the summers of 1990 and 1991 at Laurentian University (Blomme and Laws 1995). By 1995, there were 15 sightings of the released peregrines, six of which were mortalities; only seven active nest sites were identified in the Sudbury area (Blomme and Laws 1995). A second wave of releases occurred in 1992 and 1993 in an area near Killarney Provincial Park, 60 kilometers south of Sudbury; at that site, 59 birds were released (Blomme and Laws 1995). There is evidence that peregrines are breeding in the Sudbury area (Archived Birding Reports 2002; MNR 2003 pers. comm.). For example, in recent years, local naturalists have observed peregrine falcons nesting with young near lakes in Sudbury (Schoenefeld 2004 pers. comm.). Moreover, preliminary data indicate that the peregrine falcon was breeding in the City of Greater Sudbury in 2000 to 2005 (Monet and Boucher 2005).

11.4 Risk Characterization

The objective of the risk characterization was to determine the likelihood of adverse effects to terrestrial wildlife populations occurring from exposure to metals in the Sudbury area. Risks to wildlife occur when the chemical can cause one or more adverse effects, and the chemical co-occurs with or contacts an animal long enough and at a sufficient intensity to elicit the adverse effect. Risk characterization combines the information from the exposure assessment and the effects assessment. Where risk was not ruled out using direct toxicity modeling, information on local populations was used to provide a weight-of-evidence analysis of wildlife risks. The wildlife risk model was used to predict exposures for each COC on a probabilistic basis. Exposure estimates were compared to the point estimate TRV for specific chemicals and wildlife VECs. The risk was expressed as an Exposure Ratio (ER) value, calculated as follows:

$$Exposure\ Ratio\ (ER) = \frac{Exposure\ Estimate}{Toxicity\ Reference\ Value}$$

Exposure Ratios are also sometimes referred to as hazard quotients (HQ) but for this ERA, the term 'exposure ratio' was used. Predicted ERs represented the potential for adverse effects to individuals within a population or the probability of an individual animal receiving a potentially adverse exposure across the area of interest (i.e. assuming the VEC is homogeneously distributed in the study area). The best use of ERs is for ruling out risks (Fairbrother 2003; Hull and Swanson 2006), as opposed to predicting risks. The 90th percentile was recognized as a reasonable maximum exposure for an individual (U.S. EPA 2001b; Health Canada 2004) because there is only a 10% probability that an individual VEC will receive an exposure that exceeds the TRV. The 90th percentile level has been accepted by other Environment Ministries in Canada for similar sites.

The risk estimates for large home range wildlife were based on average concentrations in soil defined by the estimated confidence limits. For small home range VECs (i.e. meadow vole, short-tailed shrew, American robin, ruffed grouse), the risk estimates were not based on confidence limits about the mean, but rather on the distribution of measured data for the entire sub-area (e.g. Zone 1, Zone 2, Coniston, etc.). This was done intentionally recognizing that small home range VECs could be exposed to the entire distribution of concentrations. The risk assessment results were first reviewed to determine the probability of the ER exceeding an ER=1.0. For the purpose of this ERA, the probabilities were interpreted as follows:

- 90% or greater probability of ER less than or equal to 1.0: signified that most estimated exposures were less than the TRV (i.e. NOAEL, LOAEL, IC_{20}, etc.) indicating that adverse effects were ruled out.
- Greater than 10% probability of ER greater than 1.0: potential for adverse effects was not ruled out; however, the significance of this potential risk was judged according to the uncertainty and degree of conservatism incorporated into the risk assessment, as well as site-specific information.

Presenting the probabilities of exceeding an ER of 1.0 did not include information with regard to the magnitude of the exceedance. Therefore, where risks could not be ruled out, the 90th percentile ER was also presented to provide an indication of the magnitude of the exceedance in relation to the TRV. Detailed risk characterization results are presented as examples for Zone 1 and Copper Cliff in Sections 11.4.1 and 11.4.2, respectively. The overall results are then summarized and presented in Section 11.4.3 for each VEC and COC by ERA study zone. In many cases, risk was estimated and discussed using different TRVs as a partial sensitivity analysis to determine the significance of altering the TRV on the prediction of risk.

11.4.1 Zone 1 Results

In Zone 1, the probabilities of ERs > 1.0 were below 10% for all COCs and VECs with the exception of Se exposure to voles (Table 11.15). The dominant exposure pathways contributing to this unacceptable risk were ingestion of roots and shoots. The distribution of Se concentrations in Zone 1 soil is shown in Figure 11.6. Approximately 65% of the Se concentrations in surface soil were less than the Ontario background concentration of 1.9 mg/kg. The probabilities of individual voles receiving a particular exposure greater than the three TRVs are shown in Figure 11.7. In addition, Figure 11.7 provides the ERs associated with various percentile exposures on the secondary y-axis, for the most conservative TRV. There was a 15% probability that an individual vole would receive an exposure that exceeded the most conservative (lowest) of the three TRVs. In other words, there was an 85% probability that an individual vole would receive an exposure that was lower than the lowest TRV (Figure 11.7). The 90th percentile ER=1.1 for voles exposed to Se using the most conservative TRV. Only the most conservative TRV was exceeded. Therefore, risks to populations of voles in Zone 1 were considered low.

Table 11.15 Probability of ER > 1.0 for Zone 1

VEC	As	Cd	Co	Cu	Pb	Ni	Se
American robin	0%	0%	0%	0%	0%	0%	9%
Beaver	0%	0%	0%	0%	0%	0%	0%
Deer	0%	0%	0%	0%	0%	0%	0%
Fox	0%	0%	0%	0%	0%	0%	5%
Meadow vole	0%	0%	0%	0%	0%	0%	**15%**
Moose	0%	0%	0%	0%	0%	0%	0%
Peregrine falcon	0%	5%	0%	0%	3%	0%	8%
Ruffed grouse	0%	0%	0%	0%	0%	0%	2%
Short-tailed shrew	0%	0%	0%	0%	0%	0%	6%

Values in bold indicate that the probability is >10% that ER>1.0

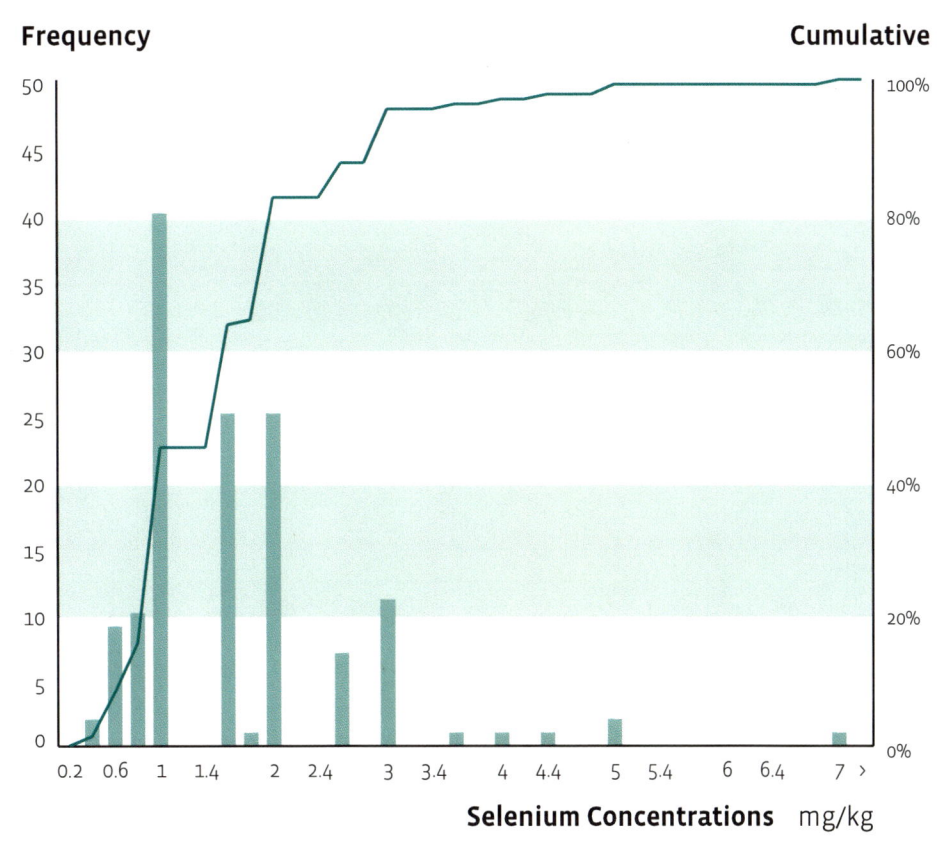

Ecological Risk Assessment: Evaluating Risks to Terrestrial Wildlife

Figure 11.6   **Distribution of Se Concentrations in Soil (0 to 5 cm) in Zone 1**

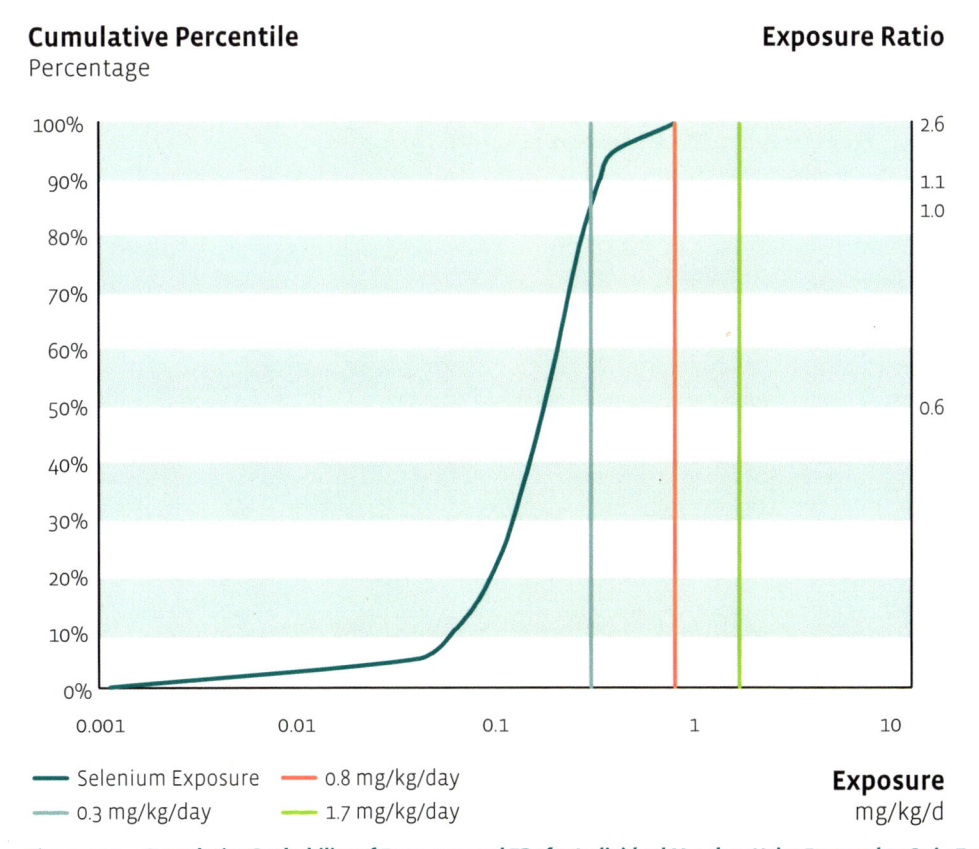

Figure 11.7   **Cumulative Probability of Exposure and ERs for Individual Meadow Voles Exposed to Se in Zone 1**

11.4.2 Copper Cliff Results

In Copper Cliff, the probabilities of ERs > 1.0 were greater than 10% for Se (all VECs) and for Cu (ruffed grouse) and Ni (meadow vole) (Table 11.16). The dominant exposure pathways contributing to unacceptable risks were: ingestion of insects for robin and shrew; ingestion of roots and shoots for vole; ingestion of mammals for fox and falcon; ingestion of shoots for grouse exposed to Se; and, ingestion of shoots and soil for grouse exposed to Cu.

Table 11.16 **Probability of ER > 1.0 for Copper Cliff**

VEC	As	Cd	Co	Cu	Pb	Ni	Se
American robin	0%	0%	0%	8%	1%	2%	**55%**
Fox	0%	0%	0%	0%	0%	0%	**16%**
Meadow vole	0%	0%	0%	4%	0%	**51%**	**74%**
Peregrine falcon	0%	7%	0%	0%	6%	0%	**20%**
Ruffed grouse	0%	0%	0%	**11%**	2%	2%	**29%**
Short-tailed shrew	0%	0%	0%	0%	0%	1%	**41%**

Values in bold indicate that the probability is >10% that ER>1.0

There was an 11% probability that an individual grouse would receive an exposure to Cu that exceeded the TRV. The 90th percentile ER = 1.1. All sample locations that may result in an ER>1.0 were residential yards, baseball diamond outfields, or schools, with one exception which was a park. Ruffed grouse are known to avoid developed areas (Endrulat et al. 2005); therefore, risk from Cu to populations of grouse in Copper Cliff was considered low.

There was a 51% probability that an individual vole would receive an exposure to Ni that exceeded the TRV; the 90th percentile ER = 2.7. There was a 74% probability that an individual vole would receive an exposure to Se that exceeded the most conservative of the three TRVs. The 90th percentile ER = 3.3 for voles exposed to Se using the most conservative TRV. The second-most conservative TRV was also exceeded, with an ER = 1.3 at the 90th percentile of exposure. In the case of voles, a significant contribution to total exposure came from consumption of roots. Both Ni and Se were found at higher concentrations in *Deschampsia* roots than in soil. *Deschampsia* concentration data were used for all plant root and shoot exposures, although this genus is known to be tolerant to acid and metals in soil (Bagatto and Shorthouse 1991; Archambault and Winterhalder 1995; Reid 2000). Therefore, metal concentrations may be higher in this genus of plant than in others used as food by wildlife.

The 90th percentile ER for Se exceeded 1.0 for several VECs (Table 11.16). The uptake models for Se from soil into dietary items were more uncertain for Copper Cliff than for other areas because the range of soil concentrations used to develop the regression equations was narrower than the range of concentrations measured in Copper Cliff soil. There was a 16% probability that an individual fox would receive an exposure to Se that exceeded the most conservative of the three TRVs. The 90th percentile ER = 1.2 for fox exposed to Se using the most conservative TRV. Only the most conservative TRV was exceeded at the 90th percentile of exposure. Therefore, risks to populations of fox from Se in Copper Cliff were low. There was a 29% probability that an individual grouse would receive an exposure to Se that exceeded the most conservative TRV. The 90th percentile ER = 1.6 for grouse exposed to Se using the most conservative TRV. At the 90th percentile of exposure, the ER = 1.1 using the second-most conservative TRV. As mentioned above, most samples were collected from residential areas in Copper Cliff, as well as schools, playgrounds and parks, and ruffed grouse are known to avoid developed areas (Endrulat et al. 2005). Therefore, risks of Se to populations of grouse in Copper Cliff were considered low.

There was a 41% probability that an individual shrew would receive an exposure to Se that exceeded the most conservative TRV. The 90th percentile ER = 1.8 for shrews exposed to Se using the most conservative TRV, and only the most conservative TRV was exceeded at the 90th percentile of exposure. There was a 20% probability that an individual falcon foraging exclusively in Copper Cliff would receive an exposure to Se that exceeded the more conservative TRV. The 90th percentile ER = 1.5 for falcons exposed to Se using the more conservative TRV. Only the more conservative TRV was exceeded at the 90th percentile of exposure. Therefore, risks to Se in falcons and shrews in Copper Cliff were considered low.

Ecological Risk Assessment: Evaluating Risks to Terrestrial Wildlife

There was a 55% probability that an individual robin would receive an exposure to Se that exceeded the most conservative of the three TRVs (Figure 11.8). The 90th percentile ER=1.9 for robins exposed to Se using the most conservative TRV. Only the most conservative TRV was exceeded at the 90th percentile of exposure. Therefore, risks to populations of robins in Copper Cliff were considered low.

Cumulative Percentile
Percentage

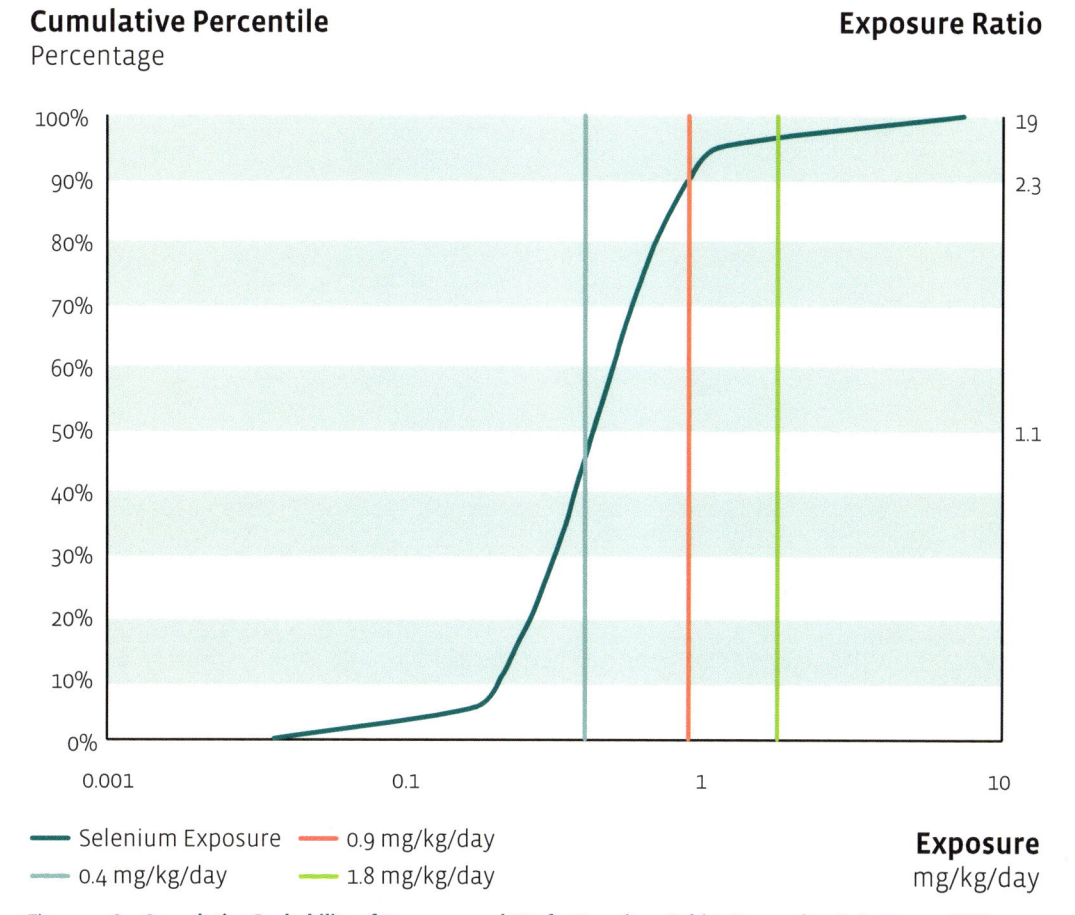

Exposure Ratio

— Selenium Exposure   — 0.9 mg/kg/day
— 0.4 mg/kg/day   — 1.8 mg/kg/day

Exposure
mg/kg/day

Figure 11.8 **Cumulative Probability of Exposure and ERs for American Robins Exposed to Se in Copper Cliff**

11.4.3 Risk Characterization Summary

The results of the risk modeling are summarized in Table 11.17, where the results represent the 90th percentile ERs and where there was a greater than 10% probability of an ER > 1.0. Alternative ERs are also provided in parentheses for ERs calculated using alternative TRVs. Modeling of direct toxicity suggested that individual short-tailed shrews and meadow voles may be at risk. In particular, voles exposed to Ni in Copper Cliff and Falconbridge, and voles exposed to Se in Copper Cliff were the primary concerns. In addition, risk to Se for robins, red fox and falcons could not be ruled out.

Table 11.17 Summary of ERs Where There Was a >10% Probability of an ER>1.0

Zone or Community and VEC	Copper ER	Nickel ER	Selenium ER*
Zone 1			
Meadow vole			1.1 (0.41)
Zone 2			
American robin			1.4 (0.61)
Meadow vole			1.7 (0.63)
Short-tailed shrew			1.3 (0.48)
Peregrine falcon			1.1 (0.58)
Zone 3			
Meadow vole			1.2 (0.44)
Sudbury Central			
None			
Coniston			
Meadow vole		1.3	1.2 (0.45)
Copper Cliff			
American robin			1.9 (1.0)
Red fox			1.2 (0.40)
Meadow vole		2.7	3.3 (1.2, 0.58)
Peregrine falcon			1.5 (0.75)
Ruffed grouse	1.1		1.6 (1.1, 0.8)
Short-tailed shrew			1.8 (0.68)
Falconbridge			
American robin			1.4 (0.62)
Meadow vole		2.5	1.9 (0.71)
Peregrine falcon			1.2 (0.6)
Short-tailed shrew			1.2 (0.45)

*ERs based on alternate TRV presented in parentheses beside the ER using the most conservative TRV.

Small mammals were predicted to be most at risk, based on the number and magnitude of exceedances of ER=1.0. However, the predicted 90th percentile risks to individual short-tailed shrews and meadow voles were all ERs ≤ 3. The ERs generally fell below 1.0 for individual shrews and voles when less conservative TRVs were used. Therefore, direct toxicity risks were considered low because: there was a low magnitude of exceedance of ER=1.0; exceedances in all but one case occurred when using only the most conservative TRV; and naturalists and researchers indicate that shrews and voles are abundant in the Sudbury area and that the Sudbury area as a whole is suitable to sustain populations of small mammals.

However, direct toxicity modeling does not address risks to small mammals or other VECs due to loss of or changes in their habitat or changes in the habitat of their predators. Habitats must provide suitable cover from predators, nest sites and food sources. Changes in plant communities that provide cover and food may result in changes in the populations or communities of small mammals and other organisms that use the habitat. In fact, habitat may have a greater influence on small mammals than direct toxicity from metals. Preliminary results showed that mammalian diversity increased in reclaimed areas (Robitaille and Linley 2006). Habitat improvements for most species would result from continued revegetation and restoration of the Sudbury area.

Modeling of direct toxicity suggested that risks to individual falcons and populations of other birds (robins, grouse) were low. There are approximately 300 species of birds known to occur in Sudbury (Whitelaw 1989). Christmas bird count data for non-migratory species showed increasing numbers of birds from 1980 to 1995 (National Audubon Society 2005). American robins are breeding in Sudbury. Grouse were extirpated from portions of the Sudbury area but the population has since recovered. Peregrine falcons were re-introduced into the Sudbury area in 1990 and 1991 and there was evidence that peregrines are breeding in the Sudbury area.

In summary, based on the available information, it was concluded that metals in soil were probably not exerting a significant direct toxic effect on VEC populations in the Sudbury area. However, previous effects of smelter emissions on habitat quality (e.g. loss of particular plant species used as food or cover) may have a continued influence on birds and mammals in the study area.

11.5 Uncertainty Analysis

There was both uncertainty and variability associated with the risk models. Uncertainty derives from a lack of knowledge. Variability arises from temporal and spatial differences in metal concentrations across the study area, and differences between individuals in body weight and food intake rates. Characterizing uncertainty and variability in an assessment is crucial to the success of the decision-making process (Moore and Bartell 2000). The method used to assess the uncertainty and variability in the exposure estimates depends on the complexity of the model, the information available, and sources of uncertainty. Potential sources of uncertainty in the ERA can be divided into one of the following categories (U.S. EPA 2001b): parameter uncertainty; model uncertainty; and, scenario uncertainty.

One of the difficult issues in assessing ecological risks is the establishment of a *priori* performance criteria for model results (Moore and Bartell 2000). In some cases, the risk assessment need only be certain that risks are not under-predicted and that the model will rarely predict the absence of risk when there is a risk (i.e. avoid a false negative). Therefore, the objective of the analysis of uncertainty for the model input parameters was to determine whether:

- Model input variables accurately reflected the natural variability in the environment with minimal uncertainty, and
- Model input variables assumed conservative values in the face of uncertainty (lack of knowledge).

To determine the relative importance of variability and uncertainty in a model input parameter, one asks the question 'Will the collection of more information significantly improve the understanding of the variability of an input variable?' At some point, the collection of additional data reaches the point of diminishing returns. When the variability in a measured parameter is well characterized for a particular area, uncertainty has been reduced to an acceptable level and no further data collection is required. The variability and uncertainty for each parameter were characterized according to the following criteria (Tables 11.18 and 11.19).

Table 11.18 Definition of Variability Qualifiers

Qualifier	Definition
Well characterized	Sample size is adequate for the area of coverage. Repeated measures confirm or fall within an order of magnitude of previous measurements. Good coverage across the study area
Moderately characterized	Sample size is moderate for the area of coverage. Only one monitoring study or sampling period. Missing coverage in certain regions across the study area
Poorly characterized	Sample size is low for the area of coverage. Only one monitoring study sampling period. Missing coverage in many regions across the study area

Table 11.19 **Definition of Uncertainty Qualifiers**

Qualifier	Definition
Low	Collection of additional data unlikely to improve understanding of centrality and variability
Medium	Collection of additional data may improve understanding of centrality and variability. Analyses assumed that site-specific conditions were not significantly different from literature studies used to parameterize models (i.e. assumptions or differences are within a factor of 10)
High	Collection of additional data likely would improve understanding of centrality and variability. Site-specific conditions are significantly different from assumptions and methodologies in the literature (i.e. differences may be greater than a factor of 10)

The variability and uncertainty in the data related to the wildlife modeling are described in the following sections for surface water, soil, medium-to-biota uptake models, dietary apportionment, wildlife exposure variables, soil bioaccessibility, model structure and TRVs.

11.5.1 Surface Water Concentrations

Based on the criteria presented in Table 11.18, surface water data in general were considered to be poorly characterized. The number of lakes sampled ($n=30$) was small and represented less than 10% of lakes located in the study area. However, concentrations of most of the COC (As, Co, Cd, Pb, Se) in water were below the laboratory MDL. As a result, the degree of variability was unknown. The uncertainty associated with surface water concentrations was considered to be medium. Because all samples were collected at the same time of year (i.e. midsummer), from a similar location within each lake (i.e. the epilimnion/metalimnion of a deep basin), concentrations may not be representative of those throughout the entire lake basin and for all seasons. In addition, data were unavailable for comparisons among years. Surface water concentrations were used to predict exposures to wildlife from drinking water and consumption of algae or other aquatic vegetation. However, uncertainty in surface water concentrations was unlikely to significantly affect the risk characterization conclusions because exposure pathways from water did not contribute significantly to wildlife exposures and risks.

11.5.2 Soil Concentrations

Soil concentrations were well characterized. The sampling number ($n > 8000$) was high for all zones and urban communities, and there was good coverage across the study area, especially in the communities of interest. Variability in concentrations of all COC was high but well characterized. The uncertainty associated with soil concentrations was considered to be low. Because metals do not degrade or volatilize, little seasonal variation was expected.

11.5.3 Medium-to-Biota Uptake Models

Four different uptake models were used in the terrestrial ERA exposure modeling: (i) surface water-to-aquatic plants, (ii) soil-to-plant, (iii) soil-to-invertebrates, and (iv) soil-to-animal uptake models. Sudbury-specific data were collected for surface water, soils, plant tissues (as food for herbivores), and grasshoppers and earthworms (representing invertebrate food sources). In general, the sample sizes were considered small resulting in significant uncertainty in these models. The variability in the site-specific uptake models was substantial in some cases, resulting in variations in predicted exposure concentrations of one to two orders of magnitude. There was also uncertainty in the use of regression models for media concentrations beyond the concentration range used to develop the models. The uncertainty or 'error' in the models would likely be reduced through collection of more information. Sudbury-specific data and uptake models were compared with literature models (e.g. BJC 1998) and indicated that Sudbury-specific uptake models were similar to what was reported in the literature.

11.5.4 Dietary Composition and Use of Surrogates

Dietary composition was variable for each wildlife VEC. The variability in dietary composition accounted for individual, geographic and seasonal differences in available food items and wildlife preferences. Extensive literature searches were conducted to locate data and information on dietary compositions. Given the extensive literature review and consultation with local biologists, dietary composition is believed to be well characterized. The wildlife exposure models often relied on surrogate models to estimate dietary exposures. The use of surrogate dietary items in the exposure model to represent the preferred dietary items of a wildlife species was judged to be reasonable and appropriate given the available data from the study area and literature. The overall uncertainty in the use of surrogate dietary items was considered to be medium to low.

11.5.5 Wildlife Receptor Variables

The wildlife exposure models included several variables (i.e. body weight, food ingestion rate, soil ingestion rate, water ingestion rate, assimilation efficiency, and gross energy). Given the extensive literature review and use of relatively large data sets or recognized allometric equations, the variability in receptor variables was considered to be well characterized. Uncertainty in most wildlife receptor variables was judged to be low. No site-specific data were available for the derivation of receptor variables.

11.5.6 Soil Bioaccessibility

The exposure model used conservative and deterministic point estimate values for bioaccessibility that were derived for the human health risk assessment (see Chapter 7). Variability in bioaccessibility was not incorporated in the assessment. The *in vitro* assays used to derive bioaccessibility estimates for the HHRA were specifically designed to mimic the gastric and intestinal phase of human digestion. The degree to which this assay would reflect wildlife digestion is uncertain. Bioaccessibility values selected for the ERA were equal to or higher than values recommended by other sources. Given the conservatism used in the derivation of bioaccessibility values and the use of site-specific soil data, the uncertainty in the bioaccessibility values for the ERA was considered to be medium.

11.5.7 Model Uncertainty

The likelihood of underestimating risks was considered to be low. This was due to the availability of site-specific data as well as the use of conservative assumptions where appropriate. The exposure model did not explicitly incorporate habitat suitability information. Therefore, it remained unknown how the chemical, physical and biological variables were inter-related to affect wildlife populations and communities. Habitat suitability is a critical factor that determines the media and concentrations a wildlife VEC will be exposed to in their home range. Spatially explicit exposure modeling techniques that account for movement of VECs between habitats of varying suitability have been described in the literature (Freshman and Menzie 1996; Hope 2000). Such techniques were not used in this assessment because of a lack of habitat suitability information for VECs in the study area.

11.5.8 Toxicity Reference Values (TRVs)

A significant source of uncertainty in characterizing risk was the use of conservative and deterministic TRVs. The TRVs used in the wildlife risk modeling were selected from the literature to be as relevant as possible (e.g. form of metal, closely related animal species), while ensuring that the effects metrics were not underestimated. Specific uncertainties related to TRVs were:

- Toxicological information directly related to the VEC of concern was often unavailable or limited in nature. Therefore, most of the TRVs were derived from data for similar or related species exposed under controlled laboratory conditions designed to maximize adverse effect (e.g. using highly bioavailable forms of metals, exposure via drinking water rather than diet). The sensitivities of laboratory animals and a wildlife species may differ because of behavioral and ecological parameters (e.g. stress from competition, seasonal changes in temperature, etc.), disease and exposure to other contaminants. Extrapolation of toxicity data to unrelated species required the use of uncertainty factors to ensure that risks were not underestimated.

- The majority of the toxicity studies reported the exposure concentrations in terms of mg/kg in diet or mg/L water. To make these exposures comparable across test species and studies, they were converted to normalized daily doses (mg/kg bw/day). In the absence of study-specific data, generic body weight values and intake rates from the literature were used to calculate the normalized daily dose.

- TRVs were chemical-specific; there were no metal-mixture TRVs that could have been used to assess responses to multiple stressors.

- TRVs were based on effects in individual organisms, and not population-level endpoints. However, assessment endpoints were related to wildlife populations, not individual organisms (with the exception of threatened or endangered species).

Despite these uncertainties, given the level of conservatism built into the TRVs and application of uncertainty factors, the results were considered to be conservative, and would not underestimate risk.

11.6 Summary and Conclusions

This chapter addressed two objectives:

- Objective #2: Evaluate risks to terrestrial wildlife populations and communities due to COC.
- Objective #3: Evaluate risks to individuals of threatened or endangered terrestrial species due to COC.

Direct risk from exposure to some COC to small mammals could not be ruled out by the modeling approach. Similarly, based on the wildlife model, risk due to Se could not be ruled out for the avian VECs using the lowest, or most conservative TRVs. However, the Se TRVs were very conservative, and the use of alternative TRVs generally ruled out risk to avian receptors. Although no formal field studies on wildlife populations were conducted as part of the Sudbury ERA, there was anecdotal information concerning wildlife abundance from area naturalists, researchers, and other groups. There was no indication from any of these sources that the abundances of the VECs, or wildlife in general, were unusually low in the study area.

Based on consideration of all of the available information, it was considered unlikely that metals in soil were exerting a significant direct toxic effect on VEC populations in the Sudbury area. However, previous effects of smelter emissions on habitat quality (e.g. loss of particular plant species used as food or cover) may be having a continued influence on birds and mammals in the study area.

11.7 Acknowledgements

The authors thank the Intrinsik coworkers who contributed to the wildlife ERA, including Scott Teed and Dwayne Moore (wildlife model), Sara Rodney (statistics), and Josephine Archbold (review of previous studies, and compilation of information on population trends).

11.8 References

Abbey J (2004) Personal communication to Cantox Environmental Inc., 13 and 14 December 2004. Ruffed Grouse Society of Canada. Email communication re: Ruffed grouse

Alkins M (2005) Personal communication to Cantox Environmental in regards to moose and deer populations in the wildlife management unit 41 on 27 September 2005 via email: melanie.alkins@mnr.gov.on.ca, North Bay District Ministry of Natural Resources Office

Allard P, Fairbrother A, Hope BK et al. (2010) Recommendations for the development and application of wildlife toxicity reference values. Integr Environ Assess Manag 6(1):28–37

Ambrose AM, Larson PS, Borzelleca JF et al. (1976) Long-term toxicologic assessment of nickel in rats and dogs. J Food Sci Technol 13:181–187

Amiro BD, Courtin GM (1981) Patterns of vegetation in the vicinity of an industrially disturbed ecosystem, Sudbury, Ontario (Canada). Can J Bot 59:1623–1639

Archambault DJ, Winterhalder K (1995) Metal tolerance in Agrostis scabra from the Sudbury, Ontario, area. Can J Bot 73(5):766–775

Archived Birding Reports (2002) Sudbury district. www.web-nat.com/bic/ont/Archives/arcsud2002.htm. Updated March 2003. Accessed 15 September 2005

Aldrich RJ, Ringer RK, Bleavins MR et al. (1982) Effects of supplemental dietary copper on growth, reproductive performance and kit survival of standard mink and the acute toxicity of copper to mink. J Anim Sci 55(2):337–343

Azar A, Trochimowicz HJ, Maxfield ME (1973) Review of lead studies in animals carried out at Haskell Laboratory: Two year feeding study and response to hemorrhage study. In: Barth D, Berlin A, Engel R et al. (eds) Environmental health aspects of lead, pp. 119–210. Proceedings, International Symposium, Amsterdam, The Netherlands, October 1972. Commission of the European Communities, Luxembourg

Bagatto G, Shorthouse JD (1991) Accumulation of copper and nickel in plant tissues and an insect gall of lowbush blueberry, Vaccinium angustifolium, near an ore smelter at Sudbury, Ontario, Canada. Can J Bot 69(7): 1483–1490

Barrett GW, Stueck KL (1976) Caloric ingestion rate and assimilation efficiency of the short-tailed shrew, Blarina brevicauda. Ohio J Sci 76:25–26

BC MELP (1998) Protocol for contaminated sites, guidance and checklist for Tier 1 ecological risk assessment of contaminated sites in British Columbia. British Columbia Ministry of Environment, Lands and Parks, Victoria, BC

Bell GP (1990) Birds and mammals on an insect diet: a primer on diet composition analysis in relation to ecological energetics. Stud Avian Biol 13:391–415

Biscaia P (2005) Personal communication to Cantox Environmental. Acting/Area Biologist, Sudbury District, Ministry of Natural Resources

BJC (1998) Empirical models for the uptake of inorganic chemicals from soil by plants. Prepared for the U.S. Department of Energy Office of Environmental Management. BJC/OR-113. September 1998. Bechtel Jacobs Company LLC, Oak Ridge, TN

Blomme CG, Laws KM (1995) Partnerships for wildlife restoration: Peregrine falcons. In: Gunn JM (ed) Restoration and recovery of an industrial region, pp. 155–168. Ontario Ministry of Natural Resources, Cooperative Freshwater Ecology Unit, Laurentian University, Sudbury, Ontario, Canada. Springer-Verlag, New York

Byron WR, Bierbower GW, Brouwer JB et al. (1967) Pathologic changes in rats and dogs from two-year feeding of sodium arsenite or sodium arsenate. Toxicol Appl Pharmacol 10:132–147

Cadman M, Eagles PFJ, Helleiner FM (1987) Atlas of the breeding birds of Ontario. Federation of Ontario Naturalists and Long Point Bird Observatory. University of Waterloo Press, Waterloo, Ontario, Canada

Cain BW, Pafford EA (1981) Effects of dietary nickel on survival and growth of Mallard ducklings. Arch Environ Contam Toxicol 10:737–745

Calder WA, Braun EJ (1983) Scaling of osmotic regulation in mammals and birds. Am J Physiol 244:R601–R606

Collopy MW (1975) Behavioural and predatory dynamics of kestrels wintering in the Arcata Bottoms. Masters thesis, Humboldt State University, Arcata, CA

Cummins KW, Wuycheck JC (1971) Caloric equivalents for investigations in ecological energetics. International Association of Theoretical and Applied Limnology, no. 15. E. Schweizerbart, Stuttgart, Germany

Demayo A, Taylor MC, Taylor JW et al. (1982) Toxic effects of lead and lead compounds on human health, aquatic life, wildlife, plants, and livestock. CRC Crit Rev Environ Control 12:257–305

Deschamps M (2006) Personal communication to Inco regarding moose populations in Wildlife Management Unit 41, as provided to Cantox Environmental in the TC Comments on Draft 1 of the Sudbury ERA. Thunder Bay District Ministry of Natural Resources

Dice LR (1922) Some factors affecting the distribution of the prairie vole, forest deer mouse, and prairie deer mouse. Ecology 3:29–47

Dobbyn J, Eger J, Wilson N (1994) Atlas of the mammals of Ontario. ISBN 1-896059-02-3. Federation of Ontario Naturalists, Don Mills, Ontario

Drozdz A (1968) Digestibility and assimilation of natural foods in small rodents. Acta Theriol 13:367–389

Eadie WR (1944) The short-tailed shrew and field mouse predation. J Mammal 25:359–364

Eadie WR (1948) Shrew-mouse predation during low mouse abundance. J Mammal 29:35–37

Edens FW, Garlich JD (1983) Lead-induced egg production decrease in Leghorn and Japanese quail hens. Poult Sci 62:1757–1763

Endrulat EG, McWilliams SR, Tefft BC (2005) Habitat selection and home range size of Ruffed grouse in Rhode Island. Northeast Nat 12:411–424 (Abstract only)

ep&t (1996) Toxicity extrapolations in terrestrial systems. Prepared for California Environmental Protection Agency, Sacramento, CA. ep&t, Corvallis, Oregon

Fairbrother A (2003) Lines of evidence in wildlife risk assessments. Hum Ecol Risk Assess 9(6):1475–1491

Freshman JS, Menzie CA (1996) Two wildlife exposure models to assess impacts at the individual and population levels and the efficacy of remedial actions. Hum Ecol Risk Assess 2(3):481–498

Gilliom RJ, Helsel DR (1986) Estimation of distributional parameters for censored trace level water quality data 1. Estimation techniques. Water Resour Res 22(2):135–146

Golley FB (1960) Energy dynamics of a food chain of an old-field community. Ecol Monogr 30:187–206

Golley FB (1961) Energy values of ecological materials. Ecology 42:581–584

Górecki A (1975) Calorimetry in ecological studies. In: Grodzinski W, Klekowski RZ, Duncan A (eds) Methods for ecological energetics. IPB handbook No. 24, pp. 275–281. Blackwell Scientific Publications, Oxford.

Grodzinski W, Wunder BA (1975) Ecological energetics of small mammals. In: Golley FB, Petrusewicz K, Ryszkowski L (eds) Small mammals: Their productivity and population dynamics. Cambridge University Press, Cambridge, MA

Hamilton WJ Jr (1941) The foods of small forest mammals in eastern United States. J Mammal 22:250–263

Health Canada (2004) Federal contaminated site risk assessment in Canada. Part I: Guidance on human health preliminary quantitative risk assessment (PQRA). Contaminated sites program. Health Canada, Ottawa, Ontario

Heinz GH, Hoffman DJ, Gold LG (1989) Impaired reproduction of mallards fed an organic form of selenium. J Wildl Manag 53:418–428

Helsel DR (1990) Less than obvious: statistical treatment of data below the detection limit. Environ Sci Technol 24(12):1767–1774

Helsel DR (2006) Fabricating data: how substituting values for nondetects can ruin results, and what can be done about it. Chemosphere 65(11):2434–9

Hope BK (2000) Generating probabilistic spatially-explicit individual and population exposure estimates for ecological risk assessments. Risk Anal 20(5): 573–589

Hull RN, Swanson SM (2006) Sequential analysis of lines of evidence – An advanced weight-of-evidence approach for ecological risk assessment. Integr Environ Assess Manag 2(4):302–311

Huybrechts T, Thas O, Dewulf J et al. (2002) How to estimate moments and quantiles of environmental data sets with non-detected observations? A case study on volatile organic compounds in marine water samples. J Chromatogr A 975:123–133

Jenkins KJ, Hidiroglou H (1989) Tolerance of the calf for excess copper in milk replacer. J Dairy Sci 72:150–156

Karasov WH (1990) Digestion in birds: Chemical and physiological determinants and ecological implications. Stud Avian Biol 13:391–415

Keller B, Heneberry J, Gunn JM et al. (2004) Recovery of acid and metal damaged lakes near Sudbury Ontario: Trends and status. Cooperative Freshwater Ecology Unit. Department of Biology, Laurentian University. Sudbury, Ontario, Canada

Koplin JR, Collopy MW, Bammann AR et al. (1980) Energetics of two wintering raptors. Auk 97:795–806

Lambou V, Lin B (1970) Hazards of arsenic in the environment, with particular reference to the aquatic environment. Federal Water Quality Administration, U.S. Department of the Interior (unpublished report)

Lijinski W, Milner JA, Kovatch RM et al. (1989) Lack of effect of selenium on induction of tumors of esophagus and bladder in rats by two nitrosamines. Toxicol Ind Health 5:63–72

Linzey DW, Linzey AV (1973) Notes on food of small mammals from Great Smokey Mountains National Park, Tennessee-North Carolina. J Elisha Mitchell Sci Soc 89:6–14

Litvaitis JA, Mautz WW (1976) Energy utilization of three diets fed to a captive red fox. J Wildl Manag 40:365–368

Long RC (2005) Bird fact sheet – American robin. Hinterland's who's who. Environment Canada. www.hww.ca/hww2.asp?id=25. Accessed 16 July 2006

Mehring AL Jr, Brumbaugh JH, Sutherland AJ et al. (1960) The tolerance of growing chickens for dietary copper. Poult Sci 39:713–719

Menzie C et al. (2000) An approach for incorporating information on chemical availability in soils into risk assessment and risk-based decision making. Hum Ecol Risk Assess 6(3):479–510

MNR (2003) Personal communication between Jan Linquist and Randy Staples at the Sudbury District MNR office. VTE statistics within 100 km radius of Sudbury based on NRVIS database, 31 July 2003

MOE (2005) Procedures for the use of risk assessment under Part XV.1 of the Environmental Protection Act. PIBs 5404e. October 2005. Ontario Ministry of the Environment, Standards Development Branch, Toronto.

Monet S, Boucher N (2005) The natural heritage background study. The Environmental Initiatives Group. The City of Greater Sudbury. www.greatersudbury.ca/content/div_planning/documents/Natural_Heritage_Background_Study.pdf. Accessed 18 September 2006

Moore DRJ, SM Bartell (2000) Estimating ecological risks of multiple stressors: Advanced methods and difficult issues. In: Foran J, Ferenc S (eds) Multiple stressors in ecological risk and impact assessment: Approaches to risk estimation, pp. 117–168. SETAC Press, Pensacola, FL

Moore DRJ, Caux PY (1997) Estimating low toxic effects. Environ Toxicol Chem 16:794–801

Moore DRJ, Sample BE, Suter GW et al. (1999) A probabilistic risk assessment of the effects of methylmercury and PCBs on mink and kingfishers along East Fork Poplar Creek, Oak Ridge, Tennessee, USA. Environ Toxicol Chem 18:2941–2953

Nagy KA (1987) Field metabolic rate and food requirement scaling in mammals and birds. Ecol Monogr 57:111–128

Nagy KA, Girard IA, Brown TK (1999) Energetics of free-ranging mammals, reptiles, and birds. Annu Rev Nutr 19:247–277

NAS (2005) Superfund and mining megasites. Lessons from the Coeur d'Alene River Basin. Committee on Superfund Site Assessment and Remediation in the Coeur d'Alene River Basin. National Academy of Sciences, National Academies Press, Washington, DC, USA. www.epa.gov/superfund/accomp/coeur/index.htm. Accessed 12 September 2007

National Audubon Society (2005) Sudbury area Christmas bird count results 1980–2005. The Christmas bird count historical results [Online]. www.audubon.org/bird/cbc. Accessed 15 September 2005

NEPI (2000) Assessing the bioavailability of metals in soil for use in human health risk assessments. National Environmental Policy Institute, Washington, DC

NRC (1980) Mineral tolerance of domestic animals. Subcommittee on mineral toxicity in animals – Agriculture and renewable resources commission on natural resources. National Research Council. National Academy Press, Washington, DC

O'Dell GD, Miller WJ, King WA et al. (1970) Effect of nickel supplementation on production and composition of milk. J Dairy Sci 53:1545

Ort JF, Latshaw JD (1978) The toxic level of sodium selenite in the diet of laying chickens. J Nutr 108:1114–1120

Owen BA (1990) Literature-derived absorption coefficients for 39 chemicals via oral and inhalation routes of exposure. Regul Toxicol Pharmacol 11:237–252

Passmore RC (1990) Mammal fact sheet – White-tailed deer. Hinterland's who's who. Environment Canada. CW69-4/7 – 1990E. www.hww.ca/hww2p.asp?id=106&cid=0. Accessed 8 October 2006

Penumarthy L, Oehme FW, Galitzer SJ (1980) Effects of chronic oral lead administration in young beagle dogs. J Environ Pathol Toxicol 3:465–490

Reid NB (2000) Tolerance of Carex scoparia to acid-generating/metal contaminated coppermine tailings, and its dependence on substrate moisture. Unpublished BSc thesis, Laurentian University, Sudbury, Ontario

Renecker LA, Hudson RJ (1985) Estimation of dry matter intake of free-ranging moose. J Wildl Manag 49:785–792

Riegert J, Dufek A, Fainova D et al. (2007) Increased hunting effort buffers against vole scarcity in an urban Kestrel Falco tinnunculus population. Bird Study 54:353–361 (Abstract only)

Robel RJ, Bisset AR, Dayton AD et al. (1979) Comparative energetics of bobwhites on six different foods. J Wildl Manag 43:987–992.

Robitaille JF (2005) Personal communication to Cantox Environmental. Laurentian University, Department of Biology

Robitaille JF, Linley RD (2006) Structure of forests used by small mammals in the industrially damaged landscape of Sudbury, Ontario, Canada. For Ecol Manag 225:160–167

Ruby MV, Schoof R, Brattin W et al. (1999) Advances in evaluating the oral bioavailability of inorganics in soil for use in human health risk assessment. Environ Sci Technol 33(21):3697–3705

Ruffed Grouse Society (2003) Ruffed grouse facts. www.ruffedgrousesociety.org/grouse-facts Accessed 10 September 2006

Sallabanks R, James FC (1999) American robin Turdus migratorius. In: Poole A, Gill F (eds) The birds of North America, No. 462. The Birds of North America Inc, Philadelphia, PA.

Sample BE, Arenal CA (1999) Allometric models for interspecies extrapolation of wildlife toxicity data. Bull Environ Contam Toxicol 62:653–663

Sample BE, Suter II GW (1994) Estimating exposure of terrestrial wildlife to contaminants. ES/ER/TM-125. Oak Ridge National Laboratory, Environmental Sciences Division, Oak Ridge, TN

Sample BE, Opresko DM, Suter II GW (1996) Toxicological benchmarks for wildlife: 1996 revision. ES/ER/TM-86/R3. Oak Ridge National Laboratory, Oak Ridge, TN

Sample BE, Beauchamp JJ, Efroymson RA et al. (1998) Development and validation of bioaccumulation models for earthworms. ES/ER/TM-220. Oak Ridge National Laboratory, Oak Ridge, TN

Santolo GM, Yamamoto JT, Pisenti JM et al. (1999) Selenium accumulation and effects on reproduction in captive American kestrels fed selenomethionine. J Wildl Manag 63(2):502–511

Schoenefeld D (2004) Personal communication to Cantox Environmental Inc., 18 February 2004. Sudbury naturalists interview re: Sudbury wildlife

Schroeder HA, Mitchener M (1971) Toxic effects of trace elements on reproduction of mice and rats. Arch Environ Health 23:102–106

Smith GJ, Heinz GH, Hoffman DJ et al. (1988) Reproduction in black-crowned night-herons fed selenium. Lake Reserv Manag 4:175–180

Stanley TR, Spann JW, Smith GJ et al. (1994) Main and interactive effects of arsenic and selenium on mallard reproduction and duckling growth and survival. Arch Environ Contam Toxicol 26:444–451

Suter II GW, Cornaby BW, Hadden CT et al. (1995) An approach for balancing health and ecological risks at hazardous waste sites. Risk Anal 15(2):221–231

Thayer GW, Schaaf WE, Angelovic JW et al. (1973) Caloric measurements of some estuarine organisms. Fish Bull 71:289–296

U.S. DOD (2003) Guide for incorporating bioavailability adjustments into human health and ecological risk assessments at U.S. Department of Defense facilities. Part 1. Overview of metals bioavailability

U.S. EPA (1992) A supplemental guidance to RAGS: Calculating the concentration term. Publication 9285.7-081. U.S. Environmental Protection Agency, Office of Research and Development, Washington, DC

U.S. EPA (1993) Wildlife exposure factors handbook. EPA/630/R-93/18a7. U.S. Environmental Protection Agency, Office of Research and Development, Washington, DC

U.S. EPA (1999a) Data collection for the hazardous waste identification rule. Section 12.0: Ecological exposure factors. U.S. Environmental Protection Agency, Office of Solid Waste, Washington, DC

U.S. EPA (1999b) Screening level ecological risk assessment protocol for hazardous waste combustion facilities. Volume I. Peer review draft. EPA530-D-99-001A. 9 August 1999. U.S. Environmental Protection Agency, Office of Solid Waste and Emergency Response, Washington, DC

U.S. EPA (2001a) Final ecological risk assessment – Coeur d'Alene Basin remedial investigation/feasibility study. May 2001. U.S. Environmental Protection Agency, Region 10. Prepared by CH2M Hill and URS Corp. U.S. Environmental Protection Agency, Washington, DC

U.S. EPA (2001b) Risk Assessment guidance for Superfund: Volume III – Part A, Process for conducting probabilistic risk assessment. December 2001. EPA 540-R-02-002. U.S. Environmental Protection Agency, Washington, DC

U.S. EPA (2004a) Housatonic river ecological risk assessment. U.S. Environmental Protection Agency, Washington, DC

U.S. EPA (2004b) ProUCL Version 3.0. Statistical software, USEPA, Technical Support Center, NERL-LV, Las Vegas, NV. www.epa.gov/nerlesd1/tsc/tsc.htm. Accessed 15 December 2008

U.S. EPA (2005) Guidance for developing ecological soil screening levels. Office of Solid Waste and Emergency Response, 1200 Pennsylvania Avenue, N.W. Washington, DC 20460. OSWER Directive 9285.7-55, November 2003 (Revised February 2005). United States Environmental Protection Agency, Washington, DC

Vogtsberger LM, Barrett GW (1973) Bioenergetics of captive red foxes. J Wildl Manag 37(4):495–500
Whitaker JO, Ferraro MG (1963) Summer food of 220 short-tailed shrews from Ithaca, New York. J Mammal 44:419
Whitelaw C (1989) Seasonal occurrence of birds in the Sudbury District. Sudbury. Unpublished report
Wiemeyer SN, Hoffman DJ (1996) Reproduction in eastern screech-owls fed selenium. J Wildl Manag 60(2):332–341

11.9 Appendix Chapter 11: Abbreviations

AE, assimilation efficiency
BCF, bioconcentration factor
BW, body weight
CEM, Centre for Environmental Monitoring
CGS, City of Greater Sudbury
COC, chemical of concern
COI, community of interest
Eco-SSLs, ecological soil screening levels
ER, exposure ratio
ERA, ecological risk assessment
FIR, food ingestion rate
FMR, free metabolic rate
GE, gross energy
GSA, Greater Sudbury Area
HHRA, human health risk assessment
HQ, hazard quotient
IC$_{20}$, inhibition concentration resulting in a 20% effect in test organism
LFH, litter fermentation humus layer
LMF, lack of model fit
LN, log normal
LOAEL, lowest observed adverse effect level
MDL, minimum detection limit
MOE, Ministry of the Environment (Ontario)
NOAEL, no observed adverse effect level
PDF, probability distribution function
RAF, relative absorption factor
SARA, Sudbury Area Risk Assessment
SIR, soil ingestion rate
TDI, total daily intake
TRV, toxicity reference value
UCLM, upper confidence level of the mean
US DOD, United States Department of Defense
US EPA, United States Environmental Protection Agency
VEC, valued ecosystem component
VTE, vulnerable, threatened, endangered (species)
WIR, water ingestion rate

12.0 Risk Management

DOI: 10.5645/b.1.12

Authors

Stephen Monet, Ph.D.
City of Greater Sudbury,
200 Brady Street, Sudbury, ON Canada P3A 5P3
Email: stephen.monet@greatersudbury.ca

Marc Butler
Xstrata Nickel,
Falconbridge, ON Canada P0M 1S0
Email: MButler@xstratanickel.ca

Glen Watson, M.Sc.
Vale,
337 Power Street, Copper Cliff, ON Canada P0M 1N0
Email: Glen.Watson@vale.com

William E. Lautenbach
City of Greater Sudbury,
200 Brady Street, Sudbury, ON Canada P3A 5P3
Email: bill.lautenbach@greatersudbury.ca

Tina McCaffrey
City of Greater Sudbury,
200 Brady Street, Sudbury, ON Canada P3A 5P3
Email: tina.mccaffrey@greatersudbury.ca

Table of Contents

12.1 Key Findings Requiring Risk Management Consideration 383
12.2 Human Health Risk Management 384
   12.2.1 Risk Reduction Strategies Regarding Pb 384
   12.2.2 Vale Risk Reduction Strategies for Ni 385
   12.2.3 Xstrata Nickel Improvements in Dust Emission Controls 387
   12.2.4 New Emission Reduction Requirements and Initiatives 389
12.3 Ecological Risk Management 389
   12.3.1 Greater Sudbury Re-greening Efforts – from 1917 to 2009 390
   12.3.2 Biodiversity Action Plan Development 394
   12.3.3 Implementing the Greater Sudbury Biodiversity Action Plan 397
12.4 Conclusions 400
12.5 Acknowledgements 400
12.6 References 400
12.7 Appendix Chapter 12: Abbreviations 402

Tables

12.1 Particulate Emission Reductions – Vale Smelter and Ni Refinery ... 386
12.2 Particulate Emission Reductions – Vale Smelter Yard and Lands ... 386
12.3 Xstrata Nickel Particulate Emission Management Programs ... 388
12.4 Woody Plant Species Planted from 1979 to 2010 ... 391
12.5 Grass/Legume Seed Mixture Used in the City's Re-greening Program ... 394

Figures

12.1 Combination of Factors Contributing to Health Risk ... 383
12.2 View of the Community of Copper Cliff and the Vale Smelter Operations ... 385
12.3 Air Monitoring Station in Sudbury ... 387
12.4 View of Community of Falconbridge and the Xstrata Nickel Smelter ... 388
12.5 (a) Photo Taken in 1981 near Coniston, Greater Sudbury, Ontario Showing the Barren Landscape; ... 393
12.5 (b) Photo Taken in 2008 at Same Location as the Photo in (a) ... 393
12.6 Development Process of the Greater Sudbury Biodiversity Action Plan ... 395
12.7 City Re-greening Crew Harvesting Forest Floor Herbaceous Plants at Highway Construction Site in 2010 ... 398

12.1 Key Findings Requiring Risk Management Consideration

Both the Human Health Risk Assessment (HHRA) and the Ecological Risk Assessment (ERA) identified issues that required attention to either reduce risk or, in the case of the natural ecosystem, to reverse historical impacts (SARA 2008a, 2009). The HHRA showed there were no unacceptable health risks predicted from any exposure pathway to As, Cu, Co and Se or to the oral/dermal pathway for Ni within the study area. Minimal risks were identified for two chemicals of concern (COC):

- Oral/dermal exposure to Pb in very localized areas in some of the communities of interest
- Ni inhalation exposure in Copper Cliff and the western portion of Sudbury Centre.

Figure 12.1 illustrates the three factors that must be present before there can be a human health risk. For human health risks to exist, there must be the presence of a chemical, presence of people and an available pathway of exposure. The following are three basic approaches that may be considered to reduce human health risks associated with a chemical; a) remove the chemical, b) remove the people; and/or c) reduce or block the exposure pathway of the chemical to the people. Risk management decisions can be directed at any or all of the three contributing factors.

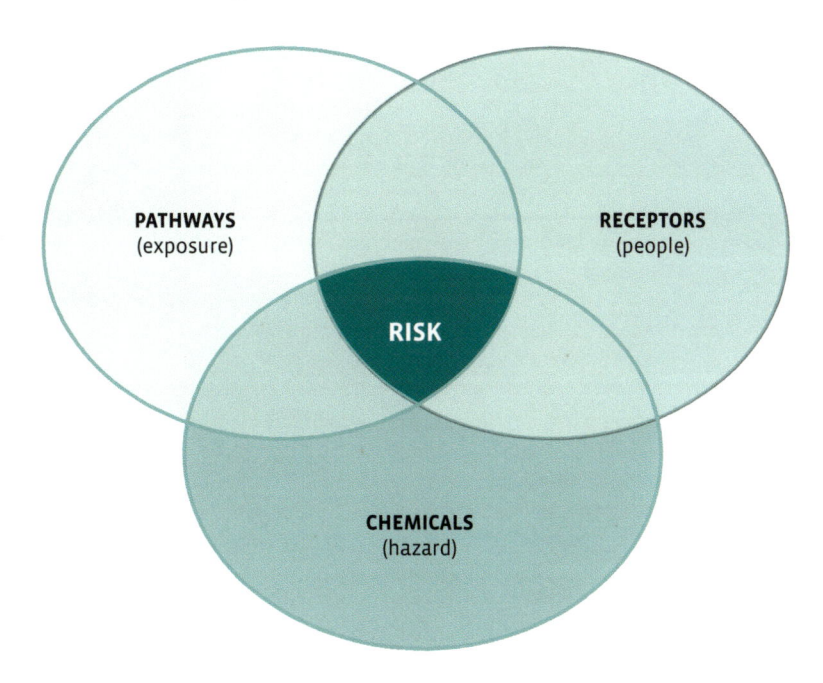

Figure 12.1 Combination of Factors Contributing to Health Risk

Both Vale and Xstrata Nickel are committed to taking all reasonable measures to reduce potential human health risks related to their operations. Risk reduction strategies are outlined in the following section to address the two areas of risk related to Pb and Ni exposure.

The ERA clearly documented that vegetation impacts due to early logging and smelting activities were still measurable over an area of at least 81,000 ha around the Greater Sudbury area. Furthermore, the ERA demonstrated that recovery of naturally sustaining diverse vegetation communities in this area is inhibited by the presence of elevated metal levels in the soil, low soil pH, low organic matter and nutrient status. All of these conditions are a direct result of prolonged smelter emissions combined with soil erosion and other interacting factors (see also Chapters 2 and 6). As a result of these findings, the two mining companies renewed their commitment to ecological restoration efforts that have been coordinated by the City of Greater Sudbury. This chapter provides an overview of some of the past and recent efforts to create a more biologically diverse environment in the impacted landscapes.

12.2 Human Health Risk Management

12.2.1 Risk Reduction Strategies Regarding Pb

The HHRA demonstrated that typical or average exposures to Pb in the environment throughout the Greater Sudbury area were within acceptable benchmarks for protection of human health (see Chapter 8). However, when maximum soil Pb levels were used in the exposure assessment, a risk of health effects for young children was found in four neighborhoods: Copper Cliff, Coniston, Falconbridge and Sudbury Centre. While the predicted risks were only marginally above the established hazard quotient (HQ) benchmark of 1.0, it was considered appropriate to derive a Sudbury-specific soil risk management level (SRML) for Pb to ensure the protection of children. The term SRML is defined as the average Chemical of Concern (COC) soil concentration within an exposure unit (EU) that corresponded to an acceptable level of risk (U.S. EPA 2001). In other words, the SRML is the exposure point concentration (EPC) in soil within a given community which would yield an acceptable level of risk.

A weight-of-evidence approach was used in the evaluation of health risk estimates and the development of the SRML for Pb as outlined previously in Chapter 8. Based on the available information and literature, a value of 400 mg/kg was recommended as the SRML for Pb for the Greater Sudbury area. The recommended SRML was not a target or clean up level above which soil must be remediated. Rather, it was a guideline to be used to initiate further investigation on a property by property basis. Approximately 5210 soil samples were collected from within the five communities of interest (COI). Of these, only nine samples (or 0.17%) contained Pb concentrations that were equal to or above the SRML of 400 mg/kg. Of the 2164 soils collected in the 0 to 5 cm surface layer, which is most accessible to humans, only three samples (0.14%) were equal to or above the lead SRML of 400 mg/kg.

To provide effective risk management measures, it is important to identify the sources of the contaminant and the exposure pathways. The study determined that the major source of oral/dermal exposure to Pb in the study area was supermarket foods. Supermarket foods represented approximately 45–55% of the total exposure to Pb by the female toddler, with minor variability between communities. This is a source that is unaffected by emissions from local metal production facilities. Household dust was generally the second largest contributor to Pb exposure (15–20%). Some of the Pb in household dust may come from outdoor soil, which in turn may be impacted by metal production emissions. However, Pb concentration levels can be elevated in household dust from historic use of Pb-based paints and other consumer products such as window blinds, gasoline and many other products. Because there are a variety of sources of Pb present in the environment, and exposures are dependent on many factors, it is important to consider all sources when implementing risk reduction measures. Soil ingestion represented a relatively small fraction of the total oral/dermal exposure of lead to area residents (from about 6% to 10% in all communities of interest). Both companies are continuing to take all reasonable measures to reduce the amount of Pb and other particulates emitted by operations. Improvements in dust control under way at Vale and Xstrata Nickel will continue to reduce Pb emissions and contributions to soil and dust within the community.

A comparison revealed that Pb levels in soils and dust in the Sudbury area were similar to levels in other older communities in Ontario. Exposure to Pb is a general community concern throughout the province and specifically in older urban communities due to historic use of Pb-based paints, leaded gasoline and plumbing (Pb pipes and Pb solder). Because of the societal nature of Pb, all of the study partners worked cooperatively on risk management activities related Pb. A significant component of this effort was a public information campaign to inform residents about how to reduce Pb exposure. A toll free line was established in Greater Sudbury for anyone with questions about any of the COC addressed in the study. If a resident had particular concerns with regard to Pb exposure on their property, they would be interviewed by a medical doctor on contract for the project. The medical doctor would determine the age of the house, possible exposures and age of family members living in the home to determine risk factors. If it was deemed appropriate, a member of the MOE (Ministry of the Environment) would collect soil samples from the front and back yards and have them analyzed for Pb. In addition, in Ontario, any person can have their blood Pb tested by their family physician to measure blood Pb levels as an indicator of exposure to Pb. If a home displayed elevated soil Pb levels and potential risk, the mining companies agreed to pay for soil removal and replacement. Two years after the conclusion of the risk assessment, soil removal and replacement was only undertaken at one property in Copper Cliff.

Vale and Xstrata Nickel also continue to take measures to reduce the amount of Pb and other particulates emitted by operations which will reduce contributions to soil and dust within the community. The following general improvements in dust control are being carried out by both companies.

- Baghouses (dust capture system): Improvements to equipment, maintenance and inspections
- Electrostatic precipitators (smelter emissions capture system): Improvements in cells, maintenance programs, stack opacity monitoring, and operator training
- Smelter dust capture and improvements to emission capture capability
- Site and roadways: Comprehensive dust reduction programs, including annual paving, road maintenance/cleaning, and application of dust suppressants
- Material handling: Implementation of best practices
- Re-greening: Implementation of re-greening programs to reduce blowing dust from exposed land surfaces.

Additional detail on emission control efforts is provided in the following two sections.

Figure 12.2 **View of the Community of Copper Cliff and the Vale Smelter Operations**

12.2.2 Vale Risk Reduction Strategies for Ni

Improvements in Dust Control – Copper Cliff Operating Facilities

Vale took immediate action to reduce dust emissions from its facilities to lower the inhalation health risks in Copper Cliff and the western portion of Sudbury Centre. Current efforts to reduce dust emissions are in addition to significant achievements made in dust reduction over the previous three decades. Efforts at reducing dust emissions have benefits in reducing not only Ni emissions, but in reducing all of the COC. Table 12.1 provides a general outline of the steps implemented to reduce Ni and other particulate emissions from the Copper Cliff operating facilities.

Table 12.1 **Particulate Emission Reductions – Vale Smelter and Ni Refinery**

Improvement activity	Key activities completed or under way
Baghouses *(Devices with fabric bags, such as a vacuum cleaner, that filter air)*	A number of activities are/were under way to reduce dust emissions by: • improving baghouse design and dust capture efficiencies • improving efficiency of monitoring and instrumentation • improving operating and maintenance programs
Electrostatic precipitators *(Devices with electrically-charged plates that attract and separate dust from air)*	A number of activities are under way to reduce dust emissions from the electrostatic precipitators, with most activities focused on upgrades to precipitators to improve design and capture efficiencies
Fluid bed roasting (FBR) *(A key part of the Copper Cliff Smelter that converts nickel sulfide feeds to nickel oxide products)*	A multi-year multi-phased project was initiated to construct improved cleaning facilities for gases originating from the FBR at the smelter. This is expected to improve dust capture efficiencies
Bulk converter off-gas capture *(A key part of the Copper Cliff Smelter that removes Fe from furnace matte)*	A prototype dual-mouth converter was installed at the #8 converter. If successful, the outcome from this project would lay the groundwork for possible future gas capture and dust removal throughout the entire converter aisle
Emission testing	Testing provides an accurate measure of dust emissions from individual sources. The outcome from this work provides focus on areas that can provide the best benefits for reducing Ni exposures

Improvements in Dust Control – Copper Cliff Smelter/Nickel Refinery Yard and Lands

A general outline of steps currently under way to reduce fugitive emissions including Ni from the Copper Cliff Smelter and Copper Cliff Nickel Refinery operations is provided in Table 12.2.

Table 12.2 **Particulate Emission Reductions – Vale Smelter Yard and Lands**

Improvement activity	Key activities completed or under way
Stockpiles of products and materials	A number of activities were initiated to reduce dusts originating from material and product stockpiles by: • applying Best Management Practices for stockpile management • applying Best Management Practices for crushing, screening and conveying • developing a land use policy to better control location of products and materials • examining the feasibility of enclosed storage facilities for stockpiling and handling of process materials
Roadways and truck haulage	Activities have been initiated to reduce dusts originating from roadways, e.g. by paving high traffic areas, improving dust control measures on roadways and applying Best Management Practices for truck haulage Since 2005, approximately 5 km of roads were paved at a cost of over $8M
Re-greening	Re-greening work continues, to reduce dusts generated from exposed soil surfaces on the nearby smelter lands Since 2006, approximately $7M was spent on re-greening efforts, including re-vegetation of approximately 30 acres of slag and the planting of approximately 1700 trees and 1000 shrubs
Slag	Improvements are being planned for reducing dust emissions from the Fisher–Wavy slag processing operations and from the slag skull area

Improvements in Ambient Air Monitoring

Before the Sudbury Soils Study, the MOE conducted ambient air monitoring of dust emissions in Copper Cliff and in Sudbury. In 2003, the companies agreed to fund an expanded ambient air monitoring program that included sampling of particulate matter less than 10 μm in diameter (PM_{10}). PM_{10} is thought to be a relevant parameter for quantifying dust that could be inhaled into a person's lungs. The SARA Group initiated the air monitoring program across the City of Greater Sudbury including 20 monitors at 10 locations (see also Chapter 6). Air samples were collected from October 2003 to September 2004 and the data were used in the HHRA. Of the 10 monitoring stations, two were placed in close proximity to Vale's Copper Cliff Smelter. After 2004, Vale continued monitoring at two of the stations used during the risk assessment by the SARA Group. In response to recommendations in the HHRA, the monitoring program was expanded in 2006 to include three additional stations. A 6th particulate monitoring station was established in Copper Cliff in September 2007. An example of an air monitor is shown in Figure 12.3. Information gained from these particulate monitoring stations will allow the company to accurately measure its performance in implementing dust control measures and reducing ambient levels of Ni in PM_{10} dusts.

In May 2007, Vale initiated additional monitoring, beyond the standard application of fixed monitoring stations that are typically used. Monitors were mounted on a truck and trailer system which allowed for a more flexible assessment of particulate emissions than could be readily determined using fixed station monitors. This system adds value to the assessment of sources and characteristics of particulate emissions from the Copper Cliff complex and the associated industrial properties.

Figure 12.3 **Air Monitoring Station in Sudbury**

12.2.3 Xstrata Nickel Improvements in Dust Emission Controls

Although no unacceptable human health risks were identified in the community of Falconbridge related to inhalation, Xstrata Nickel committed to further emission reductions as part of a continual improvement program and to comply with future emission requirements. Environmental improvements at Xstrata Nickel are addressed through its Environmental Management System (EMS), formally registered under ISO 14001 (International Standard) since 2004. Particulate emissions have been defined as a Significant Environmental Aspect within the EMS. A number of activities are aimed at addressing reductions in particulate emissions. An inventory of particulate emission sources from the plant and surroundings was conducted in 2001 and 2004. The following were identified as the most significant sources of particulates:

- stacks, vents and exhausts from various plant facilities
- material handling activities
- paved and unpaved roads
- outside erosional sources.

An assessment of each emissions source was conducted to determine its significance and to identify possible operational control measures. Formal emissions modeling was completed as part of an overall site investigation, to establish priority metals and to evaluate potential impacts within the community of Falconbridge. This work was conducted in association with the established ambient air quality monitoring, operating in the community of Falconbridge for approximately 20 years. In the Falconbridge area, total suspended particulates and PM_{10} (fine particulates) are monitored at two stations and all data collected continue to be shared with the MOE.

Figure 12.4 **View of Community of Falconbridge and the Xstrata Nickel Smelter**

For the Xstrata Nickel facilities in Falconbridge, the EMS provides the framework for managing regulatory requirements associated with site-wide particulate emissions. The company developed a series of internal inspection and monitoring points to formally measure and report on progress for a number of emission control initiatives. Table 12.3 provides a summary of these initiatives to reduce particulate emissions.

Table 12.3 **Xstrata Nickel Particulate Emission Management Programs**

Improvement activity	Key activities completed or under way
Baghouses	• Optimization of the baghouse inspection program • Installation and completion of broken bag detection systems • Replacement of specific baghouses (075–076)
Electrostatic precipitators	• Conducted a Cottrell Plant opacity/optimization study • Improved stack opacity measuring, monitoring, and reporting • Upgrading of cells in Roaster ESPs and Cottrell Plant • Implementing training program for Cottrell Plant operators and maintenance personnel

Improvement activity	Key activities completed or under way
Converter aisle emissions reduction	• Conducting concept studies for improved dust collection in converters 7 and 8 • Develop plan to collect process and converter aisle emissions; design large baghouse for entire converter aisle emissions • Installation of new hood for No. 7 converter • Replacement of hood for No. 8 converter
Outside fugitive emissions	• Paving of roads according to annual plan • Improvement of management activities on paved and unpaved roads, including operation and maintenance of new road sweeper • Selection of new dust suppressants with on-site storage • Development of 20 year slag management program, with 5 year development planning to coincide with closure • Implementation of material handling operation controls, including construction of various custom feed facilities
Site reclamation	• Establishment of site footprint for reclamation program • Completion of 5 Year Reclamation Plan • Planting of 60,000 to 100,000 trees annually • Completing 50 ha of reclamation per year • Implementing reclamation program for flux pit (i.e. aggregate operation)

12.2.4 New Emission Reduction Requirements and Initiatives

In addition to the efforts described above, Vale and Xstrata Nickel continue to address changing federal and provincial air regulations and requirements. Health protective air standards are being developed under Ontario's new regulation *Air Pollution – Local Air Quality (O. Reg. 419/05)* (MOE 2005). Both Vale and Xstrata Nickel are committed to complying with this regulation. Both companies have completed required Emission Summary Dispersion Modeling Reports, a key component of assessing emission sources and predicting ground level concentrations.

Vale's Atmospheric Emissions Reduction (AER) project aims to significantly reduce metal-containing dust and SO_2 emissions from its Sudbury smelting and refining operations. The project is in the feasibility stage and is expected to cost $1.5 to $2 billion. Xstrata's Emissions reduction plan aims to reduce sulfur dioxide and metals emissions from the smelter and materials handling operations. Vale and Xstrata Nickel are also addressing federal requirements for Pollution Prevention (known as P2 Planning). Criteria were rolled out in 2006 for the facilities, including standards set for Ni, total suspended particulates (TSP), greenhouse gases, and SO_2.

Vale and Xstrata Nickel are required to submit a variety of reports to regulatory agencies concerning their dust emissions. Annual reports are submitted as part of the National Pollutant Release Inventory (NPRI) and P2 Planning. Other reports are submitted as required by various approvals, orders and agreements with provincial and federal regulatory agencies. Public input and consultation is also conducted by both companies to inform residents of risk management procedures and emission reduction programs. Vale is committed to continuing ongoing consultation with Sudbury residents through community information meetings and participation on the Copper Cliff Liaison Committee, which meets bi-monthly and the Copper Cliff Community Action Network, which meets monthly. The company also holds an annual Open House in Copper Cliff every September, which is open to all members of the public. Xstrata Nickel has maintained ongoing consultation on air quality improvements with the Falconbridge Citizen Committee, and remains committed to ongoing consultation on a regular basis.

12.3 Ecological Risk Management

Environmental impacts associated with decades of airborne SO_2 and metal particle emissions from smelting in Greater Sudbury are described in Chapter 2. The virtual elimination of vegetation over tens of thousands of hectares has had a profound, direct ecological impact on the plant and organism communities present in the soils that subsequently eroded. Dramatic losses of vegetation cover also resulted in the loss of habitat for the numerous forest-dwelling animal species. For decades, plant species richness was reduced to mainly those species that were either metal tolerant or were able to persist on micro-sites topographically shielded from SO₂ fumigations. Animal species were reduced to those few for which the barren and semi-barren conditions provided preferred habitat. While most of the industrially generated impacts occurred within a few years, a few have taken a long time to fully manifest themselves. Red maples, for example, have taken decades to succumb to the onslaught of SO₂ fumigations and elevated metal soil levels and are now less common as a component of the vegetation than they were in the early 1980s (Amiro and Courtin 1981; Sinclair 1996).

This section starts by describing the history of Greater Sudbury's re-greening efforts and the ecological benefits that these efforts have created. Next, the development of a Biodiversity Action Plan is outlined. As the Action Plan is implemented, ecological recovery in Greater Sudbury will enter a new phase that should lead to the establishment of healthy, diverse, self-sustaining ecosystems on formerly barren lands. Social aspects of ecological recovery, including research, education and community engagement in re-greening actions and ecological monitoring are also discussed.

12.3.1 Greater Sudbury Re-greening Efforts – from 1917 to 2009

Greater Sudbury's re-greening efforts, which have been described in detail elsewhere (Lautenbach 1985; Winterhalder 1996, 2002; SARA Group 2008b), are highlighted in the following paragraphs. Early re-greening efforts by the mining companies included importing soil and landscaping previously denuded sites to improve site appearance and encourage settlement. Nickel Park in Copper Cliff and Centennial Park in Falconbridge, both established on severely damaged land, were reclaimed by this method in 1917 and 1967, respectively. INCO's (now Vale) Agricultural Department pioneered direct seeding techniques using agricultural equipment in the 1950s and both Vale and Xstrata Nickel (then Falconbridge Limited) have reclaimed thousands of hectares of tailings and other relatively flat areas using this method. Tree planting by Vale and Xstrata Nickel also began in the 1950s when loam was added to the planting holes to improve growth.

To address land reclamation on the thousands of hectares of barren areas outside of mining company property, initial steps were undertaken between the Ontario Department of Lands and Forests' Timber Branch and Laurentian University's Biology Department in 1969 under the Sudbury Environmental Enhancement Program (SEEP). The Program's mandate included determining whether soils from denuded areas could support vegetation with or without soil amendments. Tree planting experiments involving thousands of seedlings were undertaken in 1969 and 1970 on a barren site and a semi-barren site. Results of these experiments showed that trees could only survive on amended soils on barren sites and growth could be improved with amendments on semi-barren sites. Additional experiments undertaken through SEEP in 1971 clearly revealed that the most significant factor limiting plant growth on the barrens was low pH combined with elevated soil metal concentrations. Secondary factors included phosphorus and nitrogen levels depending on the site. The addition of crushed limestone to the soils raised the pH, which decreased metal uptake by the plants allowing trees to grow unimpeded. Further treatments assuring sustained tree growth included the addition of a commercial fertilizer (6-24-24) and seeding with a grass/legume mixture to ameliorate the ground-layer microclimate.

The Regional Municipality of Sudbury, at the request of the Ontario Ministry of Natural Resources, formed a multi-stakeholder committee in the mid-1970s to change Sudbury's reputation of a barren and inhospitable environment. This was accomplished by 1) vegetating barren areas around urban areas and along highways and railways, 2) making trees available for residents for planting on private property, and 3) encouraging the revegetation of vacant land around urban areas. Various trials were established through the oversight of the Tree Planting Committee using the same technique of liming, fertilizing, seeding the grass/legume mixture and tree planting. The results of these trials proved encouraging enough for the Regional Municipality of Sudbury to establish a Regional Land Reclamation Program in 1978. Initially, the Land Reclamation Program employed hundreds of people that were affected by the 1977 layoff of 3500 employees from INCO. Employing this large labor force was made possible with the assistance of various Federal and Provincial funding programs.

In the initial years of the Land Reclamation Program, liming, fertilizing, seeding the grass/legume mixture and tree planting took place primarily on barren sites along major transportation routes, including around the airport. Emphasis on these sites was intended to create an immediate and dramatic improvement for residents and visitors to Sudbury. After several years, sites along roads were reclaimed and the focus shifted to areas farther away yet still visible to travelers. Helicopters and trains were sometimes employed to carry lime bags to sites where access was difficult. Additional activities of the Land Reclamation Program included large-scale removal of woody debris for beautification purposes, native seed collection, planning and mapping of the operations, monitoring and assessment, experimental composting, and transplanting of trees, shrubs and herbaceous plants.

Risk Management

When the City of Greater Sudbury was formed in 2000 as a result of municipal amalgamation, it continued the commitment to the Land Reclamation Program, now renamed the Re-greening Program. Since 1978, the Re-greening Program has limed over 3400 hectares, fertilized and seeded over 3100 hectares, planted over 9.2 million trees, and created over 4500 temporary employment opportunities. Conifer seedlings have been planted on all of the land that received lime as well as on an additional 15,000 hectares. Calculations using a geographic information system (GIS) show, however, that 35,000 hectares of land within the original impacted zone have never been limed or planted with conifer seedlings. This does not include vast tracks of land that are considered 'behind the mining company gates' (i.e. areas for which mine closure plans exist), urban land, or rights-of-way. While not all of the remaining 35,000 hectares of land require lime, most probably need to be planted with conifer tree seedlings to at least supplement the tree seedlings that have naturally colonized some sites.

The majority (89%) of tree seedlings planted through the Re-greening Program consist of four conifer species: jack pine, red pine, white pine and white spruce. A number of additional coniferous and deciduous tree species as well as shrub species have been planted as seedlings over the years (Table 12.4).

Table 12.4 **Woody Plant Species Planted from 1979 to 2010**

Species scientific name	Species common name	Number planted	Percent of total
Trees			
Pinus banksiana	Jack pine	2,549,914	27.5
Pinus resinosa	Red pine	2,439,985	26.3
Pinus strobus	White pine	1,621,984	17.5
Picea glauca	White spruce	1,639,638	17.7
Quercus rubra	Red oak	206,755	2.2
Thuja occidentalis	Eastern white cedar	191,748	2.1
Picea mariana	Black spruce	190,089	2.1
Robinia pseudoacacia*	Black locust	122,746	1.3
Larix laricina	Tamarack	114,558	1.2
Acer (rubrum, saccharum, saccharinum, spicatum, pensylvanicum)	Maples (red, sugar, silver, mountain, and striped)	42,718	0.46
Fraxinus (pensylvanica, americana, nigra)	Ash (red, white, black)	28,809	0.31
Picea abies*	Norway spruce	20,240	0.22
Larix (decidua*, kaempferi*)	Larch (European, Japanese)	15,120	0.16
Tsuga canadensis	Eastern hemlock	8139	0.09
Abies balsamea	Balsam fir	5653	0.06
Eleagnus angustifolia*	Russian olive	5175	0.06
Betula alleghaniensis	Yellow birch	4567	0.049
Quercus macrocarpa	Bur oak	2269	0.024
Fagus grandifolia	American beech	376	0.004
Carya cordiformis	Bitternut hickory	200	0.002
Pinus nigra*	Austrian pine	95	0.001

Species scientific name	Species common name	Number planted	Percent of total
Shrubs			
Cornus sericea	Red osier dogwood	15,510	0.167
Caragana arborescens*	Siberian pea shrub	7410	0.080
Diervilla lonicera	Dwarf bush-honeysuckle	5950	0.064
Sambucus pubens	Red elderberry	3850	0.042
Amelanchier sp.	Serviceberry (Amel. Sp.)	2513	0.027
Viburnum trilobum	Highbush cranberry	2170	0.023
Pinus mugo*	Mugho pine	1950	0.021
Viburnum lentago	Nannyberry	1541	0.017
Cornus alternifolia	Alternate leaf dogwood	1415	0.015
Sambucus canadensis	Common elderberry	1178	0.013
Rhus typhina	Staghorn sumac	910	0.010
Spiraea alba	Narrow-leaved meadowsweet	896	0.010
Physocarpus opulifolius	Ninebark	803	0.009
Prunus pensylvanica	Pin cherry	800	0.009
Cornus rugosa	Round-leaved dogwood	450	0.005
Spiraea latifolia	Large-leaved meadowsweet	300	0.003
Myrica gale	Sweet gale	300	0.003
Ilex verticillata	Winterberry	300	0.003
Viburnum cassinoides	Wild raisin	290	0.003
Amelanchier sanguinea	Red-twigged serviceberry	252	0.003
Alnus viridis ssp. *crispa*	Green alder	250	0.003
Gaultheria procumbens	Wintergreen	250	0.003
Rosa palustris	Swamp rose	222	0.002
Spiraea tomentosa	Steeplebush	210	0.002
Shepherdia canadensis	Buffaloberry	200	0.002
Cephalanthus occidentalis	Buttonbush	200	0.002
Prunus virginiana	Choke cherry	200	0.002
Arctostaphylos uva-ursi	Bearberry	199	0.002
Aronia melanocarpa	Black chokeberry	28	0.0003
Prunus nigra	Canada plum	28	0.0003

*Species not native to Ontario.

Reductions in local SO_2 levels due to the construction of taller smelter stacks and improved industrial capture technologies have considerably lessened the environmental pressure allowing spontaneous colonization by some wind-dispersed plant species on soil treated with crushed limestone. Winterhalder (1996) considered the liming of soils within the barren zones as the 'trigger factor' that had an immediate detoxifying effect and created positive feedback loops throughout the system. Emission reductions along with the soil liming and fertilization also allowed the planted trees to thrive unimpeded, creating closed canopy conditions along with the colonizing birches and poplars and the relict red oaks. Figure 12.5a and 12.5b shows the dramatic results of the City's re-greening efforts.

Figure 12.5 (a) **Photo Taken in 1981 near Coniston, Greater Sudbury, Ontario Showing the Barren Landscape;**

Figure 12.5 (b) **Photo Taken in 2008 at Same Location as the Photo in (a)**

Although stands of trees now occur on many formerly barren areas, these stands typically contain few plant species, often only the species present is the grass/legume seed mix used in the initial stages of re-greening accompanied by other non-native species (Rayfield et al. 2005). In certain areas, the plants resulting from the seed mix have created dense mats of mixed grasses making natural colonization by native herbaceous species impossible. The grass/legume seed mix is entirely composed of species that are not native to Greater Sudbury, with the possible exception of Canada bluegrass (Table 12.5). This mix has been traditionally used at a rate of 40 kg/ha because of its relatively low cost and because the species establish predictably well in various soil conditions.

Table 12.5 **Grass/Legume Seed Mixture Used in the City's Re-greening Program**

Grasses (75%)	
Agrostis gigantea	Red top
Festuca rubra	Red fescue
Phleum pratense	Timothy
Poa compressa	Canada bluegrass
Poa pratensis	Kentucky bluegrass
Legumes	
Lotus corniculatus	Bird's foot trefoil
Trifolium hybridum	Alsike clover

Despite over three decades of re-greening and natural ecological recovery following large reductions in ambient SO_2 levels, the Sudbury ERA concluded that terrestrial plant communities in Greater Sudbury have been and continue to be impacted not only by the presence of elevated levels of the COC in the soil, but also by other factors including soil erosion, low nutrient levels, lack of soil organic matter, and/or low pH (SARA 2009). Historic impacts on vegetation by smelter emissions also are probably having a continued influence on re-colonization by animals due to habitat unsuitability.

Managing the residual ecological risks (i.e. remaining impacts on terrestrial plant and animal communities) required a solution that would not only be focused on specific ecological recovery actions, but would also build on the successes of the City's Re-greening Program. It was decided that the development and implementation of a Biodiversity Action Plan for Greater Sudbury would provide the solution for moving forward with local ecological recovery. Such an Action Plan would, it was thought, provide the opportunity to re-engage the community in its commitment to the re-greening challenge. Release of the Biodiversity Action Plan was timed fortuitously with the start of the International Year of Biodiversity in 2010.

12.3.2 Biodiversity Action Plan Development

The process followed in the development of the Biodiversity Action Plan involved a variety of community engagement exercises aimed at eliciting key priorities and concerns relating to local biodiversity (Figure 12.6). The process, which was initiated in the 2nd quarter of 2009, took several months to complete and resulted in the release of the final Biodiversity Action Plan by the end of that year.

March 31, 2009
Release of the Sudbury soils study's ecological risk assessment

PUBLIC INPUT

June 6, 17 and 24, 2009
Biodiversity stakeholder involvement sessions

July 7, 8 and 9, 2009
Biodiversity 'have your say' workshops

July to end of September, 2009
Biodiversity feedback questions (website)

Conducted in August, 2009
Biodiversity telephone survey

October 30 to November 30, 2009
Biodiversity action plan draft review period

December 23, 2009
FINAL BIODIVERSITY ACTION PLAN RELEASE

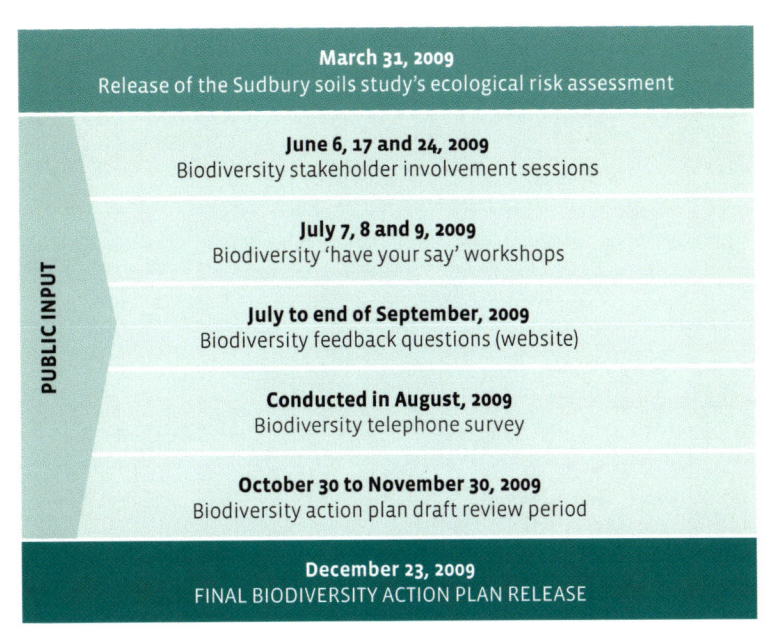

Figure 12.6 **Development Process of the Greater Sudbury Biodiversity Action Plan**

Biodiversity Stakeholder Involvement Sessions

Advertisements were placed in local newspapers asking for volunteers to be part of Biodiversity Stakeholder Involvement Sessions aimed at hearing from members of the community on key issues relating to biodiversity. Thirty people replied to the ads and all were selected to participate in three sessions, each moderated by a professional facilitator. Participating volunteers included professors and students from the local university (Laurentian University), professors from the two local colleges, teachers, a retired leader of a local mining union, homemakers, members of local naturalist groups, retired individuals, and employees of two provincial ministries. The first session was designed to immerse the participants in the topic by taking them to two sites in Greater Sudbury; one impacted by past mining and smelting and one not impacted. This half-day excursion was useful for the participants to get to know each other as well as to experience the difference between impacted and non-impacted sites. Questions were answered by invited experts as they came up but the experts were instructed not to lead the participants and to minimize any lecture-style communications to the group. It was important for the participants to reflect on the two starkly different environments based on first-hand experience, which would hopefully build common ground in the two sessions to follow.

Two additional sessions were held in the evenings, each 1 week apart. With the assistance of the facilitator, the participants were involved in various exercises designed to elicit their thoughts and impressions on what they valued in the natural environment. The facilitator kept the group working at a rapid pace by clearly structuring each exercise, dividing the participants into small groups of 6 to 8 people and ensuring that no one participant dominated group discussions. Written material emanating from the sessions was diligently recorded and transcribed into electronic format. Participants expressed that the sessions had worked very well and felt that they had been given the opportunity to express their thoughts and to be heard. Participants were then asked if they would like to be reunited several months later to get an advanced preview of the Draft version of the Biodiversity Action Plan before the document was released to the general public for comments. All participants were very enthusiastic about being given an opportunity to see if their input had been considered in the development of the Biodiversity Action Plan. Several months later when the group was reassembled, they were indeed appreciative of the attention paid to their comments and to the breadth of issues addressed by the Draft Action Plan.

Biodiversity 'Have Your Say' Workshops and Website

In addition to the intensive sessions with the Biodiversity Stakeholders, three workshops were offered to the general public on three consecutive evenings. The workshops were intended as opportunities for members of the public to express concerns, ask questions, and provide input on matters pertaining to biodiversity in the City. The workshops were professionally facilitated. Although only about a dozen people participated, the information gained from these people was useful and contributed to the development of the Action Plan.

A website was also developed for the sole purpose of obtaining community members' input on matters pertaining to biodiversity. Again, although this particular tool was not used extensively by the public, the information gained was reflected in the Biodiversity Action Plan.

Biodiversity Telephone Survey

A telephone survey was conducted by a professional polling firm, which asked 602 residents of Greater Sudbury a series of 16 questions on matters pertaining to biodiversity and the value they place on various components of biodiversity. A sample of 602 provides for a confidence interval of better than +/- 4%, 19 times out of 20 (95% level of confidence). The survey sample was developed as a stratified random sample of the seven former area municipalities (before municipal amalgamation) to ensure that a representative sample of the entire Greater Sudbury population was achieved.

This survey began by asking respondents to indicate their level of familiarity with the term 'biodiversity'. Results show that 63% of respondents 'had heard of it' (i.e. the term 'biodiversity'), were 'familiar with it' or had 'used the word in conversation'. When asked how important biodiversity was to them within the City of Greater Sudbury, 76% of respondents rated biodiversity as a 3 or higher, where 1 is 'not at all important' and 5 is 'very important'. Respondents were asked to indicate using a five-point scale, whether they felt there had been changes in the natural environment over the last 20 years in the City of Greater Sudbury. When asked which environmental action was most desirable to be undertaken, 49% of respondents selected 'lake and stream clean-up'. This was followed closely by 're-greening' (46%), 'education of the public' (44%) and 'wildlife habitat creation' (38%).

Draft Biodiversity Action Plan Review Period

The draft Biodiversity Action Plan was presented to the Biodiversity Stakeholder Involvement session participants. This offered an opportunity for participants to provide comments first-hand, many of which were incorporated into the draft that was released to the general public several days later. Following a 1 month review period, comments were received from many individuals and groups and the Biodiversity Action Plan was further modified to reflect the comments. Once finalized, the Action Plan was made available online (www.greatersudbury.ca/biodiversity) and hard copies were placed in all 12 municipal library branches and libraries of all local schools and post-secondary educational institutions. Although a final version was released, the Biodiversity Action Plan is intended to be a 'living document' to be modified as needed based on new ecological findings and community expectations.

Contents of the Biodiversity Action Plan

The Biodiversity Action Plan process resulted in the release of 'Living Landscape – A Biodiversity Action Plan for Greater Sudbury' in December 2009. The Action Plan is written for a general audience and, although comprehensive in scope, is intended to only highlight many of the issues surrounding local biodiversity. The Action Plan stresses the need to build on what has already been accomplished through the Re-greening Program and that much land has yet to be limed and trees planted. Some planted areas are now over 30 years old and the target for these stands of trees is now the development of healthy, diverse, and self-sustaining forest ecosystems. The Action Plan addresses issues relating not only to the actions needed to accomplish this, but also the need to do so in the context of watersheds, climate change and, in some instances, urban environments. The latest version of the Action Plan can be viewed at www.greatersudbury.ca/biodiversity.

Importantly, the Action Plan also addresses the social dimension by including sections on education, research and community engagement recognizing that local ecological recovery efforts reflect a strong sense of shared responsibility. Since its inception, the City's Re-greening Program has offered multiple opportunities for community participation in liming and tree planting and City personnel have offered tours of the successes of the Program to thousands of people. Re-greening efforts have also attracted considerable research from post-secondary educational institutions. Not only must these efforts continue, but these should also expand and adopt a broader and more comprehensive perspective on ecological recovery.

Finally, the Biodiversity Action Plan recognizes the sizeable and long-term financial commitments to the ecological recovery efforts by Vale and Xstrata Nickel. These commitments not only allow the Re-greening Program to step-up its activities, but also provide resources for additional educational and community engagement efforts.

12.3.3 Implementing the Greater Sudbury Biodiversity Action Plan

12.3.3.1 Restoring Diversity through the Re-greening Program

Native herbaceous and woody plant species that are representative of local forest understory are not common in Greater Sudbury's barrens and semi-barrens, even though some areas were planted with conifers and deciduous trees over three decades ago. Isolated groupings of certain understory woody species, such as beaked hazel (*Corylus cornuta*), mountain maple (*Acer spicatum*), and trailing arbutus (*Epigaea repens*) can be found on some topographically sheltered pockets that protected varying complements of plants from past SO_2 fumigations. These refugia vary in the number of species that survived, likely based on the degree of exposure to SO_2 and subsequent soil erosion. Some refugia appear to have maintained relatively intact species complements, others have maintained only a few species and still others have maintained only 1 or 2 species. Some woody species, such as prince's pine (*Chimaphila umbellata*), American yew (*Taxus canadensis*) and striped maple (*Acer pensylvanicum*) appear to have been eliminated entirely from the industrially impacted areas.

A similar pattern appears for the herbaceous plants on the barrens and semi-barrens, despite decades of growth of localized tree stands. Certain species, such as mayflower (*Maianthemum canadense*), wild sarsaparilla (*Aralia nudicaulis*) and blue-bead lily (*Clintonia borealis*), have been maintained in local refugia and are now likely spreading but numerous other species, such as pink lady's slipper (*Cypripedium acaule*), rose-twisted stalk (*Streptopus roseus*), and violets (*Viola* spp.) are either totally absent or exceedingly rare in the impacted areas.

The absence of so many woody and herbaceous plant species is of concern if the tree stands are to become healthy forests. Boreal understory vegetation is the most diverse component of the boreal plant communities (Roberts 2004) and serves as an ecosystem driver, affecting canopy succession, nutrient cycling and wildlife (see review by Hart and Chen 2006). In response to fire cycles under natural conditions, understory communities form a diverse mosaic across the boreal forest contributing to both temporal and landscape diversity (Weir et al. 2000).

12.3.3.2 Re-greening Program Operations

Planting of Deciduous Trees and Shrubs

Although various deciduous tree and shrub species have been planted over the years through the Re-greening Program, there remains much work that can be done to increase these elements of biodiversity in the developing forest stands on industrially impacted areas in Greater Sudbury. To provide coordinated efforts in seedling plug supply, the City has entered into a 5-year agreement with a propagation nursery. This agreement will allow advanced planning for plantings specified in the Five-Year Operations Plan (latest version can be viewed at www.greatersudbury.ca/regreening). City crews will also step-up seed collection activities to increase local provenance of seedlings and increase the species complement available for planting.

Forest Floor Mats

To increase the species of forest floor herbaceous plants (including bryophytes and lichens), the City's re-greening crew salvaged portions of the herbaceous layer in the summer of 2004 from a natural forest area that was to be cleared for a mine exploration operation. Forest floor vegetation mats were hand-dug to a depth of about 10 cm, placed in trays and transported into the impacted zone. The salvaged plant material was laid on two main sites on former barren land that had been limed and planted with conifers. Several years later, many of the original plant species that were salvaged have survived, thrived and, in some cases, have spread beyond the plots in which they were deposited.

Based on the positive results obtained from the small trial outlined above, the City conducted large-scale vegetation salvage operations in the summer of 2010. Thanks to a partnership with the Ontario Ministry of Transportation (MTO), forest floor mats were dug from the construction corridor of a local highway widening project and transplanted onto formerly barren sites in Greater Sudbury that had received the re-greening treatment decades earlier. From June to October, 2010, the City's re-greening crews hand dug the top 10 cm or so of forest soil (i.e. the mat) containing targeted plant species (Figure 12.7). The forest floor mats, which measured approximately 60 cm by 50 cm, were placed in trays, brought into Greater Sudbury at the end of each day, unloaded and watered. The following morning, the trays were loaded again into the carrier truck and moved to receptor sites. Forest floor mats were hand laid into 4×4 m plots that had been previously selected based on safety and accessibility for crews and appropriate environmental conditions (i.e. deciduous or mixed overstorey in stands at least 25 years old). Each plot was carefully measured, laid out with double corner markers (metal pin and metal nail), and photographed before and after the forest floor mats were placed. Over the season, 250 plots were completed, dispersed on sites strategically located throughout the formerly barren lands. In total, crews hand dug over 12,500 trays, enough to cover an entire American football field (approximately 0.4 ha).

Vegetation surveys were undertaken within all of the 4×4 m plots and their immediate surroundings before and after the forest floor mats were laid down. Plant species and their relative abundances were recorded using the Braun-Blanquet method (Kent and Coker 1992). Surveys were also undertaken for all of the plots using the same method. Additional surveys will be undertaken in the future to document vegetation changes in and around the plots. Careful establishment of vegetation transplant plots should be an attractive element to current and future researchers. Several smaller plots were also established for plants requiring special consideration for transplantation, including pink lady's slipper and wetland species such as round-leaved sundew (*Drosera rotundifolia*) and purple pitcher plant (*Sarracenia purpurea*) all of which are absent within the City's industrially impacted areas. Lady's slippers were moved in the fall and established within their own dedicated plots measuring 1.5×1.5 m or were mixed with the larger 4×4 m plots. Wetland species were moved separately and placed in low-diversity wetland sites. Several 1×1 m plots of lichens (*Cladina* sp.) that are currently very uncommon within the impacted zone were also established to determine their long-term ability to colonize from these plots. *Cladina* lichens were also collected in large paper bags, crushed into fragments of about 1 cm and dropped from planes during the aerial liming and seeding operations. All of the plots established as a result of these techniques will be monitored over time to determine success.

Although transplanting and plant species reintroductions are fraught with difficulties, and low survival and persistence are common (Fahselt 2007; Godefroid et al. 2011), transplants can help to mitigate species loss or speed up the recovery of extensively damaged areas. In a 7-year study, persistence of transplanted understory herbaceous species to a degraded, early-successional woodland in Iowa was 57% averaged across 24 species (Mottl et al. 2006). Our own anecdotal evidence with pilot plots established in 2004 indicates that, after 6 years, species persistence of transplanted species is high. Godefroid et al. (2011) provided a number of recommendations aimed at enhancing the value of plant reintroductions as a conservation tool, including 1) using a higher number of transplants (preferably seedlings rather than seeds) from non-decreasing populations, 2) clearly defining success of the reintroduction efforts, including measures of reproduction, and 3) consistent long-term monitoring after reintroduction. The forest floor mats in Greater Sudbury were transplanted from an area that would be destroyed, to an industrially damaged area that would likely take a century or more to recover naturally. None of the transplanted species are rare or threatened, thus likely do not have some of the exacting requirements that may have contributed to some of the reported failures in the literature. Finally, some of the 250 plots that were established in Greater Sudbury in 2010 will be monitored and studied for years to come to assess persistence and reproductive success of the original plants and their progeny.

Figure 12.7 **City Re-greening Crew Harvesting Forest Floor Herbaceous Plants at Highway Construction Site in 2010**

12.3.3.3 Education, Research and Community Engagement

Since its beginnings, the City's Re-greening Program has nurtured close ties with visiting dignitaries and groups, local community groups, educational institutions and the general public by offering opportunities to participate in liming and tree planting and by offering tours of the re-greening legacy to thousands of individuals in groups of various sizes. In 2002, for example, the City's Re-greening Program arranged a liming and tree planting activity for hundreds of youth that had come to Toronto for World Youth Day to be part of the visit of Pope John Paul II to Canada, but had specifically requested to take part in Greater Sudbury's re-greening efforts.

To help share its environmental recovery efforts more broadly, in 1999, the City commissioned the production and publication of a hard-cover book entitled '*Healing the Landscape*' that tells the story of the re-greening efforts (Ross 2001). This book proved so popular that a revised edition was released in 2008 in celebration of the City's 125th Anniversary. The Re-greening Program also produces annual reports and other publications available through www.greatersudbury.ca/regreening.

Recovery efforts in Greater Sudbury have been the subject of numerous research efforts published in scientific papers, books and conference proceedings. Laurentian University has regularly hosted two conferences, Sudbury Restoration and Mining and the Environment, that have included the results of many graduate students and professors on local environmental recovery. In 2002, Vale and Xstrata Nickel commissioned the Sudbury Soils Study, a 7-year study that resulted in world-class human health and ecological risk assessments.

Although the City's Re-greening Program will continue to offer opportunities for community involvement in liming and tree planting, the new emphasis on biodiversity allows for new directions in community engagement and education. In 2010, for example, at the request of a local high school, forest floor mats collected by the Re-greening crew were established on suitable sites on the school property by students. The mats will be monitored annually by current and future students with assistance from City staff. Students from other schools also participated in the establishment of some of the 250 forest floor plots in 2010 through the local chapter of Roots and Shoots, Jane Goodall Institute's worldwide initiative aimed at environmental involvement of youth.

Over the coming years, the City, in collaboration with community groups, will also attempt to broaden involvement of the community in monitoring various aspects of Greater Sudbury's ecological recovery through a number of initiatives. Two such initiatives involved monitoring for amphibians and a bird species. The FrogFind and the Whip-poor-will Survey, were initiated in 2011 (see www.greatersudbury.ca/biodiversity). The intent of these two initiatives is to introduce citizens to biodiversity monitoring through relatively simple surveys that provide useful information on the recovery of Greater Sudbury's formerly barren lands. Once involved in simple monitoring surveys, some participants may be willing to attempt more demanding monitoring efforts, such as the Cornell Laboratory of Ornithology's Project Feederwatch, Environment Canada's Marsh Monitoring Program, or many other programs coordinated by Bird Studies Canada. Monitoring through citizen science programs aimed at common and familiar species not only provides information on the species involved, but also provides an opportunity to examine values and visions that people hold for their own landscape (Devictor et al. 2010).

By involving the public in monitoring progress in ecological recovery, Greater Sudbury should, over time, build capacity in citizen science projects through the recruitment and nurturing of dedicated volunteers. Recent reviews highlight the importance of volunteer participation in ecological studies, especially for complementing more traditional research aimed at the conservation of biodiversity (Greenwood 2007; Danielsen et al. 2008; Dickinson et al. 2010). Large-scale projects can engage participants in continental or even global data-gathering networks (Bonney et al. 2009) through advances in the cyberinfrastructure (e.g. online databases, training, communications, networking, mapping, etc.) (Dickinson et al. 2010). Volunteer participation in monitoring not only contributes to biodiversity conservation but can also increase scientific literacy and promotes the reconnection between people and nature (Bonney et al. 2009; Devictor et al. 2010). By involving members of the community, a greater sense of ownership and connection with the recovery process is fostered which contributes to project sustainability (Weston et al. 2003).

12.3.3.4 Greater Sudbury Biodiversity Partnership

A separate yet complementary initiative to the Biodiversity Action Plan was the formation of an informal Greater Sudbury Biodiversity Partnership in 2009 between local offices of the Ontario Department of Natural Resources, Ontario Ministry of the Environment, the federal Department of Fisheries and Oceans, the City of Greater Sudbury, Vale, Xstrata Nickel, local naturalist groups and fishing and hunting groups, biology professors at Laurentian University, Cambrian College and Collège Boréal. The intent of the Biodiversity Partnership is to enhance opportunity for networking between these various parties on matters relating to biodiversity, which should lead to synergies for special habitat management and monitoring projects, better communication to the general public, and recruitment opportunities for local naturalist groups.

The Biodiversity Partnership will be maintained through regular breakfast meetings and e-newsletters. Already the Partnership participated actively by hosting booths at several events held in 2010 to celebrate locally the International Year of Biodiversity. A regular Greater Sudbury Biodiversity Forum open to the public is being planned starting in 2012. The Forum will provide an opportunity for undergraduates, graduates, professors, naturalists, and government staff to present research or project findings and to communicate information relating to species at risk or invasive species. The Forum should provide an important venue for recruiting new volunteers to help build local capacity for monitoring ecological recovery in Greater Sudbury.

12.4 Conclusions

The Greater Sudbury community has been actively engaged in ecological recovery for over three decades. In that time, thousands of hectares have been limed, fertilized and planted with tree seedlings. Planted conifer trees have now merged with naturally colonizing hardwood species forming a closed canopy in the older stands, allowing the City's Re-greening Program to turn its attention to increasing understory plant diversity through a variety of techniques. Although tens of thousands of hectares remain to be limed and planted, earlier efforts have now created the conditions for the next phase in the development of healthy, diverse forest ecosystems on once barren land.

Complementing the community's efforts in changing its living environment, the City will continue to provide opportunities for community-based monitoring of local ecological recovery. Monitoring by volunteers allows a deeper sense of ownership of the ongoing ecological recovery and helps build local capacity for citizen science projects. These types of projects have recently received widespread recognition for their contribution to the conservation of biological diversity. By addressing the financial, operational, educational, research, and social aspects of ecological recovery, the Greater Sudbury Biodiversity Action Plan provides a holistic model for community renewal.

12.5 Acknowledgements

The authors wish to acknowledge and thank current and past members of VETAC, the City's Re-greening Advisory Panel, for their dedicated efforts in promoting and directing local ecological recovery since 1978.

12.6 References

Amiro BD, Courtin GM (1981) Patterns of vegetation in the vicinity of an industrially disturbed ecosystem, Sudbury, Ontario. Can J Bot 59:1623–1639

Bonney R, Cooper CB, Dickinson J et al. (2009) Citizen science: a developing tool for expanding science knowledge and scientific literacy. BioScience 59:977–984

Danielsen F, Burgess ND, Balmford A et al. (2008) Local participation in natural resource monitoring: a characterization of approaches. Conserv Biol 23:31–42

Devictor V, Whittaker RJ, Beltrame C (2010) Beyond scarcity: citizen science programmes as useful tools for conservation biogeography. Divers Distrib 16:354–362

Dickinson JL, Zuckerberg B, Bonter DN (2010) Citizen science as an ecological research tool: challenges and benefits. Annu Rev Ecol Evol Syst 41:149–172

Fahselt D (2007) Is transplanting an effective means of preserving vegetation? Can J Bot 85:1007–1017

Godefroid S, Piazza C, Rossi G et al. (2011) How successful are plant species reintroduction? Biol Conserv 144:672-682

Greenwood JJD (2007) Citizens, science and bird conservation. J Ornithol 148(Suppl 1):S77–S124

Hart SA, Chen HYN (2006) Understory vegetation dynamics of North American boreal forests. Crit Rev Plant Sci 25:381–397

Kent M, Coker P (1992) Vegetation description and analysis: a practical approach. CRC Press, Boca Raton, FL, 354 pp

Lautenbach WE (1985) Land Reclamation Program 1978–1984. Regional Municipality of Sudbury. 65 pp. www.greatersudbury.ca/regreening. Accessed 28 February 2010

MOE (2005) O. Reg. 419/05. Air pollution – Local air quality. www.ene.gov.on.ca/envision/AIR/regulations/localquality.htm. Accessed 2 August 2010

Mottl LM, Mabry CM, Farrar DR (2006) Seven-year survival of perennial herbaceous transplants in temperate woodland restoration. Restor Ecol 14(3):330–338

P2 Planning. www.ec.gc.ca/NOPP/P2TUT/en/prevention/index_e.html. Accessed 15 January 2011

Rayfield B, Anand M, Laurence S (2005) Assessing simple versus complex restoration strategies for industrially disturbed forests. Restor Ecol 13(4):639–650

Roberts MR (2004) Response of the herbaceous layer to natural disturbance in North American forests. Can J Bot 82:1273–1283

Ross N (2001) Healing the landscape. Published by City of Greater Sudbury, Friesens Printers, Canada 124 p

SARA Group (2008a) Sudbury soils study volume II: human health risk assessment, part A. Prepared by SARA Group. www.sudburysoilsstudy.com. Accessed 10 February 2009

SARA Group (2008b) Sudbury soils study volume I: Background, study organization and 2001 soil survey. Prepared by SARA Group. www.sudburysoilsstudy.com. Accessed 28 February 2009

SARA Group (2009) Sudbury soils study volume III: Ecological risk assessment. Prepared by SARA Group. www.sudburysoilsstudy.com Accessed 10 March 2011

Sinclair A (1996) Floristics, structure and dynamics of plant communities on acid, metal-contaminated soils. Unpublished M.Sc. thesis, Laurentian University, Sudbury, Ontario, 95 pp

Sudbury Soils Study. www.sudburysoilsstudy.com. Accessed 8 March 2011

U.S. EPA (2001) Risk assessment guidance for superfund. Volume I: Human health evaluation manual (Part E, Supplemental guidance for dermal risk assessment). Interim review draft – for public comment. EPA/540/99/005. Office of Emergency and Remedial Response, U.S. Environmental Protection Agency. www.epa.gov/superfund/programs/risk/ragse/index.htm. Accessed 15 October 2003

Weir JMH, Johnson EA, Miyanishi K (2000) Fire frequency and the spatial age mosaic of the mixed-wood boreal forest in western Canada. Ecol Appl 10:1162–1177

Weston M, Fendley M, Jewell R, Satchell M, Tzaros C (2003) Volunteers in bird conservation: insights from the Australian Threatened Bird Network. Ecol Manage Res 4:205–211

Winterhalder K (1996) Environmental degradation and rehabilitation of the landscape around Sudbury, a major mining and smelting area. Environ Rev 4:185–224

Winterhalder K (2002) The effects of the mining and smelting industry on Sudbury's landscape. In: The physical environment of the City of Greater Sudbury, pp. 145–173. Ontario Geological Survey, Special Volume 6. Ontario Geological Survey, Greater Sudbury, Ontario

12.7 Appendix Chapter 12: Abbreviations

AER, atmospheric emissions reduction
CGS, City of Greater Sudbury
COC, chemical of concern
COI, community of interest
EMS, environmental management system
EPC, exposure point concentration
ERA, ecological risk assessment
EU, exposure unit
FBR, fluid bed roasting
GIS, geographic information system
GSA, Greater Sudbury Area
HHRA, human health risk assessment
HQ, hazard quotient
ISO, International Standards Organization
LFH, litter fermentation humus layer
MOE, Ministry of the Environment (Ontario)
MTO, Ministry of Transportation (Ontario)
NPRI, National Pollutant Release Inventory
PM$_{10}$, particulate matter passing through a 10 μm filter
SARA, Sudbury Area Risk Assessment
SEEP, Sudbury Environmental Enhancement Program
SRML, soil risk management level
TSP, total suspended particulates
VETAC, Vegetation Enhancement Technical Advisory Committee

13.0 Public Consultation and Risk Communication

DOI: 10.5645/b.1.13

Authors

Marc Butler
Xstrata Nickel
Falconbridge, ON Canada P0M 1S0
Email: MButler@xstratanickel.ca

John Hogenbirk, M.Sc
720 Beatrice Crescent, Sudbury, ON Canada P3A 5B5
Email: Jhogenbirk@laurentian.ca

Franco Mariotti
Science North
100 Ramsay Lake Road, Sudbury, ON Canada P3E 5S9
Email: mariotti@sciencenorth.ca

Trevor Smith-Diggins
360 Woolwich Street, Guelph, ON Canada N1H 3W6
Email: Trevor@smithdiggins.com

Glen Watson, M.Sc.
Vale
337 Power Street, Copper Cliff, ON Canada P0M 1N0
Email: Glen.Watson@valeinco.com

Christoher Wren, PhD
MIRARCO
935 Ramsey Lake Road, Sudbury, ON Canada P3E 2C6
Email: cwren@mirarco.org

Table of Contents

13.1 Introduction	405
13.2 Key Community Concerns	405
13.3 Consultation Framework	406
13.3.1 Communications Sub-committee	406
13.3.2 Public Advisory Committee	406
13.3.3 Independent Process Observer (IPO)	408
13.3.4 Primary Spokesperson	411
13.3.5 Communication Plan	411
13.4 Communication Activities	411
13.4.1 Mailing List	412
13.4.2 'Have Your Say' Workshops	413

- 13.4.3 Update Newsletter ... 413
- 13.4.4 Project Website ... 413
- 13.4.5 Telephone Survey ... 413
- 13.4.6 Toll-Free Phone Line and Email ... 414
- 13.4.7 Physicians Package ... 414
- 13.4.8 Individual and Group Meetings ... 414
- 13.4.9 Community Information Sessions ... 414
- 13.4.10 Media Relations ... 414
- **13.5 Risk Communication and Language of the Sudbury Soils Study** ... 415
- **13.6 Concluding Remarks** ... 418
- **13.7 Acknowledgements** ... 419
- **13.8 References** ... 419
- **13.9 Appendix Chapter 13: Abbreviations** ... 419

Tables
- 13.1 Summary of Consultation Activities and Methods ... 412

13.1 Introduction

When the Sudbury Soils Study was initiated in 2001, it was recognized that a strong public consultation and communication effort was a key element to the overall success of the study. This chapter describes the organizational framework and key activities of the public consultation and communication program.

To assist with public consultation, the Technical Committee (TC) formed a Public Advisory Committee (PAC), as well as a Communications Sub-committee (CSC), and retained the services of an Independent Process Observer (IPO). These groups, their roles and activities are described in this chapter. In addition, one primary spokesperson was identified by the TC to liaise with the media and to provide information to the public. Public consultation and community participation were important considerations throughout the study. As part of this consultative approach to communication, a strong emphasis was placed on informing the community on the human health and ecological significance of the study findings. The objectives of the consultation program were defined as follows:

- To foster ongoing public awareness and increase understanding of the goals, objectives and results of the Sudbury Soils Study
- To provide regular opportunities for public consultation and community involvement
- To address questions and concerns from all stakeholders, including identified interest groups, individuals, community and public leaders, and the media
- To carry out the above communications in clear, concise language, and reiterate messages to ensure that they are incorporated into the community's common knowledge base
- To provide members of the public, residents of the communities of interest, and other stakeholders with timely and relevant information relating to technical findings in the Sudbury Soils Study, and the role of the Sudbury Area Risk Assessment (SARA) Group.

13.2 Key Community Concerns

Members of the community identified several issues of concern during the study. These issues were raised through a variety of communications channels including PAC meetings, open houses, workshops, and through the website, telephone and email networks. As in many projects concerning environmental risk, the primary concern for community members involved the health and wellbeing of residents, particularly the children who lived in the study area. Assessing the current environmental conditions and identifying potential health risks to children were important goals of the study. This concern was addressed directly because the female toddler was the primary human receptor considered in the risk assessment. Additional concerns expressed by residents included the following:

- Residential property values may be affected by poor soil conditions
- Occupational exposure and assessment of risk to workers was not considered in the study
- A perception that the risk assessment process would be biased since the mining companies were directly involved and paid for the studies
- The scope of the study should be broader, i.e. all chemical substances present in the City should have been included in the assessment
- The risk assessment did not include a detailed community health study
- The risk assessment did not include direct bio-monitoring of human tissues such as measurement of metals in blood, hair, or other tissues
- Cumulative effects of simultaneous exposure to many chemicals or stressors was not adequately addressed in the risk assessment
- Historical exposures to the metals of concern (20–40 years earlier, when emissions were known to be much higher) were not included in the assessment.

Considerable thought, time and consideration were given to each of these issues by the study team and answers to these questions were provided to the community in different forms including newsletters, at PAC meetings, Q and A's on the website and written responses as an appendix to the human health risk assessment (HHRA) and ecological risk assessment (ERA).

13.3 Consultation Framework

This section describes the formation and function of the Communications Sub-Committee, the PAC, the IPO, and the role of the primary study spokesperson. The PAC and IPO were important components of the study 'checks and balances' put into place to ensure a fair process, and to address public concerns that the study process and results were not influenced by the mining companies funding the risk assessments.

13.3.1 Communications Sub-committee

The CSC was created to provide strategic communications support to the TC and the SARA Group. This committee was comprised of senior communications representatives from each of the member organizations of the TC. Later, a member of the PAC joined the CSC since it was felt that they may provide valuable contributions from their perspective.

Responsibility for chairing the CSC meetings rotated quarterly through each of the member organizations. Meetings were held monthly, and as required throughout the course of the study. As part of its mandate, the CSC developed and reviewed the communications plan for the study, which was implemented by the SARA Group, and approved all public information and materials before dissemination in the community. A key role of the CSC was to develop key messages, news releases and media briefings, and regular newsletters to communicate study findings or progress in plain language, and following the principles of risk communication. Communication efforts on behalf of the SARA Group were led primarily by Mr Trevor Smith Diggins, a risk communication consultant. Detailed Minutes and Actions of each CSC meeting were distributed to each TC member agency, the IPO, and the PAC.

13.3.2 Public Advisory Committee

In 2002, at the early stages of the Sudbury Soils Study, a PAC was formed as a means of soliciting public input and providing a direct link between the study team and the community. The PAC was comprised of 10–12 volunteer residents from the study area, with two seats reserved for members of Whitefish Lake and Wahnapitae First Nation communities. Advertisements were placed in local newspapers requesting applications to join the PAC. In the first few years, members were selected by the TC. In later years, applicants were selected by a panel composed of TC and PAC members. PAC member renewals were handled exclusively by the PAC. In all, 26 citizens served on the PAC. The Terms of Reference for the PAC are available on the study website: www.sudburysoilsstudy.com.

Over 7 years, the PAC held 40 public meetings for sharing information, providing a forum for making presentations, and serving as an opportunity for consultation and public involvement throughout the study process. Meetings were held in various venues throughout the study area in an attempt to generate broad-based community involvement. Meetings were advertised in local newspapers and on the Study's website. Initial PAC meetings were well attended by the public. However, public attendance decreased over the duration of the study. A number of factors could account for this trend, including apathy, fatigue, or growing confidence in the study process. Other opportunities for public interaction were also made available such as workshops or monitoring surveys to collect Sudbury specific exposure data.

Representatives of the TC and the Independent Process Observer also attended all PAC events and meetings. The PAC received regular updates on study progress from the SARA Group, the consultants conducting the risk assessments, at their bi-monthly meetings. It should be noted that, in the beginning, PAC members did not receive any reimbursements for their travel or meal expenses. This changed after the first year where meals were provided at evening meetings and during the last few years of the study PAC members were reimbursed for travel as well as any materials related to printing, i.e. reports.

The PAC played a vital role in the communications of scientific findings to the community. PAC members assisted in development of communications materials, and reviewed all communications documents developed by SARA and the CSC before public release. In this way, the PAC played an important role in the communication of study findings, conclusions, and general progress to the public.

The PAC did not comment on specific technical issues per se, as they recognized they did not possess the necessary scientific expertise. However, the PAC did request special presentations at some PAC meetings to better understand the process and to ask questions of the study team. Both scientific advisors to the TC for the study (HHRA and ERA) also made formal presentations to the PAC and were made available to the PAC to provide independent interpretation of study results.

The PAC prepared a final report on the soils study process and their involvement which is also available on the soils study website. Excerpts from this report are provided below (PAC 2009).

Successful Aspects of the Soils Study and the PAC's Role:
- The role of the PAC was to serve as a conduit for communications between the public and study proponents, and vice versa. The PAC was not the author of these communications nor was the PAC responsible for the scientific or technical direction of the study. After some initial growing pains, the PAC settled into this role
- The PAC achieved greater responsibility for its membership selection and was fully responsible for re-appointment and/or dismissal
- The PAC and IPO advocated for an open bidding process for the selection of the consultants who would be hired to conduct the study
- The PAC and IPO also advocated for an open bidding process for the selection of an independent expert panel to review the methods and results of both risk assessments
- The PAC had access to the independent scientific advisors and the Independent Expert Review Panel peer review teams as well as full access to the SARA Group and members of the TC. Information was provided when requested
- Observer status on the TC and CSC was very important for information gathering
- Although comprised of individuals having differing backgrounds, the PAC worked well as a team and everyone was able to speak freely
- Most PAC members persevered through many years of the project and maintained a high level of involvement as volunteers
- The PAC successfully cooperated with the TC, CSC and SARA Group – particularly with respect to confidential documents, which were shared with the PAC in advance of general publication or presentation. Feedback from PAC members allowed the study team to refine these documents, and to better anticipate and respond to community questions and concerns.

Less Successful Aspects of the Study Process and the PAC's Role:
- The lack of active participation by Health Canada First Nations and Inuit Health Branch (FNIHB) on the Technical Committee was disappointing and eventually stopped. As such, the PAC was unable to tell if the First Nations communities were well represented by this agency
- While the scientific advisors gave one session each on HHRA and ERA methods, the PAC may have missed opportunities for greater utilization of the scientific advisors for informing both its members and the public on risk assessment
- Professional communicators became involved when key results were to be announced. In these instances, the PAC felt it was not given sufficient time to review public communications
- Though the public was provided with opportunity to appear before both PAC and TC, this opportunity was greatly underutilized
- PAC believes that the media occasionally let the people of Sudbury down by focusing on a few critics of the study rather than on the total effort
- It may not be realistic to have expected the public to maintain a high level of interest throughout a highly technical study, particularly towards the middle and later phases of the study
- There was a perception among some PAC members that the decision makers for the TC stakeholders were not always present at key TC meetings. This may have caused delay while other TC members needed to consult with either their superiors or technical experts.

PAC Recommendations and Conclusions
- In studies such as this, all efforts must be made to achieve and maintain transparency
- The knowledge, interests and concerns of the public should be monitored in a systematic way, perhaps through random surveys

- The non-confidential components of the technical database must be made available for use by other scientists. These data are essential to providing a baseline against which existing risk assessments can be evaluated and future risk management activities may be measured. *(Editor's Note: all monitoring data and results were made available to university scientists.)*
- The rights and responsibilities of observers must be clearly defined. An issue resolution process should also be available
- The Chair and Vice-Chair of committees such as PAC should rotate every two years. Though the PAC Executive performed well, it may be too much to ask a few volunteers to take on such a task for seven or more years
- It is difficult to get volunteers to spend several years on a project. Although there may be a perceived conflict of interest, the possibility of paying stipends to members should be considered. Many government committees pay stipends to public members. At the very least, members should be adequately reimbursed for legitimate expenses related to official events
- It was encouraging that several different stakeholders could come together to produce an excellent study and a model process for future studies. While it seems likely that not every stakeholder's interests were fully met, a sufficient and meaningful study did emerge through consensus to the benefit of all citizens of the study area
- PAC believes that the science carried out by SARA for the Sudbury Soils Study was exceptional and that the use of independent expert review teams provided additional quality control and transparency
- The scientific advisors contributed to the study and served to reinforce rigor and improve understanding of the science
- The Independent Process Observer completed his task in an exceptional manner that served to strengthen the validity and transparency of the process
- It is clear that greater involvement by the public would have helped to ensure that a more representative sample of public opinion was considered throughout the study
- Finally, the PAC believes it has been a beneficial asset to the study and that the public has been well served by its volunteer members.

13.3.3 Independent Process Observer (IPO)

An Independent Process Observer (IPO) was appointed to the study as another measure to counter criticism that involvement of the mining companies, or any other stakeholder, would unfairly influence the study process and decisions. The person chosen for this role was Mr Franco Mariotti, a respected local biologist, and a member of the staff at Science North in Sudbury. The IPO played a critical role in the study and was invited to observe all meetings, and report on the process used to conduct the HHRA and ERA. The purpose of this was to ensure that decision-making was transparent to the community and that public communication was timely and effective. All member organizations of the Technical Committee contributed to a trust fund administered by a third party that reimbursed Mr Mariotti for his time and expenses.

The role of IPO was outlined as follows:

- act as an impartial observer and recorder of the process
- is independent of any bureaucracy
- maintains the right to review information and files such as minutes of meetings, terms of reference, proposals, draft reports, and final reports pertaining to the HHRA/ERA process;
- acts as an observer and, where necessary, as a facilitator to ensure that proper practice is followed with the TC and PAC
- receives comment/input/complaints from the public on matters relating to process and responds appropriately
- points out and suggests remedies for inconsistencies in procedures in consultation with committee members
- recommends process improvements to the TC and PAC to ensure effective and timely completion of work assignments, investigations, studies, and reporting

- suggests opportunities to improve the process for a more effective outcome for all parties
- prepares a quarterly written report on the overall progress and direction of the work of the committees for dissemination to the public; and
- encourages teamwork through consultation and communication.

The IPO reported that the Sudbury Soils Study was, in many ways, a ground-breaking study for Ontario and perhaps for Canada, in the following ways:

- All key members of the Sudbury Soils Study were at the decision-making table and were identified as the Technical Committee (TC)
- The terms of reference for the position of the IPO's role stressed that he was to be present as a 'watchdog' and not be connected to any member of the TC
- Several observers, such as the IPO and the PAC chair were present at most TC meetings (the decision-making body for the study) with union observers occasionally present
- All key decisions made by the TC were made by consensus. This was a crucial part of the process that ensured understanding and support for all decisions, leading to the final conclusions of the report.

The IPO successfully fulfilled this mandate, regularly attending most TC and PAC meetings, as well as scientific meetings, selected CSC meetings, and other information sessions. He reported back to the community in regular reports that described the process and highlighted any questions or issues of concern. His reports were published unedited. They were not reviewed in advance by any of the TC or study team members, before distribution. These reports were mailed to 1100 contacts on the study mailing list, and posted on the study website. During the study, Mr Mariotti produced 23 comprehensive reports including a Final Report. The following section contains excerpts from his final report and appears as a series of questions and answers.

"Technical Committee (TC) meetings were held primarily behind closed doors. Did this fact impede one of the primary purposes of the Sudbury Soils Study which was to have an open, fair and transparent process to the public?"

IPO response: It was essential that the TC could speak freely and openly about their positions. In my opinion, if these meetings were held in a setting with the public and the media attending, the atmosphere of openness and trust would have been compromised and at worst stifled!

The criticism by some that the Sudbury Soils Study process was dominated by the mining companies because of their participation at the TC is false. The presence of all of the observers mentioned above ensured that no TC member dominated or forced the opinion of any other member. The accusation that the mining companies dominated the process is not based on factual evidence whatsoever and is contrary to the findings of all of the observers.

The public was well represented on the Technical Committee by four public institutions. I attended almost all Technical Committee meetings as well as many other sub-committee meetings, public forums and open houses. My belief is that the consensus approach worked well.

In the entire 8 years, I have never observed any undue influence by any Technical Committee member over another. During that time the PAC members, union observers and I were privy to all information such as methodologies, data results and their implications. There was never discussion nor were there ever any attempts made to hold information from the public.

"Was the Sudbury Soils Study a fair and open process?"

IPO response: The process assured fair and open discussion with the ultimate decisions made for the benefit of the public's best interest in several key aspects:

- First, the scientists selected to conduct the scientific study were chosen in an open bid process that was international in scope. Advertisements for qualified bodies were circulated in Canada and the USA
- Second, the entire study was peer-reviewed by TERA, an international team of respected toxicological experts whose mandate was to provide qualified expertise to conduct a review of the study
- Third, to ensure transparency, several observers including the IPO, the PAC chair and two mining union representatives were always invited to be present at all of the Technical Committee meetings.

"Did the Sudbury Soils Study process provide ample opportunity for public engagement in the study?"

IPO response: I want to respond to this key issue in two ways. First, the public could have engaged in the study in several ways – by attending PAC meetings, by attending the first hour of all TC meetings and by attending open house events that were held several times across the Sudbury area. In the first few years, PAC meetings were held in many wards to accommodate the public so they would not have to travel far to attend. Due to the low public turnout, no attendance in some cases, the PAC meetings were eventually held at a central location. Attendance at PAC meetings was very high during the first year of the study, but soon after dwindled to just a few or none at every meeting.

This model of public participation is the traditional way in which to engage the public in a community study. Based on this model, the public was provided with the opportunity to engage in the Sudbury Soils Study process. However, due to the low public response, it is time to review this traditional approach to public engagement and this leads to my second point.

The reality of the Sudbury Soils Study, or for that matter any major risk assessment process, is that meetings are long, tedious and, frankly, at times very boring. It would have been a challenge for any member of the public to voluntarily attend all of the PAC meetings during the 8 year timeline of this study. In retrospect, other methods of public engagement may have been more meaningful. Overall, in my opinion, public engagement in the Sudbury Soils Study was poor.

"What was the role of the Scientific Advisors?"

IPO response: Two scientific advisors were used throughout the Sudbury Soils Study and their role was to guide and provide suggestions to the science methodology. In the end, I believe they were underutilized and could have contributed more positively to the public's understanding of the soils study.

"What was the role of the International Expert Peer Review scientists?"

IPO response: One of the greatest assets of the Sudbury Soils Study was the manner in which the International Expert Peer Review scientists were chosen and utilized. Peer review is essential and mandatory in any scientific study. It is essential that a research project be scrutinized by scientific peers, scientists who were not involved in the study. This standard approach to scrutinizing the science behind a study is crucial to ensuring the validity and impartiality of the study.

The method used to choose the international team of scientific reviewers for the Sudbury Soils Study is an excellent example of impartiality. To ensure a total arm's length approach, a widely distributed advertisement went out to request scientific organizations to bid on the proposal. The objective was to find and create two teams of scientists, experts in their fields. One team was responsible for reviewing the HHRA and the other the ERA.

The meetings were intense and the SARA Group scientists were grilled on the methodologies and science behind each study. The two meetings lasted one and a half days each and were held at College Boreal.

In my interactions with members of the public on the topic of the International Expert Peer Review, the significance of how the scientific advisors were chosen and utilized was poorly understood and not fully appreciated by the public at large. This part of the Sudbury Soils Study process should not be underestimated; from my perspective it was a crucial and an essential aspect in the validity of the science behind the Sudbury Soils Study.

"Were members of the public given ample opportunities to respond to the HHRA and ERA results?"

IPO response: At one of the Public Advisory Committee meetings, several members of the public expressed concern that after the release of the ERA and the HHRA, there was no opportunity for the public to engage the SARA Group scientists for additional discussions apart from by mail or e-mail. Furthermore, it was suggested that once the public had adequate time to understand the results, they should have been given the opportunity to ask questions directly of the SARA Group scientists again. I think this was an excellent idea and has its merits.

"Should the mining companies, Vale and Xstrata Nickel, have been a part of the decision-making process as members of the Technical Committee?"

IPO response: This question has been repeated several times during the 8 years of the study. The IPO made it very clear that the mining companies have to be a part of the solution. Furthermore, without a doubt, the companies that have caused the environmental degradation in Sudbury need to pay for any scientific research that aims to study its impact on human health and the environment. This responsibility has never been an issue with the companies.

13.3.4 Primary Spokesperson

Early in the study, it was determined that a primary spokesperson would be required to address media questions, provide updated information to the public at various intervals, and to act as a representative for the study. Members of the TC represented very different organizations and no individual member could speak on behalf of the TC as a whole. Although the TC members had respect for each of the other agencies and representatives, there was unwillingness to allow any single member (or agency) to represent the group. Therefore, it was decided that the Director of the SARA Group, Dr Christopher Wren, would act as the primary study spokesperson to address questions of a scientific nature and give regular updates to the public on the risk assessments. However, if members of the public or media had questions pertaining to the study process or one of the TC stakeholders organizations, a representative from that organization would step forward to address the question.

The primary spokesperson participated in more than 50 media interviews during the study. In addition, he made presentations at numerous workshops, PAC meetings and meetings with special interest groups. The various communication activities are described in more detail in Section 13.5.

13.3.5 Communication Plan

At the onset of the study, it was determined by the TC and the CSC that a concrete, iterative Communication Plan was required to identify and track communication initiatives throughout the course of the study. This plan was developed by the CSC and the SARA Group, and reviewed and updated as the study progressed. The activities outlined were implemented by the SARA Group, under the guidance of the TC and the CSC.

The Communications Plan incorporated the following goals:

- Publish regular reports to the community in the form of newspaper supplements issued approximately twice a year
- Continue to regularly update the study website, providing results and reports as they are released, as well as minutes from public meetings
- Participate in individual and group meetings with interested parties who request additional information on the study and its possible outcomes
- Conduct ongoing, proactive media relations, develop 'op ed' pieces (opinion articles of interest to the community placed in the editorial section of local newspapers), and seek editorial board meetings as needed throughout the study process
- Conduct open houses or public information sessions to describe the study process to the community (June 2003, November 2003, February 2005)
- Monitor coverage of the study in local news media
- Issue regular news updates to local media on SARA Group activities
- Release results of the HHRA and ERA at separate public information sessions, and allow public review of the reports.

Methods for implementing these initiatives are described in the following section.

13.4 Communication Activities

A number of communication initiatives were established to keep the public informed of study developments and to provide opportunities for consultation with members of the study team. These activities continued throughout the course of the study, as summarized in Table 13.1 and described in the following section.

Table 13.1 **Summary of Consultation Activities and Methods**

Activity	Frequency
TC meetings: at the beginning of each meeting the public was invited to make presentations	64 monthly TC meetings were held over the course of the study, with lower frequency near the conclusion
PAC meetings: these bi-monthly meetings were advertised and open to the public	40 PAC meetings held during course of the study
Update Newsletters: a 4-page newspaper insert was circulated to over 40,000 households	12 Update Newsletters were written and distributed
Independent Process Observer Reports	23 IPO Reports were prepared and circulated
Community Information Sessions	11 different workshops and community information sessions were held
Meetings with special interest groups and First Nations Communities	Over 30 meetings held with consultants and members of the TC
Newspaper and media articles	Over 150 media articles were reported in the local newspapers
Direct mailing to study participants	Any resident that participated in an aspect of the study (i.e. soil sampling, vegetable garden survey) was mailed their individual results with an explanation of data
Website and toll-free telephone number	A website remained active during the study; people could send in comments or questions via email. Also, a toll-free telephone number was maintained. All queries were answered within 24 h. Over 600 calls and emails were received
'Reader Friendly' Summary reports	The complex and lengthy final HHRA and ERA technical reports were summarized into 50-page 'plain-language' reports for general audiences. Copies were produced in English and French
Public release of Final reports	Results of the HHRA were presented in May, 2008, in three of the communities of interest: Sudbury Centre, Copper Cliff and Falconbridge. Results of the ERA were released at one community information session in March, 2009
Public review of Final HHRA and ERA reports. Each final report was made available to the public with review periods ranging from 90 to 180 days	A detailed response was prepared for each comment provided by the public. Both the comment and response are provided in an appendix to the reports and are now included as part of the public record

13.4.1 Mailing List

A mailing list of interested local groups and individuals was created early in the Sudbury Soils Study process, and was regularly updated throughout the study. The final list comprised approximately 1100 contacts. This list was updated to include anyone who participated in studies related to the Sudbury Soils Study. These studies included (with year in which it was conducted) the initial soil survey (2001), the residential vegetable garden survey (2003), indoor dust survey (2004), drinking water survey (2005), and the Falconbridge Arsenic Exposure Study (2004). The list was also updated with participants who attended workshops or open houses. Communications material produced during the study (IPO Reports and Update Newsletters) were distributed to all contacts on the mailing list. Confidential mailings containing specific survey results were sent to participants of individual surveys mentioned in the paragraph above. An email list was also compiled, with approximately 350 recipients. All notifications and invitations to meetings (workshops, community information sessions, TC and PAC meetings) were distributed to this group.

13.4.2 'Have Your Say' Workshops

Three 'Have Your Say' workshops were held in the communities of Copper Cliff, Coniston and Falconbridge in May, 2003. The purpose of these workshops was to obtain information from community members on their expectations for the study. The TC members used this input from the community to provide a framework for refining the study parameters. In particular, input was solicited for the selection of valued ecosystem components (VECs) for the ERA. Meetings were well advertised in the local media, and invitation letters were mailed to approximately 400 people living in the smelter communities. Over the course of three evenings, a total of 115 people attended the meetings.

Through these workshops, the study team received detailed input from the community on specific plants, animals and natural areas that should receive special attention in the study. Participants expressed the importance of the links between human health, clean drinking water, edible foods from natural areas, and recreation – particularly where children play. Participants also expressed concern about biodiversity and economic impacts on tourism, as well as concern about the health of family pets.

13.4.3 Update Newsletter

A community newsletter, entitled *Update* was published by the SARA Group and distributed in the local *Northern Life* newspaper approximately twice per year. During the study, 12 editions of the Update Newsletter were produced. Based on circulation statistics provided by the newspaper, *Update* was distributed to approximately 40,000 homes in the Greater Sudbury area. In addition, copies of the newsletter were mailed to the 1100 residents on the SARA Group mailing list, and were also available at local government offices, schools, Science North, and other community venues.

Each issue of the newsletter contained updates on study progress, and results released from related projects within the study. In addition, it provided a forum for specific groups involved with the study. Groups and professional organizations profiled in the newsletter included:

- the PAC
- TERA (responsible for the independent expert peer review process)
- the Technical Committee
- the SARA Group
- the Scientific Advisors to the TC.

The newsletter also featured a schedule of upcoming events, and contained contact information that allowed readers to provide comments directly to study team members. Newsletters published over the course of the study were well received by the community, and resulted in increased interest in the study, as evidenced by media attention and public feedback through phone and website enquiries.

13.4.4 Project Website

The website for the Sudbury Soils Study (www.sudburysoilsstudy.com) was developed early in the study process (launched March 2003). It provided general study information, as well as copies of all communication materials produced and distributed to the community (newsletters, IPO reports, news releases, articles, frequently asked questions, and related links). This online information repository also provided an archive of PAC meeting minutes, TC meeting decisions and details from community open houses and workshops. Notices of future PAC and TC meetings were posted as pop-up windows on the home page. The site also provided contact information for the study team, and an email link for direct contact.

The website is no longer active but is still accessible and contains copies of all final reports and contact information for each of the TC members.

13.4.5 Telephone Survey

A telephone survey was conducted in the fall of 2004 to determine public awareness of the study and to monitor public opinion. Information from this survey was used to determine the effectiveness of communications initiatives, and to improve efforts for the remainder of the study. A total of 606 residences participated in the survey. Of those who participated, 32% stated that they had heard of the Sudbury Soils Study. Of those who had awareness of the study, 30% stated that they had a clear idea of what the study was examining, and 83% correctly identified the statement that most closely represented the intent of the study, while 67% of respondents suggested that there was not enough information being communicated to the public, and 24% suggested there was nothing further the study team could do to encourage them to participate.

This survey provided a solid benchmark to evaluate the effectiveness of communications efforts at the halfway point in the communications program. Based on this survey, the study team continued with its ongoing initiatives. Modifications included more media invitations to meetings (through public service announcements and television ads), and increased efforts to inform school-aged children about the study.

In retrospect, it would have been useful to undertake a similar telephone survey near the conclusion of the study to obtain a better perspective of public awareness and opinion of the results. The study team regularly heard from a small vocal group of concerned residents, but it was not confirmed how widely this group represented the general public.

13.4.6 Toll-Free Phone Line and Email

A toll-free phone line and email address were set up at the beginning of the project, with contact details provided in all communication materials produced during the study. The email address was directly linked to the study website. Approximately 600 calls and emails were received between January 2003 and December 2009. All calls were answered by a member of the SARA Group within 24 h.

13.4.7 Physicians Package

People generally trust their family physician as a source of information, especially when it relates to matters such as environmental conditions that could affect personal health. To assist local healthcare workers address questions related to the risk assessment, a package of medical information related to the COCs was provided to all physicians and nurses in the Sudbury area. This information was prepared in consultation with Dr Lesbia Smith, the SARA Group's medical advisor, as well as the medical directors from Vale and Xstrata Nickel and the Sudbury & District Medical Officer of Health.

Doctors were invited to review the information provided, and to contact the SARA Group with any questions or concerns. Information updates were also provided in *The Advisory*, a quarterly newsletter produced by the Sudbury & District Health Unit (SDHU) and mailed to healthcare providers within the health unit's catchment area. Residents could also speak with Dr Lesbia Smith, and discuss medical concerns in a confidential manner. Many residents took advantage of this opportunity.

13.4.8 Individual and Group Meetings

The study team routinely attended meetings with different stakeholders, community groups, and other interested parties. The meetings were initiated by either the study team or interested party to share information on the study, as well as to solicit feedback from individuals with local knowledge and experience. Members of the SARA Group and TC participated in more than 20 meetings with special interest groups as well as more than a dozen meetings with First Nations community members. All feedback received during these meetings was used to inform the study team and make improvements to the Sudbury Soils Study.

13.4.9 Community Information Sessions

Members of the community were invited to attend a total of 11 community information sessions. Between 2002 and 2007, seven information sessions were conducted to introduce the study, present risk assessment methodology and to discuss any interim results. At these meetings, attendees were given the opportunity to review up-to-date information on the study, and to speak with scientists and other members of the study team. A brief, formal presentation also allowed for sharing of information, and was followed by a facilitated question and answer session.

13.4.10 Media Relations

The local Sudbury media displayed considerable interest in the study, conducting more than 50 interviews with team members, and publishing or broadcasting more than 150 media stories directly related to the Sudbury Soils Study. The media serves as a formal communication channel through which risk perceptions and behaviors can be influenced (Masuda and Garvin 2006).

The relationship between the study team members and the media was sometimes strained with the scientists and TC members taking exception to the newspaper headlines. It was also apparent that the media were attracted to opponents to the study, or any outspoken member of the public. Many newspaper articles, particularly in the early part of the study, focused on criticism of the mining companies' involvement, rather than the science involved. But, as one reporter maintained, "science is boring".

To provide the media with correct information, and updates on study activities, almost 40 news releases were prepared and circulated during the study. More than half these news releases were used by the local newspapers. News releases were also used as a means to generate public involvement in a particular aspect of the risk assessment. The residential garden survey was a good example. Notices were placed in local newspapers and on the radio asking residents to participate in a survey of metal levels in their home-grown vegetables.

Examples of unbalanced media coverage were apparent in the spring of 2003, when the local Medical Officer of Health (MOH) issued precautions to the residents of the town of Falconbridge regarding elevated soil arsenic concentrations. The MOH was perhaps being overly cautious in this situation, but the residents were upset and the media had a field day. Front-page headlines read 'Residents demand answers: Arsenic levels in Falconbridge's soil cause panic, anger'. An editorial appeared in the paper titled 'The danger below our feet: soil studies in Falconbridge are alarming'. The articles fuelled the situation and created additional uncertainty among the population. Later, one newspaper ran front-page headlines stating that vegetables grown in Sudbury were 'toxic'. Again, this caused angst among the population and upset many local farmers and producers. The headline was taken out of context from an external review of the HHRA and contradicted the study findings. The result of this misrepresentation of results was a public backlash against the media and the external reviewer who made the comments. When the SARA Group provided news releases that contained very positive results of the vegetable garden survey these news items were sometimes ignored or generally given low profile by the media. It is not unusual for journalists or the media to carry stories of conflict and risk while downplaying or omitting 'safety' stories thus contributing to fracture public opinion (Heath and Nathan 1990-91).

In a recent survey of almost 1,000 members of the Society of Toxicology (SOT), the toxicologists almost unanimously believed the media does a poor job covering basic scientific concepts and explaining risk (STATS 2009). In that survey, 90% of the respondents thought that media coverage of risk lacks balance and diversity, while 97% replied that the media do not distinguish good studies from bad studies.

From a study of focus groups in communities north of Boston, Mass., Scammell et al. (2009) reported that lay people often had difficulty accepting the findings of health studies that contradicted their own experiences of environmental exposure and illness. This was certainly the case in Sudbury where some of the most outspoken critics of the HHRA process and results were members, or former members, of the local workers unions.

Toward the conclusion of the study, the media reported the study results as provided through media briefings and news releases. However, articles and headlines continued to emphasize that a handful of people were skeptical of the study results. This position was emphasized, despite detailed review by the regulatory agencies, international peer review, input of the scientific advisors, and the professional qualification of the study consultants. The media also continued to report that the study cost much more than originally estimated, and took much longer to complete than originally predicted.

The study team took extra steps to meet with reporters and editors of the local newspapers to ensure they had the necessary information and 'both sides of the story' on which to report. It is difficult to measure the success of those efforts, but since the local media plays a leading role in getting information to the public, it is important to try to ensure the information that is reported is fair, accurate and as balanced as possible.

13.5 Risk Communication and Language of the Sudbury Soils Study

One of the major challenges to the study team was how to translate the results of 6 years of complex scientific study into simple language that could be understood by the general public, while remaining technically accurate. The study team also learned that scientists and the general public have differing perceptions and understanding of risk. In discussing why Americans often worry about the wrong things, Kluger (2006) states that people have a habit of worrying about possibilities, while ignoring the probabilities. He provides several clear examples of this behavior such as public angst over the mad cow pathogen which is unlikely to be present in a person's hamburger, compared with less worry over the cholesterol in the meat that contributes to heart disease, which kills 700,000 Americans annually. This train of thought was echoed by Sandman (2007) who stated that the risks that kill people and the risks that upset people are often very different.

Throughout the process, the study team was committed to providing updates and interim results to the public on a regular basis. In these presentations or written communications, it was considered important to use appropriate and consistent language to characterize the study and the related risks. This following section provides some examples of how the team communicated the final results and addressed some difficult questions such as cancer risk related to environmental conditions.

Several hundred private citizens participated in a variety of surveys such as the residential soil survey, vegetable garden survey, indoor dust survey and drinking water survey. As a condition of participation, the resident was told they would receive their individual results in a timely manner. Each participant was notified in writing of their results and some limited interpretation or context to the numerical values was provided. Press releases were also given to the media on the general interim results, again with some context provided.

At the most basic level, interim results were compared with existing regulatory guidelines or criteria. For example, if a soil or drinking water concentration was below the appropriate regulatory guideline, it was conveyed to the resident or general public that the samples were within the regulatory limits for health protection and, therefore, did not pose an unacceptable risk. In no instance did the study team say a certain level posed 'no risk' since there is no such thing as 'no risk'. The term 'safe' was also used carefully. If metal concentrations exceeded a particular guideline, it was explained that this meant further study was required, and the information would be used in the full risk assessment to better understand the total risk from all exposure pathways. It was also stated that, based on the available information, the metal levels reported in individual media (i.e. vegetables) "did not pose an immediate health risk". This phrase was used with most interim results with full support from the local MOH.

In many instances, it was necessary to explain to the public the meaning of regulatory guidelines, how they were developed and how they may be applied. If a reported COC concentration exceeded a particular guideline, some members of the public interpreted this to mean that the sample posed an immediate health risk. Ongoing education was required to respond to this misperception. Another term the study team avoided was 'hot spot', which the media and some members of the public liked to use to denote an area where the concentrations of metals were higher than surrounding properties. The phrase itself, quite intentionally, suggests a very dangerous area that should be avoided or subjected to immediate remediation. Instead, the study team would refer to these areas as having elevated metal concentrations. Similarly, when displaying maps of study results, the color red was avoided, since red is often associated with an emergency or danger, or an area to be avoided. One of the objectives of a risk assessment is to be confident not to underestimate risk, but on the other hand, it is important to not significantly overestimate risk, which may unnecessarily alarm stakeholders.

To interpret the estimated risks for the public, the terms 'acceptable' or 'unacceptable' were used in the context of regulatory guidelines. These terms are value-laden and need to be used in context. They also promoted strong reaction from some members of the public who asked "who determines what is acceptable?" This was often followed by the demand that the residents and society should be able to choose what they consider to be acceptable or unacceptable risks. Clearly, the public considered zero risk as the only acceptable alternative, which scientists do not consider plausible. The scientists were sometimes considered arrogant by members of the public in determining what is considered acceptable or not acceptable.

As in any risk assessment, it can be challenging to explain the difference between uncertainty in the risk assessment, and confidence in the results. Most risk assessment reports include a distinct section or chapter that discusses the 'limitations and uncertainties' in the study whether they are related to sample sizes, data gaps, toxicity reference values, modeling uncertainties or other technical issues. In fact, the explicit identification of these uncertainties is one of the features that make risk assessment very different from environmental assessment. However, in discussing or identifying scientific uncertainties with the general public or the media, they often mistook these limitations as a lack of confidence in the results, which was not the case. Furthermore, scientific uncertainty in the risk assessment process is different from the public's uncertainty about the study or associated health risks, as discussed by Powell et al. (2007).

The end result of the study for both the HHRA and ERA was risk characterization. As discussed in Chapter 6, the COCs were classified as either threshold (non-carcinogens) or non-threshold (carcinogens) substances. Risk is quantified differently for each of these groups of chemicals. For threshold substances, the quotient method is most commonly used, and risk is expressed as a Hazard Quotient (HQ). The HQ is determined as the Exposure divided by a Reference Value. In simplest terms, if HQ is less than one, risk is considered to be negligible and is ruled out. However, if HQ is greater than one, risk cannot be ruled out and further study may be warranted.

Explaining the HQ equation, and in particular, the significance of HQ values greater than one to the general public posed a challenge. For example, does an HQ value of 12 suggest a risk is 12 times more likely to occur than an HQ value of 1.0? We could find no satisfactory explanation or definitions in the general literature to assist in this interpretation. We felt confident that the HQ system was not linear, in that an HQ of 12 was not 12 times 'worse' than an HQ of 1.0. But, the HQ was a complex concept that was not readily understood.

Risk characterization for carcinogens consists of calculation of the Cancer Risk Level (CRL), which is defined as the predicted risk of an individual in a population of a given size developing cancer over a lifetime. The CRL is expressed as the prediction that one person per n people would develop cancer, where the magnitude of n reflects the risks to that population. The resulting estimated cancer risk can then be compared to an 'acceptable' risk of cancer, as determined by a regulatory agency. In Ontario for example, the incremental lifetime cancer risk (ILCR) level of one-in-one million (10^{-6}) is considered acceptable to the regulatory authorities. Health Canada, on the other hand, considers an ILCR of one-in-one hundred thousand (10^{-5}) as acceptable.

An ICRL refers to the contribution that a facility or activity makes to the total risk. In many situations, it is difficult to separate the actual incremental risk from the facilities (i.e. smelters) over background cancer risk. One of the key issues surrounding cancer risk is how that risk is communicated to the public. As Calman (1996) clearly states, of all of the diseases in the Western world, cancer is perhaps the most alarming. To provide a consistent framework of relative cancer risks reported in the study, we adopted terms and definitions as suggested by recognized experts in the field of risk communication (Calman 1996; Paling 2003). The terms associated with different levels of numerical risk are described below as defined by Calman (1996):

- **High:** Risks may be fairly regular events and would occur at a rate greater than 1 in 100. They may also be described as *frequent, serious* or *significant*
- **Moderate:** This term relates to a risk of between 1 in 1,000 and 1 in 100. This would apply to a wide range of medical procedures and environmental events
- **Low:** This relates to a predicted increased risk of between 1 in 10,000 and 1 in 1000. Again many risks of clinical procedures and environmental hazards fit into this broad category. Other words that might be used include *reasonable, tolerable* and *small*
- **Very Low:** This describes a risk between 1 in 100,000 and 1 in 10,000; many healthcare interventions have adverse effects that are in this range
- **Minimal:** This refers to a risk that is in the range of 1 in 1 million to 1 in 100,000 and that the conduct of normal life is not generally affected as long as reasonable precautions are taken to minimize exposure. Some policy makers consider a probability of anything less than 1 in 100,000 as *acceptable*
- **Negligible:** This describes an adverse event occurring in less than 1 per 1 million episodes. While still important to identify and monitor, such a risk would be of little concern for normal living. Other words that could be used in this context are *remote* or *insignificant*.

In fact, any cancer risks predicted in this study were either Minimal or Negligible. Using this framework had several advantages. First, the various members of the TC quickly agreed to adopt this terminology which provided consistency on wording and interpretation of results. Second, it provided language that could be understood by members of the public much better than a probability value such as 'one in a million' where the significance is difficult to grasp. Finally, the study team could confidently state that this classification system was developed by other scientists in the field of risk communication.

The key results of the HHRA and ERA are presented below exactly as they were presented to the public. The statements were supported by all stakeholders and members of the TC as well as the PAC.

HHRA conclusions

1. Based on current conditions in the Sudbury area, the study predicted little risk of health effects on Sudbury area residents associated with metals in the environment
2. There were no unacceptable health risks predicted for exposure to four of the six Chemicals of Concern studied: arsenic, copper, cobalt, and selenium
3. The risks calculated for typical exposures to lead in the environment throughout the Greater Sudbury area are within acceptable benchmarks for protection of human health. However, levels of lead in some soil samples indicate a potential risk of health effects for young children in Copper Cliff, Coniston, Falconbridge and Sudbury Centre
    - Lead levels in soils and dust in the Sudbury area are similar to levels in other older urban communities in Ontario

4. The study calculated a risk of respiratory inflammation from lifetime exposures (70 years) to airborne nickel in two areas: Copper Cliff and the western portion of Sudbury Centre
    - Respiratory inflammation has been linked to the promotion of cancer caused by other agents
    - Based on the conservative assumptions and approaches used in this risk assessment, it is unlikely that any additional respiratory cancers will result from nickel exposure over the 70-year lifespan considered in the risk assessment
    - Health risks related to nickel inhalation were not identified in the other communities of interest
5. Anglers, hunters and First Nations people who may consume more local and wild game are at no greater risk of health effects due to metals in the environment than the general population.

The results and conclusions from this risk assessment will be used as the basis for risk management decisions in the Greater Sudbury area.

The main conclusions from the ERA were as follows:
1. Terrestrial plant communities in the Greater Sudbury area have been and continue to be impacted by the Chemicals of Concern (COC) in soil
2. Terrestrial plant communities in the Greater Sudbury area are also impacted by other factors such as soil erosion, low nutrient levels, lack of soil organic matter, and/or low soil pH
3. The assessment suggests that COC originating from smelter emissions are not currently exerting a direct effect on wildlife populations in the Greater Sudbury area, nor are they predicted to in the future. However, historic impacts of smelter emissions on plant communities have affected habitat quality and, therefore, may be having a continued indirect influence on birds and mammals in the study area
4. There are very few recognized threatened or endangered species in the study area. It is unlikely that COC from the smelters are having a direct effect on these species
5. An aquatic problem formulation was developed as an information gathering and interpretation stage to focus the approach for a possible future detailed aquatic ecological risk assessment. However, given the extensive aquatic research and monitoring studies that have been conducted in this area over the past two decades, no detailed aquatic ecological risk assessment is planned at this time.

The SARA Group is confident that the ERA did not underestimate risks to plants and animals in the Greater Sudbury area. The results and conclusions from this risk assessment will be used as the basis for future risk management decisions in the Greater Sudbury area and to support activities related to the re-greening of the Greater Sudbury area landscape.

13.6 Concluding Remarks
- Communication and consultation with the public was considered a critical part of the risk assessments and was given considerable effort and resources
- The complexity and duration of the study posed ongoing communications challenges
- These challenges were, for the most part, met successfully by the diverse and numerous initiatives related to communication and transparency
- The diversity of committees and of the stakeholders within the TC, PAC and CSC was a real strength of the study and communication efforts, but not without cost in time, effort and other resources
- Traditional methods of public engagement were used and are recommended for future studies:
    - Regular and frequent communication
    - Multiple modes of communication
- Communication should be two-way with dialog between all of the parties such that members of the public can have meaningful input to the study
- Regular surveys should be conducted to assess awareness, interest, and concerns of the public in general. This would help to better understand whether issues and concerns raised by the vocal few critics are reflective of the majority of the population.

13.7 Acknowledgements

Many people were involved in the extensive communications and consultation program. Particular thanks is given to the many members of the public who voluntarily participated on the PAC. The tremendous efforts of Ms Lindsay Boyle are recognized for production and coordination of all the newsletters, reports and display material, as well as coordination of all responses to public inquiries. Input to the CSC by Ian Hamilton was particularly helpful and appreciated.

13.8 References

Calman KC (1996) Cancer: science and society and the communication of risk. BMJ 313:799–802

Heath RL, Nathan K (1990–91) Public relations' role in risk communication: Information, rhetoric and power. Public Relat Q 35(4):15–22

Kluger J (2006) Why we worry about the wrong things; the psychology of risk. Time Magazine December 4:35–41

Masuda JR, Garvin T (2006) Place, culture and the social amplification of risk. Risk Anal 26(2):437–454

PAC (2009) The Sudbury Soils Study: A commentary. The Public Advisory Committee for the Study, June 2009. www.sudburysoilsstudy.com/PAC/reports

Paling J (2003) Strategies to help patients understand risks. BMJ 327:745–748

Powell M, Dunwoody S, Griffin R, Neuwirth K (2007) Exploring lay uncertainty about an environmental health risk. Public Underst Sci 16(3):323–343

Sandman PM (2007) Avian flu, a pandemic and the role of journalists. Understanding the risk: what frightens rarely kills you. www.psandman.com/articles/NiemanReports.com

Scammell MK, Senier L, Darrah-Okike J, Brown P, Santos S (2009) Tangible evidence, trust and power: Public perception of community health studies. Soc Sci Med 68:143–153

STATS (2009) Are chemicals killing us? Toxicologists say media overstate risks. News Release, May 21, 2009. Statistical Assessment Services. Center for Health and Risk Communication at George Mason University, Washington, DC, USA

13.9 Appendix Chapter 13: Abbreviations

COC, Chemical of Concern

CRL, Cancer Risk Level

CSC, Communications Sub-committee

FNIHB, First Nations Inuit Health Branch

HQ, Hazard Quotient

ICRL, Incremental Cancer Risk Level

IPO, Independent Process Observer

MOH, Medical Officer of Health

PAC, Public Advisory Committee

SARA, Sudbury Area Risk Assessment

SOT, Society of Toxicology

TERA, Toxicological Excellence for Risk Assessment

14.0 A Comparison of Two Community Based Risk Assessments

DOI: 10.5645/b.1.14

Authors

Laura Mucklow, M.E.S.
Xstrata Nickel,
Falconbridge, ON Canada P0M 1S0
Email: LMucklow@xstratanickel.ca

Christopher Wren, Ph.D.
MIRARCO,
935 Ramsey Lake Road, Sudbury, ON Canada P3E 2C6
Email: cwren@mirarco.org

Disclaimer

The comments and recommendations in this chapter include a collection of diverse opinions expressed by the participants interviewed for this report, as well as opinions of the authors based on a review of the strengths and weaknesses of each study as described by participants. Individual comments and recommendations do not necessarily reflect the views of all participants, proponents or the regulatory agencies.

Table of Contents

14.1 Introduction 423
    14.1.1 Objectives 423
    14.1.2 Themes of Analysis 423
14.2 Background 423
    14.2.1 Port Colborne 423
    14.2.2 Greater Sudbury 424
14.3 Approach 424
    14.3.1 Document Review 424
    14.3.2 Interviews 424
    14.3.3 Report Organization 425
14.4 Theme 1: Study Organization 425
    14.4.1 Organizational Structure 425
    14.4.2 Technical Committees and Meetings 429
    14.4.3 Advisory Roles 430
    14.4.4 Project Consultants and Scope of Work 431
    14.4.5 Project Schedule and Budget 432
    14.4.6 External Peer Review Process 433
    14.4.7 Summary of Study Organization 434

14.5 Theme 2: Public Communication ... 434
    14.5.1 Public Committees and Meetings ... 435
    14.5.2 Communications Plans and Spokespeople ... 436
    14.5.3 Openness and Transparency ... 437
    14.5.4 Introducing the Study and Gathering Public Input for Problem Formulation ... 438
    14.5.5 Communication Methods ... 438
    14.5.6 Communication of Study Results ... 440
    14.5.7 Media Coverage ... 441
    14.5.8 Key Public Concerns ... 441
    14.5.9 Summary of the Public Communication Process ... 442
14.6 Conclusions ... 443
14.7 Recommendations ... 444
14.8 Acknowledgements ... 447
14.9 References ... 447
14.10 Appendix Chapter 14: Abbreviations ... 448

Figures
14.1 Port Colborne CBRA: Organizational Structure ... 426
14.2 Sudbury Soils Study: Organizational Structure ... 428

14.1 Introduction

The term, Community Based Risk Assessment (CBRA), has been applied in Ontario to risk assessment studies of large geographical areas where potential impacts extend well beyond the boundaries of any particular property. Two CBRAs were recently conducted in Port Colborne and Sudbury, Ontario. The Port Colborne Community Based Risk Assessment was conducted from 2000 to 2010 to address elevated metal concentrations of soils within the City of Port Colborne and the surrounding agricultural areas as a result of historical air emissions from a former Ni refinery. The Sudbury Soils Study was conducted from 2001 to 2009 to address elevated metal concentrations in soils within the City of Greater Sudbury and surrounding areas as a result of historical air emissions from two active Ni and Cu smelting and refining facilities and one closed smelter. To date, the Port Colborne and Sudbury studies are the largest risk assessments to be conducted in Canada. The experiences of these projects were reviewed and analyzed and recommendations were provided for future CBRAs.

14.1.1 Objectives

The objectives of this study were:

- to summarize, compare and contrast the processes used in the Port Colborne and Sudbury studies
- to comment on the strengths and weaknesses of the approaches
- to summarize the opinions of the people that played a major role in these projects, and
- based on the above, to make recommendations for future CBRAs.

14.1.2 Themes of Analysis

The purpose of this comparative analysis was not to examine the scientific and technical content of the Port Colborne and Sudbury studies. Both studies were conducted by qualified experts and were subjected to extensive peer review processes to ensure the scientific integrity of the risk assessments. While the technical content of the studies was comparable, the way in which the studies were organized, the procedures that were followed by the technical teams, and the methods of public communication varied greatly between the two risk assessments. Thus, this study focused on the following two themes:

- **Theme 1:** Study Organization
- **Theme 2:** Public Communication

14.2 Background

14.2.1 Port Colborne

Port Colborne is a small community located on the north shore of Lake Erie in the Regional Municipality of Niagara (Statistics Canada 2006). The City of Port Colborne is surrounded by active agricultural areas (JWEL 2007a; City of Port Colborne 2010). In 1918, the former Inco Ltd. (now Vale) established a Ni refinery that produced electrolytic Ni until 1984. The refinery now produces platinum-group metals and Co, and also packages and distributes Ni originating from other Vale sites. The number of local people employed was substantially reduced when the Ni refinery ceased production in 1984. The City's current economic development strategy focuses on the key priorities of small business, manufacturing, tourism and government services (City of Port Colborne 2010). According to the 2006 census, approximately 18,600 people live in Port Colborne, with approximately 9100 people making up the labor force. Only 200 people (or 2% of the labor force) were employed directly with Vale in 2010.

Historical stack and fugitive emissions from the refinery in Port Colborne (1920s to 1960s) resulted in elevated concentrations of Ni, Cu and Co in soils within the urban center of Port Colborne, as well as portions of the surrounding agricultural and natural environment. The Ontario Ministry of the Environment (MOE) conducted a series of soil studies in agricultural areas between 1975 and 1985, followed by periodic studies in non-agricultural areas, and culminating in an extensive urban soil sampling program from 1998 to 2000. These studies confirmed the presence of substantially elevated soil Ni, Cu and Co concentrations in surface soils in the Port Colborne area. The MOE, as well as the Mayor and Council of Port Colborne, wanted to understand the spatial extent and environmental implications of elevated soil metal concentrations in urban areas. Rather than conduct a series of site-specific sampling programs and risk assessments for numerous specific properties, Vale proposed initiating a CBRA process. The MOE and the City agreed to this suggestion on the condition that an extensive, open public consultation process was included. The CBRA commenced in 2000. The study area encompassed approximately 29 km², including the City and surrounding agricultural and natural areas where concentrations of one or more of the chemicals of concern (Ni, Cu, Co, As) were in excess of the MOE (1997) guidelines (JWEL 2007a). An ecological risk assessment (ERA) for the natural environment and a crop studies report for the agricultural environment were completed in 2004. The final report of the human health risk assessment (HHRA) was submitted to MOE by Vale in 2007. A final commentary report was submitted by the Public Liaison Committee (PLC) in July 2010.

14.2.2 Greater Sudbury

The City of Greater Sudbury was formed in 2001 as an amalgamation of the City of Sudbury and the surrounding towns and unincorporated townships. It is the largest city in northern Ontario by population and the largest city in Ontario on the basis of land area (City of Greater Sudbury 2010). The City of Greater Sudbury has a rich history in mining and smelting operations (SARA 2008b). Today, the mining industry, and the associated mining supply sector, continue to be a major employer of local residents. However, in recent years the economy has diversified substantially and supports a healthy mixture of business and retail services, tourist attractions, healthcare and research services, post-secondary educational institutions and government offices. According to the 2006 census, approximately 160,000 people live in Greater Sudbury, with approximately 82,000 people making up the labor force. Over 8000 people (or 10% of the labor force) were employed directly with local mining companies (Statistics Canada 2006).

In 2001, the MOE published the results of soil monitoring studies conducted in the Sudbury area between 1971 and 2000 (MOE 2001). The report identified that Co, Cu, Ni and As were elevated compared to background and frequently exceeded MOE generic soil quality criteria (MOE 1997) in soils near three historic smelting and refining centers. The 2001 MOE report made two primary recommendations:

- that a more detailed soil study be undertaken to fill data gaps, and
- that a human health risk assessment (HHRA) and ecological risk assessment (ERA) be undertaken (SARA Group 2008).

These recommendations were voluntarily accepted by Vale (formerly Inco Limited) and Xstrata Nickel (formerly Falconbridge Limited) and the Sudbury Soils Study commenced in 2001. The overall study area was very large and encompassed approximately 40,000 km², including the City of Greater Sudbury, two neighboring First Nations reserves and other outlying regions (SARA Group 2008a). The study was overseen by a collaborative Technical Committee (TC). A Public Advisory Committee (PAC) was developed to 'address the concerns of the community at large, by providing comments and input' to the TC (Sudbury Soils Study 2002b). The HHRA report was released in 2008, and the ERA was released in 2009. The PAC released a commentary report on the process in June 2009. The final report of the Independent Process Observer (IPO) was also released in June 2009.

14.3 Approach

14.3.1 Document Review

A comprehensive list of documents produced for the Sudbury CBRA was compiled based on a review of the Sudbury Soils Study website (Sudbury Soils Study 2010) and interviews with members of the Sudbury TC. For Port Colborne, a list of pertinent reports was obtained from key members of the Technical Subcommittee (TSC). Copies of these unpublished reports were obtained from Vale and the PLC.

14.3.2 Interviews

The purpose of conducting interviews was to allow the people most closely involved in each study to express their opinions about the study successes and weaknesses and to make recommendations for future CBRAs. All participants who had a defined primary role in either the Port Colborne or Sudbury CBRAs were invited to be interviewed. This included members of the technical and public committees who had a long-term involvement in the studies. The participation of all individuals in this study was voluntary. Two standard questionnaires were developed to gather opinions about the two study themes: study organization and public communication. The questionnaire pertaining to public communication was administered to all participants, while the questionnaire pertaining to study organization was administered only to participants who had some involvement with the technical committees of the projects. The questionnaires were administered verbally in personal or telephone interviews with each participant. Consent forms were completed by each participant before the interviews. Participants were given the opportunity to provide additional comments or recommendations they believed would be useful to the analysis. Approximately 25 people participated in this review, with in-depth interviews held with several participants to gain information that could not be gathered from the document review. The recommendations made in this report represent a combination of the opinions expressed by the participants interviewed and the opinions of the authors based on the information received.

14.3.3 Report Organization

The report is organized into two main sections corresponding to the two themes of analysis: study organization and public communication. Each section is further subdivided into topics. For each topic, the processes used in the Port Colborne and Sudbury CBRAs are described, followed by a commentary section. At the end of each theme section, a list of the most successful aspects and most challenging aspects of the CBRAs is provided. The conclusions section highlights the key findings of this analysis and implications for future CBRAs. Finally, based on the study findings, a number of recommendations are provided for consideration in future CBRAs.

14.4 Theme 1: Study Organization

This theme of analysis evaluated the organizational processes that were used in the Port Colborne CBRA and Sudbury Soils Study. Specifically, the projects were evaluated with respect to the following six topics:

- Organizational structure
- Technical committees and meetings
- Advisory roles
- Project consultants and scope of work
- Project schedule and budget
- External peer review process

14.4.1 Organizational Structure

The Port Colborne and Sudbury CBRAs used two different models of organization. The way in which these two studies were organized had a direct effect on the way in which the studies were carried out. The fundamental difference between the two studies is the degree of emphasis placed on the technical and public communication aspects of the CBRA. Both of these models had inherent strengths and weaknesses.

Port Colborne CBRA

The Port Colborne CBRA was designed with a strong mandate to be a completely open and transparent public process. A Public Liaison Committee (PLC) was developed to oversee the CBRA process and provide input to Vale, the MOE and the City of Port Colborne. The Technical Subcommittee (TSC) was established as an afterthought several months into the process when it was decided that technical discussions were consuming a disproportionate amount of time at PLC meetings. The TSC was a subcommittee of the PLC. The organizational structure of the Port Colborne Study is displayed in Figure 14.1 and consisted of the following groups of people:

- **Public Liaison Committee (PLC):** A committee of volunteer community members appointed by the City to solicit public input, inform the public and provide input to the CBRA process
- **Technical Subcommittee (TSC):** A subcommittee of the PLC comprised of representatives from Vale (the proponent), the City of Port Colborne, the MOE, Niagara Public Health, the project consultants and the independent consultant
- **Project Consultants:** A group of consultants that was hired by Vale to conduct the HHRA and ERA (Jacques Whitford Environmental Limited (JWEL) now part of Stantec)
- **Independent Consultant:** An individual hired to assist the PLC (Watters Environmental Group Inc. and who was also selected to be chair of the TSC)

- **Community:** The residents of Port Colborne who were invited to participate in local surveys, open houses and PLC meetings, and to observe at TSC meetings
- **External Peer Reviewers:** International experts in HHRA and ERA who conducted a technical peer review of the HHRA, ERA and Crop Studies reports.

The arrows in Figure 14.1 indicate the communication pathways among the various groups. The PLC communicated directly with all parties. PLC members rotated attendance at TSC meetings and asked questions or provided comments throughout. Similarly, TSC members were able to attend PLC meetings and address questions from the public, as directed by the PLC chair. The PLC was in a position of control over the organizational process and public aspects of conducting the CBRA, but it did not have control over the technical aspects of the project.

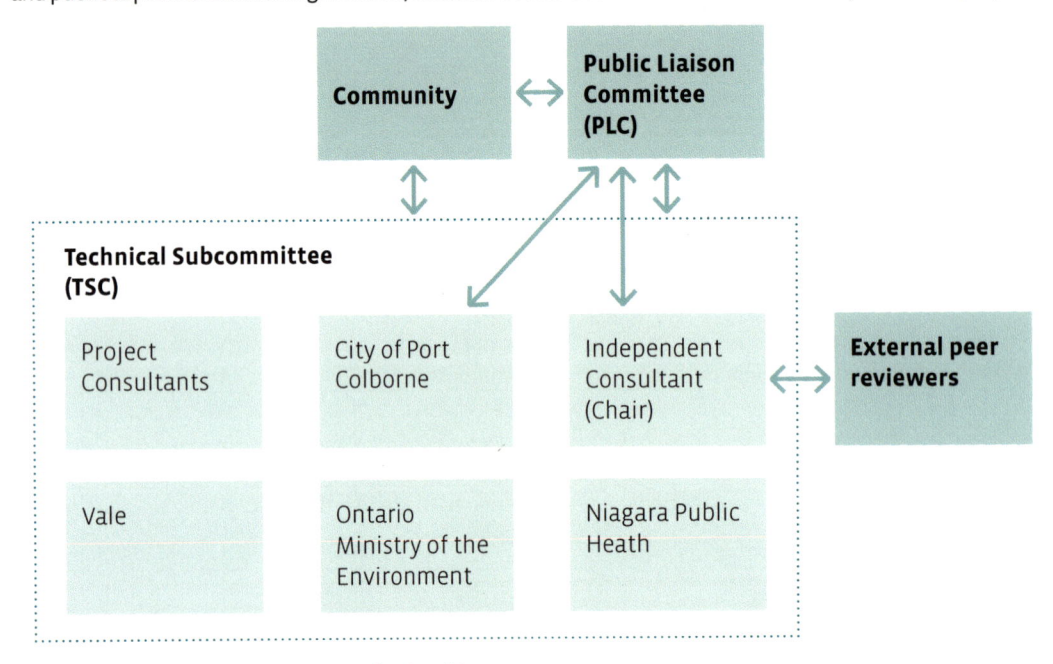

Figure 14.1 Port Colborne CBRA: Organizational Structure

The community was invited to participate in all PLC meetings and open houses, and to observe at TSC meetings. The PLC was provided with the services of an independent consultant dedicated to providing technical information and review services solely to the PLC and the City of Port Colborne. The independent consultant was selected by the PLC and was funded by Vale through a third party contractual arrangement with the City of Port Colborne.

The TSC was an assemblage of the various stakeholders involved in the process. Vale was the proponent of the CBRA that owned the former Ni refinery. The MOE was the government agency responsible for ensuring that the proponent conducted the CBRA according to the principles of the risk assessment process. Niagara Public Health and the City of Port Colborne participated to represent public health and City interests, respectively. The project consultants provided technical reports to Vale, which were then reviewed by the TSC. The independent consultant was selected to be the chair of the TSC since this was the only member of the TSC that was viewed to be truly independent at the beginning of the project. In his capacity as Chair of the TSC and consultant to the PLC, the independent consultant was the main pathway of communication between the PLC and the TSC. The independent consultant was also a liaison between the external peer reviewers and the TSC. The study was entirely funded by the proponent, Vale.

Sudbury Soils Study

In contrast to the Port Colborne CBRA, the Sudbury Soils Study was designed with a collaborative, multi-stakeholder TC at the helm, supported by a PAC to assist with public communication and facilitating information transfer between the TC and the public. The primary focus was on working collaboratively to produce a technically sound CBRA. Openness and transparency were built into the process through the use of an Independent Process Observer (IPO), PAC member observers and union observers. The organizational structure of the Sudbury Soils Study is displayed in Figure 14.2 and consisted of the following groups of people:

- **Technical Committee (TC):** A committee comprised of up to three representatives from six organizations (Vale, Xstrata Nickel, MOE, Sudbury & District Health Unit, the City of Greater Sudbury and the First Nations and Inuit Health Branch of Health Canada) who provided technical direction for the study. The TC was the primary group directing the CBRA
- **Working Group (WG):** A subcommittee of the TC, comprised of one representative of each organization on the TC (with the exception of Health Canada) who conducted the detailed day-to-day work with the Project Consultants on behalf of the TC between TC meetings and who attended all TC and PAC meetings
- **Independent Facilitator:** A third party individual retained to chair and facilitate the TC meetings and help to bring consensus for decision making
- **Communications Subcommittee (CSC):** A subcommittee of the TC, comprised of communications experts from each of the TC organizations (with the exception of Health Canada) and the project consultant team that was responsible for developing and overseeing implementation of the communications plan
- **Public Advisory Committee (PAC):** A committee of volunteer community members that provided public perspective and facilitated two-way communication between the TC and the public
- **Project Consultants:** A group of consultants hired by the TC to conduct the HHRA and ERA (the SARA Group)
- **Independent Process Observer (IPO):** An impartial observer and recorder of the process who was invited to attend all TC, WG, CSC and PAC meetings, open houses, workshops and other events to ensure that the process was transparent and that communication to the public was timely and effective
- **Unions:** Representatives of the Canadian Auto Workers and United Steel Workers Locals who observed the process on behalf of their members who were employees of Vale and Xstrata Nickel
- **Scientific Advisors (SA):** Two specialists in HHRA and ERA who reviewed the work of the project consultants, provided advice to the TC and delivered presentations to the PAC
- **Community:** The residents of Greater Sudbury who were invited to participate in open houses, local surveys, PAC meetings and a portion of TC meetings
- **External Peer Reviewers:** Panels of international experts in HHRA and ERA selected by an independent organization who conducted peer reviews of the HHRA and ERA.

The arrows in Figure 14.2 indicate the communication pathways among the various groups involved in the Sudbury Soils Study. The TC was the center of the organizational structure, maintaining communication with all other groups. The project consultants attended TC and PAC meetings as requested and submitted technical reports to the TC. The PAC communicated with the TC, the project consultants, the community and the scientific advisors. The community was invited to communicate directly with the TC, the PAC, the IPO and the project consultants. The external peer reviewers communicated with the TC collectively at Independent Expert Review Panel meetings for the HHRA and ERA. The City of Greater Sudbury representative was the liaison between the TC and TERA (Toxicological Excellence for Risk Assessment), the external organization hired by the TC to select the expert peer reviewers and organize the review process. The study was entirely funded by the two proponents, Vale and Xstrata Nickel. Financial arrangements were carried out directly between the proponents and the project consultant, but all technical reports were submitted directly to the TC.

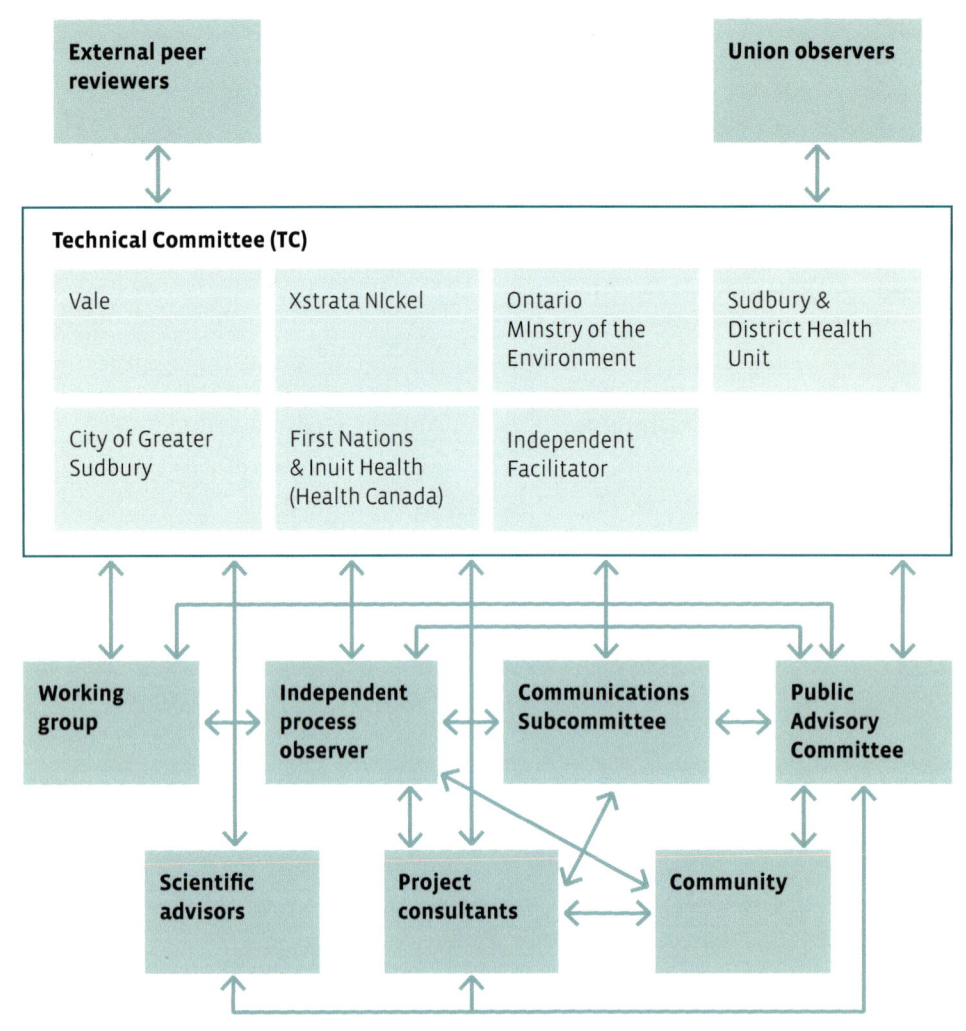

Figure 14.2 **Sudbury Soils Study: Organizational Structure**

Commentary

The differences between the organizational structures can be seen by comparing Figures 14.1 and 14.2. The Port Colborne CBRA had a relatively simple organizational structure. While the PLC was in a position of control over the organizational process and public aspects of conducting the CBRA, it did not have direct control over the technical aspects of the project. However, requests made from the PLC to other stakeholders were almost always accommodated, so even though the PLC held no formal operational control over the CBRA's technical program, it did have substantial influence over the process. Ultimate control for the technical aspects of the project resided with the proponent, Vale, with technical input from other members of the TSC and input from the PLC. The key strength of this organizational structure was the commitment to an open, public process, as evidenced by putting the PLC in a position of control over the process of the study. However, in doing this, the TSC had no control over the process and was forced to conduct all of its business in a public forum. This had a direct effect on the dynamics and productivity of the TSC (refer to Section 14.4.2 for more details).

In contrast, the Sudbury Soils Study had a much more complex organizational structure. The TC was in complete control over the technical and public communication aspects of the project. Other roles were created to ensure communication with the public (i.e. PAC) and openness and transparency in the process (i.e. IPO, union observers). The key strengths of this organizational structure were: the centralized management of the project within a cohesive TC; the creation of two subcommittees to focus on technical issues and communication issues, respectively; and the direction of project consultants by the TC as a whole, rather than by the proponents alone. A potential weakness of this organizational structure, in comparison to Port Colborne, was the subordinate position held by the PAC. This weakness was partially overcome by creation of special roles, the IPO and union observers, to ensure openness and transparency in the process.

14.4.2 Technical Committees and Meetings

Port Colborne CBRA

The TSC was essentially a working group of technical stakeholders that reported to the PLC. Although the MOE could have played a stronger regulatory role in the CBRA process, it was decided at the beginning of the project that MOE membership on the TSC would be equivalent to any other technical stakeholder. The independent consultant was selected to be the chair of the TSC. There was not consensus among participants about how the independent consultant was selected to be the chair.

A Terms of Reference (TOR) was not developed for the TSC. Participants indicated that the purpose of the TSC was to discuss technical issues only. It was agreed at the beginning of the project that the TSC would attempt to achieve consensus on decisions; however, no alternative method of decision-making or dispute resolution was established. Initially, the consensus model worked well, but over time, differences of opinion developed that could not be resolved. Therefore, the process of collective decision-making broke down. Meetings were typically held during daytime hours, usually monthly (more frequently when necessary), and typically on the same day as PLC evening meetings. Meetings were fully open to the public and media (as observers) and to PLC members (with the ability to ask questions or seek clarification on behalf of the public). While complete minutes were not kept, chairman's notes were prepared by the independent consultant that summarized the key decisions and discussion points. A total of 99 TSC meetings were held during the project (PLC 2010).

Sudbury Soils Study

The TC was established at the beginning of the project and developed the following overall vision for the CBRA:

> "A transparent process that provides a thorough scientifically sound assessment of the environmental and health risks to the Sudbury community, and effectively communicates the results so that future decisions are informed and valued" (Sudbury Soils Study 2002a)

All members of the TC, including the MOE, had an equal voice. An independent facilitator was hired to run the meetings and an administrative assistant recorded full minutes as well as one-page 'Key Progress and Decision Summaries' (posted on the study website: www.sudburysoilsstudy.com). A TOR was developed early in the process to define the purpose, reporting and decision-making process of the TC (Sudbury Soils Study 2002a). The TOR specified that all decisions would be made by consensus. If consensus could not be reached, a resolution mechanism would be adopted by the TC if and when required. It was clearly stated that the MOE and Sudbury & District Health Unit had legislative responsibilities 'that may take precedence over some matters'. The TOR also included a clause that allowed for development of working groups to undertake tasks under the direction of the TC (Sudbury Soils Study 2002a). As in Port Colborne, TC meetings were held during daytime hours, usually monthly, or more frequently when necessary. In contrast to Port Colborne, only the first hour of TC meetings was open to the public and media. A total of 62 TC meetings were held (PAC 2009; Mariotti 2009).

Commentary

All Sudbury participants agreed that the TOR for the TC was a valuable resource, and it provided clear direction for the TC throughout the study. Participants of both projects indicated that membership of the TCs was appropriate and included the key stakeholders. One participant indicated that a representative of the Ontario Ministry of Agriculture, Food and Rural Affairs would have been beneficial to the Port Colborne project because a large portion of the CBRA focused on agricultural areas. Many participants of the Sudbury Soils Study mentioned that there was not adequate representation of First Nations interests on the TC. While the First Nations and Inuit Health Branch of Health Canada was invited to participate on the TC, they did not take an active role in the project. To overcome this deficiency, the WG and project consultants consulted directly with First Nations communities via special meetings. Many participants also indicated that consistency in individual members throughout the time frame of the project was extremely beneficial for decision-making and maintaining the project schedule.

The technical committees differed at the leadership level. A chair position was established for the TSC in the Port Colborne CBRA, while a facilitator position was established in the Sudbury Soils Study. While both a chair and a facilitator can be successful, a facilitator has a greater ability to focus solely on the processes of discussion and decision-making. A chair must manage both his/her own perspective as a member of the discussion and allow for open discussion among all other members. The Sudbury model was more successful overall because the facilitator was not affiliated with any of the TC member organizations and did not represent any specific interests. In contrast, in Port Colborne, the independent consultant could not remain neutral and objective in the role of chair to the TSC because he was also representing the interests of the PLC and public.

TSC meetings in Port Colborne were completely open to the public and media. All of the public participants indicated that this was a strength of the process, while most of the technical participants viewed this as a weakness. Public participants indicated that there should be nothing to hide and, therefore, no reason to have closed door meetings. Some also indicated that open meetings provide opportunities for the public to become knowledgeable about the process of risk assessment. Technical participants indicated that it was difficult for technical stakeholders to have frank, open discussions in a public forum, especially in light of the class action lawsuit taking place. Some technical participants indicated that the TSC may have been more productive if the meetings had not been conducted in public.

In contrast, TC meetings in Sudbury were generally closed, apart from for the first hour when members of the public and media could make presentations or ask questions of the TC. Although the technical business of the TC was not conducted in public, the presence of the IPO, the chair of the PAC and union observers at TC meetings helped to ensure that the process was transparent and balanced. Discussions among members of the Sudbury TC were often very frank with members presenting contrasting opinions. TC members agreed that these types of discussions likely would not have taken place if members of the public were present. Openness and transparency is further discussed later in this chapter.

Decision-making by consensus at TSC and TC meetings added a substantial amount of time to both projects, but all participants agreed that the consensus model was the most appropriate method to use. In the best case scenario, the final product will be supported by all parties involved, as was the case in Sudbury. In a less desirable scenario, one or more dissenting stakeholders can delay the process, and consensus may not be possible. Without a previously established alternative decision-making process, there is a danger that the final product will not be supported by all parties. This was the case in Port Colborne where the independent consultant and PLC were not in agreement with the final study findings.

14.4.3 Advisory Roles

Port Colborne CBRA: Independent Consultant Role

The independent consultant role was created at the request of the City of Port Colborne to provide a neutral and objective source of technical expertise to the PLC to assist them in understanding the complex science involved in the CBRA process and to assist with communication of this information to the public. A TOR was not developed for the independent consultant. However, one clause within the TOR for the PLC and a column within the appended 'anticipated work program' for the independent consultant role included: providing ongoing professional and technical advice; providing reports and advice to the PLC as required; reviewing and advising PLC on the scope of work; monitoring progress of the CBRA, and reviewing and advising the PLC on the CBRA reports (Port Colborne CBRA 2000). In addition to the above roles, the independent consultant was selected to be the Chair of the TSC, prepare chairman's notes for the meetings, and report on the results of TSC meetings at PLC meetings. At times in the early stages of the project, the independent consultant was also selected to be a spokesperson for the TSC. The independent consultant was funded by Vale through a third party contractual arrangement with the City of Port Colborne.

Sudbury Soils Study: Scientific Advisor Role

The purpose of the scientific advisor role was to 'provide support and guidance for the TC'. Scientific advisors were expected to remain independent and conduct their work in a professional manner. They provided arm's length advice to the TC to help ensure that the study was conducted to high quality scientific standards. The advisors also made presentations to the PAC on occasion to discuss the processes of HHRA and ERA. The scientific advisors did not correspond directly with the project consultants. The scientific advisors were selected by and reported to the TC collectively, although they were paid directly by the proponents.

Commentary

While both the independent consultant and scientific advisor roles included review of information and input to the CBRA process, the primary functions of these two roles differed. The primary role of the independent consultant was to provide advice and expertise to the PLC and the public for the purpose of understanding the science, while the primary role of the Sudbury scientific advisors was to critically review the scientific aspects of the project and provide advice to the TC. The TOR for the scientific advisor's role was very detailed and clearly defined this role as one of scientific review. In contrast, the clauses within the PLC TOR pertaining to the independent consultant were vague and did not specify the depth of review and input expected.

According to PLC (2010), the independent consultant 'provided the public a better understanding of the science and their findings for the community and played a key role in representation of the interest of the public'. The role of independent consultant in the Port Colborne CBRA also had some challenges. The independent consultant was given a full member position on the TSC and was placed in a position of authority as the chair, a role in which he was expected to remain neutral and objective. At the same time, the same independent consultant was selected by and reported to the PLC. Given these circumstances, the independent consultant could not remain neutral and objective and it was likely a natural course of events that the independent consultant took on a larger role than perhaps was intended. In hindsight, a TOR for this role would have been useful. Another weakness suggested by PLC members was that the selected independent consultant did not have expertise in public communication and, therefore, could not assist the PLC with public communication of the CBRA process and results (PLC 2010).

All participants indicated that the two scientific advisors were beneficial to the Sudbury Soils Study. However, this was more likely due to the expertise, personalities and excellent communication skills of the individuals selected, rather than the roles themselves. Most agreed that these roles were not essential, but were beneficial. Many participants indicated that the scientific advisors could have been used more by the PAC to increase their level of understanding (PAC 2009; Mariotti 2009).

14.4.4 Project Consultants and Scope of Work

Port Colborne CBRA

The project consultant was selected by Vale on the basis of scientific expertise. There was no requirement for communications expertise. The project consultant had a contractual arrangement with Vale for funding and technical direction. The scope of work was developed collectively by the TSC, with input from the PLC, using the consensus model over an 8-month time frame. The deliberations during this process were intense and not all parties were fully in agreement with the final scope of work, especially with respect to the chemicals of concern. The scope of work continually expanded as the project progressed. Approximately 40 scope changes were processed by the proponent during the project to accommodate additional requests from TSC members and the PLC.

Sudbury Soils Study

A team of project consultants was selected collectively by the TC. The TC issued a Request for Proposals (RFP) and evaluated the credentials and expertise of the applicants with respect to their technical ability in HHRA and ERA as well as expertise in public communication. By consensus, the TC selected the SARA Group to conduct the HHRA and ERA and to develop and implement a communications plan. The project consultants had a contractual arrangement with the proponents for funding purposes, but technical direction was provided by the TC collectively. The initial scope of work was prepared by the TC, as outlined in the RFP. This process took almost a year to complete. Similar to the Port Colborne CBRA, the scope of work continually expanded as the project progressed. Approximately 26 scope changes were processed to accommodate additional requests from TC members and the PAC, to address external peer review comments, and to investigate new issues that arose during the study.

Commentary

The key difference between the two projects was the method of selection of the project consultants. In Port Colborne, these decisions were made solely by the proponent, Vale, although the credentials of the selected consultant were reviewed and approved by the MOE. In Sudbury, the project consultant team was selected collectively by the TC. The process used in Sudbury was viewed positively by all participants because the decision was made collectively, rather than by the proponents alone. In contrast, the process used in Port Colborne was not viewed favorably by some participants because it did not involve input from other TSC members and the PLC. The requirements for project consultants and scope of work also differed between projects. In Sudbury, the project consultant scope of work incorporated communication as a key component because communications were directed by the TC. Collective development of the scope of work was viewed as a lengthy, but necessary process by most participants.

14.4.5 Project Schedule and Budget

Port Colborne CBRA

The initial schedule for the Port Colborne CBRA indicated that the project would be completed within 18 months (Port Colborne CBRA 2000). Realistic expectations were that the study would be completed within 2 to 3 years with a proposed initial budget of $2 million. The actual CBRA has taken 11 years at a cost of approximately $20 million. An additional $10 million was spent on the community health studies. The project began in 2000. Eight months were required to establish a scope of work (finalized in November, 2000). Four reports were produced regarding the chemicals of concern (COC) for the risk assessment in 2001, followed by a re-evaluation report for Pb in 2004. Each of these reports was reviewed by the TSC, the PLC and the independent consultant. Two draft reports were released in 2003 and a final report was released in 2004 for each of the ERA and Crop Studies components. Each of these reports was reviewed by the TSC, the PLC and the independent consultant. Four draft reports of the HHRA were released from 2003 to 2005, with a final report in 2007. The PLC and the independent consultant continued to review the final report of the HHRA and hold PLC meetings, approximately monthly, until 2010. The final report of the PLC was submitted in July 2010. At the time of this writing, two of the four reports from the independent consultant to the MOE were still outstanding. According to participants, the key challenges involved in maintaining the study schedule in the Port Colborne CBRA included:

- the extensive time required to develop the original scope of work and sampling protocols using the consensus model
- obtaining as much local information as possible (resulting in additional field seasons and laboratory time)
- expanding the scope of work to address public requests, including property values and health studies
- external factors (including the lawsuit and media sensationalism)
- the time required for the TSC, PLC, independent consultant and public to review multiple draft reports, and
- the time required to communicate and build trust with the public throughout the study.

In addition to the above, the travel and administrative expenses required for the large number of meetings held throughout the project added substantially to project costs. Development of detailed sampling protocols and collection and analysis of duplicate samples (at the request of the independent consultant) also increased the budget for the Port Colborne CBRA.

Sudbury Soils Study

The Sudbury Soils Study was expected to be completed within 4 years with a proposed initial budget of approximately $2 million for the risk assessment portion. The actual CBRA was completed in 8 years (2 years to complete soil sampling and 6 years to complete the risk assessment). The total cost was approximately $12 million. The Sudbury Soil Study consisted of two phases. The first phase consisted of an extensive soil sampling and analysis program that began in 2001. Sample analyses were completed in 2003. The project consultants were retained in early 2003 to begin Phase 2. The first two volumes of the final report (study background and the HHRA) were released in 2008 and the third volume (the ERA) was released in 2009. The final PAC and IPO reports were released in June 2009. The final HHRA and ERA reports were accepted by the MOE and the project was completed. According to participants, the key challenges involved in maintaining the study schedule in the Sudbury Soils Study included:

- obtaining as much local information as possible (resulting in additional field seasons and laboratory time)
- extensive communications efforts to engage the public
- clarifying the requirements of the MOE, given the lack of guidance document or policy on CBRA
- continually expanding scope (approximately 26 scope changes to the original scope of work)
- additional work to address peer review comments and recommendations
- personnel turnover throughout the process
- extensive internal document review for clarity and language, and
- the process of making decisions by consensus.

In addition to the above, the travel and administrative expenses required for the large number of meetings held throughout the project added substantially to project costs.

Commentary

The challenges in maintaining the study schedule and budget were similar for both projects. Participants of both studies agreed that the original project schedules were ambitious and that there was not enough foresight at the beginning of the projects to clearly define the scope of work and to identify reasonable timelines. Some participants believe that an 8- to 10-year time frame for completion of a CBRA is realistic, given the size and scientific complexity of the projects and the requirement for extensive ongoing public communication. Others have suggested that the schedule could be substantially reduced if a definitive scope of work and budget were established and agreed to by all parties at the beginning of the project and then adhered to throughout the study, except for unforeseen circumstances. Decision-making by consensus at TSC and TC meetings added a substantial amount of time to both projects, but this method was supported by all participants of both projects. Perhaps the time required to obtain consensus in both studies could have been reduced if there had been an established protocol for arbitrating contentious issues, a role that could have been carried out by a neutral facilitator.

14.4.6 External Peer Review Process

Port Colborne CBRA

The decision to subject the Port Colborne CBRA to external peer review was made by the proponent, Vale, with support from other members of the TSC. Vale selected and hired the consultants to perform the external peer reviews. Once hired, the external peer reviewers reported only to the independent consultant. The peer review process for all reports was completed within 6 months. The results of the external peer review were discussed at TSC meetings and additional work was completed to address reviewer comments, where applicable. Peer reviewer comments, and associated responses from the project consultant, were provided in appendices and addendum reports to the HHRA, ERA and Crop Studies reports (JWEL 2004a,b, 2007b).

Sudbury Soils Study

The external peer review process for the Sudbury Soils Study was completely arm's-length. The chair and vice-chair of the PAC participated as observers in the process to select the external peer review organization (PAC 2005). The TC collectively selected an American organization, TERA, to coordinate the peer review process. TERA, in turn, independently selected two panels of experts in toxicology and risk assessment to review the HHRA and ERA. The panels of review experts convened a series of meetings in Sudbury to discuss the findings of their review with the TC and the project consultants. The external peer review process took approximately 1 year each to complete for the HHRA and the ERA, respectively. The peer reviewer reports were provided in appendices to the HHRA and ERA, along with the project consultant's responses to the panel's key recommendations and conclusions (SARA Group 2009).

Commentary

Technical and public participants agreed that the external peer review process was an important component of both CBRAs. An external evaluation of the technical aspects of the study increases the credibility of the study among the scientific community and can enhance public trust. According to participants of both projects, the length of time required for peer review was considered to be reasonable. One weakness of the Port Colborne external peer review process was that the external peer reviewers were selected by the proponent. The arm's-length selection process used in Sudbury provided another layer of credibility. Another weakness identified by public participants of both projects was that the public was not aware of, or did not fully appreciate, the significance of the external peer review process. This weakness was very evident at the end of the Sudbury Soils Study. A public interest group was formed and they hired two individuals to conduct separate reviews of the HHRA and ERA reports on their behalf. Perhaps, if there had been a greater understanding by this interest group of the rigor and arm's-length nature of the external peer review process, they wouldn't have needed to seek another opinion. Alternatively, they may have benefited from having the opportunity to ask their questions directly to the peer reviewers, rather than seeking other external reviewers.

14.4.7 Summary of Study Organization

Most Successful Aspects of Study Organization

According to participants, the most successful aspects of the organization of both studies were:

- the development of comprehensive scopes of work and protocols, involving input from multiple stakeholders
- collaboration of multiple diverse stakeholders
- involvement of qualified consultants
- the external peer review processes, and
- the dedication and efforts of the TSC/TC members and the volunteer PLC/PAC members.

Several other successful aspects of the Sudbury Soils Study were mentioned, including:

- agreement by consensus at the TC, which was described as being difficult, but satisfying in the end
- creation of the WG to conduct most of the detailed work and bring recommendations to the TC
- creation of the CSC to focus on the communication aspects of the project, and
- creation of the IPO role to promote openness and transparency.

Most Challenging Aspects of Study Organization

The most challenging aspect of both projects was the extremely long time it took to complete the projects. It was difficult for participants and the public to remain engaged for 8 or 10 years. The continually expanding scope and schedule of the projects was a constant challenge from both a time and budget perspective. Another common challenge was the process of coming to agreement by consensus at the TC/TSC. With the diverse mandates and responsibilities of TSC/TC members, it was not easy to come to consensus on issues. The consensus model failed in Port Colborne and no alternative methods of dispute resolution were established to guide the process to a conclusion that all TSC members could collectively support. The lack of regulatory guidance documents and approved methodologies for conducting CBRAs was a challenge for both projects. There was a general lack of clarity among participants of the expectations and role of MOE in this voluntary CBRA process. In many ways, this was a learning process for the MOE, as well as for the proponents and other stakeholders. While the final reports for the Sudbury Soils Study were accepted by the MOE, the reports for the Port Colborne CBRA were still in the review stage. The perception by some members of the public that the studies were inherently biased because they were being funded by the proponents was a constant challenge for both studies to address.

14.5 Theme 2: Public Communication

This theme of analysis evaluated the processes that were used in the Port Colborne CBRA and Sudbury Soils Study to communicate with the public. Specifically, the projects were evaluated with respect to the following eight topics:

- Public committees and meetings
- Communications plans and spokespeople
- Openness and transparency
- Introducing the study and gathering public input
- Communication methods and public interest
- Communication of results
- Media coverage
- Key public concerns and report comments.

14.5.1 Public Committees and Meetings

Port Colborne CBRA

The PLC was formed at the beginning of the study to address requests from the MOE and the City of Port Colborne for extensive public consultation throughout the CBRA. A TOR for the PLC was developed by Vale, the City and the MOE and reviewed by the PLC. The purpose of the PLC, as stated in the TOR, was to 'solicit public input; to inform the public; and to provide input' to Vale and the MOE with regard to the scope of work, preparation and conduct of the CBRA (Port Colborne CBRA 2000). The PLC's mandate included:

- receiving and reviewing all information associated with the CBRA
- providing input to Vale and the MOE Director respecting the scope of work for the CBRA
- monitoring progress of the CBRA
- reviewing and providing input on findings and recommendations of the CBRA
- providing inputs to Vale and MOE on the methods of implementing recommendations of the CBRA, and
- preparing a final report to the MOE with regard to the CBRA process (Port Colborne CBRA 2000; PLC 2010).

City Council reviewed the applications from interested members of the public and selected 7 members and one alternate to form the committee (PLC 2010). For the last 3 years of the project, only three of the original PLC members remained active. The services of an administrative assistant were provided by the City and funded by Vale. One of the PLC members volunteered to be an interim chair for the first few meetings. He was subsequently asked to remain as the permanent chair of the committee by the other members of the PLC. At the beginning of the project, the meetings were held more frequently to respond to the high level of public interest and to develop the scope of work for the project. Towards the end of the project, meetings were less frequent. The PLC meetings and monthly agendas were advertised in local newspapers and also emailed to interested members of the public. In total, 107 PLC meetings were held throughout the project. According to PLC members, these meetings provided ample opportunity for public input and engagement in the CBRA process (PLC 2010). PLC meeting minutes and associated reports were available to the public electronically, and in hard copy at the Public Library and at City Hall (PLC 2010). The PLC produced a final report at the conclusion of the study to summarize the work conducted by the PLC and the opinions and recommendations of the PLC (PLC 2010). All activities of the PLC were conducted on a voluntary basis, with no financial remuneration.

Sudbury Soils Study

In 2002, the PAC was formed to provide a linkage between the study team and the general public. An initial TOR was developed by the TC. It was subsequently revised by the PAC with approval of the TC (Sudbury Soils Study 2002b). The purpose of the PAC, as stated in the TOR, was to 'address the concerns of the community at large, by providing comments and input to the Technical Committee...' (Sudbury Soils Study 2002b). Unlike Port Colborne, the PAC was not responsible for scientific or technical review, but was offered access to study scientists to assist in understanding the scientific and technical aspects of the study. The PAC was invited to review communications materials before dissemination to the public and to provide suggestions to improve public understanding of the information (Sudbury Soils Study Website 2010). An additional role assumed by the PAC was to 'assure that the study was an open, transparent process' (PAC 2009).

The original PAC members were selected by the TC collectively. Two positions on the 10 member committee were reserved for representatives of two local First Nations. Members were asked to serve a term of at least 2 years. Once the PAC was established, it took over the responsibility of renewing the terms of existing members and selecting new members to replace those who resigned (SARA Group 2008c; PAC 2009). A chair and vice-chair were selected by PAC members. Twenty-six citizens served on the PAC during the project (PAC 2009). Meetings were held approximately bi-monthly at various locations within the community for convenient access. In the first years of the study, meetings were more frequent. A total of 40 PAC meetings were held throughout the study from 2002 to 2009 (PAC 2009). Notices were provided in local newspapers. The PAC made many recommendations to the TC throughout the study. The PAC produced five annual reports from 2003 to 2007, as well as one final commentary report in 2009. All activities of the PAC were conducted on a voluntary basis, with no financial remuneration.

Commentary

Public participants indicated that the public committees provided meaningful input to the CBRAs. The public committees brought a common sense perspective to the process and were able to provide local knowledge and historical information that was useful for designing sampling programs and collecting relevant data. While the intent expressed by technical participants in both projects was for both of these public committees to act as facilitators of information transfer to and from the public, the public committee in Port Colborne took on a much larger role in providing oversight for the entire project. In contrast, the Sudbury public committee maintained a supporting role throughout. TORs were essential to define roles and expectations.

Most participants of both studies indicated that PLC/PAC meetings provided an appropriate venue for people to present their concerns, ask questions and obtain information about the study. Meeting frequency was also considered to be adequate or more than adequate. Almost three times as many public meetings were held in Port Colborne, compared to Sudbury. More opportunities for written submissions or one-on-one discussions may have been useful. The time commitment by volunteer PLC and PAC members was extensive. It is unreasonable to expect volunteers to stay involved for a period of longer than 2 or 3 years. To maintain public trust, most participants felt strongly that membership on public committees should be voluntary (with paid expenses only). Stipends were also suggested for long-term positions (PAC 2009; PLC 2010).

Most technical and public participants in the Sudbury Soils Study indicated that review of communication materials was the most beneficial service provided by the PAC. The PAC provided useful comments and recommendations that helped to improve clarity and applicability of the materials for the local public audience.

14.5.2 Communications Plans and Spokespeople

Port Colborne CBRA

A communications plan was not developed for the Port Colborne Study. The TSC did not officially designate a spokesperson for the project. The chair of the PLC acted as the primary spokesperson for the PLC. The media were invited to attend all TSC and PLC meetings and were free to interview any member of the TSC or PLC or anyone else in attendance.

Sudbury Soils Study

In 2002, a Communications Sub-Committee (CSC) was formed to develop and oversee implementation of a communications plan. The CSC reviewed all public communications materials before dissemination. The CSC was comprised of senior communication representatives from each member organization of the TC, as well as a professional risk communicator who was a member of the project consultant team. A PAC member and the IPO were able to attend to maintain transparency of the process. A communications plan was developed by the CSC and the project consultants. It was reviewed and updated as necessary throughout the project (refer also to Chapter 13). Early in the study, it became apparent that a single spokesperson was required to address media questions, provide updated information to the public at various venues, and to act as a representative for the study itself. Because of the differences in mandates and responsibilities, no individual TC member could speak on behalf of the TC collectively. Therefore, it was decided that the director of the project consultant team would act as the primary study spokesperson. He addressed questions of a scientific nature and gave regular updates to the public about the technical aspects of the risk assessments.

Commentary

Overall, participants of the CSC in the Sudbury Soils Study indicated that everyone had a genuine interest in getting accurate information out to the public in a responsible and timely manner. One of the greatest challenges the group faced was how to communicate about risk in a way that was understandable to the public and that addressed the results of the risk assessment without causing undue alarm and without appearing to minimize the risks. All technical participants of the Sudbury Soils Study indicated that the communications plan was very useful. The major weakness of the communications plan was the minimal effort devoted to measuring the effectiveness of communication. A telephone poll was conducted in the early stages of the project that provided some information, but it would have been beneficial to conduct a second telephone poll towards the end of the project to see if any improvements had been realized. Responses were mixed about whether a communications plan would have been useful for the Port Colborne CBRA. PLC members indicated that they were not provided with any communications expertise and that their time was already overcommitted. While some participants of the Port Colborne CBRA thought that a spokesperson may have been useful, most agreed that there was not one candidate that could speak for all of the stakeholders. Opinions were mixed among participants in the Sudbury Soils Study process. Although the selected spokesperson was effective in his role, he could not speak on all issues. This presented a challenge, both for the spokesperson and the TC members. Most participants agreed that the idea of a spokesperson was good, but it was difficult to choose a candidate that could speak on all of the perspectives of the TC member organizations.

14.5.3 Openness and Transparency

Port Colborne CBRA

The Port Colborne CBRA was a completely open and transparent process. All PLC and TSC meetings were fully open to the public and media. Unlike Sudbury, special observer roles were not necessary because the public was allowed full access. Draft reports were made public and were reviewed and critiqued by the public, PLC and the independent consultant. PLC meeting minutes and TSC meeting chairman's notes were also publicly available. Public comments and questions, as well as responses from the project consultant were published in appendices to the reports.

Sudbury Soils Study

The Sudbury Soils Study model was not a completely open process, but it was made transparent by the creation of specific public observer roles, including the IPO, union observers and PAC observers. PAC meetings and the first hour of TC meetings were open to the public and media. The business portion of TC meetings, as well as WG meetings and CSC meetings, were closed. However, the IPO and PAC observers were invited to attend these meetings. Draft reports were not made public and the PAC was not responsible for technical review. The final reports were publicly released. Public comments and questions, as well as responses from the TC, were published in appendices to the reports. The IPO was provided with an honorarium of $10,000 per year. The honorarium was administered by a third party using funds provided by all member organizations of the TC (SARA Group 2008c).

Commentary

All participants of both projects indicated that openness and transparency was an essential component for trust and credibility. Most participants of the Sudbury Soils Study stressed the importance of the IPO, union and PAC observers in maintaining transparency throughout the process. The PLC did an exceptional job of providing the public with exhaustive access to the CBRA process. However, the PLC states that the Port Colborne CBRA process was not fully open at the conception stage when closed meetings were held between Vale, the City and MOE to develop the CBRA concept and TOR for the PLC (PLC 2010). Most of the technical participants of the Port Colborne CBRA suggested that technical stakeholder meetings in future CBRAs should be closed to the public, while all public participants thought they should remain open. In Sudbury, most participants favored closed technical stakeholder meetings, while a few participants recommended that at least a portion of the meetings could be open or that some meetings could be fully open and others fully closed. Technical participants of TSC meetings in Port Colborne indicated that it was difficult to have frank, open discussions and scientific debates in a public forum, especially in light of the class action lawsuit. In addition, many TSC members believed that productivity could have been enhanced if meetings had been closed. In contrast, public participants of the Port Colborne CBRA stressed the importance of maintaining open meetings to maintain public trust and increase public understanding of the risk assessment process. In Sudbury where technical meetings were closed, dynamic discussions and debate were challenging, but productive, because they helped the TC work towards consensus. The IPO also stressed the importance of having closed door TC meetings to provide members the opportunity to speak 'freely and openly' while maintaining transparency through the use of public observers (Mariotti 2009). Draft reports were publicly available in Port Colborne, but not in Sudbury. From an openness and transparency perspective, release of draft reports was favored. However, release of multiple draft reports was confusing to the public and a considerable amount of time was required to complete multiple reviews.

14.5.4 Introducing the Study and Gathering Public Input for Problem Formulation

Port Colborne CBRA

A series of open houses were held at the beginning of the project to inform the community about the project and to invite residents to participate in local surveys and data collection programs. Residents of Port Colborne were invited to participate in a variety of surveys and sampling programs to gather site-specific data for the risk assessments. These included a food consumption and time-activity survey, and sampling programs for residential soil, indoor air and dust, maple sap, home garden vegetables, local farm produce, poultry and eggs, and wild game and fish. Technical participants indicated that the PLC helped the technical team determine where to obtain local data.

Sudbury Soils Study

Three 'Have your Say' workshops were held in the communities of concern at the beginning of the risk assessment process (2003) to obtain local information from the public. Public input was solicited at four community information sessions (2002–2005) through a question and answer period following a presentation, during one-on-one conversations with study team members, and through exit survey questions and comments. Residents of Greater Sudbury were invited to participate in a variety of surveys and sampling programs to gather site-specific data for the risk assessments. These included a local food consumption survey, and sampling programs for home garden and farm produce, household dust, livestock, fish, and drinking water from private wells (SARA Group 2008c, d). The WG and project consultants delivered presentations to First Nations residents in the band communities. The project consultants also worked on two separate projects at the request of a First Nations community to sample and analyze metal concentrations in traditional food items. Because of the confidential nature of the information provided by the First Nations community, the data were not used in the CBRA. However, these studies helped to address some of the concerns expressed by the First Nations communities.

Commentary

The study team's effort to gather local information for the risk assessment was rated as good or excellent by all participants of both projects. All participants indicated that the methods used to gather local information were appropriate. No other methods were suggested.

14.5.5 Communication Methods

Port Colborne CBRA

The following methods of communication were used to update the public on study progress and results throughout the Port Colborne Study:

- Monthly PLC and TSC meetings: All PLC meetings were advertised in local newspapers. A total of 107 PLC meetings and 99 TSC meetings were held throughout the study (PLC 2010)
- Email list: An email list was maintained by the City representative for distribution of meeting notices and minutes
- Direct correspondence: Representatives provided information directly to interested members of the public
- Open houses: A total of 25 open houses were facilitated by the PLC during the study (PLC 2010)
- Project updates: Vale wrote several project updates, entitled 'An Open Letter to the Residents of Port Colborne.' These were printed in local newspapers to inform the public about study progress and results
- Individual and group meetings: The PLC offered to meet with individuals and groups in special meetings to discuss specific issues (PLC 2010)
- Media: The local media were invited to attend all PLC and TSC meetings. On a few occasions, the TSC also invited the media to conduct specific interviews to communicate important information to the public. The independent consultant and PLC submitted several informational articles and columns to the local newspapers regarding the CBRA process (PLC 2010).

No project website, toll free telephone line or central email address were established for this project.

Sudbury Soils Study

The communication methods used to update the public on study progress and results are described in Chapter 13 and are summarized here:

- PAC meetings and the first hour of TC meetings: These were open to the public and advertised in local newspapers. A total of 40 PAC meetings and 62 TC meetings were held during the project (PAC 2009; Mariotti 2009)
- Project website: The Sudbury Soils Study website (www.sudburysoilsstudy.com) provided general information about the study and contact information. It also served as a repository for documents including newsletters, risk assessment reports, IPO reports, news releases, PAC reports and meeting minutes and event/meeting notices
- Mail and email lists: Lists of approximately 1100 mail contacts and approximately 350 email contacts were compiled. Newsletters and IPO reports, as well as meeting notifications and invitations, were sent to all contacts on the mail and email lists
- Telephone and email communication: A toll-free telephone line and a project email address on the website were set up for the public to ask questions or provide comments. Approximately 550 telephone calls or emails were received between January 2003 and June 2009 (Mariotti 2009). The topic of each enquiry was logged and all enquiries were answered by a member of the project consultant team within 24 hours
- Open houses: Four open houses, also called community information sessions, were held to discuss progress and results of the study (2002–2005). Four additional community information sessions were held to present the results of the HHRA and ERA in 2008 and 2009, respectively
- Community newsletters: Twelve community newsletters were published by the project consultants, with review by the TC, CSC and PAC
- Individual and group meetings: Twenty-four meetings were held with local interest groups and First Nations (SARA Group 2008c; Mariotti 2009)
- Information packages: Packages of information were provided to local physicians nurses and healthcare providers
- Media: Members of the media were invited to TC meetings (first hour only), PAC meetings and open houses. On a few occasions, the study spokesperson submitted 'Op-Ed' pieces and 'Letters to the Editor' to local newspapers (SARA Group 2008c).

A telephone survey was conducted in the fall of 2004 to measure the effectiveness of communications initiatives in the early stages of the study. A total of 606 residents participated in the random sample survey. A follow-up telephone survey was not conducted at the conclusion of the study.

Commentary

Most participants of the Port Colborne CBRA indicated that the communication methods were appropriate, and the frequency of communication was adequate or more than adequate. All participants of the Sudbury Soils Study agreed that the communication methods were appropriate. Sudbury participants overwhelmingly believed that the 'high frequency, low intensity' communication strategy represented a strong and sincere effort to communicate with the public through diverse means. All participants of both studies indicated that the public had sufficient opportunities throughout the studies to provide input and ask questions. However, it was acknowledged that few residents exercised those opportunities in both projects. Unfortunately, the effectiveness of the communication methods in both studies was unknown because it was not measured, aside from one telephone poll in the early stages of the Sudbury Soils Study.

All participants of the Port Colborne CBRA rated public attendance at PLC meetings as high or very high at the beginning of the project (i.e. hundreds) and low or very low for the remainder of the project (i.e. 6 to 12 people; typically the same set of people). All participants of the Sudbury Soils Study rated public attendance at PAC meetings as low or very low throughout the project (i.e. typically less than 10 people). Many people in Sudbury voluntarily participated in local sampling programs, showing that they were interested in the study although they may not have attended meetings. For both studies, public attendance was higher at open houses than at PAC/PLC meetings.

14.5.6 Communication of Study Results

Port Colborne Study

The draft and final reports of the HHRA, ERA and Crop Studies were tabled at PLC meetings. The PLC would 'accept' the reports and make them available to the public electronically and at the local public library. After each report was accepted by the PLC, an open house was hosted by Vale in which the project consultants presented the findings of the study. Approximately 6 weeks later, another open house would be hosted by the PLC in which the independent consultant would present a review of the report and his evaluation of its strengths and weaknesses. The media were invited to attend all open houses. A non-technical executive summary was produced in both official languages (English and French) but complete public summary reports were not prepared.

Sudbury Soils Study

The results of the HHRA were released at community open houses in three locations in May 2008. The ERA results were released at one community open house in March 2009. The final reports were publicly released at these open houses. A press conference was held before the first open house for the HHRA and before the open house for the ERA. Reader friendly public summary reports and short summary newsletters were prepared for both the HHRA and the ERA in both official languages. Hard copies of these reports were available at the open houses. The summary newsletters were circulated with the Northern Life newspaper to approximately 40,000 homes, mailed to 1100 residents on the mailing list, and emailed to 350 individuals. Reports were made available on the study website and at the local public libraries.

Commentary

All participants of both studies agreed that a sincere effort was made to ensure that study results were communicated to the public in a manner that was easy to understand, although many also agreed that the risk assessment process and results were challenging to communicate. Public comprehension of the results was not measured in either study. There were three major differences between the two projects with respect to communication of study results. First, all draft and final reports were released to the public in Port Colborne, whereas only the final reports were available to the public in the Sudbury Soils Study. The advantage to releasing draft reports is that it makes the process of report review completely open. However, multiple draft reports with differing results can be confusing to the public and some people may lose interest during the long process of getting to the final results. Conversely, if only one message is communicated in a final report, as was done in the Sudbury Soils Study, there is less tendency for confusion among the public, but more opportunity for people to question whether they trust the process that was used to derive that message.

Secondly, two separate open houses and two sets of messages were provided for each report in Port Colborne, while one open house and one message was delivered collectively by the TC for each report release in Sudbury. While the process in Port Colborne may have been confusing to the public, the one open house format in Sudbury may not have provided a meaningful opportunity for public dialog about the study results.

Thirdly, plain language public summary documents and newsletters were prepared in Sudbury, while only executive summaries were available in Port Colborne. The public summary documents and newsletters prepared for the Sudbury Soils Study were viewed as an asset by all participants.

The issue of project closure was discussed by many participants of both studies. To date, the Port Colborne CBRA has not achieved closure because consensus was not reached during the process and the MOE has received conflicting submissions – the final risk assessment reports from Vale and the critique reports from the independent consultant and the PLC. Because closure was not achieved at the technical level, it was also not achieved with the public. In Sudbury, closure was achieved at a technical level for the risk assessment. The final reports were accepted by the MOE and the project moved on to the risk management phase. However, some participants argued that closure was not achieved among the public for the HHRA. After the final results were released, a small group of citizens challenged the results of the study and received media coverage that may have left the public with a confusing collection of messages. Some participants, including the IPO (Mariotti 2009), suggested that another public meeting may have been warranted to allow residents to engage in open discussion with the study team, and perhaps other experts, to provide public closure.

14.5.7 Media Coverage

Port Colborne CBRA

The Port Colborne CBRA received local and national media coverage. Over time, national media lost interest in the study while local media continued. All PLC and TSC meetings and open houses were open to the media and coverage was fairly regular throughout the duration of the project, until the last couple of years where media coverage was infrequent. Overall, media interest was much higher in Port Colborne than in Sudbury. Occasionally, Vale would prepare project updates that appeared in local newspapers. The independent consultant and PLC also submitted several articles and columns to the local newspapers (PLC 2010).

Sudbury Soils Study

The Sudbury Soils Study received primarily local media coverage by two local newspapers, the local CBC radio station and the local CTV news. Newspapers provided the most frequent coverage. PAC meetings and the first hour of TC meetings were open to the media. Most participants indicated that media interest was low throughout the project.

Commentary

Overall, participants of the Port Colborne CBRA viewed the national media coverage to be a weakness and the local media coverage to be a strength. Most participants agreed that local reporting tended to be more accurate and less sensational than the reporting that occurred in the national media. Some participants suggested that the initial media frenzy was a major factor leading to the high level of concern among the public in Port Colborne at the beginning of the study. In instances where there were diverse opinions about the study, some participants indicated that the national media sources tended to be biased against the study. However, most participants indicated that the local media, over time, was more likely to, but not always, fairly represent all of the opinions. While most technical participants indicated that the media reports provided enough information for the public to understand the project, some participants indicated that the public was not adequately informed through media reports alone.

In contrast, most participants of the Sudbury Soils Study indicated that local media coverage of the study was poor overall and not beneficial to the project. Media interest was low in general and coverage was sporadic, depending on events that unfolded during the study. The media largely reported about the events (e.g. open houses) and the opinions of people at these events, rather than the facts of the study. Most participants indicated that a disproportionate amount of media attention was given to vocal opponents of the project. The WG held editorial board meetings with newspapers to attempt to improve coverage, and some improvements were made, but inaccuracies still occurred. Some of the errors were addressed in 'Op-Ed' articles and 'Letters to the Editor' by the project spokesperson. After the release of study results in Sudbury, media reporting was confusing because a lot of attention was given to vocal interest groups with opposing views of the study results, and then coverage abruptly ended. Therefore, the public was left with a confusing array of articles about study results, some of which were factual and some of which were speculative.

Many participants of both projects agreed that members of the public who relied solely on media reports for information would not have received sufficient information to be adequately informed about the studies. However, the media reports did provide an opportunity for members of the public to be aware of the study and to learn where further information could be obtained.

14.5.8 Key Public Concerns

Port Colborne CBRA

A variety of comments and questions were received by the public throughout the project at PLC meetings and open houses, as well as via written submissions, email and telephone. The key public concern expressed by residents at the beginning of the project was human health, especially the health of children (and pets). There was a perception among some residents that there was a disproportionate amount of illness within the community and that this may be due, at least in part, to soil contamination. Members of the agricultural community were also concerned about potential impacts to crops and livelihoods. Later in the study, public interest shifted from health risks to the impact of soil metal levels on property values and the ethical/legal issues surrounding disclosure of soil quality results during property sale transactions. In addition to the above concerns, there was an underlying distrust of the CBRA process because the study was funded entirely by Vale and some of the major decisions in the process were made solely by Vale (e.g. selection of project consultants and external peer reviewers).

Sudbury Soils Study

As in the Port Colborne CBRA, the key public concern of the Sudbury Soils Study was human health, especially children. Farmers, gardeners and blueberry harvesters were also concerned about potential impacts to crops and livelihoods. Some residents were also concerned about impacts to property values, the health of recovering ecosystems, and the overall reputation of Sudbury (Sudbury Soils Study 2003, 2004). Some residents believed that the scope of the study was too narrow because it did not evaluate occupational exposures or the potential health risks associated with historical impacts when emissions were known to be much higher. Some residents also believed that a community health study and/or direct measurement of metal concentrations in human tissues (e.g. blood, hair) was required to address public concerns. The Sudbury Soils Study attempted to address these concerns to the extent that was possible. In addition to the above concerns, there was an underlying distrust of the CBRA process by some residents because the study was funded by Vale and Xstrata Nickel. The TC attempted to address this issue by making all decisions collectively by consensus, and by creating the IPO role with a mandate to act as an impartial public observer and recorder of the process. Despite these attempts to make the process open and transparent, some residents continued to distrust the process and this issue was a regular focus of media articles.

Commentary

Although the primary concern in both studies was human health, a number of other concerns were identified. Some concerns could not be addressed by the CBRA process (e.g. occupational exposures, property values, legal/ethical issues associated with sampling on private properties); however, simply stating this fact did not alleviate these concerns. The CBRA process does not address issues associated with property values and sampling on private properties, but it is clear from these two studies that these issues are of extreme importance to the public and that a mechanism for addressing these concerns should be developed.

14.5.9 Summary of the Public Communication Process

Most Successful Aspects of the Public Communication Process

According to participants, the most successful aspect of the public communication process in both studies was the focus on openness and transparency. While the two studies employed different methods to promote openness and transparency, all participants of both projects believed, for the most part, that this had been achieved. Participants specifically highlighted the IPO and other observer roles in the Sudbury Soils Study and the number of open public meetings in the Port Colborne CBRA as contributing substantially to the transparency. The roles of the PLC and PAC were also noted as being beneficial to the public communication processes of both projects. Technical participants praised the dedication of the volunteer participants of these committees. Both projects provided opportunities for the public to become engaged in the process.

Most Challenging Aspects of the Public Communication Process

One of the common challenging aspects expressed by participants of both projects was how to most effectively communicate complex scientific information to the public in a manner that was accurate, timely and easy to understand. In addition, choosing an appropriate study spokesperson in the Sudbury Soils Study was challenging. In Port Colborne, the class action lawsuit and media sensationalism were external factors that were challenging to deal with throughout the project. Poor and often sensational local media coverage was also a factor that was challenging for the Sudbury Soils Study. Public distrust for the proponents, government agencies, and even sometimes for the public committees, was difficult to deal with. The emergence of a public interest group with a dissenting view at the end of the Sudbury Soils Study process was challenging to handle at the conclusion of the project. Participants of both projects agreed that it was a huge challenge to get the public involved and to sustain that involvement given the length of the projects and highly technical nature of the process.

One of the key challenges of the public communication process was determining the appropriate amount and nature of public involvement. A comprehensive public communication process added value to the studies, however, effective public consultation takes time and resources. In studies such as these, determining the appropriate amount of public involvement is a balancing act. Not enough involvement may result in mistrust and opposition to the project, while too much can lead to substantially extended project timelines and stakeholder and public fatigue.

14.6 Conclusions

The experiences of these projects demonstrate that a complex scientific risk assessment can be conducted under the direction of a diverse team of stakeholders and within an open and transparent public process. Both studies used a multi-stakeholder process involving the proponents, the MOE, local health units, local municipal staff, project consultants and independent consultants, as well as committees of local citizens. The consensus model of decision-making was used as much as possible and extensive efforts were made to engage the local communities. The Port Colborne CBRA and Sudbury Soils Study used different models of organization, which had a direct effect on the way in which they were carried out. The fundamental difference between the two studies is the degree of emphasis placed on the technical and public communication aspects of the CBRA. Both of these models had inherent strengths and weaknesses.

The key strength of the Port Colborne CBRA organizational model was the focus on a completely open and transparent process, with plenty of opportunities for public involvement and engagement. However, perhaps the greatest weakness of this model was the lack of cohesiveness and control exhibited by the TSC. At the end of the project, the public was given ample opportunity to provide input, but the TSC as a whole did not support the final reports and study findings. Thus, study closure was not achieved in Port Colborne. The parties involved in the Sudbury Soils Study were fortunate to be able to learn from the early experiences of the Port Colborne CBRA and incorporated greater structure and control into the process. The key strength of the Sudbury Soils Study was the TC, who remained a cohesive group that spoke with one voice. However, the increased level of structure and control exerted by the TC may have decreased the openness and transparency of the process. While this project was not as open as the Port Colborne CBRA, all participants agreed that the project was sufficiently open and transparent. Some of the common successes of both projects included:

- the development of comprehensive scopes of work and protocols, involving input from multiple diverse stakeholders
- the dedication and efforts of the TSC/TC members and volunteer PLC/PAC members
- the commitment to making decisions by consensus
- the openness and transparency in the process
- the external peer review process
- the involvement of public committees
- the open houses for soliciting input and communicating results, and
- the opportunities for public engagement.

Some of the common challenges of both projects included:

- the long time frame for completion of the projects
- the continually expanding scope, schedule and budgets
- the time and process required for agreeing by consensus
- lack of regulatory guidance documents and approved methodologies for conducting CBRAs
- distrust of the process by some residents
- determining how to most effectively communicate complex scientific information in a manner that is accurate, timely and easy to understand, and
- public involvement and sustained engagement over the length of the projects.

The major challenges for future CBRAs will be to reduce the project schedule and budget, and sustain public engagement throughout the process. Future CBRAs would also benefit from regulatory guidance that defines expectations for technical reports and public communication, and the role of the regulatory agencies within this voluntary process. The CBRA process does not inherently address issues associated with property values and sampling on private properties, but it is clear from these two studies that these issues are of extreme importance to the public and that a mechanism for addressing these concerns ought to be developed before proceeding with a future CBRA. Finally, before embarking on a large and complex CBRA, the community should be consulted to identify their key concerns and whether these concerns are best addressed through a CBRA or through another process. These two projects have demonstrated that CBRA can be a lengthy, costly and logistically complex process, and it may or may not address the underlying concerns within the community.

14.7 Recommendations

Based on the comments received, interviews and review of available information for the Port Colborne CBRA and the Sudbury Soils Study, a series of recommendations were developed for future large scale risk assessments. The following recommendations are organized according to themes and topics as provided in the preceding text.

Organizational Structure:

- Develop an organizational structure that maintains control of the budget, schedule and study process at the technical level, but permits as much public involvement as possible
- Ensure that the organizational structure reflects the principles of openness and transparency
- Consider the development of subcommittees to facilitate technical work and public communications
- Clearly define the roles of all stakeholders as well as any advisors and process observers; and
- Clearly define project expectations (both technical and communication).

Technical Committees and Meetings:

- Develop a comprehensive Terms of Reference (TOR) with the preferred decision-making process and consideration of alternatives and mechanisms for dispute resolution
- Ensure adequate representation of key stakeholders including specific interest groups (e.g. agricultural, First Nations)
- Hire a neutral, experienced facilitator with a clear TOR
- Develop a model for openness and transparency at technical committee meetings
- Designate at least a portion of technical meetings to be conducted in private to allow for open, scientific discussions and optimal productivity
- Provide training and team building exercises to help group dynamics; and
- As much as possible, make decisions collectively and make a commitment to remain cohesive as a group throughout the project.

Advisory Roles:

- Collectively decide whether these roles are necessary and/or desirable for the specific project
- Select independent consultants in an open public process
- Develop a comprehensive TOR for the position and clearly specify the roles and responsibilities
- Maintain the independence of the advisor roles (i.e. no decision-making or chair roles); and
- Provide opportunities for the consultants or advisors to interact directly with the public.

Project Consultant and Scope of Work:

- Use an open, public bidding process to post the request for proposals for candidate consultants
- Collectively develop criteria with which to evaluate proposals in a fair and equitable manner
- Develop the scope of work collectively by consensus; and
- Consider public involvement or representation in the consultant selection process.

Project Schedule and Budgets:

- Maintain continuity of technical committee personnel throughout the project, as much as possible
- Consider an initial scope of work and budget only for the problem formulation phase, then define budgets and schedule for subsequent detailed risk assessment
- Establish a process at the beginning of the project for how changes to the scope of work can be made and how the additional work will be funded

- Limit the number of draft reports and maintain strict timelines for internal (and external) review
- Establish a smaller working group to research issues and make recommendations to the technical committee for approval
- Consider managing funds through a third party to limit the perception of bias due to proponent funding; and
- Deliberate on the importance of collecting additional data relative to whether it will increase confidence in the study results and what effect it will have on project timelines.

External Peer Review Process:
- An external peer review process is highly recommended for future CBRAs
- Select external peer reviewers in an open process by collective agreement of the technical committee
- Consider the input of PAC members or public interest groups in the selection of external peer reviewers
- Keep the process as arm's-length as possible to enhance public trust; and
- Provide opportunities for the public to interact directly with the external peer reviewers, and emphasize the value of this component of the project in public communications.

Public Committees and Meetings:
- Involve the public at the conception stage and seek their input on member selection processes
- Consider alternative methods for recruitment of public committee members to encourage broad representation that reflects the makeup of the community
- Maintain the voluntary nature of the public member positions but reimburse expenses for travel, meals and child care; consider stipends for long-term positions and provide recognition
- Set a specific term of 2 or 3 years for membership, with option to renew
- Limit the size of the public committee to a manageable number (6 to 8)
- Develop a clear and comprehensive TOR for the public committee
- Allow public committee members to review and make changes to their TOR, within reason
- Allow public committee members to be involved in other aspects of the project as observers (such as at technical stakeholder meetings, during the expert review process, at seminars and workshops)
- Ensure that the general public understands the mandate of the public committee and its limitations
- Consider hiring a professional facilitator or a rotating chair position
- Consider less frequent public committee meetings and more open information sessions; and
- Create opportunities for written submissions or one-on-one discussions for residents who are not comfortable raising their questions during public meetings.

Communication Plans and Spokespeople:
- Develop a communication plan and principles (e.g. openness, transparency, accuracy, timeliness) at the beginning of the project
- Involve qualified risk communication experts in development and delivery of the communications plan;
- Create a communications subcommittee to focus solely on communication issues
- Involve the public committee in the development or review of communications plans
- Consider having a central spokesperson and whether having a spokesperson is necessary; and
- Choose a neutral spokesperson with good communication skills, and that will be viewed by the public as a credible representative of the project.

Openness and Transparency:
- Strive to make the process as open and transparent as possible, while maintaining the ability of the technical stakeholders to get the work completed in a timely and scientifically-defensible manner
- Consider the trade-offs of having open or closed technical stakeholder meetings
- If any meetings are closed to the public, involve public observers in the process
- Clearly define the rights and responsibilities of observers and a process for any issues that may arise
- If the public committee is involved with scientific review, provide draft reports that are as complete as possible; and
- If the public committee does not have responsibility for scientific review, there is no need to release draft reports.

Gathering Public Input:
- Actively solicit the advice of the public committee, local residents and special interest groups when gathering local information and determining where to obtain local samples
- Advertise widely to maximize participation of the community in local surveys and sampling programs; and
- Show how public input was incorporated into the risk assessments.

Communication Methods and Public Interest:
- Create an engaging and user-friendly project website and upload all relevant materials
- Maintain a physical collection of documents in the local public library
- Consider a wide variety of communication methods to reach the public at large
- Create alternative mechanisms for direct communication between the public and public/technical committee members, rather than just attending meetings
- Investigate the use of social media (e.g. Facebook, Twitter) for public engagement;
- Develop a plan for measuring the effectiveness of communication methods (at least twice during the study). Use the results to improve the communication strategy; and
- Explore novel approaches for engaging the public in future CBRAs, such as participatory workshops, consensus conferences or focus groups in which citizens become more educated about the science and play a more active role in the process.

Communication of Results:
- Consider the need to release draft reports to the public
- Use an open house format with a range of public interaction methods
- Allow time for the public to read reports after the open house and provide another opportunity for public dialog to achieve closure (e.g. a second open house, public forum, experts debate); and
- Consider creating public summary documents and/or project highlight summary sheets.

Media Coverage:
- Invite the local media to attend all public meetings, open houses and other events during the project;
- Use the local media to advertise all public events
- Request editorial board meetings at the beginning of the project to provide local media with some background information and lists of key contacts for technical and public spokespeople
- Continue to meet on a regular basis with the media to provide them with accurate and timely information
- Actively participate in media reporting by submitting educational articles or 'Letters to the Editor' authored by the technical or public committees; and
- Prepare and widely distribute study-specific communications materials (e.g. newsletters) so that the media are not the only source of information for the public at large.

Key Public Concerns:
- Conduct a comprehensive consultation program at the beginning of the project to determine community concerns and project expectations
- Consider evaluating risks to pets if this is a key concern of the public
- Explain why a CBRA process is not designed to address occupational exposures
- Clearly communicate the differences between risk assessments and community health studies
- Consider bio-monitoring programs or community health studies if this would address public concerns; and
- Investigate the legal and ethical issues associated with collecting environmental samples on private properties and adequately inform property owners of their legal obligation for disclosure, and their right to refuse sampling.

14.8 Acknowledgements

We would like to acknowledge the contributions of the Steering Committee for this project, which included: Dr Beverley Hale, University of Guelph; Marc Butler, Xstrata Nickel; Glen Watson, Dr Mike Dutton and Dr. Bruce Conard, Vale; Dave McLaughlin and Brian Cameron, Ontario Ministry of the Environment; Dr Stephen Hall, formerly Laurentian University, now at Curtin University, Australia; Dr Graeme Spiers, Laurentian University. We would also like to thank the many people in Sudbury and Port Colborne who volunteered their time to participate in interviews, gather documents, and provide information that made this project possible. Funding for this study was provided by the Canadian Network of Toxicology Centres (University of Guelph), Vale and Xstrata Nickel.

14.9 References

City of Greater Sudbury (2010) Keyfacts: City of Greater Sudbury. www.greatersudbury.ca/keyfacts. Accessed 2 August 2010

City of Port Colborne (2010) City of Port Colborne (municipal website). www.portcolborne.ca. Accessed August 2010

JWEL (2004a) Community based risk assessment Port Colborne Ontario. Ecological risk assessment – Natural environment. Volume 1 – Main report. Appendix A: Jacques Whitford responses to independent third party peer review comments. Jacques Whitford Environmental Limited, Toronto, 71 pp

JWEL (2004b) Community based risk assessment Port Colborne Ontario. Ecological risk assessment – Natural environment. Volume 1 – Main report. Appendix B: Jacques Whitford responses to public comments. Jacques Whitford Environmental Limited, Toronto, 44 pp

JWEL (2007a) Port Colborne community based risk assessment. Human health risk assessment. Final report. Volume 1. Jacques Whitford Environmental Limited, Toronto, 310 pp

JWEL (2007b) Port Colborne community based risk assessment. Human health risk assessment. Final report. Volume VI. Report review comments and public notices. Jacques Whitford Environmental Limited, Toronto, 1605 pp

Mariotti F (2009) Independent process observer quarterly report #23. June 2009. 18 pp

MOE (1997) Guideline for use at contaminated sites in Ontario. Ontario Ministry of the Environment, Toronto, Ontario

MOE (2001) Metals in soil and vegetation in the Sudbury area (survey 2000 and additional historic data). Ontario Ministry of the Environment, Toronto, Ontario

PAC (2005) Public Advisory Committee's (PAC) 2004 Annual Report. PAC, 8 pp

PAC (2009) The Sudbury Soils Study: A commentary. June 2009. PAC, 20 pp

PLC (2010) The final Public Liaison Committee report on the Port Colborne community based risk assessment. 8 July 2010. PLC, 23 pp (plus appendices)

Port Colborne CBRA (2000) Terms of reference of the Public Liaison Committee for the community based risk assessment for soils contaminated in the Port Colborne area. Unpublished report, Port Colborne, Ontario

SARA Group (2008a) Sudbury area risk assessment, Volume I, Chapter 1.0: Introduction to the Sudbury soils study. SARA Group, 18 pp www.sudburysoilsstudy.com. Accessed 15 September 2010

SARA Group (2008b) Sudbury area risk assessment, Volume I, Chapter 2.0: History of the Sudbury smelters. SARA Group, 70 pp www.sudburysoilsstudy.com. Accessed 15 September 2010

SARA Group (2008c) Sudbury area risk assessment, Volume I, Chapter 6.0: Communications and public consultation program. SARA Group, 15 pp www.sudburysoilsstudy.com. Accessed 15 September 2010

SARA Group (2008d) Sudbury area risk assessment, Volume I, Appendices F, G, H, K, L and M. SARA Group www.sudburysoilsstudy.com. Accessed 12 August 2010

SARA Group (2009) Sudbury area risk assessment, Volume II, Appendix P: IERP comments and SARA Group responses. SARA Group, 66 pp www.sudburysoilsstudy.com. Accessed 12 August 2010

Statistics Canada (2006) 2006 Census. Community profiles for Greater Sudbury and Port Colborne. www.www12.statcan.gc.ca/census-recensement/2006/rt-td/index-eng.cfm and www.investinontario.com/ Accessed 22 July 2010

Sudbury Soils Study (2002a) Terms of Reference for the Technical Committee of the Sudbury Soils Study. www.sudburysoilsstudy.com. Accessed 10 August 2010

Sudbury Soils Study (2002b) Public Advisory Committee Terms of Reference. www.sudburysoilsstudy.com. Accessed 12 August 2010

Sudbury Soils Study (2003) Have Your Say Workshop Summary Report. June 2003. 4 pp www.sudburysoilsstudy.com. Accessed 12 August 2010

Sudbury Soils Study (2004) Public Open House Summary Report. February 2004. 9 pp www.sudburysoilsstudy.com. Accessed 21 August 2010

Sudbury Soils Study (2010) www.sudburysoilsstudy.com. Accessed 12 August 2010

14.10 Appendix Chapter 14: Abbreviations

CBRA, community based risk assessment

COC, chemicals of concern

CSC, Communications Subcommittee (subcommittee of the TC; Sudbury CBRA)

ERA, ecological risk assessment

HHRA, human health risk assessment

IERP, Independent Expert Review Panel (Sudbury CBRA)

IPO, Independent Process Observer (Sudbury CBRA)

MOE, Ministry of the Environment, Ontario

PAC, Public Advisory Committee (Sudbury CBRA)

PLC, Public Liaison Committee (Port Colborne CBRA)

RFI, Request For Information

RFP, Request For Proposals

SA, Scientific Advisors (Sudbury CBRA)

SARA, Sudbury Area Risk Assessment

TC, Technical Committee (Sudbury CBRA)

TERA, Toxicology Excellence for Risk Assessment (peer review organization; Sudbury CBRA)

TOR, Terms of Reference

TSC, Technical Subcommittee (Port Colborne CBRA)

WG, Working Group (subcommittee of the TC; Sudbury CBRA)

Index

A

acceptable daily intakes 176
acid-metal tolerant 300, 301
acid precipitation 233, 234, 235
acid rain 10, 11
acid-tolerant species 326, 328
additivity 235, 236
Agency for Toxic Substances and Disease Registry 178, 194, 195
air filter samples 136, 138, 139, 140
air monitoring 387
air monitoring program 119, 162, 220, 241, 243
air-monitoring program 40
Air Quality Criterion 121
Aluminum Toxicity 336
American robin 348, 352, 354, 361, 362, 363, 364, 365, 366, 368, 370, 376, 378
antagonism 236
area-wide risk assessment 99, 100
artificial soil 305, 311
assessment endpoint 268
assessment endpoints 257, 268, 269, 270
atmospheric deposition 53, 63, 64, 65, 69

B

background concentrations 75, 76, 86
background levels 52, 54, 60
baghouse 386, 388, 389
barren 262, 263
barrens 42, 43, 45, 47, 263, 390, 397
Barrens 258
beavers 364
bioaccessibility 18, 119, 130, 131, 132, 133, 134, 151, 152, 153, 154, 216, 226, 227, 228, 243, 245, 246, 247, 252, 350, 372, 373
Bioaccessibility 157
bioavailability 18, 113, 119, 130, 132, 133, 134, 151, 152, 153, 154, 176, 178, 179, 226, 228, 235, 245, 246, 287, 290, 319, 323, 329, 334, 336, 337, 338
bioconcentration factors 351
biodiversity 263, 264, 269, 280, 301, 302, 326, 327, 328, 339
Biodiversity Action Plan 337, 381, 382, 390, 394, 395, 396, 397, 399, 400
biomonitoring 243
bladder cancer 198
blood lead level 225, 232
blood Pb 226, 228, 229, 230, 231, 232, 233, 244, 384
Boreal forest 306
broad plant inventory 297, 326
Bulk density 282, 287, 290, 320, 325
bulk mineral analysis 138

C

cancer potency 177, 199
cancer risk 209, 214, 220, 222
Cancer Risk Level 157, 193
cancer slope factors 130, 177
canonical correspondence analysis 284, 318
carcinogenic 103, 176, 177, 178, 197, 202
cation exchange capacity 319, 320, 338
cation-exchange capacity 288
central tendency estimate 103, 104, 159, 165
certified reference material 289
Certified Reference Material 142
chemicals of concern 99, 209
Chemicals of Concern 18, 49, 51, 344, 345
cigarette smokers 242
City of Greater Sudbury 10, 11, 19, 51, 53, 70, 423, 424, 427, 447
City of Greater Sudbury, 259, 267, 273
coarse woody material 298
Coeur d'Alene ERA 350, 361
communications plan 427, 431, 436, 445
Communications Subcommittee 427
Communities of Interest 97, 101, 102, 159, 260
Community Based Risk Assessment 423
community health studies 432, 447
community health study 99, 442, 447
community information sessions 438, 439
conceptual model 112
Coniston 71, 72, 74, 77, 79, 85, 87, 88, 94, 101, 110
consensus 427, 429, 430, 431, 432, 433, 434, 437, 440, 442, 443, 444, 446
conservative TRV 361, 366, 368, 369, 370
Copper Cliff 71, 72, 75, 77, 79, 80, 81, 82, 83, 85, 86, 92, 93, 101, 110
Critical Hazard Quotient 192
crop studies 424

D

deficiency 200
de minimis 177, 193, 216, 251
dermal exposure pathways 209
dermal sensitization 233
deterministic 19, 345, 373
dietary consumption rates 105
distance from the smelter 284, 322, 330, 332
diversity 262, 267, 268
dolomitic limestone 325
dose–response 99, 176, 177
drinking water 99, 101, 104, 107
drinking water supply 163

E

earthworm 305, 306, 307, 310, 321, 326, 328, 329
earthworm reproduction 306, 307
earthworms 263, 268
ecological integrity 301, 302, 326, 328
ecological recovery 11, 41, 390, 394, 396, 399, 400
Ecological Soil Screening Level 361
ecosite 297, 298, 326
edible wild plant 142
effects assessment 345, 360, 365
Effects Assessment 257
Electron Microprobe Analysis 138, 141
emission controls 29
endangered 259, 263, 266, 267, 268, 269, 272
endangered terrestrial species 345, 374
enrichment factor 64
Environment Canada 279, 286, 301, 304, 306, 310, 337, 339
essential elements 238
essential macronutrients 288
estimated total daily intake 214, 219
ethical issues 442, 447
exposure assessment 120, 122, 123, 127, 141, 144, 153, 158, 159, 163, 164, 165, 167, 168, 169, 194, 195, 201, 345, 365
Exposure Assessment 257
Exposure Factors 103, 106, 107
exposure limits 19, 176, 177, 178, 192, 209, 234, 235
exposure modeling 345, 346, 347, 350, 354, 372, 373
exposure pathways 10, 18, 19, 99, 101, 107, 108, 111, 112, 270, 384
exposure point concentration 159, 174, 176, 209, 225
Exposure Ratio 365
exposure unit 194
external peer review 431, 433, 434, 443, 445
External peer review 425
external peer reviewers 426, 427, 433, 442, 445
External Peer Reviewers 426, 427

F

Falconbridge 71, 72, 73, 77, 79, 80, 85, 88, 89, 94, 101, 110, 118, 119, 120, 121, 123, 124, 127, 128, 135, 138, 139, 141, 144, 145, 146, 148, 149, 150, 151
fingerprints 136, 141
First Nation 97, 102, 105, 106, 107, 108, 112, 114, 209, 212, 224, 241

First Nations 10, 11, 102, 124, 424, 427, 429, 435, 438, 439, 444
food concentrations 169, 170
food consumption 438
food consumption survey 105, 109, 119, 123, 124
forest damage 42, 47
forest floor plots 399
for threshold-response 176
free metabolic rates 347
fugitive dust 52
fugitive dusts 220, 243
fugitive emissions 386, 389, 423

G

gastric phase 131, 132, 133
genotoxic 177
genotoxicity 198
goldenrod 305, 306, 307, 321, 328
grasshoppers 347, 349, 356, 357, 372
Greater Sudbury Area 100
Great Lakes St. Lawrence Forest 303

H

Hanmer 71, 72, 85, 89, 90, 91, 101, 111
hazard assessment 158, 176
Health Canada 99, 103, 104, 105, 114, 162, 170, 172, 176, 178, 179, 193, 195, 196, 197, 220, 221, 226, 227, 241, 242, 244, 245, 246, 249, 427, 429
historical smelter emissions 52
home ranges 351, 353, 354
human receptor 102, 103
humus 288, 313, 324
hunters/anglers 211

I

IEUBK model 103, 109, 111
impact rank 282, 284, 294
Inco 30, 31, 34, 36, 37, 42, 46, 47
incremental lifetime cancer risk 177, 193
incremental lifetime cancer risks 209
Independent Process Observer 424, 426, 427
indirect pathways 107, 108
indoor dust 10, 18, 103, 110, 114, 119, 126, 127, 129, 131, 135, 136, 138, 141, 151, 153, 160, 161, 176, 200
industrial barrens 45
inhalation pathway 119
inhalation unit risks) 220
inorganic arsenic 170, 195, 196, 197, 198, 199, 200, 202
inorganic As 134, 149, 150, 170, 176, 213, 214
Integrated Exposure Uptake Biokinetic 162
in vitro bioaccessibility 132

K

key public concern 441, 442

L

landscape degradation 41
Laurentian University 49, 51, 56, 69, 117, 147, 151, 255, 259, 260, 263, 264, 265, 266, 273, 274, 390, 395, 399, 401
lichens 297, 300, 301

life stages 103
lifestyle factors 111
lifetime cancer risk 214
lifetime composite receptor 210
line of evidence 66, 68
lines of evidence 269, 272, 280, 316, 332, 337, 339, 340
litter decomposition 267, 269, 313, 314, 315, 329, 339
local fish 119, 147
local foods 105, 123, 124, 210, 219, 223
loss of dry weight 315
lowest observed adverse effect level 360

M

macronutrients 294
market basket 210, 213, 214, 216, 217, 223, 226, 241, 245
market basket estimated daily intake 169
market basket foods 173, 174
market basket surveys 170
media coverage 440, 441, 442
Media coverage 434
metals mixture 235
method detection limit 76
microbial activity 306, 314, 315
micronutrients 288, 294, 325
Montana 161
Monte Carlo analysis 345, 346, 347, 351, 360
moose 348, 351, 354, 363, 364, 374, 375, 377
Moose 264, 265, 266
multimedia 159
Multiple linear regression 284
multiple linear regression analysis 318, 320
multi-stakeholder 426, 443
Mycorrhizal fungi 267

N

National Audubon Society 364, 371, 377
natural recovery 41, 43, 45
Niagara Public Health 425, 426
Ni dermatitis 237
Ni inhalation 219, 223, 224, 244
Ni subsulfide 134, 138, 139, 220
non-cancer endpoints 214, 219
non-cancer risk 214
non-carcinogenic 176
non-threshold chemicals 177
no observed adverse effect level 360
Norilsk smelter 81, 86
northern short-tailed shrew 346, 348
northern wheatgrass 305, 306, 308, 310, 321, 328, 329
no threshold 177

O

occupational exposures 233, 235, 442, 447
O'Donnell roast yard 27, 29
Ontario background level 74
Ontario Ministry of Natural Resources 364, 375
Ontario Ministry of the Environment 10, 20, 49, 51, 69, 70, 120, 142, 161, 168, 195, 198, 199, 200, 210, 226, 250, 251, 258, 274, 399

Ontario soil quality guideline 79, 82
Ontario Typical Range 218
open houses 426, 427, 438, 439, 440, 441, 443, 446
organic carbon content 74
organic matter 293, 294, 298, 306, 310, 315, 320, 321, 323, 325, 328, 329, 336, 338, 340
Ottawa 161, 196, 197, 200

P

parent material 73, 74, 75, 76
particulate matter 27, 29, 31, 32, 37, 39, 40, 42, 119, 120, 122, 126, 134, 135, 152, 153, 387
pedon layers 287
peregrine falcon 348, 352, 365
Peregrine Falcon 263, 264
peregrine falcons 365
pH-amended soils 304, 305, 306, 307, 308, 337
phytotoxicity 307, 339
pica 233, 236, 237, 248
plant assemblage composition 322
plant community assessment 281, 297, 301, 302, 317
plant species richness 300
PM2.5 39, 109, 119, 120, 135, 139
PM10 39, 109, 110, 118, 119, 120, 121, 122, 135, 139, 140
point source 73, 83, 84
Port Colborne 161, 170, 172, 173, 197, 198, 230, 248, 250, 421, 422, 423, 424, 425, 426, 428, 429, 430, 431, 432, 433, 434, 435, 436, 437, 438, 439, 440, 441, 442, 443, 444, 447, 448
potable drinking water 125
predictive risk assessment 280
preliminary remediation goal 225
preschool child 103
prevailing wind 286
primary spokesperson 436
probabilistic 19, 20
probabilistic. 345
probabilistic distributions 345
probability distribution functions 103
problem formulation 99, 100
Problem formulation 18
Problem Formulation 18, 255, 257, 259
Project schedule 425
property values 432, 441, 442, 443
Public Advisory Committee 424, 427, 447, 448
public committees 424, 436, 442, 443, 446
public communication 423, 424, 425, 426, 428, 431, 433, 442, 443
Public Liaison Committee 424, 425, 447, 448

Q

QA/QC 56, 126, 142
quality assurance/quality control 56

R

reasonable maximum exposure 159
reasonable worst case 111
reasonably maximally exposed 103, 104
receptor characteristics 159

receptors of concern 99, 100
reference concentration 178
reference doses 130, 176, 178
reference sites 281, 284, 285, 286, 290, 292, 293, 294, 295, 299, 301, 302, 304, 306, 307, 308, 310, 311, 312, 314, 315, 324, 325, 329, 330, 331, 335, 336, 337
refinery dust 201
regional background 74, 77, 83
regional survey 53, 54, 64
re-greening 258, 259, 280, 285, 305, 323, 324, 327, 329, 337, 385, 386, 390, 393, 394, 396, 397, 399
relative absorption factor 130, 350
remediation 59
remediation goals 193, 202
remote sensing techniques 330
residential gardens 143
residential soil samples 135
retrospective risk assessment 280
reverberatory furnaces 29, 31
risk characterization 158, 193, 195, 345, 365, 366, 372
Risk Characterization 257
risk communication 445
risk management 11, 19, 100, 101, 106, 107, 111, 112, 167, 169, 193, 210, 218, 223, 225, 237, 239, 240, 243, 249, 257, 263, 384, 389
Risk Management 381, 383, 384, 389
roast yards 27, 29, 30, 35, 41, 42, 44
Rouyn-Noranda, Quebec 315, 339

S

sampling protocols 432
SARA Group 240, 241, 245, 251
scaling factors 361
scanning electron microscope 68
scanning electron microscopy 138
schools 127, 128
scientific advisors 427, 430, 431
scope changes 431, 432
Screening Level Risk Assessment 97, 113
seedling emergence 306
selenosis 179
self-sustaining forest ecosystem 296
semi-barrens 42, 43, 262, 397
Semi-Barrens 258
sensitivity analysis 226, 239, 244, 245
Sequential leach analysis 135
sintering 31, 32, 38
Site Ranking 278, 279, 295, 296, 301, 303, 310, 312, 313, 315, 316, 317, 322, 330, 333
site-specific risk assessment 99
slope factors 216, 228, 229
small mammals 347, 353, 356, 358, 370, 371, 374, 376, 377
smelter emissions 10, 11, 383, 385, 394
SO_2 280, 323, 335, 336
soil acidification 84
soil chemistry 284, 301, 307, 318, 320, 326, 332, 335
soil fertility 287, 288, 294, 300, 313, 319, 320, 321, 322, 323, 329, 340
soil fertility factor 319, 320

soil horizons 324
soil ingestion rate 226, 236, 241, 348, 373
soil ingestion rates 233, 236
soil intake rates 162
soil intervention level 101
soil microbes 267
soil organic matter 288, 294, 339
soil pH 60, 61, 62, 280, 281, 285, 288, 290, 293, 303, 304, 305, 306, 307, 308, 310, 311, 323, 325, 328, 329, 336, 337
soil properties 294, 310, 321, 322, 323, 338
soil quality guidelines 51, 69, 133, 142
soil remediation 213, 214
soil risk management level 218, 225, 384
soil risk management levels 193
soil toxicity 284, 303, 304, 306, 310, 311, 318, 320, 323, 329, 332, 336, 337
soluble forms of Ni 134
spatial distribution 77, 92
speciation 18, 99, 113, 119, 120, 134, 135, 136, 138, 139, 141, 151, 152, 153, 154, 216, 220, 234, 243
springtail 305
stack samples 68
stack sampling 38
stack spheres 68
Standard Operating Procedures 286
Standards Council of Canada 289
stratified sampling 260
stressed plant community 328
study organization 424, 425
study schedule 432, 433
Sudbury and District Health Unit 148
Sudbury basin 25, 26, 34, 40, 41, 47, 75
Sudbury Center 72, 91, 92
Sudbury Centre 101, 111
Sudbury Centre West 118, 120, 121, 122, 135, 139, 140, 141
Sudbury & District Health Unit 10, 427, 429
sulfur dioxide 11, 12, 30, 36, 84, 233
Sulfur dioxide 29, 36, 38
supermarkets foods 169
super stack 42, 43
Super Stack 37
surface water data 372
surface wipe sampling 127
surrogate models 349, 372

T

tailings 29
Technical Committee 10, 18, 19, 100, 134, 259, 424, 427, 435, 448
Technical Subcommittee 424, 425
technogenic load factor 321
Terms of Reference 429, 444, 448
terrestrial plant communities 280, 335
terrestrial wildlife 259, 268, 272, 274
terrestrial wildlife populations 345, 365, 374
tolerable daily intakes 176, 197
Total Diet Study 170, 172, 173, 196

total metal concentration 287, 322
total organic content 56
total suspended particulate 37, 39
toxicity endpoints 321, 329
toxicity reference values 19, 209
Toxicity reference values 360
Toxicity Reference Values 133, 259
toxicological criteria) 178, 193
toxicological endpoints 320, 340
toxicological profile 195
trace mineral analysis 138
Trail, British Columbia 78, 82, 83, 85, 94
transplanting 390, 398, 400
TSP 37, 39, 109, 119, 120, 140
Typical GSA resident 112
typical Ontario resident 102, 105
Typical Ontario resident 112
Typical Ontario Resident 159, 174, 175, 209, 213

U

unacceptable risk 366
uncertainty 10, 18, 19, 209, 223, 233, 234, 236, 237, 238, 239, 240, 241, 242, 243, 244, 248, 249, 250, 251, 280, 281, 304, 330, 334, 335, 336, 337, 345, 346, 348, 351, 353, 354, 361, 366, 371, 372, 373, 374
uncertainty factor 234
uncertainty factors 176, 177
undisturbed soils 73, 83
union observers 426, 428, 430, 437
upper confidence limit of the arithmetic mean 351
upper confidence limit of the mean 159
uptake models 349, 357, 358, 368, 372
urban soil survey 53, 60, 70
Urinary Arsenic Study 118, 148
urinary As 213, 216, 243
U.S. EPA 159, 160, 162, 163, 165, 166, 176, 177, 178, 179, 194, 201, 202

V

Vale 9, 10, 12, 23, 24, 27, 28, 29, 30, 31, 35, 36, 37, 39, 40, 42, 46, 48, 51, 63, 68, 77, 85, 126, 128, 134, 135, 139, 141, 151, 259, 267, 381, 382, 383, 384, 385, 386, 387, 389, 390, 396, 399, 423, 424, 425, 426, 427, 428, 430, 431, 433, 435, 437, 438, 440, 441, 442, 447
valued ecosystem components 345
Valued Ecosystem Components 255, 256, 257, 266
vegetable garden survey 134, 142
vegetation salvage 397

W

water available metals 284
water leach extraction. 292
water leach metal 287, 324
weight-of-evidence 106, 210, 211, 214, 216, 219, 220, 223, 225, 226, 232, 239, 243, 244, 272, 280, 316, 338, 339, 340
weight-of-evidence approach 134, 150
wild game 105, 107, 112, 212, 215, 219, 223, 224, 241
wildife dietary items 356
wildlife exposure model 259, 260, 346, 349, 353, 356, 357
wild mushrooms 143, 144

workshops 427, 438, 445, 446
World Health Organization 178, 196, 202, 203
World Youth Day 399

X

XANES spectroscopy 138, 154
Xstrata 77, 78, 80, 85
Xstrata Nickel 9, 10, 12, 23, 24, 29, 32, 33, 35, 36, 38, 39, 40, 42, 48, 51, 53, 68, 120, 126, 134, 148, 149, 259, 267, 381, 382, 383, 384, 387, 388, 389, 390, 396, 399, 421, 424, 427, 442, 447

Symbols

95% UCLM 110, 111